# Sustainable Environmental Analysis of Soil and Water

# Sustainable Environmental Analysis of Soil and Water

Guest Editors

**Solmaz Fathololoumi**
**Asim Biswas**

MDPI

Basel • Beijing • Wuhan • Barcelona • Belgrade • Novi Sad • Cluj • Manchester

*Guest Editors*
Solmaz Fathololoumi
School of Environmental Sciences
University of Guelph
Guelph, ON
Canada

Asim Biswas
School of Environmental Sciences
University of Guelph
Guelph, ON
Canada

*Editorial Office*
MDPI AG
Grosspeteranlage 5
4052 Basel, Switzerland

This is a reprint of the Special Issue, published open access by the journal *Sustainability* (ISSN 2071-1050), freely accessible at: https://www.mdpi.com/journal/sustainability/special_issues/249QPI60H9.

For citation purposes, cite each article independently as indicated on the article page online and as indicated below:

| Lastname, A.A.; Lastname, B.B. Article Title. *Journal Name* **Year**, *Volume Number*, Page Range. |
|---|

**ISBN 978-3-7258-7356-2 (Hbk)**
**ISBN 978-3-7258-7357-9 (PDF)**
**https://doi.org/10.3390/books978-3-7258-7357-9**

# Contents

# About the Editors

**Solmaz Fathololoumi**

Solmaz Fathololoumi is a postdoctoral researcher at the School of Environmental Science, University of Guelph. She completed her doctoral degree in Soil Physics and Conservation, and her academic trajectory has been recognized through national awards for outstanding graduate students, as well as multiple international awards, highlighting her distinction as a highly talented and high-achieving researcher. Her research focuses on digital and precision agriculture, integrating soil and water management with digital soil mapping, remote sensing, and modeling of agricultural and environmental systems. Dr. Fathololoumi combines fundamental soil science with modern data-driven approaches and advanced analytical methods to address environmental challenges, enhance agricultural productivity, and support ecosystem resilience. She contributes to multiple projects, developing practical and innovative solutions for soil and water sustainability. Her work bridges traditional soil science with cutting-edge technologies, providing actionable insights for sustainable land and water management and promoting resilient agri-environmental systems.

**Asim Biswas**

Asim Biswas is a Professor at the University of Guelph, Canada. He holds the Canada Research Chair in Digital Agriculture and the OAC Chair in Soils and Precision Agriculture, and serves as Graduate Program Coordinator at the School of Environmental Science. He completed his Ph.D. at the University of Saskatchewan, Canada, an M.Sc. at the University of Agricultural Sciences–Bangalore, India, and a B.Sc. at Bidhan Chandra Agricultural University, West Bengal, India. Recognized internationally as a leading authority in sustainable soil and water management, Prof. Biswas has developed innovative, data-driven approaches that integrate soil sensors, satellite imagery, yield data, and advanced computational algorithms to enhance agricultural productivity, resource efficiency, and environmental stewardship. He leads multidisciplinary research teams and major international projects, bridging fundamental soil science with advanced digital technologies to address global challenges in precision agriculture and natural resource management. Prof. Biswas has held prominent leadership roles, including President of the Canadian Society of Soil Science and Chair of the Soil Physics and Hydrology Division of the Soil Science Society of America, and he currently chairs the Proximal Soil Sensing Working Group of the International Union of Soil Sciences. A Fellow and member of the Royal Society of Canada College of New Scholars, he contributes to advancing the discipline through his editorial leadership as Associate Editor for eight journals and guest editor of numerous Special Issues. His work establishes him as a highly influential scientist whose research, mentorship, and collaborations have contributed to advancing sustainable and resilient agri-food systems in Canada and worldwide.

# Preface

This Reprint is based on the Special Issue, entitled "Sustainable Environmental Analysis of Soil and Water", published in *Sustainability*. It brings together fifteen peer-reviewed contributions that collectively examine the dynamic interactions between soil and water systems under natural processes and increasing human pressures. The subject of this Reprint centers on advancing scientific understanding of how soil and water resources can be analyzed, modeled, and managed in a sustainable and integrated manner.

The scope of this Reprint is interdisciplinary and methodologically diverse. The included studies address soil mechanical behavior, matric suction effects, hydrodynamics and sedimentation processes, watershed scale erosion modeling, groundwater hydrochemical assessment, climate-based precipitation optimization, and environmental contamination associated with mining activities. In parallel, several contributions focus on sustainable land and agricultural management, including tillage impacts on soil's physical, hydraulic, microbial, and enzymatic properties, soil-conditioning strategies to enhance crop performance, and scenario-based decision frameworks for resource allocation. Together, these works reflect the multifaceted nature of soil and water systems across different climatic and geomorphological contexts worldwide.

A defining feature of this Reprint is the integration of experimental investigations, field observations, simulation studies, remote sensing applications, neural network models, regression-based optimization techniques, and multi-criteria decision-making approaches. This methodological breadth demonstrates how contemporary environmental research increasingly relies on data driven tools and quantitative modeling to improve spatial and temporal assessment accuracy.

The aim and purpose of this Reprint are to synthesize current scientific advances that contribute to the sustainable stewardship of soil and water resources. The motivation behind assembling this collection arises from the growing recognition that soil and water degradation, erosion, contamination, and hydrological instability pose significant challenges to agricultural productivity, ecosystem services, and public health. By presenting case studies and analytical approaches within a unified framework, this Reprint seeks to encourage integrative thinking and foster collaboration across soil science, hydrology, environmental engineering, and sustainability research.

This Reprint is addressed to researchers, graduate students, environmental scientists, soil and water specialists, hydrologists, and professionals engaged in sustainability, environmental modeling, and natural resource management. It is intended to serve as a rigorous scientific reference that bridges methodological innovation with practical environmental applications.

**Solmaz Fathololoumi and Asim Biswas**
*Guest Editors*

sustainability

MDPI

*Article*

# Effect of Matric Suction and Drying-Wetting Cycles on the Strength of Granite Residual Soil in Fujian Pumped Storage Power Station Slopes, China

**Xiudong Xie *, Zhidong Xie, Chenyang Wang and Yan Su**

College of Civil Engineering, Fuzhou University, Fuzhou 350108, China; xzd0918@163.com (Z.X.); bingru1747407@163.com (C.W.); suyan@fzu.edu.cn (Y.S.)
* Correspondence: zlxxd@163.com

**Abstract**

The stability of bank slopes in pumped storage power stations is crucial, particularly in regions where frequent water level fluctuations occur. This study aims to investigate the degradation mechanism of bank slope under such fluctuating conditions, focusing on granite residual soil from the pumped storage power stations in Fujian, China. To explore the effects of drying-wetting cycles and matric suction on soil shear strength, drying and wetting cycles were conducted with unsaturated triaxial shear tests. The results revealed that the shear parameter strengthening effect occurs when the matric suction increases from 50 kPa to 200 kPa. Moreover, during the first five drying-wetting cycles, soil shear strength decreased sharply, with cohesion and internal friction angle decreasing by approximately 15.4% and 11.2%, respectively. This degradation trend stabilized in the later cycles. Scanning Electron Microscopy (SEM) analysis of the soil microstructure showed an evolution from a dense structure to a penetrating cavity during the cycles. This change reflects that the strength degradation characteristics of granite residual soils are controlled by the synergistic effects of structural and frictional mechanisms, manifesting as initial degradation followed by stabilization. Additionally, by fitting the nonlinear characteristics of the experimental data, shear strength evolution functions for matric suction and drying-wetting cycles were established, revealing the effect of these factors on strength degradation. These findings provide a theoretical basis for the stability analysis of bank slopes in pumped storage power stations, offering insights into soil behavior under fluctuating water levels.

**Keywords:** unsaturated soil; granite residual soil; matric suction; drying-wetting cycle; strength evolution

## 1. Introduction

Granitic residual soils are widely distributed in Fujian, China, covering approximately 30% to 40% of the province's total land area. These soils primarily consist of debris generated by the physical and chemical weathering of granite [1]. The unique mountainous terrain and abundant water resources in Fujian provide favorable conditions for the development of pumped storage power stations [2,3]. Currently, more than ten storage power stations are being operated. These stations regulate grid stability by performing peak shaving and valley filling. However, as the stations continue to operate, the soils along the reservoir shorelines are subjected to periodic water level fluctuations, undergoing prolonged drying and wetting cycles. This process alters the bonding and arrangement of soil


https://doi.org/10.3390/su18020748

particles, affecting the pore structure of the soil [4]. Over time, the continuous wetting and drying cycles gradually degrade the water retention capacity and strength characteristics of the unsaturated shoreline soils, as shown in Figure 1. This deterioration may not only reduce soil stability but also potentially trigger slope instability, leading to landslides and other geological hazards, which can impact the safe operation of pumped storage stations. Addressing this issue has become a critical engineering challenge affecting the safe and stable operation of these stations.

**Figure 1.** Slope instability at Zhouning pumped storage power station in Fujian, China.

The impact of drying and wetting cycles on soil properties has been extensively investigated. Early studies found that soil strength varied under drying and wetting cycles, even when the matric suction was the same [5,6]. Based on experimental research, drying and wetting cycles were regarded as influencing the saturation of soils, thereby altering their shear strength [7,8]. Furthermore, drying and wetting cycles affect expansive soils, causing fatigue behavior in the soil, which reduces its swelling capacity [9]. Experimental evaluations have shown that the fissure rate increases with cycles, with the most significant changes occurring during the initial cycles [10]. Repeated cycles modify the structure and shrinkage characteristics of unsaturated soils, leading to a decrease in suction and an increase in compressibility at the same moisture content [11]. This structural alteration also significantly affects the soil–water characteristic curve, resulting in different soil strengths under the same matric suction during drying and wetting cycles. The soil–water characteristic curve can be conducted on remolded soils subjected to varying numbers of cycles [12]. It found that the soil's water retention properties deteriorated most significantly after the first cycle, stabilizing after three cycles. Although existing studies have explored the effects of drying and wetting cycles on soil properties, there is still a lack of research on the influence of cycles on the evolutionary characteristics of granitic residual soils and their relationship with the matric suction of the bank slopes in pumped storage reservoir areas.

Existing research has systematically explored the effects of drying and wetting cycles on soil microstructure, strength, water retention, permeability, and soil–water characteristic curves, confirming the adverse effects of cycles on unsaturated soils [13,14]. These effects are particularly evident in terms of shear strength, structural stability, and fissure evolution, with significant degradation trends observed [15–17]. Such degradation can compromise the integrity of soil, leading to issues like reduced load-bearing capacity and increased susceptibility to erosion or failure. Due to the rapid fluctuations in water levels of pumped storage power stations, previous studies have shown that such variations can trigger significant slope deformations, posing a threat to operational safety [18,19]. These findings highlight the need for a deeper understanding of the soil's behavior under dynamic conditions. Therefore, investigating the strength characteristics of granite residual soil under long-term regulation of pumped storage power stations is particularly crucial.

 https://doi.org/10.3390/su18020748

However, the strength evolution and underlying mechanisms of granitic residual soils, which are skeletal and non-expansive, under the combined influence of matric suction and repeated cycles, remain insufficiently explored.

Considering the topographical characteristics of Fujian and the regulation features of pumped storage power stations, this study focuses on the synergistic effect of matric suction and drying-wetting cycles on granite residual soil. The innovation lies in identifying the nonlinear degradation mechanism under coupled conditions and analyzing its correlation with the evolution of microstructural features. First, by conducting drying and wetting cycle experiments with triaxial tests while controlling matric suction, the research systematically investigates the effects of matric suction and the number of cycles on the shear strength parameters of the soil. Then, strength evolution functions are developed to reveal the strength degradation mechanisms of granitic residual soils under the influence of cycle effects and matric suction. The findings of this study provide experimental data and theoretical support for the long-term stability and protective design of pumped storage reservoir slopes.

## 2. Materials and Methods

### 2.1. Sample Preparation

Remolded soil samples were obtained via trenching from the shore slope of a pumped storage power station in Fujian. To investigate the effects of matric suction and drying-wetting cycles on the shear strength of the granite residual soil under controlled conditions, it was essential to minimize variability caused by natural soil heterogeneity. Although remolding disturbs the natural fabric and crack structure, it ensures uniform initial conditions for all samples. The basic physical properties of the soil samples were tested in the laboratory according to the standard for geotechnical testing method of China (GB/T 50123-2019) [20], and the results are presented in Figure 2 and Table 1.

**Figure 2.** Grain size distribution of the soil sample.

Based on the test results, the particle size distribution curve indicates a continuous gradation without significant gaps. The curvature coefficient ($C_c$) is 1.29, and the uniformity coefficient ($C_u$) is 14.00, classifying the soil as well-graded. The fine particle content (<0.075 mm) is relatively high at 19.05%, with no gravel (>2 mm) present. According to the local standard of Fujian, the soil sample is classified as residual cohesive soil. The specific


https://doi.org/10.3390/su18020748

gravity of the soil particles is 2.71, suggesting a composition dominated by minerals such as quartz and feldspar. The results of the compaction test show a maximum dry density of 1.57 g/cm$^3$ and an optimum moisture content of 21%, demonstrating favorable compaction characteristics.

**Table 1.** The physical properties of the soil sample.

| Property | Value |
|---|---|
| Natural moisture content (%) | 32.30 |
| Optimum moisture content (%) | 21.00 |
| Natural unit weight (g/cm$^3$) | 1.90 |
| Maximum dry density (g/cm$^3$) | 1.57 |
| Curvature coefficient | 1.29 |
| Uniformity coefficient | 14.00 |
| Specific gravity | 2.71 |

The granite residual soil was processed by crushing, drying, and grinding into a fine powder, which was passed through a 2 mm sieve to obtain the desired soil sample. The sample was then prepared by adjusting the moisture content to the predetermined level and placed in a sealed bag for 24 h to allow for uniform distribution of moisture throughout the soil. To achieve a target dry density of 1.41 g/cm$^3$, samples were prepared using the layer-by-layer compaction method to ensure homogeneity in their physical properties. Equivalent in situ stresses of 50 kPa for shallow layers and 150 kPa for deeper layers were applied during compaction to maintain consistency between the sample structure and the in situ soil. The sample specifications were set as a sample with a diameter of 39.1 mm and a height of 80 mm, intended for non-saturated triaxial testing. This preparation aims to ensure that the sample is representative and the experimental results are reliable, while accurately reflecting the mechanical behavior of the soil during testing.

### *2.2. Experimental Process of Unsaturated Soil Samples*

The experimental process is shown in Figure 3. It mainly includes field sampling, drying and wetting cycle, and triaxial test. According to relevant studies, the first drying and wetting cycle has the most significant impact on the sample, and after 5 to 7 cycles, the soil-water characteristics and mechanical properties of the sample tend to stabilize [21,22]. Therefore, 10 cycles were conducted to examine the effects of varying drying and wetting cycles on the sample's properties in this experiment. The experimental procedure involves the following steps: First, the target moisture content was aimed at achieving full saturation in the wetting phase. According to the testing procedure outlined in the standard for soil testing methods (GB/T 50123-2019), the vacuum saturation method was used to saturate the samples. The degree of saturation was determined by comparing the sample weight before and after saturation, with a saturation greater than 95% considered fully saturated. Then, the sample is subjected to 8 h of drying at 20 °C and relative humidity 50% ± 5% to reduce moisture content and simulate soil exposure during a falling water level. The process of drying and wetting cycles is shown in Figure 3. This experimental approach simulates the impact of water level fluctuations on soil samples, providing a solid basis for studying the soil-water characteristics and mechanical properties of soils under drying and wetting cycles.



https://doi.org/10.3390/su18020748

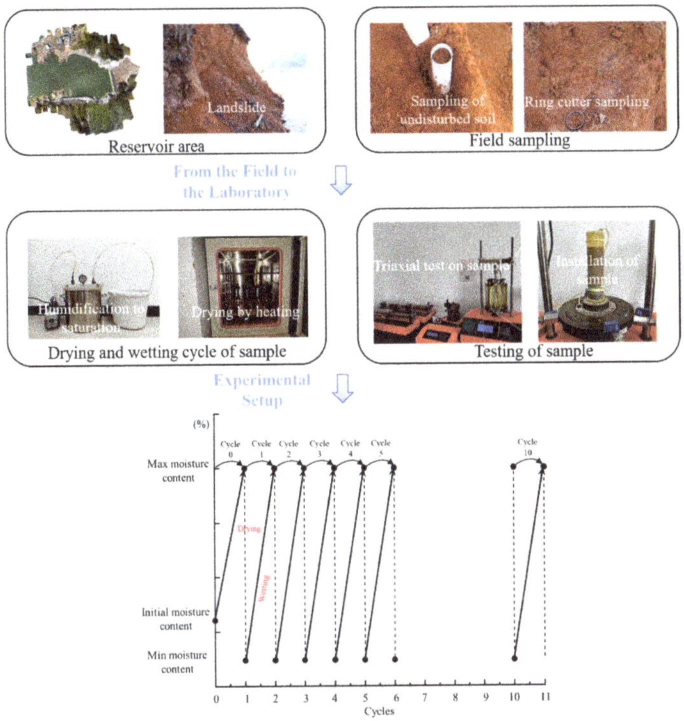


**Figure 3.** Sample preparation and experimental process of unsaturated soil.

To investigate the strength characteristics of the soil samples, a stress–strain controlled triaxial apparatus for unsaturated soils was used in the experiments. The equipment consisted of four main components: a dual-chamber pressure chamber, a pressure control system, a total volume measurement system, and a data acquisition system. The dual-chamber pressure chamber features an internal and external chamber structure, connected by ducts to a confining pressure controller. This design allows for the simultaneous application of equal internal and external confining pressures, preventing deformation of the inner chamber under high confining pressure. The pore air pressure, pore water pressure, and confining pressure are controlled by the pressurization system, while the axial pressure is applied through an axial shear controller, ensuring precise control over the axial stress during the shearing [23].

The total volume measurement system is an independent system capable of continuously and accurately measuring the changes in volume over extended periods, providing reliable data for volume change analysis during the test. The data acquisition system consists of computer software, relevant sensors, and a data acquisition instrument. The software handles real-time data collection and processing, while the sensors measure various physical quantities (e.g., pressure, volume). The acquisition instrument is responsible for storing and transmitting data. This system ensures the efficient and accurate recording of experimental data and allows for real-time monitoring of changes during the test, providing support for subsequent data analysis.

The procedure of triaxial testing is divided into three main stages: suction equilibrium, isostatic consolidation, and constant-rate shear. Each stage is designed to precisely control the moisture state and stress conditions of the soil sample, ensuring reliable experimental results. In the suction equilibrium stage, the sample's moisture state is adjusted until it


https://doi.org/10.3390/su18020748

reaches suction equilibrium. The criterion for achieving matric suction equilibrium is that the pore water drainage from the sample should not exceed 0.1 $cm^3$ within 24 h. This process ensures that the moisture distribution and suction in the sample are stabilized, providing a solid foundation for the subsequent consolidation and shear testing.

In the isostatic consolidation stage, after achieving suction equilibrium, the suction is maintained constant while a net confining pressure is applied to the sample. This pressure causes the sample to drain and gradually consolidate. The consolidation is considered complete when the drainage of pore water is less than or equal to 0.1 $cm^3$, similar to the criterion used in the suction equilibrium stage, ensuring that the sample has reached a stable, consolidated state. In the constant-rate shear stage, the suction and net confining pressure are kept constant while the sample is subjected to shear at a constant rate of 0.005 mm/min, allowing for simultaneous drainage and axial strain development. The shear rate of 0.005 mm/min is selected based on pre-experimental results. Under this rate, the soil sample drains sufficiently, the strength parameters are stable, and it conforms to the specification of soil test (GB/T 50123-2019). The shear test is terminated when the axial strain reaches 15%, simulating the stress–strain behavior of soil in engineering applications and providing essential data on the sample's strength characteristics. Throughout the experiment, stress and moisture conditions are controlled to ensure the consistency of the test data.

### 2.3. Test Condition Setting of the Granite Residual Soil Strength Characteristics

To investigate the effects of cycle and matric suction on the strength characteristics of granite residual soil, two experimental groups were designed to explore the relationship between these factors and the shear strength of granite residual soil. The scheme and the specific parameters for the triaxial tests are provided in Table 2.

**Table 2.** Test scheme and parameters.

| Groups | Cycle | Matric Suction (kPa) | Pore Water Pressure (kPa) | Pore Air Pressure (kPa) | Confining Pressure (kPa) | Net Confining Pressure (kPa) |
|---|---|---|---|---|---|---|
| 1 | 0 | 50<br>100<br>150<br>200 | 20 | 70<br>120<br>170<br>220 | 170/270/370<br>220/320/420<br>270/370/470<br>320/420/520 | 100/200/300 |
| 2 | 0<br>1<br>5<br>10 | 50 | 20 | 70 | 170/270/370 | 100/200/300 |

Group 1 involves consolidated drained tests under different matric suction conditions (50, 100, 150, 200 kPa) without cycling. In this group, the pore air pressure is maintained at a fixed value (20 kPa), and matric suction is controlled by varying the pore water pressure. During the test, matric suction is measured using the tensiometer method, with an accuracy of ±2 kPa. Data is recorded during the suction equilibrium stage (24 h drainage $\leq$ 1 $cm^3$). The confining pressure is adjusted according to the relationship between confining pressure and pore water pressure to ensure a consistent net confining pressure. Three samples are tested for each matric suction gradient, totaling 12 samples. This group aims to examine the impact of matric suction on the shear strength of granite residual soil.

Group 2 consists of consolidated drained tests under a fixed matric suction of 50 kPa and varying drying and wetting cycles. Since matric suction is fixed, both the pore air pressure and pore water pressure are kept constant at 20 kPa and 70 kPa, respectively. For



https://doi.org/10.3390/su18020748

each cycle gradient, three samples are tested, resulting in a total of 12 samples. The choice of 10 cycles is based on two factors: first, due to the low degree of weathering of the parent rock in Fujian, the shear strength parameters of granite residual soil continue to change after 5 cycles and stabilize only by the 10th cycle. Second, considering that pumped storage power station slopes experience long-term operational hydraulic fluctuations, 10 cycles provide a more reasonable simulation of the long-term environmental effects. This group investigates the influence of drying and wetting cycles on the mechanical properties of granite residual soil. The experimental design aims to investigate the effects of matric suction and drying and wetting cycles on the strength characteristics of granite residual soil, offering valuable insights for the disaster prevention of pumped storage power stations.

Moreover, a scanning electron microscope (Quanta 250, Thermo Fisher Scientific, Oakwood, OH, USA) was used to observe the microstructure of soil samples after drying and wetting cycles in this study, as shown in Figure 4. According to the instrument's sample requirements, the original and cycled soil samples were first cut in half, and representative small specimens (with dimensions controlled at 10 mm × 10 mm × 5 mm) were taken. Next, the processed samples were fixed on the sample holder and introduced into the sample chamber. Finally, observations were conducted in high vacuum mode at an accelerating voltage of 25 kV. Scanning Electron Microscope (SEM) images of the soil samples were collected at different cycles, which reveal the degradation mechanisms of soil strength at the microscopic scale.

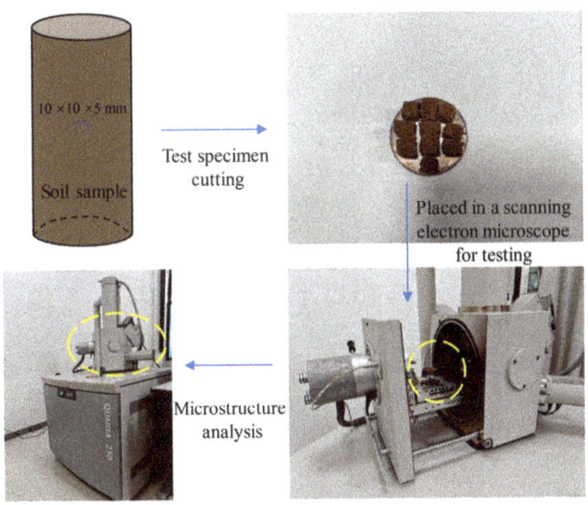


**Figure 4.** Procedure for Scanning Electron Microscopy testing of soil samples. (Blue cuboid is SEM test sample cut from the soil sample. Yellow circle is the position of the test sample within the SEM equipment).

## 3. Results

### 3.1. Influence of Matric Suction on the Strength Characteristics of Granite Residual Soil

To investigate the influence of matric suction on the strength characteristics of granite residual soil, the stress–strain relationships under varying matric suctions and net confining pressures were analyzed. Figure 5 illustrates the stress–strain relationships of granite residual soil under varying matric suctions and net confining pressures. The samples exhibit typical strain-hardening behavior without distinct peak points. In the later stages of deformation, the shear strength stabilizes, indicating good plastic deformation capacity. This suggests that granite residual soil is able to maintain its load-bearing capacity while undergoing deformation [24].


https://doi.org/10.3390/su18020748

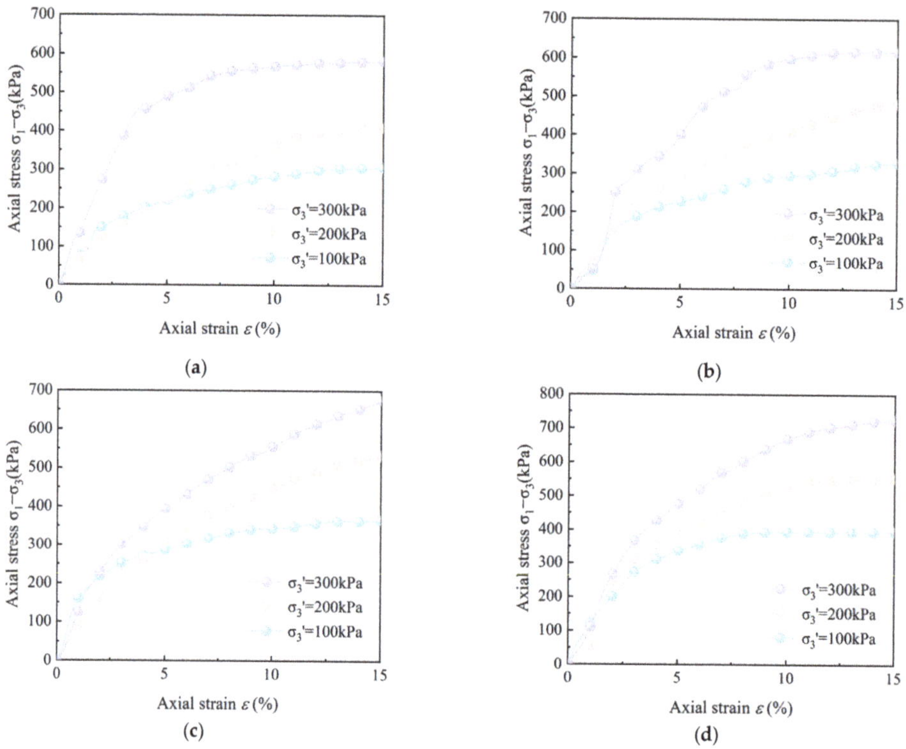


**Figure 5.** Stress–strain curves under different matric suctions: (**a**) Matric suction: 50 kPa; (**b**) Matric suction: 100 kPa; (**c**) Matric suction: 150 kPa; (**d**) Matric suction: 200 kPa.

Under low matric suction conditions, the initial stiffness of the samples is relatively low, and the deviatoric stress increases gradually, indicating that the soil structure is relatively loose with weak cementation between particles. The loose interparticle bonding leads to significant redistribution of stress during deformation, resulting in a slower increase in stress. As confining pressure increases, the slope of the stress–strain curve slightly rises, showing that the external confinement contributes to strengthening the shear resistance of the soil. This suggests that the external pressure helps maintain the soil's structural stability and enhances its shear resistance.

Under high matric suction conditions, the overall stiffness of the samples significantly increases, and the rate of deviatoric stress growth accelerates. The slope of the stress–strain curve becomes steeper, indicating enhanced shear resistance. This is attributed to the increased matric suction, which enhances the adhesive forces between particles and improves the structural stability of the soil. As a result, the sample exhibits greater resilience in the initial stages, requiring more deformation to reach failure. The higher matric suction strengthens the interparticle bonds, improving the shear strength and making the soil more stable under stress and delaying failure.

Furthermore, the rate of axial strain development slightly slows down with increasing matric suction, indicating that higher matric suction delays failure onset. This indicates that under high matric suction conditions, the soil has a longer adaptation time, allowing it to better resist external disturbances and delay the point at which stress reaches failure. Thus, the stress–strain response of granite residual soil is more sensitive under low suction conditions, with more pronounced deformation and greater susceptibility to disturbance. In contrast, granite residual soil exhibits higher strength, greater resistance to disturbance,


https://doi.org/10.3390/su18020748

and enhanced stability. These findings provide valuable insights into the mechanical behavior of granite residual soil under different moisture conditions.

As shown in Figure 6, the Mohr circles for granite residual soil at varying matric suction clearly demonstrate the relationship between shear strength parameters and matric suction. Both cohesion and internal friction angle increase with the matric suction, but the rate of increase differs across suction ranges. Cohesion increases rapidly from 50 kPa to 100 kPa, while the growth rate slows significantly as matric suction rises from 150 kPa to 200 kPa. In contrast, the internal friction angle increases slowly from 50 kPa to 100 kPa range but accelerates significantly from 100 kPa to 150 kPa.

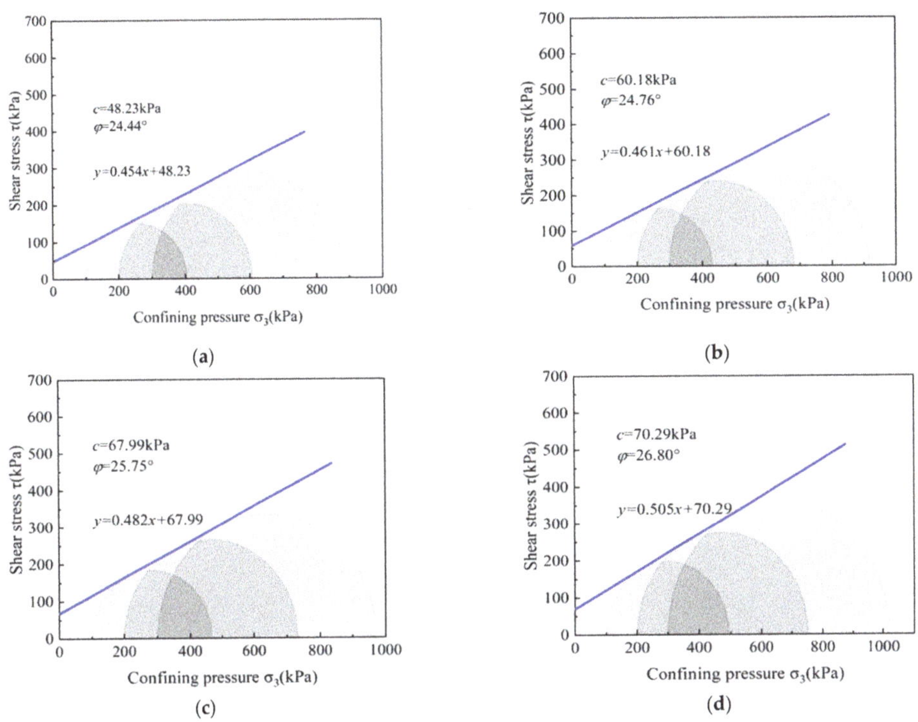


**Figure 6.** Strength envelope under different matric suctions: (**a**) Matric suction: 50 kPa; (**b**) Matric suction: 100 kPa; (**c**) Matric suction: 150 kPa; (**d**) Matric suction: 200 kPa.

These observations indicate that matric suction has a considerable influence on the mechanical properties of granite residual soil, particularly on cohesion [25]. When the soil is relatively dry, higher matric suction enhances the molecular attraction between soil particles, which in turn increases cohesion. A similar study mentioned that matric suction can form a water film on the particle surfaces, strengthening the attraction between particles and further enhancing the cohesion of the soil [26]. This enhancement effect is particularly obvious when the suction is low, because the initial bond is weak at this time, and a small increase in suction can bring significant strength growth.

However, matric suction has a relatively limited effect on the internal friction angle of granite residual soil. While matric suction changes the contact state between soil particles, the size of the internal friction angle is primarily governed by the shape and roughness of the soil particles. Even though matric suction improves the bonding between particles through water film formation, these physical characteristics have a smaller influence on the internal friction angle. Thus, the internal friction angle is more influenced by the inherent


https://doi.org/10.3390/su18020748

properties of the soil particles, such as their shape, size, and surface roughness, while matric suction plays a secondary role.

Thus, matric suction significantly enhances the strength characteristics of granite residual soil, particularly its cohesion. At lower matric suction, the cohesion is relatively low, but as matric suction increases, cohesion improves gradually. However, the internal friction angle is influenced more by particle characteristics than by matric suction. Understanding these mechanisms is essential for evaluating the mechanical behavior of granite residual soil under varying environmental conditions.

To quantify the influence of the above mechanisms, experimental data were analyzed, revealing a clear quadratic curvilinear trend. The physical basis of this trend is that matric suction influences both cohesion and internal friction angle in a nonlinear manner, with their values changing as suction increases. To further investigate this relationship, a function fitting analysis was performed to examine the influence of matric suction on the shear strength parameters, as shown in Figure 7. The optimal fitting model was selected based on the root mean square error (*RMSE*) and the coefficient of determination ($R^2$). The fitted polynomials are shown in Equations (1) and (2). The results indicate that cohesion follows an approximate quadratic polynomial relationship with matric suction, with a fitting goodness of $R^2 = 0.9999$ and an *RMSE* of 0.018. Similarly, the internal friction angle also exhibits a quadratic polynomial relationship with matric suction, with $R^2 = 0.9876$ and *RMSE* = 0.201.

$$c = 30.72 + 0.40S - 0.0010S^2 \tag{1}$$

$$\varphi = 24.69 - 0.0086S + 0.000098S^2 \tag{2}$$

where $c$ is the cohesion, $\varphi$ is the internal friction angle, $S$ is the matric suction.

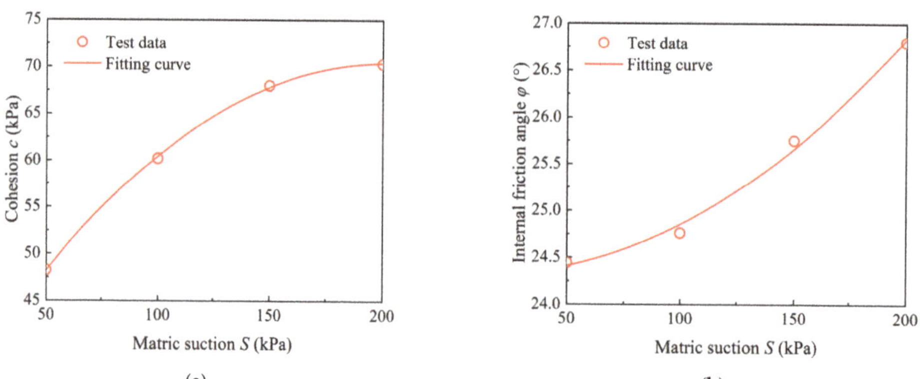


**Figure 7.** Relationship between shear strength parameters and matric suction. (**a**) Cohesion and matric suction; (**b**) Internal friction angle and matric suction.

The fitting results reveal that both cohesion and internal friction angle of granite residual soil show a quadratic polynomial relationship with matric suction. This suggests that as matric suction increases, the shear strength parameters—cohesion and internal friction angle—initially increase rapidly and then gradually approach saturation. Specifically, at lower matric suction (<100 kPa), both cohesion and internal friction angle increase significantly with increasing matric suction, showing a pronounced strengthening effect. However, as matric suction further increases, the changes in cohesion and internal friction angle become less pronounced, eventually reaching saturation.

 https://doi.org/10.3390/su18020748

This trend is captured by the quadratic term in the model, which reflects the diminishing contribution of matric suction as it increases. Physically, this effect corresponds to the stabilization of the capillary water film thickness in the soil, beyond which the suction effect weakens. In the early stages of water uptake, the capillary water films between soil particles are thin, and the suction effect is significant as the matric suction increases. As the matric suction reaches a certain threshold, the thickness of the capillary water film stabilizes, and the suction effect diminishes.

Thus, the increase in matric suction significantly enhances the shear strength of the soil, particularly cohesion and internal friction angle, during the early stages of suction increase. However, as suction increases to a certain threshold, the strengthening effect approaches saturation. This phenomenon reveals the mechanical response characteristics of soil under the effects of capillary water films.

Based on experimental observations and theoretical analysis of the strength characteristics of granite residual soil, Fredlund and Xing highlighted that a single effective stress state variable cannot fully capture the mechanical behavior of unsaturated soils [27]. To address this, they proposed separately considering net confining pressure and matric suction, using $\varphi^b$ to represent the influence of matric suction. As a result, they developed a shear strength equation for unsaturated soils, as shown in Equation (3).

$$\tau_f = c' + (\sigma - \mu_a)tan\varphi' + (\mu_a - \mu_w)tan\varphi^b \tag{3}$$

where $\tau_f$ is the shear strength of the unsaturated soil, $c'$ is the effective cohesion, $\sigma$ is the confining pressure, $\mu_a$ is pore air pressure, $\mu_w$ is the pore water pressure, $\varphi'$ is the effective internal friction angle, $\varphi^b$ is the angle with respect to matric suction.

By setting the matric suction to 0 kPa and substituting it into the equation, the effective cohesion $c'$ and effective internal friction angle $\varphi'$ of granite residual soil in a saturated state are obtained as 30.72 kPa and 24.69°, respectively. Substituting the values into Equation (3), the shear strength equation for unsaturated granite residual soil in the study area can be expressed as Equation (4).

$$\tau_f = 30.72 + (\sigma - \mu_a) \times 0.46 + (\mu_a - \mu_w) \times 0.18 \tag{4}$$

where $\tau_f$ is the shear strength of the unsaturated soil, $\sigma$ is the confining pressure, $\mu_a$ is pore air pressure, $\mu_w$ is the pore water pressure.

### 3.2. Strength Response of Granite Residual Soil Under Drying and Wetting Cycles

Figure 8 illustrates the stress–strain relationships of granite residual soil under three different net confining pressures, with varying numbers of drying and wetting cycles. The samples exhibit typical strain-hardening behavior, with no distinct peak points. The stress–strain curves flatten out in the later stages, demonstrating some plastic deformation capacity. This indicates that the soil sample can endure significant stress while undergoing deformation, with a certain degree of plastic flow before failure, showing its ability to deform without immediately breaking.

The stress–strain response of the soil is significantly influenced by drying and wetting cycles. Untreated samples exhibit high initial stiffness and a steeply rising stress curve, reflecting well-preserved structural integrity, characterized by dense particle packing and strong interparticle cementation. After 1 cycle, the curve flattens slightly, accompanied by a reduction in both initial stiffness and maximum deviatoric stress. This indicates the onset of structural deterioration, marked by the development of microcracks and weakened cementation. Following 5 cycles, a pronounced downward shift in the curve is observed, with a slower stress increase, signifying substantial structural damage where shear resis-



https://doi.org/10.3390/su18020748

tance becomes governed mainly by frictional forces and particle rearrangement rather than cementation. By 10 cycles, the curve stabilizes, displaying a hardening trend with further reduced initial stiffness. This suggests the soil has reached a residual state, wherein its mechanical behavior becomes more stable but at a considerably lower strength level.

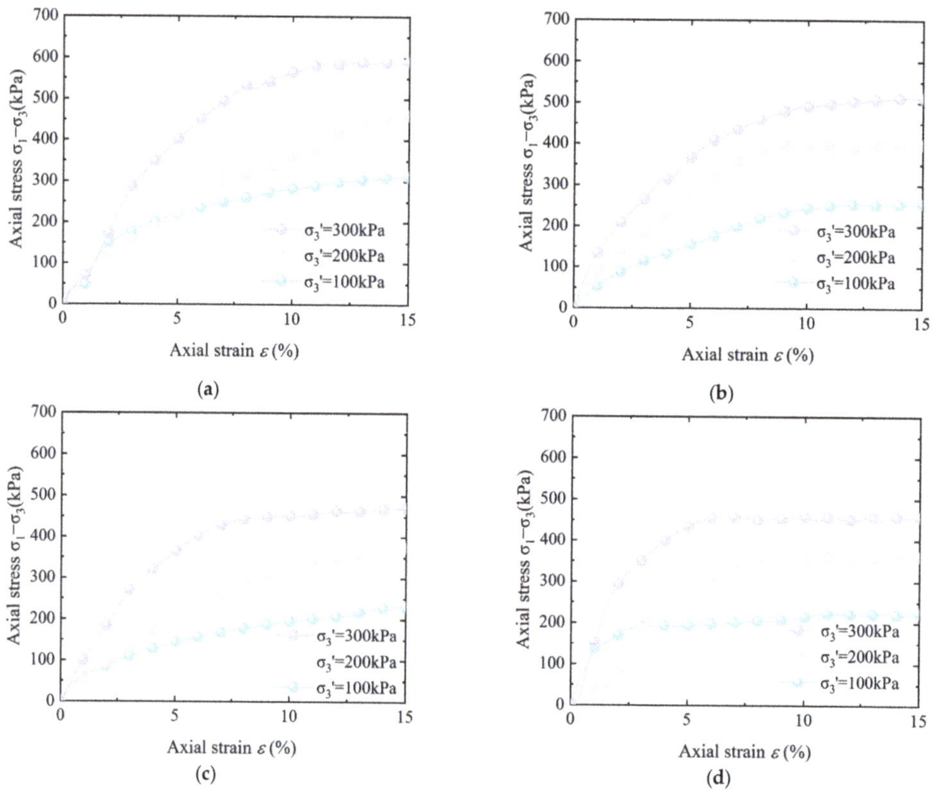


**Figure 8.** Stress–strain curves under drying and wetting cycles: (**a**) Cycle 0; (**b**) Cycle 1; (**c**) Cycle 5; (**d**) Cycle 10.

These changes are associated with the fluctuations in moisture content that occur during drying and wetting cycles. The absorption and loss of water affect the bonding strength and structural stability of the soil particles. As the cycles progress, the bonding force between particles weakens, leading to a gradual decrease in shear strength and rigidity. Additionally, the drying and wetting process damages the capillary water film, further impacting the soil's mechanical properties. Previous studies have also recognized that moisture content plays a crucial role in the drying and wetting of soil [28,29].

The shear strength parameters of granite residual soil exhibit general degradation with successive drying-wetting cycles, characterized by a pronounced reduction in cohesion and a slight decrease in the internal friction angle (Figure 9). This trend unequivocally indicates that structural degradation is the dominant mechanism governing the soil's response to these cycles. The pattern of strength reduction in granite residual soil is notably different from the typical behavior observed in cohesive soils such as loess and red clay. Unlike loess or red clay, which undergo reversible expansion and contraction due to interlayer water between mineral particles, the weathering products of the parent rock in granite residual soil are mainly composed of inert minerals, such as quartz and feldspar. These minerals are chemically stable and less prone to swelling or shrinkage due to changes


https://doi.org/10.3390/su18020748

in moisture. Therefore, the failure mode of granite residual soil is primarily driven by particle rearrangement and the loosening of the skeletal structure during the drying and wetting cycles, rather than by the reversible expansion and contraction of interlayer water between minerals.

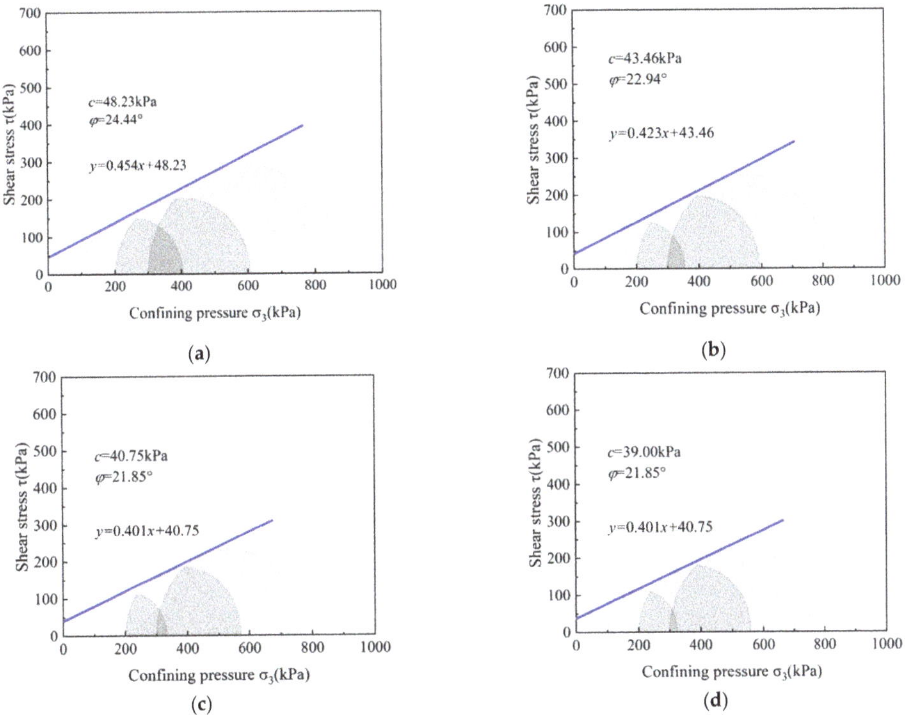


**Figure 9.** Strength envelope under drying and wetting cycles: (**a**) Cycle 0; (**b**) Cycle 1; (**c**) Cycle 5; (**d**) Cycle 10.

The skeletal framework of granite residual soil is primarily composed of coarse particles, and the cementation between particles is relatively weak. As a result, during the drying and wetting cycles, the particles are more prone to displacement and rearrangement, leading to a gradual reduction in the overall strength and stiffness of the soil. In contrast, typical cohesive soils like loess and red clay consist mainly of fine particles and clay minerals, which exhibit reversible swelling and shrinkage due to the absorption and release of water, affecting their mechanical properties.

The variation of shear strength parameters with the number of cycles is shown in Figure 10. After the first wetting-drying cycle, the cohesion decreases by 9.9%, while the internal friction angle decreases by 6.5%, accounting for 48.5% and 39.1% of the total reduction, respectively. This stage shows the largest rate of decrease, indicating that drying and wetting cycles cause significant disruption to the internal microstructure of the sample. As the number of cycles increases to 5, the cumulative reductions in cohesion and internal friction angle are 15.4% and 11.1%, respectively. After 10 cycles, the cumulative reductions in cohesion and internal friction angle reach 19.1% and 12.3%, respectively.


https://doi.org/10.3390/su18020748

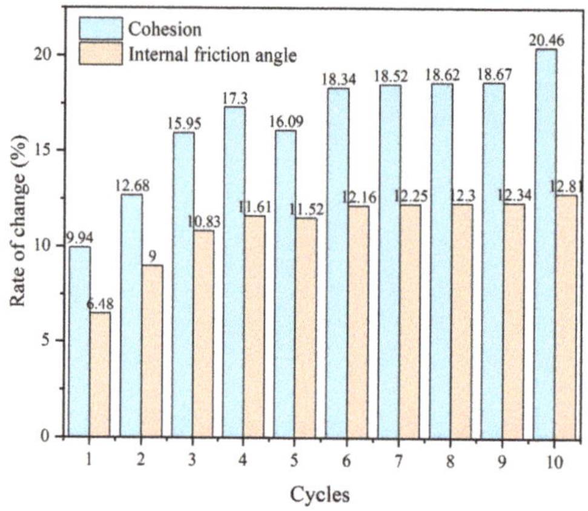


**Figure 10.** The variation trend of shear strength parameters with the cycle.

Overall, the strength parameters exhibit a rapid decline followed by a slow stabilization, suggesting that the structural damage caused by drying and wetting cycles reaches a certain upper limit. Once the particles reach a new equilibrium, the strength parameters stabilize. This change can be explained by the initial destruction of the cemented structure in the soil, followed by particle redistribution driven by drying-wetting cycles, which leads to the formation of a newly developed frictional skeleton structure, gradually slowing down the degradation [30,31]. This reflects a strength degradation feature controlled by the synergistic effect of structural and frictional mechanisms. It can be concluded that the strength evolution of granite residual soil follows a dual-stage control mechanism of structural degradation and frictional stabilization.

Although the nonlinear strength evolution observed in this study is similar to the strength changes seen in other soils after drying-wetting cycles, the underlying mechanisms are fundamentally different. For example, strength fluctuations in expansive soils are caused by significant swelling and shrinkage of clay minerals due to water absorption and loss [32], which leads to the formation of cracks. In contrast, the strength evolution of the granite residual soil in this study results from the combined effects of enhanced matric suction and degradation induced by drying-wetting cycles.

To quantify the impact of drying and wetting cycles on the shear strength parameters, a functional fitting was performed on the data presented in Figure 11. The optimal model was selected based on the coefficient of determination ($R^2$) and the root mean square error (*RMSE*). The variation of both cohesion and internal friction angle with the number of cycles was best captured by a two-term exponential function. The exponential form was selected to reflect the progressive, non-linear nature of damage accumulation in unsaturated granite residual soil under drying and wetting cycles, with rapid initial decay followed by gradual stabilization. The fitting for cohesion $c$ yielded an excellent $R^2$ of 0.9997 and an *RMSE* of 0.070. Similarly, the fitting for internal friction angle was also highly accurate, with an $R^2$ of 0.9996 and an *RMSE* of 0.044. These results demonstrate the excellent performance of the two-term exponential function. The corresponding fitting equations are presented in Equations (5) and (6).

$$c = 42.71e^{-0.0091n} + 5.47e^{-1.62n} \tag{5}$$

$$\varphi = 22.11e^{-0.0026n} + 2.44e^{-0.96n} \tag{6}$$

where $c$ is the cohesion, $\varphi$ is the internal friction angle, $n$ is the number of cycles.


https://doi.org/10.3390/su18020748

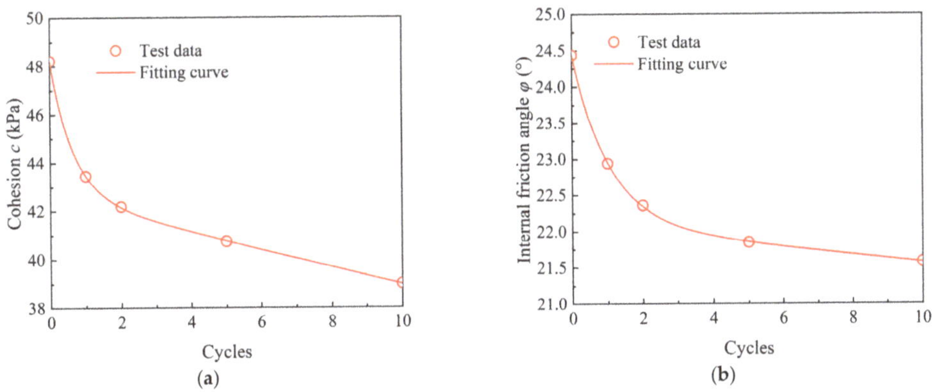


**Figure 11.** Relationship between shear strength parameters and cycles: (**a**) cohesion and cycles; (**b**) internal friction angle and cycles.

*3.3. Microstructure Evolution of Granite Residual Soil During Drying and Wetting Cycles*

To investigate the effect of drying and wetting cycles on the microstructure, samples were collected from the central section of both the original state and those subjected to 1, 5, and 10 cycles. SEM was conducted at a magnification of 300×, and the results are presented in Figure 12. The evolution of pore structure with increasing cycle numbers can be clearly observed.

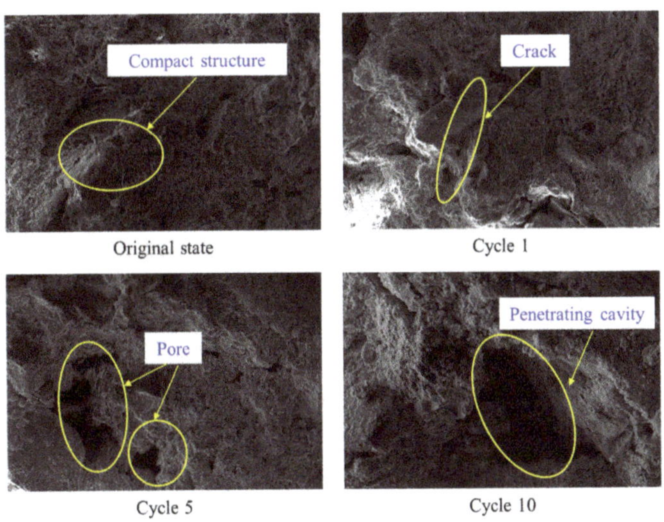


**Figure 12.** Microstructural evolution of soil samples under drying and wetting cycles observed by 300× Scanning Electron Microscopy magnification.

In the original state, the soil matrix exhibits a densely packed configuration, with particles bonded through intergranular contacts, resulting in a structurally intact and homogeneous fabric. Following the first drying-wetting cycle, the swelling and shrinkage of clay minerals induce initial microcracks within the soil structure. These incipient cracks locally disrupt the interparticle connections, leading to a slight reduction in bond strength and the onset of fabric disturbance.

As the cycle reaches 5 times, the repeated hydraulic and mechanical stresses cause the microcracks to propagate and gradually interconnect. This process not only enlarges


https://doi.org/10.3390/su18020748

existing cracks but also promotes the coalescence of smaller pores into larger and more continuous pore spaces. Consequently, the soil fabric transitions from an initially dense network to a visibly loosened structure, with increased porosity and reduced contact area between particles.

After extended cycling (10 cycles), the accumulated damage becomes more severe. The cracks and pores further expand and merge, eventually forming penetrating cavities. These discontinuities substantially weaken the structural integrity of the soil sample, thereby diminishing its load-bearing capacity. Thus, the observed progression from a dense fabric to microcracks, then to pore enlargement, and finally to penetrating cavity formation, clearly demonstrates a progressive deterioration mechanism. These SEM-based microstructural observations provide direct evidence linking the cyclic drying-wetting to the progressive degradation of macroscopic mechanical properties, thereby elucidating the physical basis for the loss of strength in granite residual soil under drying and wetting cycles.

## 4. Discussion

Although previous studies have extensively investigated the influence of matric suction [33] and drying-wetting cycles [34] as isolated factors on granite residual soil, the present work focuses on their synergistic coupling effect, which reveals a more complex and engineering-relevant mechanism. A key innovation of this study is the identification of a nonlinear strengthening-degradation mechanism under these coupled conditions, along with an analysis of its correlation with the evolution of microstructural characteristics. In comparison to the strength increase with matric suction often reported in single-factor control tests, these experiments under coupled drying-wetting cycles demonstrate a nonlinear trajectory: initial suction increase provides significant enhanced cohesion and friction angle via capillary forces, and cycling drives a progressive degradation of the soil fabric as evidenced by SEM observations of crack and pore evolution. This degradation ultimately limits the strengthening effect and leads to a strength plateau after several cycles. This critical phenomenon is not captured by single-factor analyses.

Moreover, the evolution of matric suction in granite residual soil plays a crucial role in the soil's shear strength and stability [35]. Compared to sandy soils or clayey sands, which typically show a shear strength increase with matric suction that stabilizes after relatively few drying-wetting cycles, granite residual soil exhibits a more prolonged and nonlinear evolution of shear strength [36,37]. It should be noted that the strength characteristics of granite residual soils are primarily controlled by the degree of weathering of the parent rock. For example, compared to the highly weathered granite residual soil in Guangdong, the soil samples from Fujian in this study, due to the relatively lower degree of weathering of their parent rock, exhibit higher initial cohesion and structural integrity [38,39].

In this state, the increased matric suction significantly enhances the cohesion and internal friction angle of the granite residual soil. Capillary forces improve the bonding between soil particles, thereby enhancing the structural stability of the soil. This effect is most pronounced at lower matric suction values (<100 kPa), where both cohesion and internal friction angle show significant increases. However, once the matric suction exceeds this threshold, the rate of increase in shear strength begins to slow, indicating that as the soil approaches saturation, the effect of suction diminishes, and the increase in soil strength starts to stabilize. Specifically, granite residual soil at higher matric suction demonstrates better elasticity, maintaining its load-bearing capacity more effectively while exhibiting slower strain rates during deformation [40]. This suggests that at higher matric suction, the soil can resist external disturbances for a longer period, delaying failure. Similarly, the influence of capillary force on shear strength in unsaturated granular soil was analyzed by three-dimensional discrete element simulation [41]. It was also found that apparent



https://doi.org/10.3390/su18020748

cohesion increases with matric suction, although its rate of increase declines at higher suction levels.

Therefore, it is crucial to analyze the impact of matric suction on granite residual soil in conjunction with the geological background of parent rock weathering, especially around pumped storage power stations, in order to accurately predict the soil's mechanical behavior. Slope gradients at pumped storage power stations in Fujian are typically designed between 1:2 and 1:3. The variation pattern of matric suction during the instability process can be summarized in Figure 13.

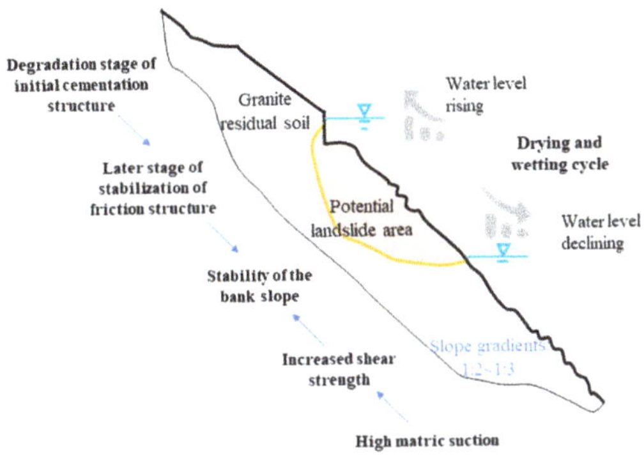


**Figure 13.** The evolution of matric suction in bank slope failure of a pumped storage power station in Fujian.

In the initial stage, the primary cementitious structure of the soil undergoes progressive damage. Subsequently, driven by drying-wetting cycles, particle redistribution occurs, leading to the formation of a new skeletal structure, which gradually moderates the degradation. The engineering contribution of this study lies in proposing a function to quantify the strength evolution of granite residual soil under the coupled conditions of drying-wetting cycles and matric suction. A high-precision exponential function can then be established. Moreover, a conceptual model for nonlinear evolution is developed, reflecting that the strength evolution of granite residual soil follows a dual-stage mechanism of structural degradation and frictional stabilization. Compared to previous empirical methods that simplified the impact of drying-wetting cycles as a strength reduction coefficient [32], this model provides an evolutionary pathway for the long-term behavior of bank slopes. In engineering practice, the number of drying-wetting cycles can be estimated based on local climate and reservoir operation patterns, allowing the model to predict strength parameters at corresponding stages.

Based on the analysis of the influence of matric suction and shear strength characteristics of granite residual soil, the following guidelines can be proposed for the bank slopes of the pumped storage power station in Fujian. (i) Control of matric suction. The findings suggest that higher matric suction improves the soil's load-bearing capacity and stability. (ii) Consideration of drying and wetting cycles. Soil stabilization techniques should be applied to maintain the integrity of the soil during these cycles. (iii) Monitoring of soil mechanical behavior. Continuous monitoring of soil mechanical properties, especially cohesion and internal friction angle, allows for timely adjustment of slope management practices. Despite the valuable insights provided by this study, there are several limitations that should be addressed in future research. The experimental data offer important findings on the strength characteristics of granite residual soil under matric suction and drying-wetting cycles. However, the study only considers a limited range of suction levels and initial dry density. Future work could investigate the interaction of environmental

 https://doi.org/10.3390/su18020748

factors, such as temperature fluctuations and rainfall, with matric suction and soil shear strength to provide a more comprehensive understanding of slope stability in pumped storage power stations.

## 5. Conclusions

This study analyzes the strength characteristics of granite residual soil under the influence of matric suction and drying-wetting cycles through experiments. The results provide fundamental data and theoretical support for the stability analysis and long-term safety assessment of the bank slopes in pumped storage power station reservoirs. The main conclusions are as follows. (i) Matric suction significantly enhances the shear strength of unsaturated granite residual soil. As matric suction increases from 50 kPa to 200 kPa, both cohesion and internal friction angle increase nonlinearly. (ii) By examining the evolution of the soil microstructure from a compact structure to penetrating cavities, it reveals a dual-stage deterioration pattern of shear strength parameters during drying and wetting cycles. In the first five cycles, rapid degradation occurs, with cohesion decreasing by approximately 15.4% and internal friction angle decreasing by about 11.2%. After five cycles, they enter a steady stage. (iii) The developed two-term exponential function considers shear strength and cycle number, enabling quantitative predictions of strength reduction under drying and wetting cycles.

**Author Contributions:** Conceptualization, X.X. and Z.X.; Methodology, Z.X. and C.W.; Writing—original draft, Z.X.; Resources, X.X. and Y.S.; Writing—review and editing, X.X. and Z.X.; Funding acquisition, X.X. and Y.S.; Visualization, Z.X. and C.W.; Supervision, X.X. and Y.S. All authors have read and agreed to the published version of the manuscript.

**Funding:** This research was funded by the Key Science and Technology Project of China's Ministry of Water Resources, grant number: SKS-2022151.

**Institutional Review Board Statement:** Not applicable.

**Informed Consent Statement:** Not applicable.

**Data Availability Statement:** The original contributions presented in the study are included in the article, further inquiries can be directed to the corresponding author.

**Conflicts of Interest:** The authors declare no conflicts of interest.

## References

1. Liu, X.; Zhang, X.; Kong, L.; Wang, G.; Lu, J. Disintegration of Granite Residual Soils with Varying Degrees of Weathering. *Eng. Geol.* **2022**, *305*, 106723. [CrossRef]
2. Kong, Y.; Kong, Z.; Liu, Z.; Wei, C.; Zhang, J.; An, G. Pumped Storage Power Stations in China: The Past, the Present, and the Future. *Renew. Sustain. Energy Rev.* **2017**, *71*, 720–731. [CrossRef]
3. Bai, R.; Jin, L.; Fan, Y.; Zhang, X.; Yin, R.; Wei, Y. Analyzing Water Resources with Sector Characteristics in Fujian Province Utilizing S2-DR-BIMCCP Model Integrated with the Gini Coefficient. *J. Clean. Prod.* **2025**, *490*, 144737. [CrossRef]
4. Yang, Y.; Li, Z.; Chen, Y.; Zhang, Y.; Lu, L. Periodic Flooding Alters Ecological Processes and Carbon Metabolism Efficiency of Riparian Soil Microbial Communities in the Three Gorges Reservoir Area, China. *J. Environ. Manag.* **2025**, *376*, 124534. [CrossRef] [PubMed]
5. Tang, C.-S.; Wang, D.-Y.; Shi, B.; Li, J. Effect of Wetting–Drying Cycles on Profile Mechanical Behavior of Soils with Different Initial Conditions. *Catena* **2016**, *139*, 105–116. [CrossRef]
6. Fan, Y.; Zheng, M.; Wu, J. A Study on the Shear Strength Characteristics and Microscopic Mechanism of Coal-Bearing Soil under Dry-Wet Cycles. *Front. Earth Sci.* **2023**, *10*, 1096980. [CrossRef]
7. Yu, M.; Gui, Y.; Li, B. Modelling Soil-Water Retention Curves Subject to Multiple Wetting-Drying Cycles: An Approach for Expansive Soils. *Comput. Geotech.* **2024**, *171*, 106335. [CrossRef]
8. Sun, Z.; Wu, C.; Qing, Y.; Li, S. Influence of Wetting-Drying Cycles on the Shear Strength Behavior of Electroosmotically Treated Soft Clay. *Sci. Rep.* **2025**, *15*, 5748. [CrossRef]

 https://doi.org/10.3390/su18020748

9. Liu, S.; Huang, X.J.; Gan, L.; Zhang, Z.B.; Dong, Y.; Peng, X.H. Drying-Wetting Cycles Affect Soil Structure by Impacting Soil Aggregate Transformations and Soil Organic Carbon Fractions. *Catena* **2024**, *243*, 108188. [CrossRef]
10. Gao, Z.; Kong, L.; Yan, J.; Wang, S. Effects of Drying-Wetting Cycles on Small-Strain Stiffness Characteristics of Fissured Clay. *J. Rock Mech. Geotech. Eng.* **2025**, *17*, 4618–4631. [CrossRef]
11. Zhuo, Z.; Cai, W.; Zhu, C.; Tang, C.-S.; Roksana, K. Morphology Characterization of Unsaturated Soils under Drying-Wetting Cycles: Crack Opening and Closure. *Acta Geotech.* **2024**, *19*, 7287–7306. [CrossRef]
12. Miao, L.; Jing, F.; Houston, S.L. Soil-Water Characteristic Curve of Remolded Expansive Soils. In Proceedings of the Fourth International Conference on Unsaturated Soils, Carefree, AZ, USA, 2–6 April 2006; pp. 997–1004. [CrossRef]
13. Wen, T.; Shao, L.; Guo, X.; Zhao, Y. Experimental Investigations of the Soil Water Retention Curve under Multiple Drying-Wetting Cycles. *Acta Geotech.* **2020**, *15*, 3321–3326. [CrossRef]
14. Xu, X.; Liu, D.; Xian, Z.; Yang, F.; Jian, W.; Xu, X.; Huang, J. Influence of Drying-Wetting Cycles on the Water Retention and Microstructure of Residual Soil. *Geofluids* **2022**, *2022*, 9948658. [CrossRef]
15. Zhang, Q.; Zhang, H.; Wang, L.; Li, Q.; Yu, H. Effect of Structural Characteristics on the Stability of Multi-Weak Rock Slopes Considering the Spatial Variability of Geotechnical Parameters. *Sci. Rep.* **2024**, *14*, 30618. [CrossRef] [PubMed]
16. Hu, B.W.; Hu, Q.Z.; Liu, Y.M.; Tao, G.L. Research on the Improvement of Granite Residual Soil Caused by Fly Ash and Its Slope Stability under Rainfall Conditions. *Appl. Sci.* **2024**, *14*, 3734. [CrossRef]
17. Chen, C.; Yang, H.; Song, K.; Liang, D.; Zhang, Y.; Ni, J. Dissolution Feature Differences of Carbonate Rock within Hydro-Fluctuation Belt Located in the Three Gorges Reservoir Area. *Eng. Geol.* **2023**, *327*, 107362. [CrossRef]
18. Lin, J.; Zhu, Y.; Hou, J.; Pan, B.; Chu, W.; Huang, W. Equivalent Continuum Coupling-Based Slope Stability Analysis of Zhouning Pumped Storage Power Station. *Geofluids* **2021**, *2021*, 9926751. [CrossRef]
19. Wei, S.; Ji, F.; Ding, J.; Guan, M.; Lu, Y. Influence of Extremely Rapid Cyclic Reservoir Water Level Fluctuations on Bank Slope Stability: Insights from Model Testing and Numerical Simulation of a Pumped Storage Power Station Slope. *Eng. Geol.* **2025**, *357*, 108380. [CrossRef]
20. *GB/T50123-2019*; Standard for Geotechnical Testing Method. China Architecture and Building Press: Beijing, China, 2019.
21. Diel, J.; Vogel, H.-J.; Schlüter, S. Impact of Wetting and Drying Cycles on Soil Structure Dynamics. *Geoderma* **2019**, *345*, 63–71. [CrossRef]
22. Shah, S.S.A.; Turrakheil, K.S.; Naveed, M. Impact of Wetting and Drying Cycles on the Hydromechanical Properties of Soil and Implications on Slope Stability. *Atmosphere* **2024**, *15*, 1368. [CrossRef]
23. Wu, T.Y.; Xia, C.B.; Chen, H.; Xu, X.C.; Pan, D.M.; Chen, J. An Improved Double Chamber Triaxial Volume Change Measurement System for Unsaturated Soils. *Front. Built Environ.* **2025**, *10*, 1518953. [CrossRef]
24. Radhey, S.S.; Longjie, H.; Sonal, S. Developments in Measurement of Volume Change in Triaxial Testing of Unsaturated Soils. *Adv. Unsaturated Soil Seepage Environ. Geotech.* **2006**, 93–101. [CrossRef]
25. Abd, I.A.; Fattah, M.Y.; Mekkiyah, H. Relationship between the Matric Suction and the Shear Strength in Unsaturated Soil. *Case Stud. Constr. Mater.* **2020**, *13*, e00441. [CrossRef]
26. Shan, X.; Chen, H.; Ma, B.; Yu, Q.; Wang, Z.; Wang, Q. Analysis of Factors and Mechanisms Influencing the Matric Suction and Water-Holding Capacity of Carbonate Saline Soil. *Water* **2025**, *17*, 469. [CrossRef]
27. Fredlund, D.G.; Xing, A. Equations for the Soil-Water Characteristic Curve. *Can. Geotech. J.* **1994**, *31*, 521–532. [CrossRef]
28. Wang, S.; Liu, Y.; Wu, C.; Li, Y.; Zhang, R.; An, S.; Jiang, H.; Yao, Z. Investigation of Water Retention and Volumetric Deformation Characteristics of Soils Subjected to Wetting-Drying Cycles. *Bull. Eng. Geol. Environ.* **2025**, *84*, 155. [CrossRef]
29. Li, B.; Dai, L.; Tan, Y.; Liu, C.; Tian, B.; Yang, J. Influence of Multiple Wetting-Drying Cycles on Water Retention and Pore Structure of Silt Loam from Reservoir Hydro-Fluctuation Belt: Experiments and Modeling. *Bull. Eng. Geol. Environ.* **2025**, *84*, 104. [CrossRef]
30. Lu, Y.; Zhu, W.; Ye, G.; Zhang, F. A Unified Constitutive Model for Cemented/Non-Cemented Soils under Monotonic and Cyclic Loading. *Acta Geotech.* **2022**, *17*, 2173–2191. [CrossRef]
31. Bai, B.; Zhou, R.; Cai, G.; Hu, W.; Yang, G. Coupled Thermo-Hydro-Mechanical Mechanism in View of the Soil Particle Rearrangement of Granular Thermodynamics. *Comput. Geotech.* **2021**, *137*, 104272. [CrossRef]
32. Zhou, R.; Wang, B.; Han, S.; Wang, D.; Zhang, F. Mechanisms of Crack Development and Strength Deterioration in Compacted Expansive Soils under Controlled Wetting-Drying Conditions. *Eng. Fail. Anal.* **2024**, *159*, 108133. [CrossRef]
33. Deng, X.; Zhu, Z.; Lin, X.Y.; Yang, X.Q. Effect of Matrix Suction on the Shear Strength Characteristics of Reinforced Granite Residual Soil. *Front. Mater.* **2023**, *10*, 1284722. [CrossRef]
34. An, R.; Zhang, X.; Kong, L.; Liu, X.; Chen, C. Drying-Wetting Impacts on Granite Residual Soil: A Multi-Scale Study from Macroscopic to Microscopic Investigations. *Bull. Eng. Geol. Environ.* **2022**, *81*, 447. [CrossRef]
35. Wu, D.; Liu, H.; Wang, C.; Xu, X.; Liu, X.; Wang, Q. The Interaction Effect of Particle Composition and Matric Suction on the Shear Strength Parameters of Unsaturated Granite Residual Soil. *Arab. J. Sci. Eng.* **2022**, *47*, 12453–12467. [CrossRef]


 https://doi.org/10.3390/su18020748

36. Awad, M.; Sasanakul, I. Hysteresis Effects on Shear-Induced Suction and Strength of Unsaturated Clayey Sand and Silt. *Geotech. Geol. Eng.* **2025**, *43*, 523. [CrossRef]
37. Bozyigit, I. Influence of Clay Mineralogy on Undrained Cyclic Strength of Sand with Fines. *Eng. Sci. Technol. Int. J.* **2024**, *59*, 101872. [CrossRef]
38. Tan, D.; XU, X.; Wang, L.; Xu, J.; Shi, Q. Deformation Evolution and Failure Mechanism of Rainfall-Induced Granite Residual Soil Landsliding Event in Northern Guangdong, China. *Landslides* **2025**, *22*, 925–941. [CrossRef]
39. Wang, R.; Li, H.; Chen, Z.; Liu, F.; Wei, M.; Liu, F.; Wang, Q.; Hu, C. Strength and Mechanism of Granite Residual Soil Strengthened by Microbial-Induced Calcite Precipitation Technology. *Appl. Sci.* **2023**, *13*, 8863. [CrossRef]
40. Kholghifard, M.; Ahmad, K. Effect of Matric Suction and Density on Yield Stress, Compression Index and Collapse Potential of Unsaturated Granite Soil. *KSCE J. Civ. Eng.* **2021**, *25*, 2847–2854. [CrossRef]
41. Shen, Z.; Jiang, M.; Thornton, C. Shear Strength of Unsaturated Granular Soils: Three-Dimensional Discrete Element Analyses. *Granul. Matter* **2016**, *18*, 37. [CrossRef]




 https://doi.org/10.3390/su18020748

*sustainability*

MDPI

*Article*

# Application of a Multi-Component Conditioner as a Sustainable Management Practice for Enhancing Soil Properties and *Hordeum vulgare* L. Growth and Yield

Jacek Długosz [1], Karol Kotwica [2], Ewelina Przybyszewska [3] and Anna Piotrowska-Długosz [1,*]

[1] Department of Biogeochemistry, Soil Science, Irrigation and Drainage, Bydgoszcz University of Science and Technology, Bernardyńska 6/8 Street, 85-029 Bydgoszcz, Poland; jacekd@pbs.edu.pl

[2] Department of Agronomy, Bydgoszcz University of Science and Technology, S. Kaliskiego 7 Street, 85-796 Bydgoszcz, Poland; kotwica@pbs.edu.pl

[3] NaturalCrop Poland Ltd., KEN Avenue 57/2, 02-797 Warsaw, Poland; e.przybyszewska@naturalcrop.com

* Correspondence: apiotr@pbs.edu.pl

**Abstract:** The purpose of this study was to evaluate how a multi-component soil conditioner consisting of zeolite, calcium carbonate, potassium humate, and *Ascophyllum nodosum* extract affects selected soil properties (physical, chemical, and water-related properties, as well as microbial and enzymatic properties) and the growth and grain yield of spring barley (*Hordeum vulgare* L.). To achieve the goal, one-year research experiments were conducted at three conventionally tilled sites, which were situated on farms across three geographically separate regions in the Kuyavian–Pomeranian Region of Midwestern Poland. Most of the chemical properties, namely, total organic C, total N, pH in KCl, cation exchangeable capacity (CEC), as well as exchangeable (Mg, Ca, K, and Na) and available (Mg, K, and P) forms of nutrients, were not significantly affected by the conditioner or sampling time. Independent of the study location, the percentage of macropores in total porosity (TP) and dissolved nitrogen content (DNt) determined in July were considerably greater in the soil treated with *Solactiv* compared to the reference soil. Bulk density (BD), in turn, showed the opposite tendency, also suggesting the positive effect of the studied conditioner. At all study sites, application of the conditioner significantly reduced the percentage of micropores in total porosity (TP) (by 17%), while significantly increasing the content of macropores in TP (15%) and enhancing the percentage of available and readily available water capacity (8.5% and 14%). No clear changes in the results of C and N form and enzymatic activity were noted. The activities of DHA and FDAH behave differently in each study site, making it difficult to draw clear conclusions. The cellulase was the only enzyme that was significantly and positively affected by *Solactiv* at all study sites and for both sampling times. The values of dry matter of roots and plants, barley root length and surface, and barley grain yield were considerably greater in soil amended with *Solactiv* compared to the reference soil. Because some important soil and plant properties showed a positive response toward the tested conditioner, despite the low dose used, further studies should be conducted at a larger scale, focusing on different soils and plants.

**Keywords:** spring barley; organic–mineral conditioner; zeolite; soil physicochemical properties; water properties; soil enzymatic activity

## 1. Introduction

Agricultural productivity and crop yields are closely associated with soil health and fertility [1]. Unfortunately, currently soil resources worldwide face significant threats from overuse, degradation, and permanent loss due to the unbalanced soil management practices and changes in land use, as well as extreme weather and climate change and events (droughts and floods) [2,3]. Widely used agricultural practices such as intensive tillage and the application of high doses of synthetic fertilisers and pesticides, combined with minimal use of organic manure and cover crops, contribute substantially to ongoing soil deterioration as manifested by loss of organic matter, reduction in microbial diversity, and soil compaction and erosion [4]. Changes in soil moisture are important for the general health of soil because they significantly affect its overall biological activity, organic matter transformation, and nutrient cycling [5].

Poland has a significant area of agricultural land, though this is systematically decreasing. Moreover, the quality of agricultural soils in Poland is also a topic of concern, since over 40% of the country's farmland comprises poor-quality sandy soils with limited agricultural potential [6]. These soils contain insufficient organic carbon and clay content, resulting in low cation exchange capacity (CEC) [7], making them highly vulnerable to drought conditions and nutrient loss through leaching. Additionally, approximately 80% of cultivated land shows varying degrees of acidification, and four million hectares require lime treatment [7].

In response to global and progressive soil deterioration and fertility loss, different soil additives and conditioners targeting specific soil-related problems are being developed and applied to limit soil degradation and improve soil quality worldwide [8–14]. Conditioners are natural or synthetic materials added to soil to improve its chemical, physical, microbial, or biochemical properties [8]. The primary aims of their application are to increase soil fertility by providing nutrients to plants; improve soil structure by enhancing soil aeration and drainage and diminishing excessive soil compaction; mitigate soil alkalinity or salinity by adjusting soil pH; and improve soil biological activity by activating soil microorganisms and their enzymes [11–20]. Therefore, soil conditioners should be dedicated to specific soil types; for example, use materials like compost, mulch, or biochar to improve sandy or coarse soils by enhancing water retention and fertility [21]. For heavy clay soils that hold too much water, materials such as gypsum or other conditioners that improve drainage and aeration are more suitable. Lime-based products have been successfully used in acid soils to reduce acidity [22]. Conditioners containing humic acid salt have been used to enhance root development and plant growth [15,23], while hydrogel and zeolite formulations have been applied to improve water retention [24]. A large group of commonly used conditioners comprises microbial preparations containing beneficial bacteria and fungi to improve soil biological activity [25,26].

Soil conditioners are often applied in relatively high doses (e.g., zeolites and biochar are applied in the dose of 5–10 t $ha^{-1}$), which can have an adverse environmental effect, caused by conditioner compounds or their degradation products, and generate higher costs of plant cultivation, e.g., [27]. Using appropriate doses helps avoid excessive applications that could be uneconomical or lead to negative environmental impacts. An unexplored scientific gap in current soil conditioner application is the application of multiple conditioners as mixtures, thus reducing individual application rates by possible complementarities derived from the interaction of diverse components, e.g., [12,17]. Integrated approaches that combine soil conditioners with other strategies can enhance the positive feedback and the effectiveness of the conditioner over time. For example, combining biochar and compost can improve water retention, microbial activity, and nutrient sorption more effectively than either material alone [17]. Similarly, organic conditioners mixed with bio-organic fertilisers and micronutrients boost microbial communities and nutrient availability, resulting in better crop growth [9,10]. Other mixed conditioners, such as gypsum and biochar, can have a significant synergistic effect in improving salt-affected soils by reducing exchangeable sodium and increasing microbial biomass [13]. Additionally, the effect of low application rate is also important from an economic point of view.

In this study, we used *Solactiv*, a commercially available, innovative multi-component mineral–organic soil conditioner containing four synergistic components: (1) zeolite-clinoptilolite for enhanced soil sorption and water retention, (2) calcium carbonate for pH stabilisation, (3) potassium humate (originating from leonardite), which is rich in humic acids for improving soil health and plant growth. The literature reports that each of these components has exhibited a significant effect on soil and/or plant properties [18,28]. Zeolites, due to their intrinsic adsorption capacity, offer multiple benefits for soil environments by improving water retention in sandy soils and increasing porosity in clay-rich soils. Additionally, soils amended with zeolites showed increased CEC and enhanced nutrient retention [29]. They retain nutrients and serve as carriers that slowly release mineral nutrients after application, consequently improving yields and quality of crops [13,30–32]. Potassium humates, which are salts of humic acid originating from lignite brown coal, are considered to be a potential soil conditioner that works by improving SOM content, soil porosity, and water-holding capacity (WHC) as well as cation exchange capacity (CEC) [33,34]. Other studies have indicated that potassium humates activated plant growth by promoting root development, increasing nutrient uptake efficiency, and improving stress tolerance [15,35]. Positive responses in plant properties were also noted for the application of seaweed extracts (with spp. *Ascophyllum*, *Fucus*, and *Laminaria* being the most commonly used genera) as biostimulants, wherein both solid (granules) and liquid (drip and foliar spray) forms of application were used [19,36]. The application of seaweed extracts, being a source of phytohormones, polysaccharides and minerals, positively affected seed germination [11], shoot and root growth [37], nutrient use efficiency [11,38] reduction in biotic and abiotic stress [18,39,40] and improvements in crop yield and quality [41,42] and soil properties, e.g., soil microbial and enzymatic properties [43,44].

In the only study on the effect of *Solactiv* on soil properties carried out in soil under maize cultivation [45], the enzymatic activity increased significantly, while no such positive response was found for the amount of C and N determined in microbial biomass. Additionally, the application of the studied conditioner enhanced the content of available K (by about 11%), as well as improving soil water storage by increasing available water capacity (AWC). Because of the promising results obtained in the above-mentioned studies carried out under maize cultivation, further research related to other important agricultural plants is expected. Additionally, while most research on soil conditioners and their components demonstrates benefits for plant and soil properties under controlled laboratory conditions, e.g., [46], fewer studies have examined their effectiveness in the actual field environment, e.g., [47,48]. Accordingly, this study evaluated the field-based effectiveness of the selected conditioner under spring barley cultivation. We hypothesised that the conditioner with a multi-component formulation, besides a low dose used, would significantly improve soil properties (mainly microbial and biochemical variables, nutrient availability, and water properties) in an intensively managed field. We also hypothesised that the conditioner would increase spring barley growth and yield. To investigate the assumed hypotheses, this study aimed to evaluate how *Solactiv* affects a range of soil characteristics and spring barley growth and productivity across field trials at three study sites.

## 2. Materials and Methods

### 2.1. Description of the Experimental Fields and Soil Sampling

The influence of *Solactiv* on a range of soil features and spring barley growth and yield was investigated in three annual trials under conventional tillage practices. All research locations were situated on farms in three distinct mesoregions in the Kuyavian–

Pomeranian Region (Midwestern Poland): Agriculture Cooperative *Zjednoczeni* in Janocin (52°36′24.6″ N 18°25′11.9″ E), an individual farm in Kobylnica (52°35′10.4″ N 18°26′46.3″ E), and an individual farm in Samsieczynek (53°13′36.4″ N 17°40′57.9″ E) (Figure 1).

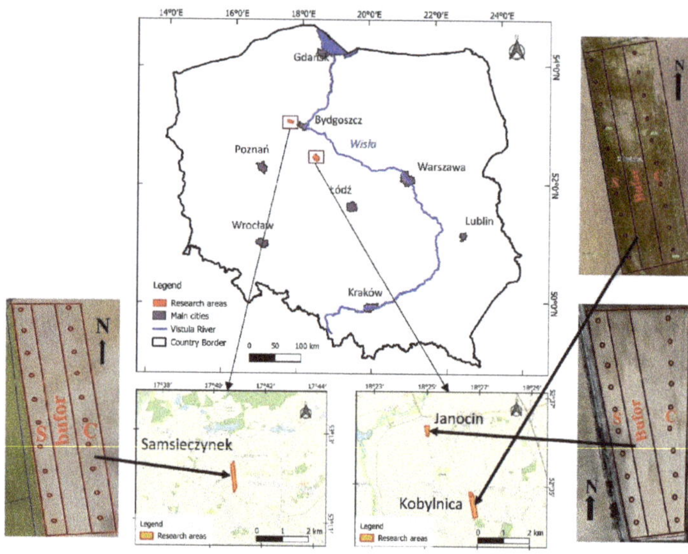


**Figure 1.** Location of research sites and sampling scheme: S—soil amended with *Solactiv*, and C—control soil.

The research sites are situated in a temperate zone characterised by variable climate conditions, where North Atlantic maritime air masses meet continental air masses from the east. This convergence results in considerable daily and annual fluctuations in meteorological patterns. Weather data for the sampling periods were collected from nearby meteorological stations positioned close to the research locations (specifics are provided in Figure 2). Based on the IUSS Working Group WRB classification system, the examined soils were categorised as Haplic at Janocin, Haplic Luvisols at Kobylnica, and Albic Luvisols at Samsieczynek [49]. The basis for this classification was the finding of the presence of an agric diagnostic horizon in all profiles studied, and also the albic horizon in the soil from the Samsieczynek site. The soils across all locations developed from glacial till deposits. The upper soil layers at every study site have a sandy loam texture. Detailed information about the applied mineral fertilisation is given in Table S1.

An experimental area was established at each study site and separated into two equal sections (fields). One experimental section was managed with a tested conditioner at a rate of 300 kg ha$^{-1}$ following the manufacturer's suggested application rate (no additional benefits were observed at higher rates), while the other section served as an untreated control. The soil conditioner was applied prior to spring barley sowing and incorporated into the soil to a depth of 12 cm using cultivation equipment.

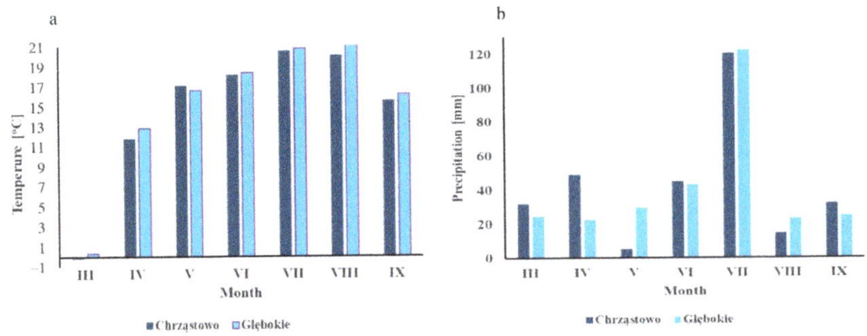


**Figure 2.** Average air temperature (**a**) and precipitation (**b**) at the experimental sites. Notes: Chrząstowo weather station is located nearest to the Samsieczynek study site, while Głębokie weather station is located nearest to the Janocin and Kobylnica study sites.

The tested conditioner contains the following components: (1) zeolite—50%, (2) calcium carbonate [$CaCO_3$]—47.5%, (3) potassium humate powder—2%, and (4) the extract of *Ascophyllum nodosum*—0.5%. The zeolite applied in the *Solactiv* production originated from Western Slovakia and is supplied by the company Zeocem, Inc. (Bystré, Slovakia). Although clinoptilolite is the primary mineral in the zeolite used (84%), cristobalite, illite, and feldspars were also found in its composition. The specific surface area of the applied zeolite, estimated using the Brunauer–Emmett–Teller (BET) technique, ranged from 31.4 to 35.4 $m^2$ $g^{-1}$, indicating that this particular zeolite has a relatively moderate surface area compared to some other zeolites [50]. Two primary raw materials were used for conditioner production. First, *Ascophyllum nodosum* seaweed extract was acquired from Atlantic Ocean waters near Ireland through the Irish company BioAtlantis Ltd. (Tralee, Ireland). The extract's organic matter (8–12%) includes alginate (27%), fucoidan (11%), phlorotannin (1%), and mannitol (1%). Potassium humate powder (95%, with 60–65% humic acids) from *Humic Growth Solution*, Inc. was also used as a pure material for conditioner production. It contained 60–65% humic acid derived from premium North American sources of weathered, oxidised sub-bituminous coal rich in humic compounds. This study used potassium humate with 95% purity, sourced from Humic Growth Solution, Inc. (Jacksonville, FL, USA), as another basic component for conditioner production. This component was extracted from leonardite—high-quality oxidised lignite (brown coal) sourced from the richest natural deposits located in North America, particularly in the western regions.

Bulk density and water-related properties were determined once (after barley harvesting), while the changes in other soil properties were assessed in samples collected twice: before barley sowing (prior to *Solactiv* application) and shortly after barley harvest. Twenty soil samples each were collected from both treated and control fields at each location. A hand auger was used to collect soil material from the Ap horizon (down to a depth of approximately 25–30 cm) at systematic intervals throughout the field zone (30 m × 10 m). At each grid point, ten individual sub-samples were randomly gathered within a two-metre radius and then combined to form the composite samples. To determine soil enzymatic activity and soil microbial biomass C and N, the soil samples were placed in plastic containers (which permitted gas exchange) and chilled to 4oC in order to minimise any changes in the populations of microorganisms. The microbial and enzymatic activity was determined within two weeks based on fresh soil samples. The soil samples used to determine the physicochemical properties of the soil were air-dried and sieved (2 mm).

### 2.2. Determination of Soil Physicochemical Properties

The following analyses were performed on dry and sieved soil samples using the methodology commonly accepted in soil science [51]: the particle size was assessed based on the Casagrande method using the modification by Prószyński, and the sand fraction content was evaluated using the sieving method [52]. Total organic carbon (TOC) and total nitrogen (Nt) concentrations were assessed by dry combustion using a CN analyser –Vario Max CN (Elementar Analysensysteme GmbH, Langenselbold, Germany) [53]. Dissolved organic carbon and nitrogen (DOC and DNt) were extracted with 0.004 M $CaCl_2$ for one hour at a 1:10 soil–solvent ratio and determined using a Multi N/C 3100 analyser (manufactured by Analityk, Jena, Germany) [53]. Each analysis was performed in triplicate, and mean values were presented. The obtained results were expressed as mg C (N) $kg^{-1}$ dry soil weight. Atomic absorption spectrometry (AAS) using a Philips PU 9100X apparatus (Eindhoven, The Netherlands) was employed to measure available magnesium and potassium levels after extraction with 0.0125 M $CaCl_2$ (for Mg) [54] and the Egner–Riehm DL method [55] (for K and P). The potassium content in the solution was determined by Atomic absorption spectrometry (AAS) using a Philips PU 9100X apparatus. Phosphorus was determined by the colorimetric method, and was measured using a spectrophotometer UV Vis Evolution 220 (Thermo Scientific, Waltham, MA, USA). The contents of exchangeable nutrients (Mg, Ca, K, Na) were analysed in 0.1 M $BaCl_2$ solutions [56]. Soil pH was determined potentiometrically in 1 M KCl (1:2.5 soil-to-solution ratios) [57], and hydrolytic acidity (Hh) was measured in 1M $CH_3COONa$ [58]. Moisture content was obtained gravimetrically after 24 h drying at 105 °C [59]. The effective cation exchange capacity (CEC) was presented as a sum of the Hh and basic saturation (BS) values [56]. Bulk density and water retention measurements were conducted on cylindrical soil samples with four replicates each [60]. The total porosity (TP) was calculated using the following formula: $TP = (Sw - So) \cdot Sw^{-1} \cdot 100$ (%) [60]. Properties related to the water retention were determined using low- and high-pressure chambers (pF range 0–2.7 and pF range 3.0–4.2, respectively). Volumetric water capacities (VWC) were evaluated at soil water potentials of 98.1 hPa (pF 2.0), 981.0 hPa (pF 3.0), and 15,547.9 hPa (pF 4.2) [60]. Pore volume categories and corresponding water capacities were calculated in the following way: macropores (as the difference between total porosity and VWC at pF 2.0), micropores (Wvol. at pF 4.2), and mesopores representing potentially useful water retention or available water capacity (AWC) (as the difference between VWC at pF 2.0 and at pF 4.2). Additionally, the readily available water capacity (RAWC) (the difference between VWC at pF 2.0 and VWC at pF 3.0) and small pore available water capacity (SAWC) (the difference between VWC at pF 3.0 and VWC at pF 4.2) were determined as the components of available water capacity (AWC) [60].

### 2.3. Soil Enzymatic and Microbial Biomass Content

The method of Thalmann et al. [61] was used to evaluate soil dehydrogenase activity (DHA). Briefly, a soil sample (1 g) was mixed with 1% TTC solution (triphenyltetrazolium chloride) and TRIS buffer (100 mM, pH between 7.4 and 7.8) and left in a shaking incubator for 24 h at 30 °C. After incubation, acetone was added to extract the reaction product (triphenyl formazan—TPF). The soil suspensions were then filtered, and the optical density of the clear filtrate was measured spectrophotometrically at 546 nm using a spectrophotometer UV Vis Evolution 220 (Thermo Scientific, Waltham, MA, USA). The DHA was presented as mg of TPF released per kg dried soil (mg TPF $kg^{-1}$ 24 $h^{-1}$). The total soil hydrolysis activity was determined by the method outlined by Adam and Duncan [62]. Soil samples with natural moisture content were mixed with a phosphate-buffer solution (60 mM, pH 7.6) that contained fluorescein diacetate salt as a substrate. After a one-hour incubation at 37 °C, the reaction was stopped by adding a methanol–chloroform solution (1:2 ratio). The soil mixture was then subjected to centrifugation (3000 rpm) using Beckman GS-6R centrifuge (Beckman Coulter Life Science, Indianapolis, IN, USA), and the absorbance of clear supernatant was measured colorimetrically at 490 nm wavelength using a spectrophotometer UV Vis Evolution 220 (Thermo Scientific, Waltham, MA, USA). The level of the FDAH activity was quantified as milligrams of fluorescein generated per kilogram of soil per hour (mg F $kg^{-1}$ $h^{-1}$). Cellulase (CEL) activity was measured following the procedure described by Schinner and von Mersi [63]. This involves incubating field-moist soil samples with the acetate buffer (50 mm, pH 5.5) and carboxymethylcellulose solution (serving as a substrate) for 24 h at 50 °C. The hydrolysis products (reducing sugars) generated during the incubation period were quantified spectrophotometrically at 690 nm wavelength using a spectrophotometer UV Vis Evolution 220 (Thermo Scientific, Waltham, MA, USA). The CEL activity levels were expressed as milligrams of glucose liberated per kilogram of dry soil per hour (mg Glu $kg^{-1}$ $h^{-1}$).

The content of microbial biomass carbon and nitrogen (MBC, MBN) was evaluated using the chloroform fumigation–extraction protocol outlined in Brookes et al. [64] and Vance et al. [65], respectively. The selected method assumes fumigation of soil samples with chloroform to lyse microbial cells, followed by extraction of the released C and N. Soil samples (25 g) with controlled moisture content (50%) were exposed to ethanol-free chloroform ($CHCl_3$) at 25 °C for a 24 h period. After incubation time, the chloroform was removed through successive evacuation using a vacuum pump. Both fumigated samples and untreated control soils were then processed with 0.5 M $K_2SO_4$ solution (using a 5:1 ratio) for half an hour and examined for soluble carbon concentrations according to the protocol proposed by Vance et al. [65]. In turn, the content of total nitrogen was evaluated using the method developed by Bremner and Mulvaney [66]. The correction factor of 0.38 (kEC) [65] was applied to adjust for the recovery of MBC, while the factor of (0.54 = kEN) [67] was used to calculate the real values of MBN content. The difference in extractable C and N between fumigated and control samples, after using the appropriate correction factors, gives the correct results of microbial biomass content.

### 2.4. Plant Characteristics

At each experimental location, plant biomass was collected for analysis during the earing stage of spring barley (BBCH 55–56) from five measurement points within each experimental replication. At a single point, a soil clod containing plants and their roots (20 cm × 20 cm × 25 cm) was dug up. Immediately after the plants were dug up, two individuals were separated from the soil in a water wash and then weighed as a whole (shoot and root) to determine their fresh weight. The plants were then divided into aboveground and belowground parts and weighed again to determine the fresh weight of the shoot and root parts. To determine the dry weight of the whole spring barley plants, from their aboveground parts and from their roots, fresh plant biomass obtained from each experimental site was dried at 105 °C and then weighed [68]. Root analysis determined the total root length, root area, and average root diameter. A portion of the roots of individual spring barley plants, previously prepared for dry weight analysis, was used for the analysis. A digital scanning system was used in the study, so the prepared roots were stained with a methylene blue solution to enhance contrast before scanning. During the analysis, each root sample was evenly distributed in a water layer on a transparent tray and scanned at 200 dpi using an Epson Expression 836 L system scanner (Seiko Epson Corporation, Nagano, Japan). The roots were analysed for total root length, root system area, and average root diameter using WinRHIZO v. 5.0 software from Regent Instruments Inc., Quebec City, QC, Canada. Spring barley grain was harvested from the experimental units of individual research sites using a Wintersteiger Classic plot combine harvester equipped with a cutting table with a belt feeder with a working width of 1250 mm (in Janocin and Kobylnica). Grain harvested from individual experimental units was weighed, and its actual moisture content was determined. Grain yield was expressed in t/ha at 15% water content.

### *2.5. Statistical Calculations*

The Shapiro–Wilk test (Statistica v. 13.3, TIBCO Software Inc., based in Palo Alto, CA, USA) revealed that the measured variables did not follow a normal distribution, so the data were subjected to logarithmic transformation. Since the transformation successfully improved distributional normality, all subsequent analyses were performed with the transformed dataset. We used one-way analysis of variance (ANOVA) to evaluate how the application of conditioner affected soil and barley characteristics as compared to the untreated controls, as well as to compare the results obtained from the two sampling times. The results of these analyses were the basis for performing a comprehensive ANOVA synthesis for all three study sites. In this synthesis, the study location served as a random effect, while conditioner application was the constant (mixed model). When ANOVA revealed statistically significant treatment effects, Tukey's HSD post hoc test was used to identify specific differences between treatment means with significance determined at $p < 0.05$. To assess differences between soil samples, Principal Component Analysis (PCA) was performed on the mean values of all studied variables. Sample ordination was based on the first two principal components (PC1 and PC2). Dataset variability was quantified using the coefficient of variation (CV%), which was categorised following Wilding's [69] classification system: low variability (CV = 0–15%), moderate variability (CV = 16–35%), and high variability (CV > 36%).

## 3. Results

### *3.1. Soil Chemical, Physical, and Water-Related Properties*

Because no significant changes were obtained for chemical properties between study sites, the data presented in Table 1 are mean values for all study sites. Most of these properties, i.e., Corg, Ntot, pH in KCl, S, ECEC, basic saturation (BS), exchangeable forms of Mg, K, and Na, as well as available forms of P and Mg, were not significantly affected by conditioner application or sampling time.

The exception was the available K and exchangeable Ca content. The first of them was significantly higher in soil taken in March than in July, and the decline was approximately 11%, while the second one was 10% higher in soil collected after barley harvest than in soil taken in March. Further, the hydrolytic acidity (Hh) was significantly higher in soil destined for and treated with the conditioner than in control soil, and this relationship held for both sampling dates. The contents of Corg and Ntot were in the ranges 8.20–8.99 mg $kg^{-1}$ and 1.07–1.13 mg $kg^{-1}$, respectively. The soil pH in KCl was slightly acidic and ranged between 6.24 and 6.51. The concentration of Pavail was in the high class of its content (67–88 mg $kg^{-1}$) while the content of available Mg and K was in the average class according to PN-R-04023 [70]. Most of the chemical properties exhibited a moderate or small variability, as indicated by CV values between 7.6 and 33.6%. Only the Hh and exchangeable forms of Mg (in control soil), Ca (in both fields), K and Na (July, control soil) have a high variability (CV = 36.3–75.0%).

The studied soils are composed of 39.2% silt (from 0.05 to 0.002 mm) and 5.1% clay (< 0.002 mm) and displayed a moderate variability (Table 2).

**Table 1.** Soil chemical properties as determined by *Solactiv* and sampling month (mean values for the study locations).

| Property | Sampling Month | Control | | Solactiv | |
|---|---|---|---|---|---|
| | | Mean ± SE | CV [%] | Mean ± SE | CV [%] |
| Corg [g kg$^{-1}$] | * III | 8.69 (±0.48) | 19.2 | 8.99 (±0.60) | 23.2 |
| | ** VII | 8.20 (±0.26) | 10.8 | 8.92 (±0.54) | 21.0 |
| Ntot [g kg$^{-1}$] | III | 1.07 (±0.03) | 11.0 | 1.12 (±0.05) | 15.4 |
| | VII | 1.09 (±0.03) | 8.7 | 1.13 (±0.04) | 10.9 |
| pH in 1M KCl | III | 6.34 (±0.15) | 12.0 | 6.24 (±0.19) | 15.2 |
| | VII | 6.51 (±0.15) | 11.2 | 6.35 (±0.18) | 14.1 |
| Hh [cmol kg$^{-1}$] | III | 0.79 (±0.10) B | 63.0 | 1.16 (±0.18) A | 73.9 |
| | VII | 0.84 (±0.11) B | 66.5 | 1.10 (±0.17) A | 75.0 |
| CEC [cmol kg$^{-1}$] | III | 9.43 (±0.61) | 31.5 | 10.3 (±0.65) | 31.2 |
| | VII | 11.3 (±1.41) | 31.4 | 12.2 (±0.71) | 28.8 |
| Basic saturation (BS) [%] | III | 89.4 (±1.8) | 9.6 | 86.0 (±2.2) | 12.7 |
| | VII | 91.1 (±1.4) | 7.6 | 89.4 (±1.7) | 9.3 |
| Mg-exchan [cmol kg$^{-1}$] | III | 0.48 (±0.04) | 43.0 | 0.49 (±0.03) | 30.0 |
| | VII | 0.49 (±0.04) | 42.5 | 0.48 (±0.03) | 33.6 |
| Ca-exchan [cmol kg$^{-1}$] | III | 7.58 (±0.64) | 41.2 | 7.97 (±0.76) | 46.8 |
| | VII | 9.41 (±0.72) | 37.4 | 9.95 (±0.78) | 38.6 |
| K-exchan [cmol kg$^{-1}$] | III | 0.34 (±0.02) | 27.9 | 0.37 (±0.01) | 14.0 |
| | VII | 0.33 (±0.03) | 36.7 | 0.38 (±0.01) | 15.5 |
| Na-exchan [cmol kg$^{-1}$] | III | 0.23 (±0.01) | 13.6 | 0.29 (±0.01) | 21.4 |
| | VII | 0.21 (±0.02) | 44.8 | 0.25 (±0.02) | 31.3 |
| P-avail [mg kg$^{-1}$] | III | 63.4 (±1.3) | 10.2 | 67.1 (±2.8) | 20.3 |
| | VII | 66.0 (±2.1) | 15.4 | 70.0 (±3.4) | 24.1 |
| K-avail [mg kg$^{-1}$] | III | 155.8 (±9.7) a | 30.5 | 158.1 (±7.1) a | 21.9 |
| | VII | 138.6 (±7.7) b | 27.4 | 139.6 (±5.5) b | 19.2 |
| Mg-avail [mg kg$^{-1}$] | III | 53.1 (±3.0) | 27.3 | 56.9 (±2.7) | 23.5 |
| | VII | 57.9 (±3.4) | 28.4 | 59.6 (±2.4) | 19.9 |

Notes: SE—standard error; CV—coefficient of variation; III—March, before barley sowing (* *Solactiv* was to be used); VII—July, after the harvest of barley (** *Solactiv* was used); Corg—organic carbon; Ntot—total nitrogen; Hh—hydrolytic acidity; BS—basic saturation; CEC—cation exchange capacity; exchan—exchangeable forms; avail—available forms. Different capital letters indicate significant differences between conditioner and control locations at the same sampling time ($p < 0.05$, Tukey HSD Test). Different lowercase letters indicate significant differences between March and July within the same soil management ($p < 0.05$, Tukey HSD Test). Means without letters are not significantly different.

**Table 2.** Physical and water-related variables as determined by the *Solactiv* used (mean values for the research locations).

| Property | Control | | Solactiv | |
|---|---|---|---|---|
| | Mean ± SE | CV [%] | Mean ± SE | CV [%] |
| Clay [%] | 5.0 (±0.3) A | 22.0 | 5.2 (±0.3) A | 21.3 |
| Silt [%] | 38.5 (±2.9) A | 26.3 | 39.9 (±2.4) A | 21.2 |
| Bulk density | 1.61 (±0.0) B | 1.1 | 1.57 (±0.01) A | 2.8 |
| Total porosity | 35.2 (±0.30) A | 3.6 | 36.2 (±0.29) A | 3.4 |
| Macropores | 7.22 (±0.34) B | 19.7 | 8.51 (±0.45) A | 22.3 |
| AWC | 17.2 (±0.25) B | 6.2 | 18.8 (±0.17) A | 3.8 |
| RAWC | 9.51 (±0.17) B | 7.7 | 11.01(±0.17) A | 6.4 |
| SAWC | 7.64 (±0.32) A | 18.0 | 7.66 (±0.29) A | 16.0 |
| Micropores | 10.8 (±0.26) A | 10.4 | 8.94 (±0.28) B | 13.4 |

Notes: SE—standard error; CV—coefficient of variation; AWC—available water capacity; RAWC—readily available water capacity; SAWC—small pore available water capacity. Different capital letters indicate significant differences between conditioner and control locations ($p < 0.05$, Tukey HSD Test). Means without letters are not significantly different.

Independent of the study location, bulk density (BD) was considerably lower in the samples collected from the area with the *Solactiv* compared to the control. When study sites were considered separately, this tendency was only true for one study site (Samsieczynek) (Table S2). The difference in total porosity (TP) between the fields with and without the tested conditioner was statistically insignificant, irrespective of the study location, as well as for soil samples taken from Janocin and Kobylnica (Tables 2 and S2). Only in the soil sampled in Samsieczynek was TP markedly greater in soil amended with conditioner as compared to the control soil (Table S2). Based on the mean values for all study sites, the percentage of macropores in TP was significantly higher in soil amended with conditioner compared to the control soil. This was also true for soil samples analysed separately in two study sites (Kobylnica and Samsieczynek) (Tables 2 and S2). At all study sites, the application of conditioner significantly reduced (by 17%) the percentage of micropores in TP, while significantly enhancing the percentage of AWC and RAWC (Table S2). In turn, the percentage of SAWC did not present significant differences between soil amended with conditioner and control soil derived from all three study sites. The soil's physical and water-related features showed little and, in some cases, moderate variability (CV = 0.6–22.0%) (Table S2).

### 3.2. Carbon and Nitrogen: Dissolved Forms and Microbial Biomass

The dataset of soil C and N was discussed in two ways: based on means calculated for the study sites (to check the general tendency of changes) and for each study site independently (Tables 3 , S3 and S4).

**Table 3.** The content of carbon and nitrogen (dissolved forms and microbial biomass) in response to the conditioner application and sampling month (mean values across the study sites).

| Property | Sampling Month | Control | | Solactiv | |
|---|---|---|---|---|---|
| | | Mean ± SE | CV [%] | Mean ± SE | CV [%] |
| DOC | * III | 101.1 (±4.0) Aa | 13.7 | 98.4 (±2.8) Aa | 9.8 |
| | ** VII | 91.2 (±2.3) Ab | 8.7 | 91.0 (±2.8) Ab | 10.6 |
| DNt | III | 23.6 (±0.9) Aa | 12.8 | 23.3 (±1.0) Aa | 15.5 |
| | VII | 18.2 (±0.5) Bb | 18.2 | 21.1 (±1.0) Aa | 16.1 |
| MBC | III | 118.3 (±7.9) Bb | 32.8 | 133.5 (±6.8) Aa | 24.8 |
| | VII | 131.3 (±8.8) Aa | 32.8 | 135.9 (±5.5) Aa | 19.9 |
| MBN | III | 19.7 (±1.33) Bb | 33.1 | 23.1 (±1.1) Aa | 22.5 |
| | VII | 21.5 (±1.6) Aa | 35.5 | 23.1 (±1.0) Aa | 22.0 |

Notes: SE—standard error; CV—coefficient of variation; III—March, before barley sowing (* *Solactiv* was to be used); VII—July, after the harvest of barley (** *Solactiv* was used); DOC—dissolved organic carbon (mg $kg^{-1}$); DNt—dissolved organic nitrogen (mg $kg^{-1}$); MBC—microbial biomass carbon (mg $kg^{-1}$); MBN—microbial biomass nitrogen (mg $kg^{-1}$). The significance of lowercase and capital letters is explained in the caption of Table 1.

When the experimental sites were considered together, no significant differences in DOC concentration were noted between experimental fields (conditioner, control), while such a difference was found for DNt content determined in July (Table 3). For each of the study sites considered separately, the differences in the dissolved forms of C and N (DOC and DNt) between fields with and without conditioner were not significant on each sampling date, except for DOC content tested in soil from Samsieczynek (Table S3). The above observation was also true in the case of DNt for most of the differences between sampling times, although its content in the control soil was significantly higher in soil collected in March than in July. In the case of soil with *Solactiv* applied, this observation was only true in soil from the Janocin site. Taking into account the average for all research sites, the content of MBC and MBN in March was significantly higher in control soil than in soil destined for and treated with the conditioner, while after barley harvest (July), values of these properties were not significantly differentiated between the two experimental fields (control vs. soil treated with conditioner) (Table S3). The content of MBC and MBN determined for the Janocin study site was significantly higher in soil from the *Solactiv* field compared to the control soil (for both March and July). The same relationship was also noted for soil from Samsieczynek (March only) and Kobylnica (July only). The opposite tendency was found for soil collected from Samsieczynek in July (Table S4).

Because the layout of the basic dataset presented in Tables 3, 4, S3 and S4 does not show the precise effect of the applied conditioner on C and N forms, we have presented the level of differences between results obtained in July and March for the same field (control, conditioner) as well as variability between data obtained for conditioner and control soils for the same sampling month (March, July) (Figures 3 and 4). This approach allows for a more accurate assessment of the influence of the conditioner (a significant increase or smaller decrease in content of the studied properties as compared to the control).

**Table 4.** Enzymatic activity as determined by the used conditioner and sampling month (mean values across the study locations).

| Property | Sampling Month | Control | | Solactiv | |
|---|---|---|---|---|---|
| | | Mean ± SE | CV [%] | Mean ± SE | CV [%] |
| FDAH | * III | 21.1 (±0.66) Ba | 15.3 | 24.4 (±0.89) Aa | 17.9 |
| | ** VII | 19.4 (±0.59) Bb | 15.0 | 22.4 (±0.47) Aa | 10.2 |
| DHA | III | 7.28 (±1.00) Ba | 67.2 | 9.28 (±1.05) Aa | 55.5 |
| | VII | 8.08 (±1.01) Ba | 61.4 | 9.20 (±0.67) Aa | 35.6 |
| CMC | III | 5.17 (±0.20) Bb | 18.8 | 6.19 (±0.17) Ab | 13.4 |
| | VII | 8.73 (±0.19) Ba | 10.6 | 11.0 (±0.30) Aa | 13.6 |

Notes: SE—standard error; CV—coefficient of variation; III—March, before barley sowing (* *Solactiv* was to be used); VII—July, after the harvest of barley (** *Solactiv* was used). FDAH—fluorescein diacetate hydrolysis activity (mg F $kg^{-1}$ $h^{-1}$); DHA—dehydrogenase activity (mg TPF $kg^{-1}$ 24 $h^{-1}$); CEL—cellulase activity (mg Glu $kg^{-1}$ $h^{-1}$). The significance of lowercase and capital letters is explained in the caption of Table 1.

When analysing variability in DOC between July and March, a large decrease in its content was found in soil taken from the *Solactiv* field at the Janocin study site (16.6 mg $kg^{-1}$) compared to the control field (1.0 mg $kg^{-1}$). A similar tendency was also observed for DOC content at Samsieczynek, while the range of results was smaller.

The beneficial effect of the studied conditioner on DOC content was only noticed for soil from Kobylnica (Figure 3a). The decline in DNt concentration between July and March in control soil was 4.9 mg $kg^{-1}$, whereas in soil with *Solactiv* it was only 0.8 mg $kg^{-1}$, suggesting a positive effect of the conditioner on this property in soil taken from Kobylnica (Figure 3c). Additionally, differences in DNt content between July and March were also found in soil from Samsieczynek, and these favoured the *Solactiv* field. Considering the level of differences between results obtained in July and March for the same field (control and conditioner), we found the positive effect of the applied conditioner on DOC and DNt content in two of the three study sites (Figure 3b,d). However, this influence had a different nature as regards the study site. In soil from Kobylnica, a positive impact was found for both DOC and DNt content, whereas in the soil originating from Samsieczynek, a lower decrease in DOC content was determined in July than in March, and the positive impact of *Solactiv* was found for DNt concentration (4.1 mg $kg^{-1}$).

There were no clear patterns for the content of MBC and MBN in the study sites regarding differences between results obtained in July and March for the same field (control and conditioner), as well as variability between data obtained for conditioner and control soils for the same sampling month (March and July) (Figure 4a–d).

Thus, in both approaches, no significant changes in these properties were found for soil collected from Janocin (Figure 4a–d), whereas in soil from Kobylnica, a lower decrease in MBC and MBN content was noted in soil taken from the field with *Solactiv* compared to the control (Figure 4a,c). In soil from Samsieczynek, in turn, large increases in the content of these properties were received in the control soil compared to the amended soil (76–MBC and 87–MBN) (Figure 3a,b). Additionally, the above properties exhibited a large reduction in contents in soil from Samsieczynek collected in July (compared to control soil), suggesting the negative influence of the studied conditioner on soil microbial C and N content (Figure 4b,d).

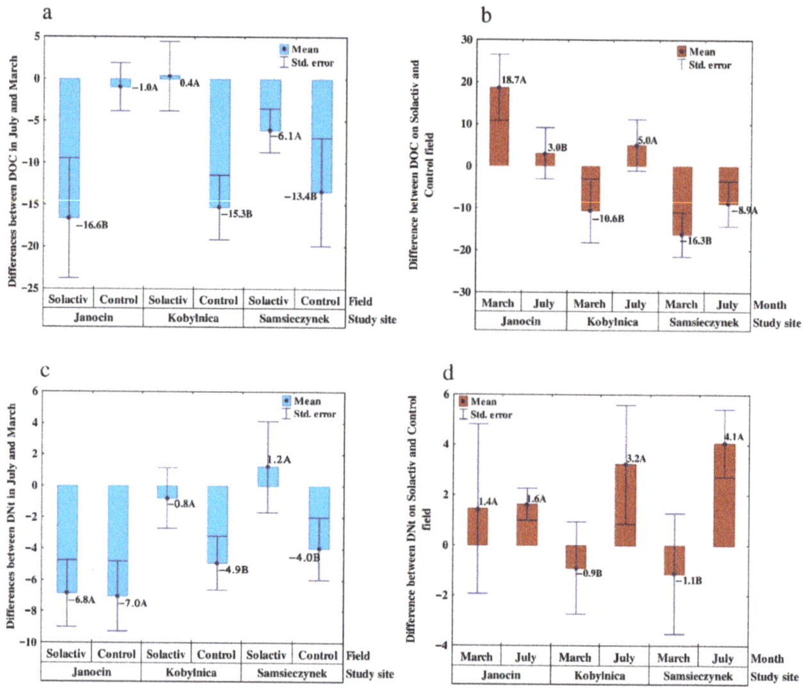


**Figure 3.** The level of differences between DOC and DNt results (mg kg$^{-1}$) obtained in July and March for the same field (control vs. Solactiv) (**a**,**c**), as well as variability between data obtained for conditioner and control soils for the same sampling month (March or July) (**b**,**d**). Different letters between means for conditioner sites and control sites (obtained from the difference between results from July and March) indicate significant differences at $p < 0.05$ according to the Tukey HSD Test (**a**,**c**). Different letters between means for March and July (obtained from the difference between results for control or *Solactiv* and control) indicate significant differences at $p < 0.05$ according to the Tukey HSD Test (**b**,**d**).

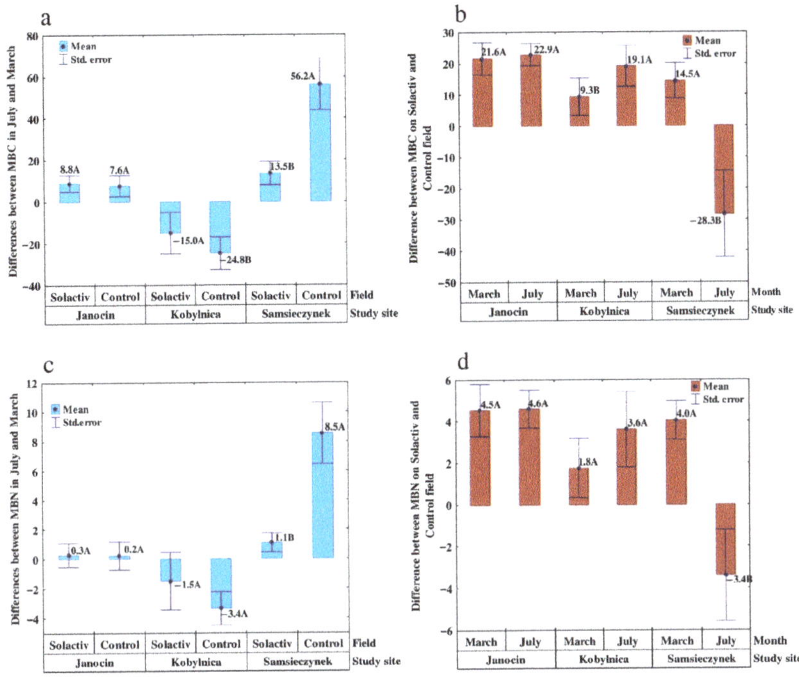


**Figure 4.** The level of differences between MBC and MBN results (mg $kg^{-1}$) obtained in July and March for the same field (control vs. Solactiv) (**a**,**c**), as well as variability between data obtained for conditioner and control soils for the same sampling month (March and July) (**b**,**d**). The significance of and capital letters is explained in the caption of Figure 3.

### 3.3. Soil Enzymatic and Microbial Properties

The results of enzymatic activity were analysed twofold: without division by study sites (to check the general tendency) (Table 4) and for each location separately (Table S4, Figures 5 and 6).

When the study locations were considered together, all the studied enzymes revealed significantly higher activity in the soil that had been dedicated for *Solactiv* application (March) as well as in the soil after conditioner application (July) compared to the control soils (Table 4). The above observation was also true in most cases (enzyme and location) when the study sites were analysed separately (Table S4). The exception was the activity of FDAH and DHA determined in soil collected at Samsieczynek in July, as well as the activity of CEL evaluated for Janocin on both sampling dates and for Kobylnica in March (Table S4). Despite the fact that the dissimilarity between enzymatic activity in the soil taken from both sites (*Solactiv* and untreated plots) before and after the application of conditioner was statistically proven, they showed similar patterns, suggesting no conditioner effect on these variables.

Since the basic data presented in Tables 4 and S4 did not clearly indicate either a significant influence (positive or negative) or a lack of influence of *Solactiv* on enzymatic properties, the level of differences between results obtained in July and March for the same field (control vs. *Solactiv*) as well as the variability between results received for conditioner and control soils for the same sampling time (March and July) were also presented (Figures 5 and 6). The actual significant and positive impact of *Solactiv* was found only in the case of CEL activity. Accordingly, the differences in the CEL activity between July and March were significantly higher in soil taken from the *Solactiv* field compared to the control soil, and this pattern was observed at two of the three research locations (Figure 4a). Additionally, the CEL activity given as a difference between *Solactiv* and control fields was significantly higher in July than in March, which further confirms the positive effect of *Solactiv* on the CEL activity at all study sites (Figure 5b).

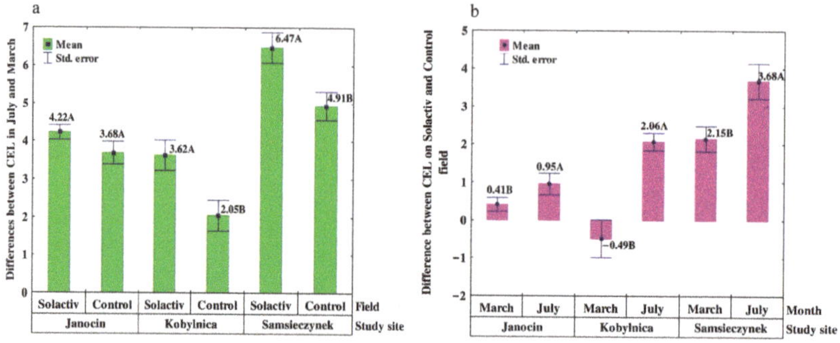


**Figure 5.** The level of differences between CEL activity (mg Glu $kg^{-1}$ $h^{-1}$) obtained in July and March for the same field (control vs. Solactiv) (**a**), as well as variability between data obtained for conditioner and control soils for the same sampling month (March and July) (**b**). Different letters between means for conditioner sites and control sites (obtained from the difference between results from July and March) indicate significant differences at $p < 0.05$ according to the Tukey HSD Test (**a**). Different letters between means for March and July (obtained from the difference between results for control or *Solactiv* and control) indicate significant differences at $p < 0.05$ according to the Tukey HSD Test (**b**).

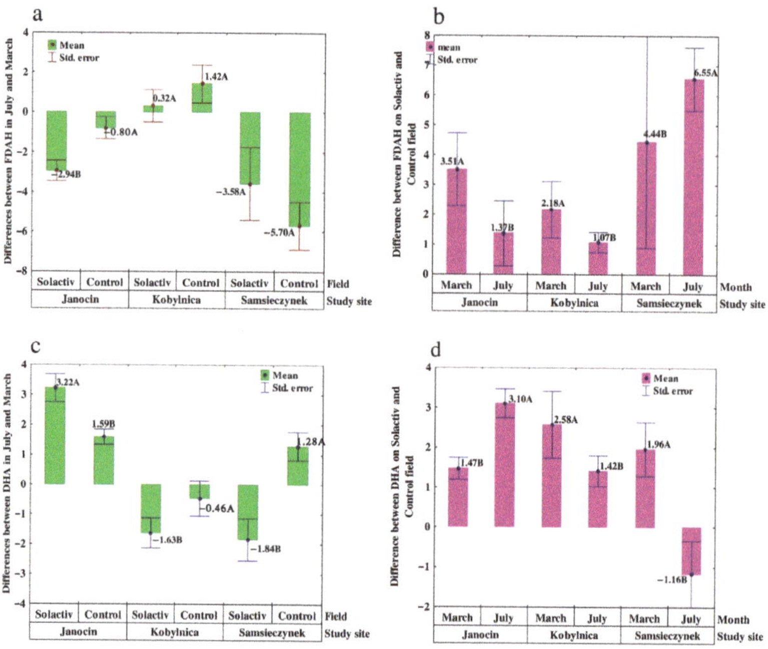


**Figure 6.** The level of differences between FDAH and DHA (mg F $kg^{-1}$ $h^{-1}$ and mg TPF $kg^{-1}$ 24 $h^{-1}$, respectively) obtained in July and March for the same field (control vs. Solactiv) (**a**,**c**), as well as variability between data obtained for conditioner and control soils for the same sampling month (March and July) (**b**,**d**). The significance of and capital letters is explained in the caption of Figure 3.

Considering the differences between the results obtained in July and March, the application of the conditioner significantly decreased the FDAH level in soil collected in Janocin; meanwhile, in soil taken from the two other locations, there were no significant changes in FDAH level between the field with *Solactiv* applied and the control site (Figure 6a). Additionally, the difference in FDAH level between *Solactiv* and control soils was significantly higher in March than in July in Janocin and Kobylnica, while the opposite trend was noted for the Samsieczynek study site (Figure 6b).

There were no clear differences in DHA due to the application of the *Solactiv* at any of the experimental sites (Figure 6c,d). Analysing the differences between results achieved in July and March has shown that the application of conditioner significantly increased the DHA at the Janocin study site, while for the other two study sites, the tendency was opposite (Figure 6c). Based on the differences between experimental fields (conditioner and control), the DHA was significantly higher in the soil studied in July (Janocin), while the opposite trend was noted for the other study sites (Figure 6d).

### 3.4. Properties Relating to the Growth and Yield of Spring Barley

Dry matter of roots and plants (the sum of surface and subsurface parts) was considerably greater in soil amended with conditioner compared to the reference soil. This was true for both considered study sites, as well as for the mean values calculated for both study sites (Figure 7).

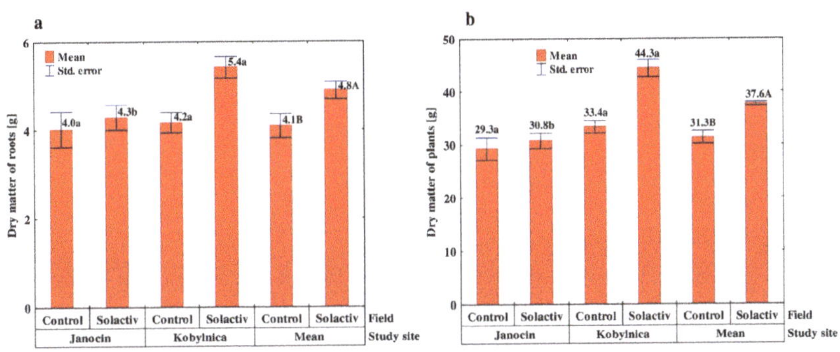


**Figure 7.** Dry matter of (**a**) barley's roots and (**b**) all plants (g). Different lowercase letters between means for conditioner sites and control sites for the given study location indicate significant differences at $p < 0.05$ according to the Tukey HSD Test. Different capital letters between means for conditioner sites and control for all study locations indicate significant differences at $p < 0.05$ according to the Tukey HSD Test.

The influence of *Solactiv* was greater in soil collected from Kobylnica than from Janocin (results were, respectively, 22 and 25% higher). The same trend was found for the values of barley root length and surface (Figure 8).

The yield of barley grain was also significantly higher in soil amended with conditioner than in control soil at both study sites, though the differences were not too high (2.4% for Janocin and 12.7% for Kobylnica) (Figure 9).

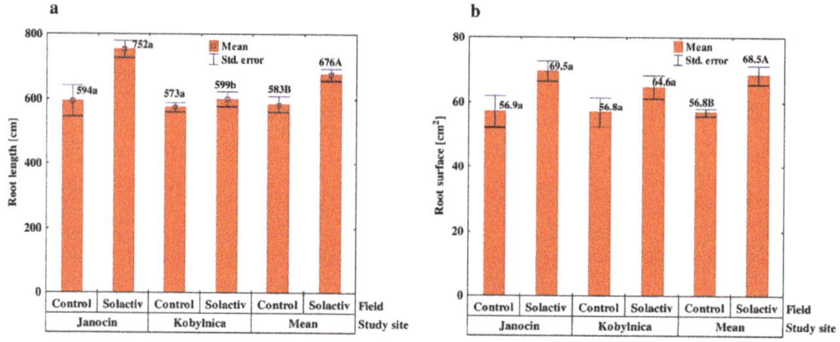


**Figure 8.** Barley's root (**a**) length and (**b**) surface (g). The significance of lowercase and capital letters is explained in the caption of Figure 7.

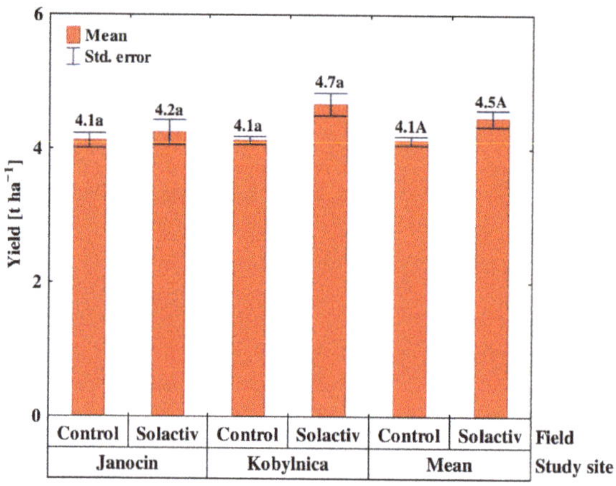


**Figure 9.** Yield of barley grain (t $ha^{-1}$). The significance of lowercase and capital letters is explained in the caption of Figure 7.

### 3.5. Principal Component Analysis

Based on PCA, three principal components were identified that accounted for 94.8% of the total variance, most of which (81.3%) were explained by PC1 and PC2 (Figure 10a).

The first principal component (PC1) had the largest contribution to the description of the total variability (47.7%). Chemical properties (e.g., CEC, available K, and Mg) and the content of different C and N forms (e.g., Corg, Ntot, MBC, MBN) were most negatively correlated (>−0.700) with PCA 1. PCA 1 had the largest contribution to the variance explanation (90%) of the following variables: exchangeable and available Mg, Ntot, clay, MBC, and MBN. For these properties, the contribution of PCA 2 to the variance explanation was small (Table S6). PCA 2, on the other hand, was correlated with the water properties and explained from 65 to 95% of the variability of these properties. The FDAH and RAWC were the only variables whose variance was not predominantly explained by either PCA 1 or PCA 2 (Table S6). PCA 3 dominated in explaining the variance of these variables (FDAH—62% and RAWC—83.1%). PCA of the cases showed that the only significant differentiating factor was the location of the experimental fields (Figure 10b). PCA 1 differentiated the results obtained in Kobylnica and Samsieczynek, which were negatively correlated with this PCA, and the results obtained in Janocin, which were positively correlated with PCA 1. PCA 2, on the other hand, differentiated the results from the Kobylnica site (negative correlation) and the results from the Samsieczynek site (positive correlation).

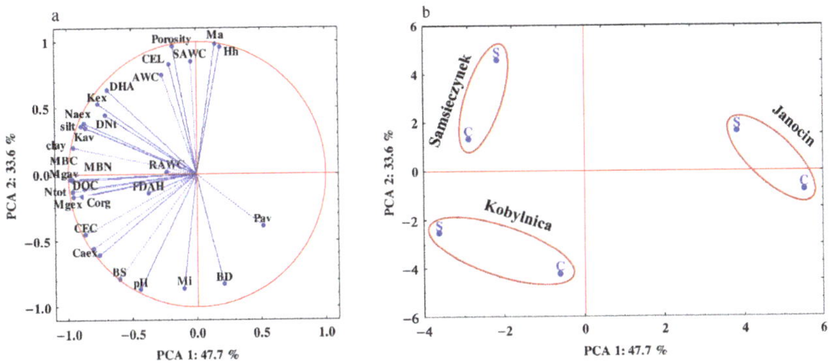


**Figure 10.** Principal Component Analysis derived from the studied soil properties: (**a**) plot of the first two principal components (PCs) for the assessed soil variables; Corg—total organic carbon, Ntot—total nitrogen, DHA—dehydrogenase activity, FDAH—fluorescein sodium salt hydrolysis, CEL—cellulase avtivity, DOC—dissolved organic carbon, DNt—dissolved nitrogen, MBC—microbial biomass carbon, MBN—microbial biomass nitrogen, AWC—available water capacity; RAWC—readily available water capacity; SAWC—small pore available water capacity, Ma—macropores, Mi—micropores, Pav—available phoshprus, Kav—available potassium, Mgav—available magnesium, Kex—exchangeable potassium, Naex—exchangeable sodium, Caex—exchangeable calcium, Mgex—exchangeable magnesium, CEC—cation exchange capacity (**b**) Principal Component Analysis of the properties determined in the individual study sites (Janocin, Kobylnica, and Samsieczynek); S—sites with *Solactiv*, C—control sites.

## 4. Discussion

### *4.1. Selected Physicochemical Properties*

It is well known that a significant increase in the total carbon and nitrogen is usually noticed after several years of ongoing treatment with different, e.g., [71]. Accordingly, the confined duration of this investigation (one year) could explain why the content of Corg and Ntot was not significantly differentiated between the objects with and without the tested *Solactiv*. Because of the presence of zeolite–clinoptilolite in the conditioner, we expected a significant increase in the soil available K. It is obvious that zeolites are known for their high capacity to exchange cations [72]. Actually, clinoptilolite, as the most abundant natural zeolite [73], which often serves as a potassium source and can regulate potassium release from fertilisers, thereby enhancing their efficient utilisation by plants [19]. High cationic exchange capacity of clinoptilolite was confirmed by Józefaciuk et al. [74], who used this zeolite in their field-based research and observed that zeolite-enriched soil had a significantly larger surface area than the combined surface areas of the original soil and zeolite alone. Previously, Długosz et al. [45] found that application of Solactiv, which contains 50% of clinoptilolite, increased the available K content by over 11%. Unexpectedly, in this study, the content of available K was not significantly affected by the conditioner. The lack of a significant effect of zeolite from *Solactiv* on the available K content may be due to the relatively high content of potassium in soil (medium class according to the abundance classes [70] and potassium fertilisation applied in the study) (Table S1).

The applied conditioner had a positive effect on some water-related properties, which could have been due to the presence of zeolite in the tested *Solactiv*. The porous nature of zeolites allows them to store water, which can be desorbed to form a moisture layer on their surface, creating an optimal habitat that promotes microorganism growth and activity and enabling plants to better tolerate water stress [75]. The addition of zeolite to sandy soil enhanced the plant-available water content by 50% [76]. In this study, the percentage of available water capacity (AWC) and readily available water capacity (RAWC) also increased, but the percentage values were lower (8.5% and 14.0%, respectively), which may have been caused by a relatively low dose of zeolite applied (150 kg ha$^{-1}$). The greater increase in the percentage of AWC (by about 10–67%) and RAWC (by about 15–111%) was found in the study of de Campos Bernardi et al. [73]. These substantial improvements in AWC and RAWC were achievable due to the initially low water retention values in the studied Entisols and the high zeolite application rates employed (reaching up to 100 g kg$^{-1}$ of soil). The reduction in bulk density (BD) that we observed in conditioner-treated fields can also be attributed to the zeolite–clinoptilolite component within the conditioner. This finding was also supported by other researchers [77], who conducted a study on wheat grown in two distinct soil types: sandy loam and silty loam. Their results demonstrated dose-related bulk density reductions in both soil types. Accordingly, the values of BD in sandy loam decreased sequentially by about 10%, 15%, 34% and 41%, respectively, while in silty loam by about 18%, 22%, 37% and 41%, respectively. Additionally, the values of total porosity in both soil types were also increased [77]. Furthermore, the pot experiment conducted by Ravali et al. [27] showed that zeolite treatment substantially decreased soil bulk density, reaching the lowest measurement of 0.97 Mg m$^{-3}$ at the 7.5 t ha$^{-1}$ application rate. The zeolite component in the tested *Solactiv* probably contributed to the observed changes in the soil pore distribution: a 15% increase in macroporosity and a 17% reduction in microporosity within the total porosity were noted. Such a tendency was also supported by Githinji et al. [78], who documented macroporosity increases by 10 m$^3$ m$^{-3}$ (from 15 to 25 m$^3$ m$^{-3}$) following clinoptilolite addition to sandy soil.

### 4.2. Soil Enzymatic and Microbial Properties

The obtained data on microbial biomass and enzymatic properties only partially point to their positive response to the applied conditioner. Key factors such as soil pH, organic matter, and microbial biomass content, as well as the range of physicochemical properties that play pivotal roles in affecting soil enzymatic activity, were not significantly changed by the applied conditioner, which was confirmed by the lack of significant correlations between these properties. Of the three tested enzymes, only cellulase activity responded positively to the applied conditioner and was significantly increased by the tested conditioner at two out of three study sites. We guessed that zeolites, characterised by their porous structure and ion-binding capacity, could significantly affect cellulase activity. Zeolites can adsorb cellulase enzymes onto their surface, potentially affecting the enzyme conformation and activity [79,80]. The pore structure of zeolites can influence which molecules can access the active site of the enzyme [81]. Thus, microporous zeolites can restrict the diffusion of large molecules, potentially hindering cellulase activity. However, some zeolites, like those with meso- and macropores, can enhance diffusion and improve catalytic performance. One example of such a zeolite is the clinoptilolite used in this study, which is characterised by high porosity, with voids constituting from 24 to 32% of its volume.

In turn, no clear trend was found in the activity of dehydrogenases; thus, these enzymes were significantly activated by the *Solactiv* only at one study site, while at the two other sites the conditioner decreased their activity. Finally, there were no significant changes in the FDAH level in any study location. What is more, the conditioner did not significantly affect the amount of microbial biomass (measured by the C and N content). Therefore, the hypothesis that microbial and enzymatic properties would be improved by *Solactiv* under spring barley cultivation could be accepted only partially. According to the literature data, interactions between *Solactiv* components and microbial and enzymatic properties are complex and depend on various factors, e.g., [12,20,24,28,31,36,38,44]. In the study of Wolny-Koładka et al. [82], determination of the DHA in soil under spring wheat and spring oilseed rape cultivation indicates the significant reduction in this property for five different mineral–organic treatments (NPK; NPK + 3% or 6% lignite and 3% zeolite–vermiculite; NPK + 3% or 6% leonardite and 3% zeolite–vermiculite) compared to the control soil, even though the number of determined microorganisms (bacteria, mould fungi, *Azotobacter* spp., and actinomycetes) increased by 116–1600%. Consequently, the authors concluded that fertilisation based on mineral and organic additions in the form of lignite and leonardite, combined with a zeolite–vermiculite, effectively increased the number of the studied microbial groups. Dehydrogenase activity acts as a measure of the total microbial oxidative metabolism, revealing a pattern of growth and activity of microbial communities, and is not always related to the amount of microbial biomass [83]. The literature data [84,85] indicated that the activity of microbial populations measured with dehydrogenase enzymes can respond to the application of a zeolite in various ways, and the dose of zeolite is often crucial. In the study carried out by Doni et al. [84] in grapevine cultivation, an increase in soil dehydrogenase activity in the range 8–110% was observed after relatively high doses of zeolite application (5 and 10 t ha$^{-1}$) after 6 months from initial treatments. In addition [84], even though the content of total organic carbon remained constant, there was a reduction in humic substances in soils treated with zeolite. This finding indicated that zeolite addition enhanced microbial activity in the soil, which in turn activated the breakdown of existing organic carbon stocks through microbial mineralization processes. The influence of seaweed-based biofertilizers on soil biochemical properties has been well-documented in the literature, e.g., [86–88]. Due to their rich composition of organic compounds, including carbohydrates, proteins, enzymes, vitamins, and fatty acids, seaweeds are believed to substantially influence soil enzymatic activity. Previously, Wang et al. [87] observed that seaweed fertilisers applied at different concentrations altered the soil microbial dynamics and enzymatic activity, with a notable effect on the dehydrogenase and protease function. Similarly, Chen et al. [88] found significant increases in the enzymatic activities of dehydrogenases, cellulases, nitrate reductase, and urease in maize rhizosphere soil a short time after the seaweed fertiliser use. Previous studies have also established that seaweed extract applications lead to increased soil bacterial activity [88,89]. However, our results did not support these findings, as the soil conditioner we tested showed no clear effect on the content of carbon and nitrogen in the soil microbial biomass, along with soil dehydrogenase activity (DHA).

The varying effects of *Solactiv* on individual enzyme activities observed in this study can be attributed to the wide range of enzymatic reactions, each with distinct optimal environmental requirements, varying levels of activity, sensitivity, and differences in soil substrate availability for specific enzymatic reactions. This is especially evident for FDA hydrolysis activity, as the breakdown of fluorescein diacetate (enzymatic substrate) is catalysed by a range of intracellular enzymes, including esterases, lipases, and proteases, that are found in metabolically active microorganisms [90]. The absence of significant variation in microbial and enzymatic properties between *Solactiv*-managed and control soils may be attributed to the limited experimental duration and the complex interactions between the various components of the soil conditioner.

### 4.3. Plant Properties

Despite the low dose used, the applied conditioner significantly increased dry matter of roots and shoots, barley root length and surface as well as barley grain yield (independent of the study location and when study sites were considered separately), which is particularly important for agricultural production. We assumed that all of the *Solactiv* components might have a positive impact on barley growth and yield. Previously, a positive influence of the individual components of *Solactiv* or their potential synergistic effects has been frequently observed, but usually, higher doses were applied [17,91]. In the first study year conducted by Szatanik-Kloc et al. [92] on a poor soil with zeolite amended at doses from 1 to 8 t/ha, an increase in spring wheat yield was noted. Research by Hassan and Mahmoud [93] showed that zeolite-containing amendments applied to the sandy and loamy sand soils increased the maize yield by 14.3 t ha$^{-1}$. Moreno et al. [17] showed a positive effect of zeolites supplemented with leonardite at a dose of 75 t ha$^{-1}$ on plant growth and yield of barley as compared with the compost treatments, either with zeolite or without. Because of the wide range of beneficial properties, this mixture has become significant for improving soil physical and chemical properties and thereby enhancing plant growth and yield. The positive effects of *Ascophyllum nodosum* extract on plant growth and productivity across different cultivated species have been reported [41,88,94,95]. In their study on strawberries and carrots, Alam et al. [94,95] demonstrated that *Ascophyllum nodosum* extract enhanced plant growth and production with increased rhizosphere microbial activity. According to Chen et al. [88], application of seaweed fertiliser resulted in significant improvements in maize seedling development, with significant seedling height and above-ground fresh mass. This effect may be attributed to the presence of biostimulants, including 2% abscisic acid, 3% adenine, and 5% indoleacetic acid [96].

Potassium humate, a humic acid salt obtained from lignite brown coal, is characterised by high concentrations of carboxylic and phenolic groups. Humic acid affects plant development through both direct and indirect mechanisms, with the indirect effects enhancing the soil's physical, chemical, and biological properties [15]. The results of the greenhouse trial [97] demonstrated that the humic acids (originating from leonardite) applied at a dose of 1.5 g per 1 kg of soil had a significant effect on potato growth (55%) and tuber yield (66.0%) in contrast to the control. However, numerous studies have demonstrated minimal or no beneficial effects when humates are applied at standard recommended rates (typically 5–10 kg ha$^{-1}$ for K humate), suggesting these application levels are likely insufficient for achieving effectiveness [98,99]. Similarly, the potassium humate concentration in Solactiv applied in this research (6 kg ha$^{-1}$) appeared insufficient to achieve the assumed efficiency, as such low application rates likely do not provide adequate levels of the bioactive compounds [45].

## 5. Conclusions

Despite the low single dose used (only 300 kg ha$^{-1}$), the applied conditioner significantly increased some important soil and plant properties, which can be attributed to its complex and unique nature. Therefore, the use of *Solactiv* represents a potentially valuable strategy for advancing sustainable agricultural practices through enhancing some physical and water-related soil properties as well as selected plant properties. Additionally, the effect of a low application rate is also important from an economic point of view. Nevertheless, inconclusive findings regarding soil enzymatic, microbial, and chemical variables indicate that further research involving multi-site barley cultivation studies and scaled-up field trials is essential to across multiple sites and commercial-scale applications are needed to confirm its effectiveness. It should be stressed that the *Luvisols* used in this study have moderate fertility and relatively high nutrient levels, which may have obscured the true potential of the soil conditioner. Future studies should test the long-term impact of *Solactiv* on various agricultural systems (e.g., no-till), different soil types, especially the low-fertility sandy soils with limited productivity, to obtain a clearer picture of how well it affects soil fertility, water, and nutrient retention in realistic field-scale conditions. Furthermore, environmental parameters such as temperature and rainfall patterns at various study locations should also be considered to validate the study's findings. Because of the clear, positive response of cellulase activity in this study, it is worth assessing the range of cellulolytic enzymes as affected by the applied conditioner in different agricultural systems. Finally, research should also focus on developing a new, more effective *Solactiv* formula by incorporating new components, such as polyacrylamide, biochar, and organic composts.

**Supplementary Materials:** The following supporting information can be downloaded at: https://www.mdpi.com/article/10.3390/su172210169/s1, Table S1. Mineral fertilization used in the experimental fields. Table S2. Physical and water-related properties as affected by the applied conditioner for individual study sites. Table S3. The content of dissolved carbon and nitrogen as affected by the applied conditioner and sampling month. Table S4. The content of microbial biomass carbon and nitrogen as affected by the applied conditioner and sampling month. Table S5. Enzymatic activity as affected by the applied conditioner and sampling month. Table S6. Loading scores of the variable for PCA.

**Author Contributions:** Conceptualization, J.D.; Formal analysis, J.D. and A.P.-D.; Investigation, J.D., A.P.-D. and K.K.; Methodology, A.P.-D. and K.K.; Project administration, J.D. and E.P.; Visualisation, A.P.-D. and E.P.; Writing—original draft, J.D., A.P.-D. and K.K.; Writing—review and editing, A.P.-D. and E.P. All authors have read and agreed to the published version of the manuscript.

**Funding:** This research was funded by The Polish Agency for Enterprise Development, grant number ID POIR.02.03.02-06-0002/17, as well as by NaturalCrop Poland Ltd.

**Institutional Review Board Statement:** Not applicable.

**Informed Consent Statement:** Not applicable.

**Data Availability Statement:** The data used to support the findings of this study are available from the authors upon request.

**Acknowledgments:** We are most grateful to the Experimental Station for Cultivar Testing in Chrząstowo and Głębokie (Poland) for providing the meteorological data. Much gratitude is due to Tim Brombley for proofreading the article.

**Conflicts of Interest:** Author Ewelina Przybyszewska was employed by the company NaturalCrop Poland Ltd. The remaining authors declare that the research was conducted in the absence of any commercial or financial relationships that could be construed as a potential conflict of interest.

## References

1. Xing, Y.; Wang, X.; Mustafa, A. Exploring the link between soil health and crop productivity. *Ecotoxicol. Environ. Saf.* **2025**, *289*, 117703. [CrossRef]
2. Furtak, K.; Wolińska, A. The impact of extreme weather events as a consequence of climate change on the soil moisture and on the quality of the soil environment and agriculture—A review. *Catena* **2023**, *231*, 107378. [CrossRef]
3. Oishy, M.N.; Shemonty, N.A.; Fatema, S.I.; Mahbub, S.; Min, E.L.; Raisa, M.B.H.; Anik, A.H. Unravelling the effects of climate change on the soil-plant-atmosphere interactions: A critical review. *Soil Environ. Health* **2025**, *3*, 100130. [CrossRef]
4. Kopittke, P.M.; Menzies, N.W.; Wang, P.; McKenna, B.A.; Lombi, E. Soil and the intensification of agriculture for global food security. *Environ. Int.* **2019**, *132*, 105078. [CrossRef] [PubMed]
5. Chen, Q.; Li, G.; Li, Q.; Zhou, Z.; Feng, Y.; Luo, Y.; Tan, H.; Tian, X. Soil moisture determines the consistency of organic matter decomposition in field and lab test patterns. *Catena* **2024**, *246*, 108485. [CrossRef]
6. Krasowicz, S.; Matyka, M.; Madej, A. The rational use of Polish soils as a challenge for science, advice, and agricultural practice State Research Institute in Puławy. *Acta Sci. Pol. Oecon.* **2023**, *22*, 17–25. [CrossRef]
7. Ochal, P.; Jadczyszyn, T.; Jurga, B.; Kopiński, J.; Matyka, M.; Madej, A.; Rutkowska, A.; Smreczek, B.; Łysiak, M. Environmental aspects of soil acidity in Poland. In *Studies and Reports of Institute of Soil Science and Plant Cultivation—State Research Institute*; Report Prepared as a Part of the Task 2.2; Institute of Soil Science and Plant Cultivation–State Research Institute: Puławy, Poland, 2017; p. 47.
8. Shinde, R.; Sarkar, P.; Thombare, N.; Naik, S. Soil conservation: Today's need for sustainable development. *Agric. Food E-Newsl.* **2019**, *1*, 175–183.
9. Thakur, P.; Wadhwa, H.; Shubham; Kaushal, S. Soil conditioners: Refinement of soil health for better tomorrow. *Curr. J. Appl. Sci. Technol.* **2023**, *42*, 1–9. [CrossRef]
10. Triphati, S.; Tiwari, T.; Sachan, R. Soil Conditioners: Substances that Enhance Physical Properties of Soil. In *Current Research and Review in Soil Science*; Bright Sky Publications: Delhi, India, 2023; Volume 3, pp. 19–29.

11. Rakkammal, K.; Maharajan, T.; Ceasar, S.A.; Ramesh, M. Biostimulants and their role in improving plant growth under drough and salinity. *Cereal Res. Commun.* **2023**, *51*, 61–74. [CrossRef]
12. Cataldo, E.; Fucile, M.; Manzi, D.; Masini, C.M.; Doni, S.; Mattii, G.B. Sustainable soil Managment: Effects of clinoptilolite an organic compost soil application on eco-physiology, quercitin, and hydroxylated, methoxylated anthocyanins on *Vitis vinifer Plants* **2023**, *12*, 708. [CrossRef]
13. Kukowska, S.; Szewczuk-Karpisz, K. Managment of the soil environment using biochar and zeolite in various combination Impact on soil condition and economical aspects. *J. Soils Sediments* **2025**, *25*, 77–102. [CrossRef]
14. Galamini, G.; Ferretti, G.; Rosinger, C.; Huber, S.; Diaz-Pines, E.; Faccini, B.; Keiblinger, K.M. Potential for agricultural recycling c struvite and zeolites to improve soil microbial physiology and mitigate $CO_2$ emission. *Geoderma* **2025**, *453*, 117149. [CrossRef]
15. Kumar, D.; Singh, A.P.; Raha, P.; Rakshit, A.; Singh, C.M.; Kishor, P. Potassium humate: A potential soil conditioner and plar growth promoter. *Int. J. Agric. Environ. Biotechnol.* **2013**, *6*, 441–446. [CrossRef]
16. Kołodziej, B.; Sugier, D.; Bielińska, E. The effect of leonardite application and various plantation modalities on yielding an quality of roseroot (*Rhodiola rosea* L.) and soil enzymatic activity. *J. Geochem. Explor.* **2013**, *129*, 64–69. [CrossRef]
17. Moreno, J.L.; Ondoño, S.; Torres, I.; Bastida, F. Compost, leonardite, and zeolite impacts on soil microbial community unde barley crops. *J. Soil Sci. Plant Nutr.* **2017**, *17*, 214–230. [CrossRef]
18. Shukla, P.S.; Mantin, E.G.; Adil, M.; Bajpai, S.; Critchley, A.T.; Prithiviraj, B. Ascophyllum nodosum-based biostimulant Sustainable applications in agriculture for the stimulation of plant growth, stress tolerance, and disease Managment. *Front. Plan Sci.* **2019**, *10*, 655. [CrossRef]
19. Ahmed, M.; Ullah, H.; Piromsri, K.; Tiisarum, R.; Chaum, S.; Datta, A. Effect of an Ascophyllum nodosum seaweed extrac application dose and method on growth, fruit yield, quality, and water productivity of tomato under water-deficit stress. *S. Afr. Bot.* **2022**, *151*, 95–107. [CrossRef]
20. Liu, Q.; Cui, H.; Yang, W.; Wang, F.; Liao, H.; Zhu, Q.; Qin, S.; Lu, P. Soil conditioner improves soil properties, regulates microbia communities, and increases yield and quality of *Uncaria rhynchophylla*. *Sci. Rep.* **2024**, *14*, 13398. [CrossRef]
21. Dueñas, J.F.; Kunze, E.; Li, H.; Riling, M.C. Soil conditioner mixtures as an agricultural Managment alternative to mitigat drought impacts: A proof of concept. *Nat. Hazards Earth Syst. Sci.* **2025**, *25*, 1377–1386. [CrossRef]
22. Goulding, K.W.T. Soil acidification and the importance of liming agricultural soils with particular reference to the Unitec Kingdom. *Soil Use Manag.* **2016**, *32*, 3. [CrossRef]
23. Ouni, Y.; Ghnaya, T.; Montemurro, F.; Abdelly, C.; Lakhdar, A. The role of humic substances in mitigating the harmful effects o soil salinity and improve plant productivity. *Int. J. Plant Prod.* **2014**, *8*, 353–374.
24. Wu, Q.; Xia, D.; Chen, G.; Sun, T.; Song, Y. Effects of zeolite on drought resistance and water—nitrogen use efficiency in padd rice. *J. Irrig. Drain. Eng.* **2019**, *145*, 04019024. [CrossRef]
25. Piotrowska, A.; Długosz, J.; Zamorski, R.; Bogdanowicz, P. Changes in some biological and chemical properties of an arable soi treated with the microbial biofertilizer UGmax. *Pol. J. Environ. Stud.* **2012**, *21*, 455–463.
26. Mayer, J.; Scheid, S.; Widmer, F.; Fließbach, A.; Oberholzer, H.R. How effective are "Effective micro-organisms (EM)"? Result from a field study in temperate climate. *Appl. Soil Ecol.* **2010**, *46*, 230–239. [CrossRef]
27. Ravali, C.; Rao, K.J.; Anjaiah, T.; Suresh, K. Effect of zeolite on soil physical and physico-chemical properties. *Int. Ref. Peer Rev Index. Q. J. Sci. Agric. Eng.* **2020**, *10*, 776–781.
28. Mondal, M.; Biswas, B.; Garai, S.; Sarkar, S.; Banerjee, H.; Brahmachari, K.; Bandyopadhyay, P.K.; Maitra, S.; Brestic, M.; Skalick M.; et al. Zeolites enhance soil health, crop productivity and environmental safety. *Agronomy* **2021**, *11*, 448. [CrossRef]
29. Ferretti, G.; Rosinger, C.; Diaz-Pines, E.; Faccini, B.; Coltorti, M.; Keiblinger, K.M. Soil quality increases with long-term chabazite zeolite tuff amendments in arable and perennial cropping systems. *J. Environ. Manag.* **2024**, *354*, 120303. [CrossRef]
30. Bhattarai, B.; Neupane, J.; Dhakal, S.P.; Nepal, J.; Gnyawali, B.; Timalsina, R.; Poudel, A. Effect of biochar from different origin or physio-chemical properties of soil and yield of Garden Pea (*Pisum sativum* L.) at Paklihawa, Rupandehi. Nepal. *World J. Agric Res.* **2015**, *3*, 129–138.
31. Ergolu, N.; Emekci, M.; Athanassiou, C.G. Applications of natural zeolites on agriculture and food production. *J. Sci. Food Agric* **2017**, *97*, 3487–3499. [CrossRef]
32. Sangeetha, C.; Baskar, P. Zeolite and its potential uses in agriculture: A critical review. *Agric. Rev.* **2016**, *37*, 101–108. [CrossRef]
33. Demirer, T. Effect of leonardite application on leaf nutrient content and fruit chemical parameters of cherry (*Prunus avium* L.). *J Plant Nutr.* **2019**, *42*, 2532–2538. [CrossRef]
34. Dinçsoy, M.; Sönmez, F. The effect of potassium and humic acid applications on yield and nutrient contents of wheat (*Triticum aestivum* L. var. Delfii) with same soil properties). *J. Plant Nutr.* **2019**, *42*, 2757–2772. [CrossRef]
35. Alharbi, K.; Rashwan, E.; Hafez, E.; El-Dein Omara, A.; Mohamed, H.H.; Alshaal, T. Potassium Humate and Plant Growth Promoting Microbes Jointly Mitigate Water Deficit Stress in Soybean Cultivated in Salt-Affected Soil. *Plants* **2022**, *11*, 3016 [CrossRef]

36. Margal, P.B.; Thakare, R.S.; Kamble, B.M.; Patil, V.S.; Patil, K.B.; Titirmare, N.S. Effect of seaweed extracts on crop growth and soil: A review. *J. Exp. Agric. Int.* **2023**, *45*, 9–19. [CrossRef]
37. Khan, W.; Palanisamy, R.; Critchley, A.T.; Smith, D.L.; Papadopoulos, Y.; Prithiviraj, B. *Ascophyllum nodosum* extract and its organic fractions stimulate Rhizobium root nodulation and growth of *Medicago sativa* (Alfalfa). *Commun. Soil Sci. Plant Anal.* **2013**, *44*, 900–908. [CrossRef]
38. Banjare, L.; Banwasi, R.; Jataw, G.K.; Shrivastav, L.K. Effect of seaweed extract on yield and nutrient uptake of rice in a vertisol. *Pharma Innov. J.* **2022**, *11*, 2193–2198.
39. Repke, R.A.; Ribeiro Silva, D.M.; Camilo dos Santos, J.C.; de Almeida Silva, M. Increased soybean tolerance to high-temperature through biostimulant based on *Ascophyllum nodosum* (L.) seaweed extract. *J. Appl. Phycol.* **2022**, *34*, 3205–3218. [CrossRef]
40. Frioni, T.; Vander-Weide, J.; Palliotti, A.; Tombesi, S.; Poni, S.; Sabbatini, P. Foliar *vs.* soil application of *Ascophyllum nodosum* extracts to improve grapevine water stress tolerance. *Sci. Hortic.* **2021**, *277*, 109807. [CrossRef]
41. Naz, S.; Din Muhammad, H.M.; Ramzan, M.; Sadiq, B.; Ahmad, R.; Ali, S.; Alsahli, A.A.; Altal, A. Seaweed application enhanced the growth and yield of pea (*Pisum sativum* L.) by altering physiological indices. *J. Soil Sci. Plant Nutr.* **2023**, *23*, 6183–6195. [CrossRef]
42. Ronga, D.; Biazzi, E.; Parati, K.; Carminati, D.; Carminati, E.; Tava, A. Microalgal biostimulants and biofertilisers in crop productions. *Agronomy* **2019**, *9*, 192. [CrossRef]
43. Siwik-Ziomek, A.; Szczepanek, M. Soil extracellular enzyme activities and uptake of N by oilseed rape depending on fertilization and seaweed biostimulant application. *Agronomy* **2019**, *9*, 480. [CrossRef]
44. Villa e Vila, V.; Rezende, R.; Marques, P.A.A.; Wenneck, G.S.; Crepaldi de Faria Nocchi, R.; de Souza Terassi, D.; Barion Alves Andrean, A.F.; Matumoto-Pintro, P.T. Seaweed extract of Ascophyllum nodosum applied in tomato crop as a biostimulant for improving growth, yield and soil fertility in subtropical condition. *J. Appl. Phycol.* **2023**, *35*, 2531–2541. [CrossRef]
45. Długosz, J.; Piotrowska-Długosz, A.; Kotwica, K.; Przybyszewska, E. Application of multi-component Conditioner with clinoptilolite and *Ascophyllum nodosum* extract for improving soil properties and *Zea mays* L. growth and yield. *Agronomy* **2020**, *10*, 2005. [CrossRef]
46. Őztürk, H.S.; Türkmen, C.; Erdogan, E.; Baskan, O.; Dengiz, O.; Parlak, M. Effects of a soil conditioner on some physical and biological features of soils: Results from a greenhouse study. *Biores. Technol.* **2005**, *96*, 1950–1954. [CrossRef]
47. Ebrahimi, E.; Asadi, G.; von Fragstein und Niemsdorff, P. A field study on the effect of organic soil conditioners with different placements on dry matter and yield of tomato (*Lycopersicon esculentum* L.). *Int. J. Recycl. Org. Waste Agric.* **2019**, *8*, 59–66. [CrossRef]
48. Jeřábková, J.; Salaš, P.; Burgová, J. The Influence of the Application of Soil Conditioners on the Temperature and Moisture of the Soil Environment. *Inż. Mineral.* **2024**, *1*, 1–7. [CrossRef]
49. IUSS Working Group WRB. World reference base for soil resources. In *International Soil Classification System for Naming Soils and Creating Legends for Soil Maps*, 4th ed.; International Union of Soil sciences (IUSS): Vienna, Austria, 2022.
50. Gorbulewski, K.; Fronczak, J.; Leszczyńska, M. Specific surface area-basic parameter of reactive material's characteristics. *Sci. Rev. Eng. Environ. Sci.* **2008**, *17*, 122–130. (In Polish)
51. Soil Survey Staff. *2014 Soil Survey Staff, Keys to Soil Taxonomy*, 11th ed.; USDA Natural Resources Conservation Service: Washington, DC, USA, 2010.
52. Domżał, H.; Słowińska-Jurkiewicz, A.; Turski, R. *Przewodnik do Ćwiczeń z Gleboznawstwa z Elementami Geologii i Mechaniki Gleb*, 1st ed.; Wydawnictwo Akademii Rolniczej: Lublin, Poland, 1976; pp. 12–16. (In Polish)
53. Gonet, S.S.; Dębska, B.; Pakula, J. *The Content of Dissolved Organic Carbon in Soils and Organic Fertilisers*, 1st ed.; PTSH: Wrocław, Poland, 2002. (In Polish)
54. *PN-R-04020:1994*; Chemical-Agricultural Analysis of Soil—Determination of Available Magnesium Content. PKN Press: Warsaw, Poland, 2018.
55. Egner, H.; Riehm, H.; Domingo, W.R. Studies concerning the chemical analysis of soils as background for soil nutrient assessment. II. Chemical extracting methods to determinate the phosphorous and potassium content of soil. *Kungl. Lantbr. Ann.* **1960**, *26*, 199–215. (In German)
56. *PN-ISO 11260.2018*; Soil Quality—Determination of Effective Cationic Exchange Capacity and Base Saturation with Barium Chloride. PKN Press: Warsaw, Poland, 2018.
57. Van Reeuwijk, L.P. *Procedures for Soil Analysis*, 6th ed.; ISRIC Technical Paper 9; International Soil Reference and Information Centre, Food and Agriculture Organization of the United Nations: Wageningen, The Netherlands, 2002; p. 14.
58. Jaremko, D.; Kalembasa, D. A comparison of methods for the determination of cation exchange capacity of soils. *Ecol. Chem. Eng. S.* **2014**, *3*, 487–498.
59. *PN-ISO 11465.1999*; Soil quality—Determination of Soil Dry Matter and Soil Water Content Expressed as Dry Matter of Soil—Gravimetric Method. PKN Press: Warsaw, Poland, 1999.

60. Walczak, R.; Ostrowski, J.; Witkowska-Walczak, B.; Sławiński, C. Spatial characteristics of water conductivity in the surface level of Polish arable soils. *Int. Agrophys.* **2002**, *16*, 239–247.
61. Thalmann, A. Zur Methodik der Bestimmung der Dehydrodgenaseaktivitat im Boden mittels Triphenyltetrazolium-chlorid (TTC). *Landwirtsch. Forsch.* **1968**, *21*, 249–258.
62. Adam, G.; Duncan, H. Development of a sensitive and rapid method for the measurement of total microbial activity using fluorescein diacetate (FDA) in a range of soils. *Soil Biol. Biochem.* **2001**, *33*, 943–951. [CrossRef]
63. Schinner, F.; von Mersi, W. Xylanase-, CM-cellulase- and invertase activity in soil: An improved method. *Soil Biol. Biochem.* **1990**, *22*, 511–515. [CrossRef]
64. Brookes, P.C.; Landman, A.; Pruden, G.; Jenkinson, D.S. Chloroform fumigation and the release of soil nitrogen: A rapid extraction method to measure microbial biomass nitrogen in soil. *Soil Biol. Biochem.* **1985**, *17*, 837–842. [CrossRef]
65. Vance, E.D.; Brookes, P.C.; Jenkinsen, D.S. An extraction method for measuring soil microbial biomass C. *Soil Biol. Biochem.* **1987**, *19*, 703–707. [CrossRef]
66. Bremner, J.M.; Mulvaney, C.S. Nitrogen—Total. In *Methods of Soil Analysis. Part 2. Chemical and Microbiological Properties*; Page, A.L., Miller, R.H., Keeny, D.R., Eds.; SSSA: Madison, WI, USA, 1982; pp. 595–624.
67. Joergensen, R.G.; Mueller, T. The fumigation-extraction method to estimate soil microbial biomass: Calibration of the $k_{EN}$ value. *Soil Biol. Biochem.* **1996**, *28*, 33–37. [CrossRef]
68. *PN-88/R-04013.1994*; Chemical-Agricultural Analysis of Plants. Determination of Air-Dry and Dry Mass. Polish Committee: Warsaw, Poland, 1996.
69. Wilding, L.P. Spatial variability: Its documentation, accommodation, and implication to soil surveys. In *Soil Spatial Variability*; Nielsen, D.R., Bouma, J., Eds.; PUDOC: Wageningen, The Netherlands, 1985; pp. 166–194.
70. *PN-R-04023.1996*; Agrochemical Soil Analysis—Determination of Available Phosphorus in Mineral Soils. Polish Committee: Warsaw, Poland, 1996.
71. Sherrod, L.A.; Peterson, G.A.; Westfall, D.G.; Ahuja, L.R. Cropping intensity enhances soil organic carbon and nitrogen in a no-till agroecosystem. *Soil Sci. Soc. Am.* **2003**, *67*, 1533–1543. [CrossRef]
72. Morante-Carballo, F.; Montalván-Burbano, N.; Carrión-Mero, P.; Espinoza-Santos, N. Cation exchange of natural zeolites: Worldwide research. *Sustainability* **2021**, *13*, 7751. [CrossRef]
73. De Campos Bernardi, A.C.; Polidoro, J.C.; de Melo Monte, M.B.; Pereira, E.I.; Ribeiro, C.; Ramesh, K. Enhancing nutrient use efficiency using zeolites minerals—A review. *Adv. Chem. Eng. Sci.* **2016**, *6*, 295–304. [CrossRef]
74. Józefaciuk, G.; Szatniak-Kloc, A.; Ambrozewicz-Nita, A. The surface area of zeolite-amended soils exceeds the sum of the inherent surface areas of soil and zeolite. *Eur. J. Soil Sci.* **2018**, *69*, 787–790. [CrossRef]
75. Tatlier, M.; Munz, G.; Henninger, S.K. Relation of water adsorption capacities of zeolites with their structural properties. *Microporous Mesoporous Mater.* **2018**, *264*, 70–75. [CrossRef]
76. Palanivell, P.; Ahmed, O.H.; Susilawati, K.; Majid, N.M.A. Mitigating ammonia volatilization urea in waterlogged condition using clinoptilolite zeolite. *Int. J. Agric. Biol.* **2015**, *17*, 149–155.
77. Abdel-Hassan, A.N.; Abdullah Radi, A.M. Effect of zeolite on some physical properties of wheat plant growth (*Triticum aestivum* L.). *Plant Arch.* **2017**, *18*, 2641–2648.
78. Githinji, L.J.M.; Dane, J.H.; Walker, R.H. Physical and hydraulic properties of inorganic amendments and modelling their e ects on water movement in sand-based root zones. *Irrig. Sci.* **2011**, *29*, 65–77. [CrossRef]
79. Xu, C.; Sun, L.; Tong, S.; Quyang, J.; Gu, X. Cellulase immobilization on zeolitic imidazolate frameworks for boosting cellulose hydrolysis at high solids loading. *Ind. Crops Prod.* **2023**, *206*, 117693. [CrossRef]
80. Sun, L.; Xu, C.; Tong, S.; Gu, X. Enhancing cellulose hydrolysis via cellulase immobilization on zeolitic imidazolate frameworks using physical adsorption. *Bioprocess Biosyst. Eng.* **2024**, *47*, 1071–1080. [CrossRef]
81. Jampana, S.R.; Jia, L.; Ramarao, B.V.; Kumar, D. Experimental investigation of the adsorption and desorption of cellulase enzymes on zeolite-β for enzyme recycling applications. *Bioprocess Biosyst. Eng.* **2021**, *44*, 495–505. [CrossRef]
82. Wolny-Koładka, K.; Jarosz, R.; Marcińska-Mazur, L.; Mierzwa-Hersztek, M. Effect of mineral and organic additions on soil microbial composition. *Int. Agrophys.* **2022**, *63*, 131–138. [CrossRef]
83. Burns, R.G.; DeForest, J.L.; Marxsen, J.; Sinsabaugh, R.L.; Stromberger, M.E.; Wallenstein, M.D.; Weintraub, M.N.; Zoppini, A. Soil enzymes in a changing environment: Current knowledge and future directions. *Soil Biol. Biochem.* **2013**, *58*, 216–234. [CrossRef]
84. Doni, S.; Gispert, M.; Peruzzi, E.; Macci, C.; Mattii, G.B.; Manzi, D.; Masini, C.M.; Grazia, M. Impact of natural zeolite on chemical and biochemical properties of vineyard soils. *Soil Use Manag.* **2020**, *37*, 1–11. [CrossRef]
85. Shivakumara, M.N.; Rangaish, K.M.; Subbarayappa, C.T.; Chamegowda, T.C.; Thimmegowda, M.N.; Ramaiah, M. Effect of zeolite and fertilizer application on soil microbial biomass and enzyme activity in finger millet. *Int. J. Curr. Microbiol. App. Sci.* **2019**, *8*, 1939–1957. [CrossRef]
86. Nabti, E.; Jha, B.; Hartmann, A. Impact of seaweeds on agricultural crop production as biofertilizer. *Int. J. Environ. Sci. Technol.* **2016**, *14*, 1119–1134. [CrossRef]

87. Wang, M.; Chen, L.; Li, Y.; Chen, L.; Liu, Z.; Wang, X.; Qin, S. Responses of soil microbial communities to a short-term application of seaweed fertilizer revealed by deep amplicon sequencing. *Appl. Soil Ecol.* **2018**, *125*, 288–296. [CrossRef]
88. Chen, Y.; Li, J.; Huang, Z.; Su, G.; Li, X.; Sun, Z.; Qin, Y. Impact of short-term application of seaweed fertilizer on bacterial diversity and community structure, soil nitrogen contents, and plant growth in maize rhizosphere soil. *Folia Microbiol.* **2020**, *65*, 591–603. [CrossRef]
89. Xavier James, V.C.; Pushpa Thiraviam, A.G.; Al-Dosary, M.A.; Hatamleh, A.A.; Bukhari, N.A.; Arokiyaraj, S.; Kalaiyarasi, M. Evaluation of nutrient composition and biostimulant properties of seaweeds for improving soil microbial population and tomato plant growth. *BioResources* **2025**, *20*, 1431–1451. [CrossRef]
90. Hernández, I.Z.; Zamopra-Natera, J.F.; Garcia, P.M.; Ramirez, E.; Trujillo, N. Biological activity in soils treated with green manures of *Lupinus* spp. (*Leguminosae*) using the hydrolysis of fluorescein diacetate method (FDA) in Jalisco, Mexico. *Hortic. Int. J.* **2020**, *4*, 203–206. [CrossRef]
91. Ippolito, J.A.; Tarkalson, D.D.; Lehrsch, G.A. Zeolite soil application method a ects inorganic nitrogen, moisture, and corn growth. *Soil Sci.* **2011**, *176*, 136–142. [CrossRef]
92. Szatanik-Kloc, A.; Szerement, J.; Adamczuk, A.; Józefaciuk, G. Effect of low zeolite doses on plants and soil physicochemical properties. *Materials* **2021**, *14*, 2617. [CrossRef] [PubMed]
93. Hassan, A.Z.A.; Mahmoud, A.W.M. The combined effect of bentonite and natural zeolite on sandy soil properties and productivity of some crops. *Topclass J. Agric. Res.* **2013**, *1*, 22–28.
94. Alam, M.Z.; Braun, G.; Norrie, J.; Hodges, D.M. Ascophyllum extract application can promote plant growth and root yield in carrot associated with increased rootzone soil microbial activity. *Can. J. Plant Sci.* **2013**, *94*, 337–348. [CrossRef]
95. Alam, M.Z.; Braun, G.; Norrie, J.; Hodges, D.M. Effect of Ascophyllum extract application on plant growth, fruit yield and soil microbial communities of strawberry. *Can. J. Plant Sci.* **2013**, *93*, 23–36. [CrossRef]
96. Lötze, E.; Homan, E.W. Nutrient composition and content of various biological active compounds of three South African-based commercial seaweed biostimulants. *J. Appl. Phycol.* **2016**, *28*, 1379–1386. [CrossRef]
97. Hartz, T.K. Humic substances generally ineffective in improving vegetable crop nutrient uptake or productivity. *Hortic. Sci.* **2010**, *45*, 906–910. [CrossRef]
98. Little, K.R.; Rose, M.T.; Jackson, W.R.; Cavagnaro, T.R.; Patti, A.F. Do lignite-derived organic amendments improve early-stage pasture growth and key soil biological and physicochemical properties? *Crop Pasture Sci.* **2014**, *65*, 899–910. [CrossRef]
99. Akimbekov, N.; Qiao, X.; Digel, I.; Abdieva, G.; Ualieva, P.; Zhubanova, A. The effect of leonardite-derived amendments on soil microbiome structure and potato yield. *Agriculture* **2020**, *10*, 147. [CrossRef]

*sustainability*

MDPI

*Article*

# Soil Erosion Modeling of Kinmen (Quemoy) Island, Taiwan: Toward Land Conservation in a Strategic Location

**Yu-Chieh Huang, Kieu Anh Nguyen and Walter Chen ***

Department of Civil Engineering, National Taipei University of Technology, Taipei 10608, Taiwan; rosenguyen@ntut.edu.tw (K.A.N.)
* Correspondence: waltchen@ntut.edu.tw; Tel.: +886-(2)-27712171 (ext. 2628)

**Abstract**

Kinmen Island, historically known as Quemoy, holds significant historical and geopolitical importance due to its strategic location in the Taiwan Strait, just a few kilometers from the Chinese mainland. This study presents the first comprehensive assessment of soil erosion and deposition on Kinmen, providing a scientific foundation for future land conservation and sustainable development initiatives. It also addresses the underrepresentation of small-island environments in soil erosion modeling by adapting the Revised Universal Soil Loss Equation (RUSLE) and Unit-Stream-Power-based Erosion Deposition (USPED) models for coarse-textured soils under limited rainfall conditions, offering insights into the reliability and limitations of these models in such contexts. The rainfall–runoff erosivity factor ($R_m$) was derived from precipitation data at four stations, while soil samples from ten locations were analyzed for the Soil Erodibility Factor ($K_m$). Topographic factors, including the Slope Length and Steepness (LS) and the Topographic Sediment Transport ($LS_T$) factors, were computed from a 20 m DEM (Digital Elevation Model), and the Cover-Management Factor (C) was obtained from land use classification. The RUSLE estimated a mean soil erosion rate of 2.17 Mg ha$^{-1}$ year$^{-1}$, while the USPED results varied with parameterization, ranging from 0.87 to 3.79 Mg ha$^{-1}$ year$^{-1}$ for erosion and 1.39 to 6.51 Mg ha$^{-1}$ year$^{-1}$ for deposition. The results were compared with studies from the neighboring Fujian Province and contextualized through two field expeditions. This pioneering research advances the understanding of erosion and deposition processes in a strategically located island environment and supports evidence-based policies for land conservation, contributing to SDG 15 (Life on Land) and SDG 13 (Climate Action).

**Keywords:** RUSLE; USPED; rainfall erosivity; soil erodibility; cover-management factor; soil conservation

## 1. Introduction

Agriculture has been indispensable to human civilization, with food availability central to societal advancement. However, soil erosion poses a major threat to agricultural sustainability [1]. Traditional plow-based practices can accelerate erosion rates beyond natural soil formation [2], depleting nutrient-rich topsoil, reducing crop yields, and increasing sedimentation in rivers and lakes. These impacts degrade water quality and harm fragile ecosystems.

Soil erosion modeling is a key tool for analyzing and predicting erosion patterns. A global review identified 435 models developed for this purpose [3], with the Revised

Universal Soil Loss Equation (RUSLE) family being the most widely used. Applications of RUSLE have had substantial impacts worldwide. In Europe, large-scale assessments have informed soil conservation policy [4]. In China, recent work demonstrated opportunities to reduce cropland erosion through crop switching and improved management [5]. In Brazil, high-resolution mapping advanced the understanding of soil erodibility at a national scale [6]. At the global level, projections of land use and climate change impacts highlighted future erosion risks and conservation priorities [7].

In Taiwan, soil erosion research has focused on watersheds and reservoirs, with USLE/RUSLE-based studies reporting high mean soil erosion rates. For example, the Shihmen Reservoir watershed has been estimated at roughly 82–127 Mg ha$^{-1}$ year$^{-1}$ [8–12], while the Zhuoshui River Basin averaged 122.9 Mg ha$^{-1}$ year$^{-1}$ [13]. Across Taiwan, a conservative estimate is 200.7 Mg ha$^{-1}$ year$^{-1}$ [14].

Numerous erosion studies worldwide have used USLE/RUSLE for their simplicity and broad applicability. However, several reviews have identified recurring problems in their application, including inconsistent estimation of model factors and the use of non-standard or empirically derived relationships that reduce result reliability and reproducibility [15,16]. Similar issues have been noted for Unit-Stream-Power-based Erosion Deposition (USPED), where the interpretation of erosion and deposition signs has occasionally been reversed [17,18]. These challenges underscore the importance of careful parameterization and transparent methodology. In this study, RUSLE and USPED are applied to Kinmen Island with an emphasis on methodological rigor, as further detailed in the Materials and Methods Section and later evaluated in the Discussion Section.

This study examines soil erosion on Kinmen Island, an offshore island of Taiwan where no previous research has applied RUSLE or USPED. Smaller islands like Kinmen remain underrepresented in global studies, making this case particularly important. Globally, small islands are often overlooked in soil erosion research despite their ecological fragility, limited monitoring infrastructure, and high exposure to climate extremes. The primary objective of this study is to quantify soil erosion on Kinmen using RUSLE [19] and compare results with USPED [20,21], which estimates both erosion and deposition. Unlike prior applications in Taiwan's watersheds [22], no such work exists for Kinmen. As the Agency of Rural Development and Soil and Water Conservation (ARDSWC) lacks relevant data, we collected new field measurements, providing a critical baseline for understanding soil erosion processes on small islands. This study addresses the following research questions: (i) What are the soil erosion rates on Kinmen as estimated by RUSLE? (ii) How do these results compare with USPED, which accounts for both erosion and deposition? (iii) What insights can these findings provide for understanding soil erosion processes in small-island environments?

## 2. Materials and Methods

This section first introduces the study site, Kinmen Island, and the models employed in this research: RUSLE and USPED. To provide a systematic overview, Table 1 summarizes the datasets used for calculating erosion factors and for field observation, including their sources and resolution or accuracy. The detailed procedures for deriving each factor and applying the models are described in the following subsections. Analysis was performed using R 4.3.3, GRASS GIS 8.3.1, and ArcGIS 10.4.1 to compute erosivity, erodibility, and other factors.

### 2.1. Study Site

Kinmen Island, located west of Taiwan Island, comprises two main islands: Kinmen Main Island and Little Kinmen (Lieyu) (Figure 1). Shaped like a dumbbell, the island covers approximately 131.7 km$^2$ and features rolling hills, coastal lowlands, and laterite platforms formed from granite gneiss [23–25]. The eastern part consists of gneiss hills shaped by long-term weathering and erosion, distinct from the shorter-term processes studied here. Taiwu Mountain, the highest point, reaches 247.6 m above sea level. The western part is mainly laterite platforms, low hills, and lowlands, with low gneiss hills in the southwest [26]. Island elevations are shown in Figure 2.

**Table 1.** Summary of datasets used for erosion factor calculation and field observation.

| Data Type/Factor | Source | Resolution/Accuracy | Notes |
|---|---|---|---|
| Rainfall data ($R_m$) | Central Weather Administration (CWA) | 1 min data (2018–2022), aggregated to 15 min intervals | Four stations on Kinmen; some data gaps in 2021 due to equipment failure |
| Soil samples ($K_m$) | Field surveys (10 sites), laboratory analysis | Particle size distribution by ASTM sieve/hydrometer methods; in situ permeability (limited) | 18 samples analyzed; universal kriging interpolation |
| DEM (LS, $LS_T$) | Ministry of the Interior | 20 m resolution DEM | Used to calculate LS and $LS_T$ factors |
| Land use/land cover (C, P) | National Land Surveying and Mapping Center | General 2021 dataset, spatial resolution not specified | C values adapted from Shihmen Reservoir study, modified for Kinmen |
| Field surveys (observation) | Authors' fieldwork (July 2023) | Locations recorded with handheld GPS; erosion depth measured with mm precision | 24 locations inspected, 6 sites with erosion depth measurements |

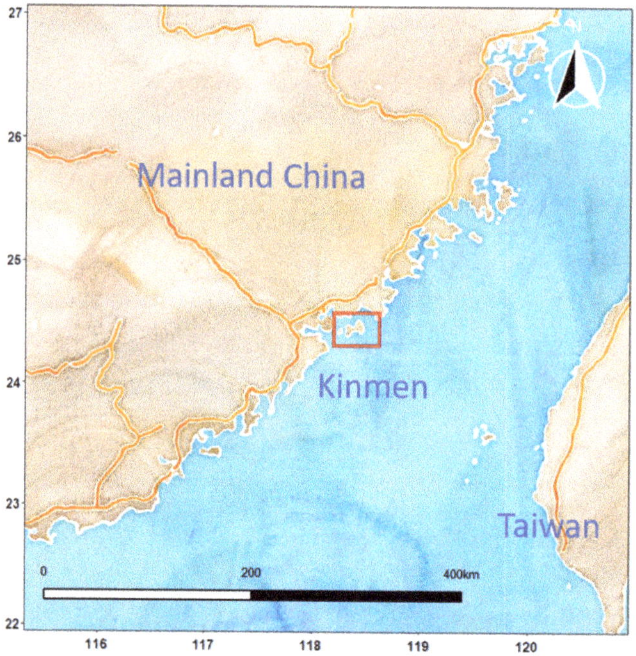


**Figure 1.** Location of the study area: Kinmen Island, situated in the Taiwan Strait between Mainland China and Taiwan. The red box highlights the island's position relative to the surrounding coastlines.

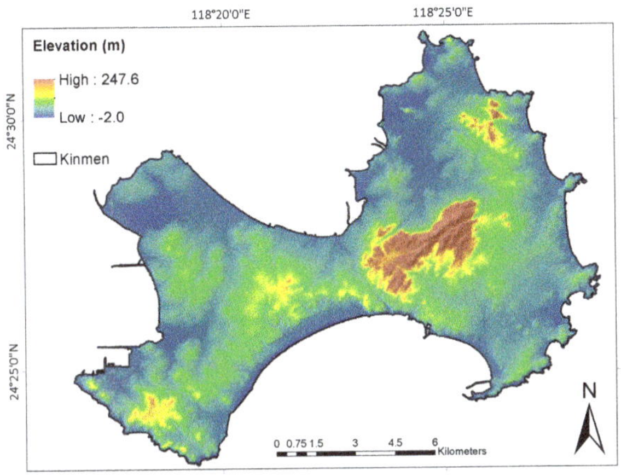


**Figure 2.** Topographic map of Kinmen Island showing elevation, locations of rainfall monitoring stations, and sampling sites for the soil erodibility factor ($K_m$).

Historically known as Quemoy, Kinmen still bears scars from the 1958 Second Taiwan Strait Crisis, when its strategic location a few kilometers from mainland China placed it at the center of heavy artillery bombardment. The conflict began on 23 August 1958, with tens of thousands of shells fired in a single day, and intermittent shelling continued until 1979. The crisis drew global attention and was described by the U.S. Secretary of State Christian Herter as the "first serious nuclear crisis" [27], paralleling more recent conflicts such as the Russia–Ukraine war.

Agriculture historically dominated Kinmen's economy, but the island also maintained a substantial military presence until 1992, which supported local livelihoods. The lifting of military restrictions initiated a shift toward tourism [28]. The Offshore Islands Development Act of 2000 promoted peaceful cross-strait relations and tourism through the "Mini Three Links" initiative [29], launched in 2001 and expanded in 2011, enabling independent travel between Kinmen and mainland China. Tourism growth has spurred infrastructure expansion, increasing pressure on natural resources and creating the challenge of balancing economic growth with ecological conservation.

Kinmen's infertile, sandy soils and limited natural resources make it vulnerable to environmental degradation and climate change. According to the USDA Soil Taxonomy, the main soils are Ultisols and Entisols [30]. Ultisols include Paleustults, Haplustults, and Plinthustults—highly weathered and acidic soils typical of humid subtropical regions. Entisols, such as Ustipsamments and Ustorthents, are young, weakly developed soils common in sandy or disturbed areas. The island's coarse-grained soils have low water retention and fertility due to extensive weathering [31]. Excessive cultivation without sufficient vegetation cover has further accelerated erosion, exposing barren laterite [30]. Kinmen was designated a low-carbon development county in 2012 [31], reflecting its sustainability commitment.

This research supports soil and water conservation on Kinmen by analyzing erosion and deposition rates and their spatial patterns. The findings will aid sustainable soil management and align with SDG 15, which emphasizes the conservation of terrestrial ecosystems and combating land degradation [32].

### 2.2. *The RUSLE and USPED Models*

The RUSLE model calculates soil erosion as follows:

$$A = R_m \cdot K_m \cdot L \cdot S \cdot C \cdot P \tag{1}$$

where

$A$ = computed soil loss per unit area (Mg ha$^{-1}$ year$^{-1}$).
$R_m$ = rainfall–runoff erosivity factor (MJ mm ha$^{-1}$ h$^{-1}$ year$^{-1}$).
$K_m$ = soil erodibility factor (Mg hour MJ$^{-1}$ mm$^{-1}$).
$L$ = slope length factor (dimensionless).
$S$ = slope steepness factor (dimensionless).
$C$ = cover management factor (dimensionless).
$P$ = support practice factor (dimensionless).

The USPED model estimates soil erosion and deposition by first computing the sediment flow at transport capacity:

$$T = R_m \cdot K_m \cdot LS_T \cdot C \cdot P \tag{2}$$

where

$T$ = sediment flow at transport capacity (Mg m ha$^{-1}$ year$^{-1}$).
$R_m$ = rainfall–runoff erosivity factor (MJ mm ha$^{-1}$ h$^{-1}$ year$^{-1}$).
$K_m$ = soil erodibility factor (Mg hour MJ$^{-1}$ mm$^{-1}$).
$LS_T$ = topographic sediment transport factor (m$^2$ m$^{-1}$).
$C$ = cover management factor (dimensionless).
$P$ = support practice factor (dimensionless).

USPED modifies the USLE/RUSLE framework by replacing LS with $LS_T$, calculated from the upslope contributing area per unit width:

$$LS_T = U^m \cdot (\sin\beta)^n \tag{3}$$

where

$U$ = upslope contributing area per unit width (m$^2$/m).
$\beta$ = angle of the slope.
$m$ = empirical coefficient.
$n$ = empirical coefficient.

Different $m$ and $n$ values are used to represent sheet erosion ($m = n = 1$) and rill erosion ($m = 1.6$, $n = 1.3$). In this study, we used $m = 1.3$ and $n = 1.2$ following [33] and also tested $m = 1.6$, $n = 1.3$ for rill erosion.

Finally, USPED derives erosion and deposition by computing the divergence of sediment flow [34,35]:

$$\mathrm{ED} = \nabla \cdot (T s_0) = \frac{\partial (T\cos\alpha)}{\partial x} + \frac{\partial (T\sin\alpha)}{\partial y} \tag{4}$$

where

ED = soil erosion or deposition (Mg ha$^{-1}$ year$^{-1}$).
$\alpha$ = aspect of the topography or the direction of flow.
$s_0$ = ($\cos\alpha$, $\sin\alpha$) = unit vector in the steepest slope direction.

A positive divergence indicates soil erosion, where sediment transport capacity increases and soil detachment rises to meet it; a negative divergence indicates deposition [17,18]. As noted in the Introduction, many past studies have mistakenly reversed these signs, leading to incorrect interpretations of erosion and deposition patterns.

Building on these considerations, this study also sought to address several common pitfalls reported in the broader soil erosion modeling literature. First, the rainfall–runoff erosivity factor (Rm) was calculated directly following the original RUSLE manual [19], using high-frequency (1 min) rainfall data aggregated to 15 min intervals, rather than relying on empirical equations that can produce biased estimates. The detailed procedure is described in the section on Rm determination. Second, the soil erodibility factor (Km) was determined from laboratory analysis of field-collected soil samples, providing physically based values instead of depending solely on literature data. Further explanation is provided in the section on Km determination. Third, in applying USPED, we carefully adhered to the correct interpretation of erosion as positive divergence and deposition as negative divergence, thereby avoiding the sign-reversal errors noted in some previous studies. The modeling framework and its implications are elaborated in the subsequent methodological and discussion sections. These steps strengthen the reliability and reproducibility of the modeling framework applied to Kinmen Island.

### 2.3. Determination of the $R_M$ Factor

Unlike the case for Taiwan Island [14], the government does not provide $R_m$ or $K_m$ values for Kinmen. To avoid the errors associated with inaccurate empirical $R_m$ equations [16], we calculated $R_m$ directly using the methodology in the RUSLE manual [19]. One-minute interval rainfall data, purchased from the government, were aggregated to 15-min intervals for ease of processing.

The $R_m$ computation involves multiple steps. First, rainfall events are defined by segmenting the storm data. Under RUSLE criteria, an event is significant if total precipitation exceeds 12.7 mm or if 6.35 mm accumulates within 15 min. Storms are separated if rainfall during a six-hour period remains below 1.27 mm. Next, the total storm energy ($E$) is calculated (Equation (5)) using $e_m$ from Equation (6). RUSLE specifies that $EI_{30}$ values—the product of total storm energy and the maximum 30 min intensity—are additive for individual storms. These $EI_{30}$ values are calculated for each event and summed (Equation (7)) and then divided by the number of years $N$ to obtain mean annual $R_m$.

For this study, rainfall data from four Central Weather Administration (CWA) stations (Figure 2) were used. Due to budget limits, only five years (2018–2022) of one-minute data were acquired. Two stations (C0W140 and C0W150) suffered equipment failures in 2021, so their $R_m$ values were averaged over four years, excluding that year.

$$E = \sum_{r=1}^{p} (e_m)_r \Delta V_r \quad (5)$$

$$e_m = \begin{cases} 0.119 + 0.0873 \log_{10}(i_m) & \text{if } i_m \leq 76 \text{ mm h}^{-1} \\ 0.283 & \text{otherwise} \end{cases} \quad (6)$$

$$R_m = \frac{\sum_{i=1}^{j} (EI_{30})_i}{N} \quad (7)$$

where

$E$ = total storm energy (MJ ha$^{-1}$).
$I_{30}$ = maximum 30 min intensity (mm h$^{-1}$).
$(EI_{30})_i$ = $EI_{30}$ for storm $i$ (MJ mm ha$^{-1}$ h$^{-1}$).
$j$ = number of storms in an $N$ year period.
$e_m$ = unit kinetic energy (MJ ha$^{-1}$ mm$^{-1}$).
$i_m$ = rainfall intensity (mm h$^{-1}$).

$(e_m)_r = e_m$ for increment $r$ (MJ ha$^{-1}$ mm$^{-1}$).
$\Delta V_r$ = rainfall depth for the $rth$ increment of the storm hyetograph (mm).

It should be noted that the above equations, while directly derived from [19], are not presented explicitly in that reference and require careful interpretation of the methodology. After computing $R_m$, values were spatially interpolated across Kinmen Island using the Inverse Distance Weighting (IDW) method.

### 2.4. *Determination of the* $K_M$ *Factor*

In this study, the calculation of $K_m$ followed the empirical formula described in the RUSLE manual [19]. As the original formula yields values in imperial units, results were converted to metric units by multiplying by the factor 0.1317, as shown in Equation (8).

$$K_m = \frac{0.1317}{100} \times \Big(2.1 \times M^{1.14} \times (10^{-4}) \times (12 - OM) + 3.25 \times (s - 2) + 2.5 \times (p - 3)\Big) \quad (8)$$

where

$OM$ = percent organic matter.
$s$ = soil-structure code.
$p$ = profile permeability class.
$M$ = (percent modified silt) × (percent silt + percent sand).
= (%silt + %very fine sand) × (100 − %clay).

Although empirical equations are often used to estimate the soil erodibility ($K_m$) factor in large-area studies, they may produce unrealistic values exceeding the theoretical upper limit of 0.1 (Mg h MJ$^{-1}$ mm$^{-1}$) [15]; therefore, this study directly determined $K_m$ through field sampling and laboratory particle-size analysis, ensuring that all $K_m$ values are physically measured and regionally representative rather than derived from empirical equations.

A total of 18 soil samples were collected from 10 locations on Kinmen Island and analyzed in the laboratory to determine $K_m$. As detailed in Table S1 of the Supplementary Materials ("Particle-size distributions and related properties for Km estimation"), sites 1, 2, 3, 4, 6, 7, 9, and 10 each include two samples (e.g., 1.1 and 1.2), while sites 5 and 8 include only one sample due to insufficient soil quantity for complete grain-size analysis after transport to the laboratory in Taiwan. Figure 3 illustrates the island's geological formations and $K_m$ sampling locations. Surface soils exhibit marked heterogeneity even over short distances; therefore, sampling aimed to capture representative conditions. The soils primarily consist of deposits formed by aeolian, fluvial, marine, or oceanic processes [23,36], laterite gravel, the Kinmen Formation, and weathered rock fragments. Particle size distribution was first assessed using sieve analysis down to a No. 270 sieve (0.053 mm). Finer fractions were analyzed with the hydrometer method to estimate silt and clay content. Both methods followed ASTM (American Society for Testing and Materials) standards.

In situ permeability tests were also performed, but access restrictions from landowners limited site selection, and infiltration rates failed to reach steady state. Soil-structure codes were unavailable because some samples were inadvertently discarded. Given that $s$ ranges from 1 to 4 and $p$ from 1 to 6, average values of $s = 2.5$ and $p = 3.5$ were adopted. The soil samples collected on Kinmen contained 85–99% sand, which provides important context for these parameter choices. In the RUSLE formulation, the soil-structure code ($s$) is defined as 2 for fine granular and 3 for medium/coarse granular soils; therefore, selecting

$s = 2.5$ represents a reasonable midpoint for the predominantly sandy soils observed. Likewise, the profile permeability class ($p$) is coded as 3 for moderate permeability and 4 for slow-to-moderate permeability. Considering the limited infiltration observations together with the sandy texture, $p = 3.5$ was adopted as a balanced representation. Prior studies indicate that the influence of $p$ and $s$ on $K_m$ may be minor [37].

The resulting $K_m$ values were interpolated using Universal Kriging to produce a spatial map for Kinmen Island from the ten sampling locations as it accounts for both spatial autocorrelation and data trends. For $R_m$, Universal Kriging was unsuitable due to too few data points to fit a variogram model, so IDW was used instead.

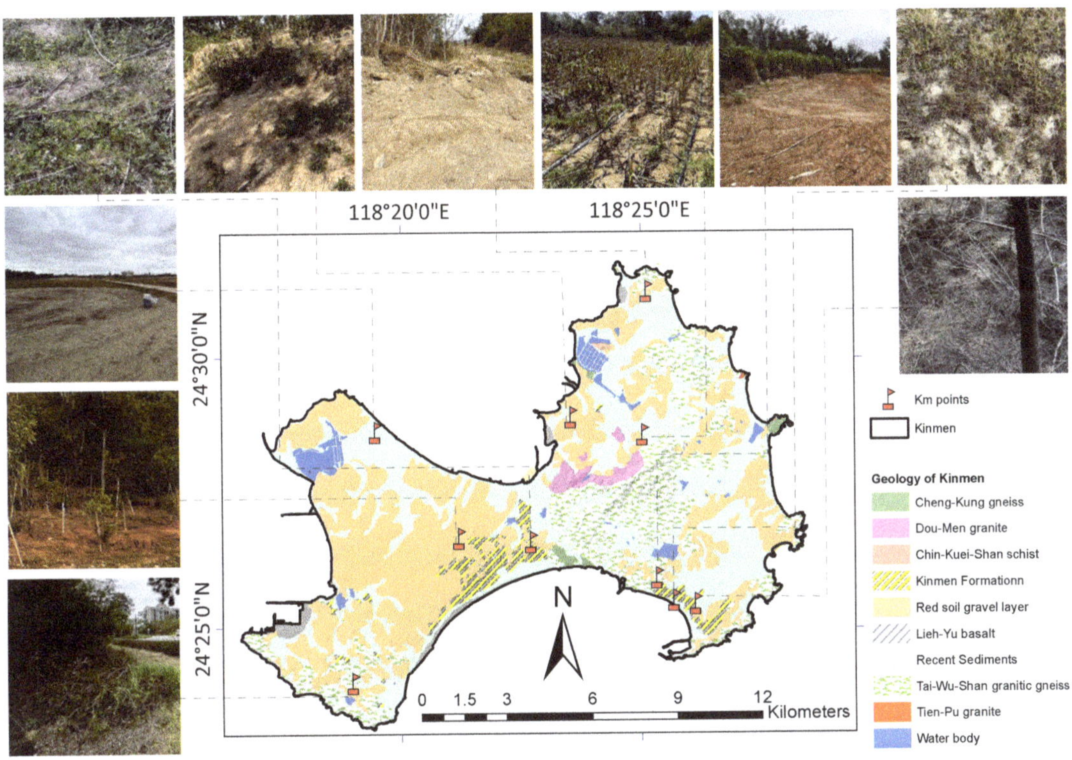


**Figure 3.** Geological map of Kinmen with the locations of the $K_m$ sampling points, accompanied by field photographs depicting the corresponding site conditions.

### 2.5. *Determination of the* LS *and* $LS_T$ *Factors*

The key difference between the RUSLE and USPED models lies in their treatment of the LS and $LS_T$ factors. For the RUSLE, this study employs the Desmet and Govers equation [38] to compute the LS factor from the upslope contributing area, using the field-based LS modules in SAGA. Notably, Chen and Nguyen [39] were the first to apply this method across the entire island of Taiwan.

In the USPED model, the LS factor is replaced by $LS_T$, which evaluates topographic sediment transport based on the upslope contributing area per unit width. Recommended values of $m$ and $n$ account for different erosion types: predominantly sheet erosion, a combination of sheet and rill erosion, or predominantly rill erosion. Typical ranges are $m = 1.0$–$1.6$ and $n = 1.0$–$1.3$, with higher values indicating prevailing rill erosion and more turbulent flow, where erosion increases sharply with water volume. These parameters can also be fitted to field data.

Both sheet and rill erosion occur on Kinmen Island. In this study, we tested two cases: (1) $m = 1.3$, $n = 1.2$ for mixed sheet and rill erosion; and (2) $m = 1.6$, $n = 1.3$ for predominantly rill erosion.

The LS and $LS_T$ factors were computed from a 20 m resolution DEM (Digital Elevation Model) provided by the Ministry of the Interior (Figure 2). Kinmen's topography includes negative elevation values due to its geographic setting and tidal fluctuations, where some areas fall below average sea level during high tide. These values are essential for accurately representing terrain morphology and its influence on slope and erosion, making their inclusion critical for realistic sediment transport modeling.

### 2.6. *Determination of the* C *and* P *Factors*

In this study, the 2021 Land Use and Land Cover (LULC) data purchased from the National Land Surveying and Mapping Center were used to determine the C values. The conversion of LULC data to C factors was initially based on a reference table developed for the Shihmen Reservoir watershed in Taiwan [40]. This table was then expanded to include LULC types specific to Kinmen Island that were absent in the Shihmen Reservoir dataset, as shown in Table 2.

Due to the lack of detailed field data on soil conservation measures, this study follows a conservative approach widely adopted in soil erosion research, assuming the P factor to be 1 for Kinmen.

**Table 2.** LULC categories and associated C values, modified from [40].

| C Value | LULC Types |
|---|---|
| 0.000 | Aquaculture, Airport, General Road, Road-related facilities, Commercial port, Fishing port, River, Irrigation canal, Reservoir, Lake, Water storage pond, Waterway sandbar, Embankment, Water gate, Pumping station, Groundwater extraction well, Other water facilities, Flood prevention road, Retail and wholesale, Service industry, Residential area, Residential area used for both residential and industrial purposes, Residential area used for both residential and commercial purposes, Residential area used for both residential and other purposes, Manufacturing industry, Warehouses, Religious facilities, Funeral facilities, Under construction, Others, Government agencies, Kindergarten, Elementary school, Junior high school, College/University, Medical and healthcare, Social welfare facilities, Meteorological facilities, Electricity, Gas, Tap water, Gas station, Environmental facilities, Legally designated cultural assets, General cultural facilities, Other cultural facilities, Amusement places, Sports facilities, Reef |
| 0.003 | Park and green space |
| 0.010 | Livestock and poultry houses, Agricultural production facilities, Agricultural marketing and processing facilities, Broad-leaved Forest, Bamboo Forest, Bamboo-broad-leaved mixed forest |
| 0.030 | Grassland |
| 0.050 | Pasture, Landslide area |
| 0.100 | Paddy field |
| 0.160 | Orchard |
| 0.228 | Rainfed field |
| 0.500 | Beach |
| 0.600 | Unused land |
| 0.700 | Earth and stone works and related facilities, Land undergoing anthropogenic modifications |
| 0.800 | Construction waste and soil storage and treatment facilities |

### 2.7. On-Site Field Observations

The soil erosion modeling results were not formally validated through monitoring data as no erosion pins or long-term measurement infrastructure were installed on Kinmen Island. Instead, on-site field surveys were conducted to provide observational evidence and contextual understanding of erosion conditions within the study area. Two field surveys were conducted in July 2023. During the first visit to Kinmen, 24 locations were inspected to evaluate local conditions and capture photographic documentation. In the second survey, erosion depth measurements were recorded at six specific sites. From the 24 locations visited, representative examples of sheet erosion and rill erosion were identified and presented. Figure 4 shows only the six locations in green where photographs are provided in the Results section and the six locations in red where erosion depths were measured. These observations illustrate the presence and characteristics of erosion features on Kinmen, but they were not intended as a systematic validation of the model predictions.

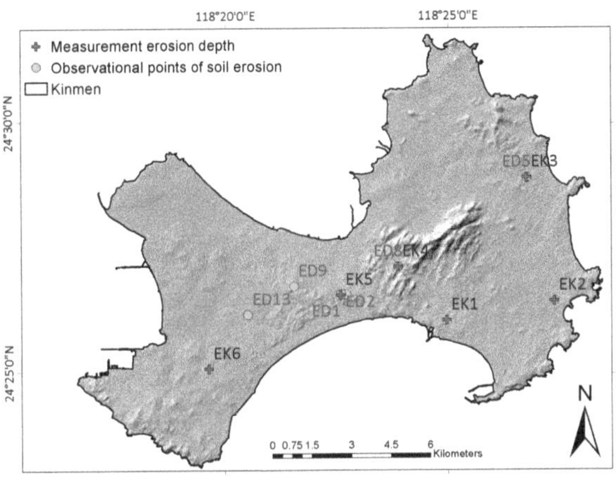


**Figure 4.** Representative locations of rill and sheet erosion on Kinmen Island, indicating sites where erosion depth was measured (in red) and sites of field observations of soil erosion (in green).

## 3. Results

This study's findings are presented in three parts. First, the erosion factors derived from the RUSLE and USPED models are analyzed. Second, soil erosion estimates from both models are compared. Finally, field measurements of erosion depth are used to provide context for the model results and offer insights into on-site erosion processes.

### 3.1. Quantitative Assessment of Erosion Factors Using RUSLE and USPED

Figure 5 shows the spatial distribution of the key erosion and deposition factors, alongside a SPOT 7 satellite image from 23 August 2022. The $R_m$ and $K_m$ maps were interpolated from rainfall station data and soil sampling points, respectively. The northwest exhibits the highest $K_m$ values, while some of the lowest $R_m$ values are found in the northeast. Conversely, the southeast displays relatively high $R_m$ values but lower $K_m$ values.

The LS and $LS_T$ maps reflect the island's generally flat terrain, except around Taiwu Mountain, the highest peak. The C factor is highest in the north, indicating sparser vegetation. Together, these spatial patterns highlight the interaction between climate, soil properties, topography, and land cover in shaping erosion and deposition on Kinmen Island.

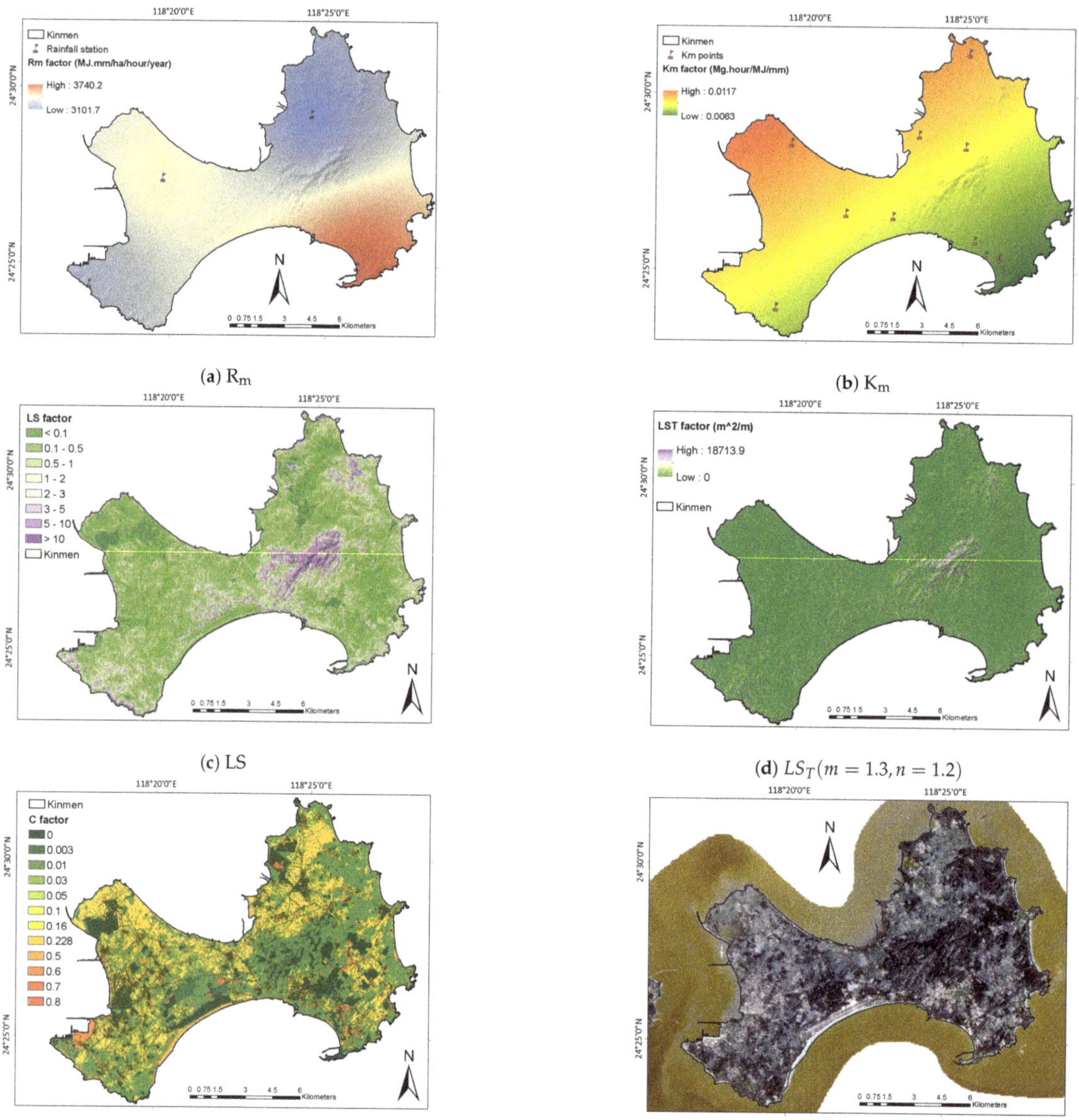


(**a**) $R_m$ (**b**) $K_m$ (**c**) LS (**d**) $LS_T(m = 1.3, n = 1.2)$ (**e**) C (**f**) SPOT 7 image (23 August 2022)

**Figure 5.** Factors of the RUSLE and USPED models: (**a**) $R_m$, (**b**) $K_m$, (**c**) $LS$, (**d**) $LS_T$, (**e**) $C$, and (**f**) SPOT 7 image of Kinmen Island.

Table 3 summarizes the average, maximum (Max), and minimum (Min) values of these factors. The $R_m$ factor averages 3352 MJ mm ha$^{-1}$ hour$^{-1}$ year$^{-1}$, ranging from 3102 to 3740, showing notable spatial variability. The $K_m$ factor averages 0.0090 Mg hour MJ$^{-1}$ mm$^{-1}$, with a range of 0.0063–0.0117, reflecting differences in soil erodibility. The detailed particle-size data used to compute the $K_m$ values, together with the per-station $R_m$ values and their averages, are provided in the Supplementary Materials. The LS factor averages 0.95, with values from 0.03 to 29.85. The topographic sediment transport factor ($LS_T$) averages 13.20 for $m = 1.3$, $n = 1.2$, and 60.57 for $m = 1.6$, $n = 1.3$. The cover-management factor (C) averages 0.12, with values from 0 to 0.80, indicating variation in vegetation cover across the island.

**Table 3.** Average, maximum, and minimum values of erosion factors for the RUSLE and USPED models.

| Factor | Average | Max | Min |
|---|---|---|---|
| $R_m$ (MJ mm ha$^{-1}$ h$^{-1}$ year$^{-1}$) | 3352 | 3740 | 3102 |
| $K_m$ (Mg hour MJ$^{-1}$ mm$^{-1}$) | 0.0090 | 0.0117 | 0.0063 |
| LS | 0.95 | 29.85 | 0.03 |
| $LS_T$ (m$^2$/m) $m = 1.3$, $n = 1.2$ | 13.20 | 18,713.92 | 0 |
| $LS_T$ (m$^2$/m) $m = 1.6$, $n = 1.3$ | 60.57 | 210,962.56 | 0 |
| C | 0.12 | 0.80 | 0 |

### *3.2. Comparison of Soil Erosion Estimates from RUSLE and USPED*

Figure 6 shows the spatial patterns of erosion and deposition across Kinmen Island. The RUSLE output (Figure 6a) indicates widespread low to moderate erosion rates, with most of the island showing values between 0 and 3 Mg ha$^{-1}$ year$^{-1}$ and 3 and 6 Mg ha$^{-1}$ year$^{-1}$. Scattered hotspots are found primarily in cultivated areas, sparsely vegetated land, and some coastal slopes, reflecting the strong influence of the cover-management (C) factor on local erosion intensity. In contrast, the generally flat topography of Kinmen limits the LS/$LS_T$ contribution, which helps to explain the dominance of low erosion values compared with more mountainous regions of Taiwan.

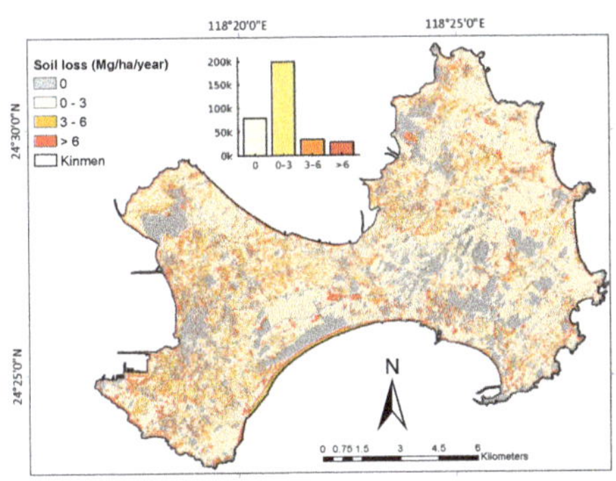


(**a**) RUSLE

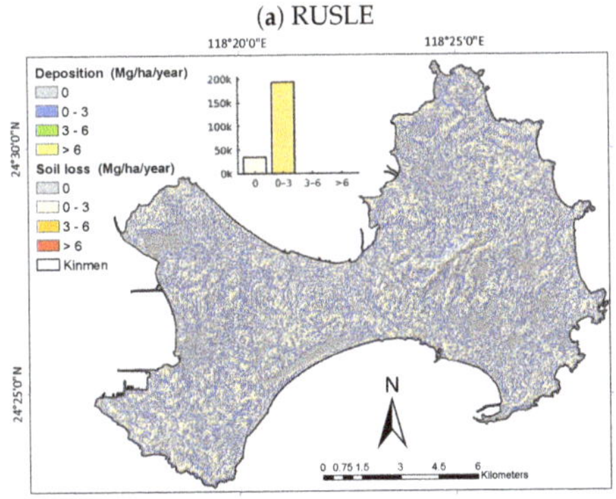


(**b**) USPED (m = 1.3, n = 1.2)

**Figure 6.** *Cont.*

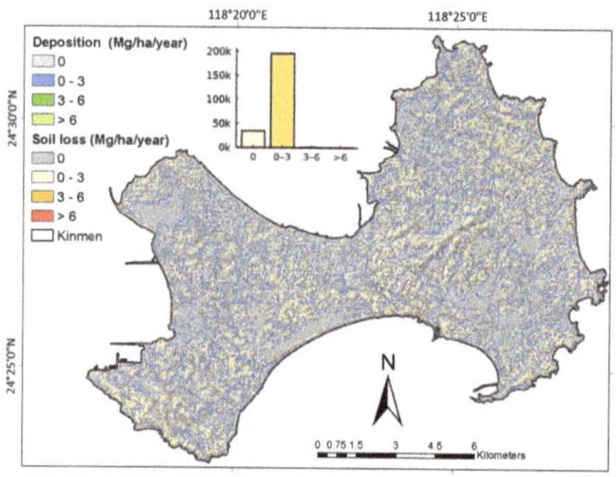


(c) USPED (m = 1.6, n = 1.3)

**Figure 6.** Distribution of soil erosion and deposition from the (**a**) RUSLE model, (**b**) USPED model (m = 1.3, n = 1.2), and (**c**) USPED model (m = 1.6, n = 1.3).

The USPED outputs (Figure 6b,c) reveal more complex spatial dynamics, with yellow, orange, and red indicating erosion and blue and green indicating deposition. This dual-process representation highlights areas of sediment redistribution that RUSLE does not capture. The USPED results are also more sensitive to parameter choices, with higher $m$ and $n$ values producing larger predicted erosion and deposition magnitudes. The observed patterns suggest that localized land use interacts with gentle slope gradients to determine whether a site functions predominantly as an erosion source or a deposition sink.

Table 4 summarizes the corresponding soil erosion and deposition rates from the two models. The RUSLE model estimates an average erosion rate of 2.17 Mg ha$^{-1}$ year$^{-1}$ but does not account for deposition. The USPED model, by contrast, provides both erosion and deposition estimates: for $m = 1.3$ and $n = 1.2$, erosion is 0.87 Mg ha$^{-1}$ year$^{-1}$ and deposition is 1.39 Mg ha$^{-1}$ year$^{-1}$; for $m = 1.6$ and $n = 1.3$, erosion increases to 3.79 Mg ha$^{-1}$ year$^{-1}$ and deposition to 6.51 Mg ha$^{-1}$ year$^{-1}$. In both scenarios, the total amounts of erosion and deposition are of similar magnitude, reflecting USPED's mass-balance framework. The USPED map for the sheet erosion case ($m = n = 1$) is provided in the Supplementary Materials, where the total erosion and total deposition amounts for different $(m, n)$ parameterizations are also reported. In addition, the island area fractions derived from RUSLE and USPED (for multiple $(m, n)$ values) are included in the Supplementary Materials.

These contrasts highlight the complementary strengths of the two models: RUSLE offers a simpler estimate of overall erosion magnitude, whereas USPED reveals the spatial interplay between erosion and deposition and accounts for the balance between the two processes, making it particularly valuable for understanding landscape-scale sediment redistribution. By linking the modeled outcomes to controlling factors such as land cover and topography, the comparison also underscores how management interventions (e.g., vegetation cover improvements) could mitigate erosion risks even in relatively flat island settings.

**Table 4.** Summary of soil erosion and deposition on Kinmen Island based on the RUSLE and USPED models.

| | RUSLE | USPED (m = 1.3, n = 1.2) | USPED (m = 1.6, n = 1.3) |
|---|---|---|---|
| Erosion (Mg ha$^{-1}$ year$^{-1}$) | 2.17 | 0.87 | 3.79 |
| Deposition (Mg ha$^{-1}$ year$^{-1}$) | - | 1.39 | 6.51 |

### 3.3. Observed Field Evidence of Soil Erosion Depth

Figure 7 presents an overlay of RUSLE-derived soil erosion estimates on a Pléiades satellite image (15 November 2022). The colored grid cells indicate modeled annual soil loss, with higher values marking potential erosion hotspots. Two field observation sites (ED1 and ED2; see also Figure 8) are also shown on the image. The overlay reveals that some modeled hotspots coincide with bare soil or disturbed land visible in the satellite image, while others do not. This discrepancy is expected because RUSLE represents long-term average annual soil erosion, whereas the Pléiades image reflects land-surface conditions at a single moment in time. Accordingly, the figure should be viewed as providing observational evidence and contextual understanding of erosion conditions on Kinmen, rather than a quantitative validation of the model outputs. The southern shoreline and Kinmen Airport are also visible, where potential erosion can be observed along coastal stretches near built-up areas.

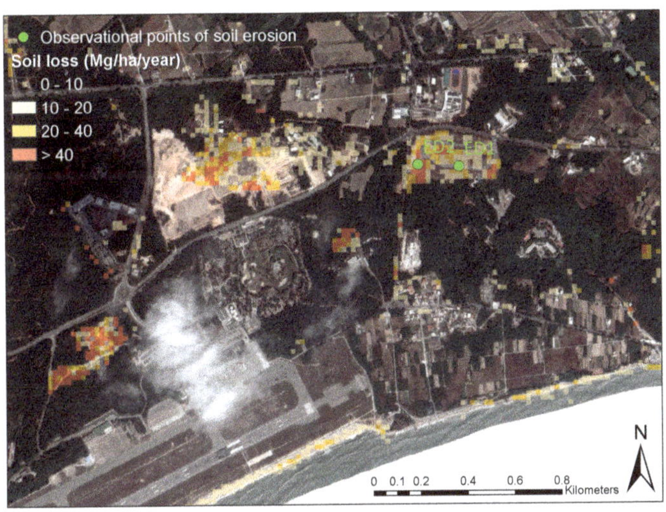


**Figure 7.** Overlay of RUSLE-derived soil erosion estimates on a Pléiades image (15 November 2022). The color scale shows annual soil loss in Mg ha$^{-1}$ year$^{-1}$, with higher values indicating potential hotspots. Two field observation sites (ED1 and ED2; see Figure 8) are marked on the image. The overlay highlights locations where modeled hotspots align with bare or disturbed land visible in the satellite image, as well as areas where such agreement is absent.

The locations of soil erosion depth measurements are shown in Figure 4, with corresponding photographs in Figure 8. As noted earlier, ED1 and ED2 are also shown in Figure 7, where their positions relative to modeled erosion hotspots can be seen. Rill erosion was observed at ED1, ED2, ED8, and ED13, while sheet erosion occurred at ED5 and ED9. Sheet erosion was marked by topsoil loss and the presence of soil pedestals—small, elevated patches of soil protected by stones or rock fragments that shield them from raindrop impact. Unlike Taiwan Island, where intense typhoons quickly erode such features, Kinmen's milder rainfall allows them to persist, making sheet erosion more easily observed.

Rill erosion, characterized by channels formed by concentrated surface runoff, was most apparent on slopes adjacent to dirt roads or sidewalks. Rainfall-driven runoff in these areas removes substantial amounts of soil, illustrating the erosive force of concentrated water flow.

(**a**) ED1 (**b**) ED2 (**c**) ED5 (**d**) ED8 (**e**) ED9 (**f**) ED13

**Figure 8.** The locations of on-site observation of soil erosion on Kinmen Island, corresponding to the six marked locations in Figure 4: (**a**) ED1, (**b**) ED2, (**c**) ED5, (**d**) ED8, (**e**) ED9, and (**f**) ED13. Each image depicts varying degrees of soil erosion observed during field visits. ED1 and ED2 are also shown in Figure 7, where they are displayed in relation to modeled erosion hotspots and the Pléiades satellite image.

At each of the six measurement sites (Figure 4), a ruler with millimeter precision was used to record soil pedestal height from the base of a protecting stone to the ground surface. Multiple readings per site were averaged. Table 5 lists the mean cumulative erosion depths and standard deviations. "Cumulative" refers to the fact that these depths have accumulated over many years, with the initial onset of erosion unknown.

It is important to note that these field measurements were conducted as observational surveys to provide context about erosion processes on Kinmen, rather than as a formal validation of the RUSLE or USPED model predictions. Using Universal Kriging, the spatial mean cumulative erosion depth for Kinmen Island was estimated at 42.9 mm—exceeding model predictions from both RUSLE and USPED. Because the onset of erosion is uncertain and may have been influenced by typhoon-related events, these measured depths are not considered reliable for quantitative model evaluation. In contrast, the model predictions fall within a more reasonable range based on empirical experience and validated studies in the neighboring Fujian Province. This issue and its implications are further discussed in Section 4.1.

**Table 5.** Cumulative erosion depth measurements at six locations on Kinmen Island. "Number of measurements" refers to the count of readings taken at each site and, therefore, has no physical unit.

| Sheet Erosion | EK1 | EK2 | EK3 | EK4 | EK5 | EK6 |
|---|---|---|---|---|---|---|
| Number of measurements | 9 | 3 | 11 | 5 | 7 | 18 |
| Mean cumulative erosion depth (mm) | 69.2 | 32.7 | 59.1 | 23.6 | 26.9 | 53.6 |
| Standard deviation (mm) | 21.9 | 2.5 | 15.4 | 3.0 | 8.8 | 16.3 |

## 4. Discussion

The application of the RUSLE and USPED models provided valuable insights into the extent and spatial distribution of soil erosion and deposition on Kinmen Island. In the following sections, we examine why measured erosion depths exceed model predictions and compare our findings with data from the nearby Fujian Province to place Kinmen's erosion dynamics in a regional context.

### *4.1. Lack of Monitoring, Erosion Onset, and Annual Soil Loss Estimation*

As noted in Section 3.3, the field erosion depth measurements should be understood as observational context rather than a formal validation reference. Unlike many regions, Kinmen Island has no history of systematic soil erosion monitoring. No erosion pins or long-term devices were installed, which prevents rigorous evaluation of the model predictions. In such cases, direct field measurements of erosion depth, as performed here, provide one of the few feasible means to quantify erosion. However, while these measurements are valuable, they are highly uncertain and likely influenced by assumptions about erosion onset related to rainstorms or typhoon events. For this reason, model-based estimates from RUSLE and USPED remain more credible for representing average island-wide conditions.

Kinmen's long-term mean annual precipitation is about 1100 mm year$^{-1}$ [41], but the 2018–2022 rainfall data used in this study average only 745.9 mm year$^{-1}$, reflecting a recent absence of typhoon landfalls. Typhoons likely play a major role in driving erosion by delivering intense rainfall that can remove soil pedestals and reset erosion surfaces. Therefore, identifying typhoons with direct impacts on Kinmen was critical to constraining the onset of erosion.

Based on the Central Weather Administration's typhoon database, four notable events were identified:

1. Typhoon Lupit (2–9 August 2021).
2. Tropical Depression WP242018 (22–26 August 2018).
3. Typhoon Guchol (4–8 September 2017).
4. Typhoon Meranti (15 September 2016; direct landfall on Kinmen).

Among these, Typhoon Meranti in 2016 caused the most severe impacts according to available damage reports and is hypothesized to mark the beginning of the current erosion phase. Figure 9 shows the typhoon's path across the study area. It should be emphasized that typhoons are not treated here as a separate research subject since their rainfall contribution is already represented in the $R_m$ factor. They are mentioned only because the absence of long-term erosion monitoring on Kinmen makes it impossible to determine when the observed erosion features first developed. In this context, we postulate that soil pedestals and similar features persist between major storm events and are reset when such events occur. Assuming erosion began after 15 September 2016, the period to our field survey dates (24–26 July 2023) spans 2503 days.

The mean cumulative erosion depth for Kinmen, derived from field measurements (Table 5), is 42.9 mm. This translates to an average annual erosion depth of 6.3 mm $year^{-1}$, or 87.58 Mg $ha^{-1}$ $year^{-1}$, assuming a soil unit weight of 1.4 Mg $m^{-3}$. These estimates are considerably higher than expected based on empirical experience and may not represent typical long-term rates. In contrast, the RUSLE prediction of 2.17 Mg $ha^{-1}$ $year^{-1}$ and the highest USPED prediction (3.79 Mg $ha^{-1}$ $year^{-1}$ for $m = 1.6$, $n = 1.3$) fall within a more reasonable range and align better with comparable studies in neighboring Fujian Province.

The disparity may stem from several factors. First, field measurements capture localized hotspots of erosion not fully represented in spatially averaged models. Second, because the exact onset of erosion is unknown, the assumption that it began after Typhoon Meranti may overestimate annual rates. Third, local variations in soil properties, land use, and vegetation cover may differ from model parameter assumptions. Ultimately, while field erosion depths provide important site-specific insight in the absence of monitoring, they should not be treated as a validation dataset, and the RUSLE and USPED outputs are considered to provide more reasonable island-wide averages.

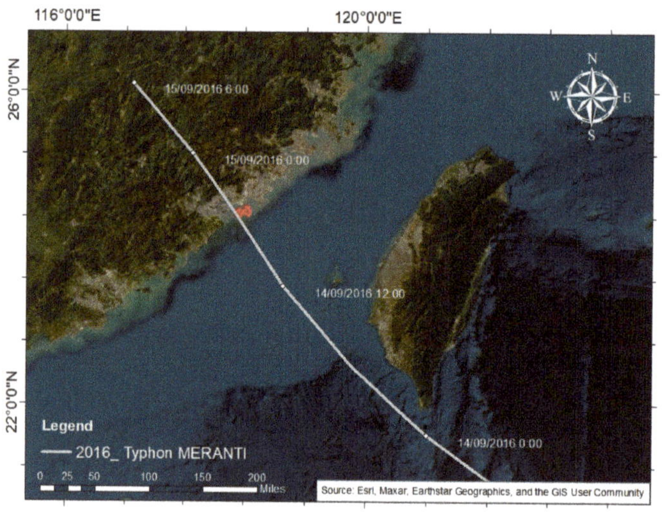


**Figure 9.** Track of Typhoon Meranti in September 2016, showing its trajectory and timing before and after making landfall on Kinmen (shown in red).

### 4.2. Comparison with Neighboring Fujian Province

The comparison between our RUSLE and USPED results and field-observed erosion phenomena shows a weak correlation. To contextualize our findings, we examined studies from the neighboring Fujian Province, including those that focus on individual erosion factors, landscape change, and ecosystem services. A synthesis of these studies is presented in Table 6.

**Table 6.** Summary and comparison of soil erosion studies conducted in neighboring Fujian Province. Some studies focus solely on individual factors, while others do not provide average or range values. Original figures from the studies are presented without rounding.

| Location | Model | Avg. Annual Precip. (mm) | $R_m$ (MJ mm $ha^{-1}$ $hr^{-1}$ $yr^{-1}$) | $K_m$ (Mg hr $MJ^{-1}$ $mm^{-1}$) | LS | C | P | Soil Erosion (Mg $ha^{-1}$ $yr^{-1}$) | Ref. |
|---|---|---|---|---|---|---|---|---|---|
| Kinmen | RUSLE | 745.9 | 3352 | 0.009 | 0.95 | 0.12 | 1 | 2.17 | This study |
| Fujian Province | USLE | - | - | 0.0329 | - | - | - | - | Orig. in [42], cited in [43] |
| Fujian Province | USLE | - | - | 0.0251 (non-standard USLE) | - | - | - | - | [43] |
| Fujian Province | USLE | - | - (Emp. Eq.) | 0.1626–0.2708 (Figure 3, no unit) | - | - | - | 80%: 5–25 | [44] |
| Fujian Province | USLE | 1614.01 | 6771.80–14,828.23 (non-standard RUSLE) | - | - | - | - | - | [45] |
| Upper Minjiang River Basin, Fujian Province | RUSLE | 1692–2677 (Figure 3a) | 289.4–460.6 (Figure 3b) * | 0–0.216 (Figure 4b) ** | 0–3 (Figure 5b) | 0–0.31 (Figure 6a) | 0–1 (Figure 6b) | 1.3783–6.9227 | [46] |
| Hetian Town, Fujian Province | USLE | 1750.8 | - | 0.22 (Emp. Data) | - | 0.18 (Masson pine forest) | - | - | [47] |
| Changting County, Fujian Province | RUSLE | 1737 | - (Emp. Eq.) | - (non-standard RUSLE) | - | - | - | 3.38 (2020) | [48] |
| Tingjiang Watershed, Fujian Province | InVEST-SDR (based on USLE) | 1450–2200 | - (Emp. Eq.) | - (non-standard USLE) | - | - | - | 993.6–12,085.01 (2020, Figure 5) | [49] |
| Pingtan Island, Fujian Province | RUSLE | - | 636.81–777.14 (2020, Figure 9) | 0–0.185 (Figure 9) ** | 0–229.65 (Figure 9) | 1.263–2.489 (2020, Figure 9) *** | 0.2–1.71 (Figure 9) | 0–8099 (Figure 9) | [50] |

* The $R_m$ factor was calculated using an incorrect R equation [16]. ** The $K_m$ factor exceeds the maximum possible value of 0.1 [15]. *** The C factor exceeds the maximum possible value of 1.

Although the literature provides useful background, it also reveals methodological issues similar to those discussed in the Introduction Section [15,16]. As summarized in Table 6, many studies did not follow standard USLE/RUSLE procedures for calculating the rainfall erosivity factor ($R_m$), often substituting empirical equations for high-resolution rainfall data. Such empirical relationships, as noted in earlier reviews, can produce substantial inaccuracies. Comparable challenges exist for the $K_m$ factor: where some reported values exceed known physical limits, others lack proper units, or were derived using non-standard methods. These inconsistencies make direct, quantitative comparison with our Kinmen results unreliable.

The regional comparison is, therefore, helpful but necessarily limited. While it highlights the prevalence of methodological shortcuts in Fujian studies, it also demonstrates the difficulty of synthesizing results when factors are derived inconsistently. The broader implication is that meaningful cross-regional comparisons require standardized methods and transparent reporting of data sources and procedures.

The study by [48], which reported a 2020 RUSLE-based erosion rate of 3.38 Mg $ha^{-1}$ $year^{-1}$ validated with hydrological station data, is a notable exception and aligns closely with our findings. However, the broader variability in methods and data quality across the Fujian studies prevents firm conclusions.

This underscores the importance of our work: Kinmen represents an isolated island setting where soil erosion modeling has not previously been conducted and where we have applied fully documented, standard-compliant methods using actual high-frequency rainfall and measured soil properties. By addressing the methodological pitfalls identified in the past literature, our study provides a more reliable reference point for understanding erosion dynamics in small-island environments and contributes to a more standardized basis for comparison with neighboring regions such as Fujian.

### 4.3. Management Implications for Kinmen

The findings of this study also have practical relevance for soil and water conservation planning on Kinmen Island. Given that overall erosion rates are relatively low to moderate but localized hotspots occur in cultivated and sparsely vegetated areas, targeted management is recommended. Conservation practices such as maintaining ground cover through cover crops, mulching, or residue retention could effectively reduce soil loss in farmland. Contour farming and small-scale terraces may further help minimize rill development on slopes near agricultural plots and roads. In coastal areas where erosion risk is higher, the establishment of vegetative buffer strips and reinforcement of vulnerable slopes would provide additional protection. These site-specific measures, combined with the development of long-term monitoring infrastructure, would support sustainable land management and align with broader soil conservation goals in Taiwan and beyond.

### 4.4. Study Limitations and Considerations

It is worth emphasizing that this study directly addressed several methodological pitfalls frequently encountered in soil erosion modeling. By calculating the rainfall–runoff erosivity factor (Rm) from high-frequency rainfall records in accordance with the RUSLE manual, we avoided the biases that can arise from simplified empirical equations. The soil erodibility factor (Km) was derived from laboratory analysis of field-collected samples, ensuring physically based values instead of relying on generalized or literature-based parameters. In applying USPED, we also adhered to the correct interpretation of positive divergence as erosion and negative divergence as deposition, thereby eliminating the sign-reversal errors that have appeared in past applications. These measures enhance the reliability and reproducibility of the results and provide a stronger reference point for small-island environments such as Kinmen.

While these steps improved methodological reliability, this study also evaluated soil erosion and deposition patterns on Kinmen Island using the RUSLE and USPED models, with comparisons to studies in neighboring Fujian Province. The results reinforce the importance of standardized modeling procedures to enable meaningful cross-regional comparisons. The relatively low erosion rates observed are consistent with Kinmen's lower rainfall and gentle topography, which limit runoff and erosive energy. While these findings provide a useful baseline for local conservation strategies, several methodological constraints warrant consideration for future modeling in similar environments.

First, the $R_m$ factor was derived from high-temporal-resolution rainfall data following the original RUSLE methodology. This improves process fidelity but depends on costly and limited datasets. In this case, only five years of data (2018–2022) were available, during which rainfall was below the long-term average. This temporal limitation may bias the erosivity estimate toward lower values, underscoring the need for extended datasets or the integration of synthetic rainfall series to capture interannual variability.

Second, the relatively low $K_m$ factor likely reflects both methodological and logistical constraints. Difficulties in achieving steady infiltration rates during field permeability testing, coupled with restricted access imposed by landowners and budgetary limitations, which constrained soil sampling to 18 samples from 10 locations in this pilot study, may have led to an underestimation of soil erodibility. Future studies should address this limitation by expanding the number of sampling sites and improving spatial coverage across diverse soil and land-cover types to better capture regional-scale variability. This highlights the importance of improving in situ measurement protocols and ensuring representative sampling, especially in areas with limited accessibility.

Third, the cover-management (C) and support practice (*P*) factors were derived from national land-use data. Because no detailed records of conservation practices (e.g., contouring, terracing, and strip cropping) are available for Kinmen, a uniform $P = 1$ was assumed. While this simplifies conditions and may overestimate erosion in managed lands, it avoids introducing speculative values. We acknowledge this as a limitation and emphasize the need for more detailed management information in future studies.

Fourth, the resolution of the DEM used in this study (20 m) may constrain the accuracy of topographic factor estimation. Higher-resolution datasets (e.g., 5 m DEMs or LiDAR-derived products) would better capture microtopographic variability and improve both LS and $LS_T$ factor calculations. As no finer DEM is currently available for Kinmen, we acknowledge this as a limitation and recommend that future work incorporate higher-resolution elevation data when feasible.

Fifth, although field surveys identified extensive sheet and rill erosion, the observed annual rate of 87.58 Mg ha$^{-1}$ exceeded both RUSLE and USPED predictions. This discrepancy suggests potential misestimation of erosion onset timing—possibly linked to Typhoon Meranti in 2016—or underrepresentation of extreme events in the model inputs. The absence of long-term erosion monitoring hampers retrospective calibration. Incorporating erosion pins or similar field-based monitoring tools could provide continuous validation data and help reconcile model–field mismatches.

A further limitation relates to scaling. The erosion depths obtained from field surveys represent localized point conditions, whereas the RUSLE and USPED models produce distributed estimates across the landscape. Directly comparing point-based measurements with gridded model outputs may, therefore, introduce mismatches and uncertainty. For this reason, the field observations in this study are interpreted as qualitative context to illustrate erosion processes, rather than as quantitative validation of model predictions.

Overall, these limitations do not diminish the study's value as the first quantitative erosion assessment for Kinmen Island, but they do indicate critical avenues for methodological refinement. Future work should prioritize longer climatic records, more robust soil sampling strategies, and permanent monitoring infrastructure. Such measures would strengthen model calibration, improve predictive reliability, and enhance transferability of results to other data-scarce, low-relief island environments.

In addition, this study was designed as a preliminary benchmark for Kinmen. However, it did not separately analyze the effect of farmland on erosion rates or delineate erosion probability zones. These tasks would require more detailed land-use stratification, longer-term monitoring, and probabilistic modeling. We note them here as important directions for future research to complement the baseline results presented in this study.

In summary, although data and logistical constraints introduced uncertainties, the corrective measures implemented in this study—such as the direct calculation of $R_m$, laboratory-based determination of $K_m$, and proper USPED interpretation—represent a meaningful step toward more reliable erosion modeling in small-island contexts.

## 5. Conclusions

This study applied two soil erosion models—the RUSLE and the USPED model—to assess soil erosion and deposition on Kinmen Island. It represents the first comprehensive evaluation of soil erosion in this region. As no rainfall erosivity data were available for the island, we compiled rainfall records from weather stations and calculated the $R_m$ factor according to the $EI_{30}$ equation specified in the RUSLE manual. The $K_m$ factor was determined through field sampling and laboratory analysis of soils from Kinmen Island. USPED was also applied using divergence signs to distinguish erosion and deposition, thereby avoiding common sign-reversal errors.

According to the RUSLE model, most of Kinmen Island exhibits low-to-moderate erosion rates (average 2.17 Mg ha$^{-1}$ year$^{-1}$), with scattered hotspots occurring mainly in cultivated land, sparsely vegetated areas, and some coastal slopes. These spatial patterns emphasize the influence of land cover and land use on erosion processes, while the island's gentle topography and relatively low rainfall help restrict the occurrence of severe erosion compared with Taiwan Island. By contrast, the USPED results, which vary with parameter settings, indicate somewhat higher erosion and deposition values, underscoring the model's greater sensitivity to topographic controls.

The main methodological contributions of this study are as follows: (i) use of high-frequency rainfall data to calculate the $R_m$ factor in accordance with the RUSLE manual, (ii) laboratory-based determination of the $K_m$ factor from field samples rather than relying solely on literature values, and (iii) careful implementation of USPED with correct interpretation of divergence signs. These refinements strengthen reproducibility and comparability across regions.

Because Kinmen lacks systematic erosion monitoring, temporal evolution of erosion could not be quantified, but field observations confirm that sheet and rill erosion remain visible. This data gap underscores the importance of future long-term monitoring to capture storm-driven dynamics and provide quantitative validation of model predictions. Despite these limitations, the findings provide valuable insights into the island's erosion dynamics and establish a foundation for soil conservation strategies in similar environments, contributing to the achievement of Sustainable Development Goals (SDGs), particularly SDG 13: Climate Action and SDG 15: Life on Land.

**Supplementary Materials:** The following supporting information can be downloaded at https://www.mdpi.com/article/10.3390/su172210052/s1.

**Author Contributions:** Conceptualization, W.C.; Data Curation, W.C., K.A.N. and Y.-C.H.; Funding Acquisition, W.C.; Investigation, W.C., K.A.N. and Y.-C.H.; Methodology, W.C.; Project Administration, W.C.; Resources, W.C.; Software, W.C., K.A.N. and Y.-C.H.; Supervision, W.C.; Validation, W.C., K.A.N. and Y.-C.H.; Visualization, K.A.N. and Y.-C.H.; Writing—Original Draft, W.C. and K.A.N.; Writing—Review and Editing, W.C., K.A.N. and Y.-C.H. All authors have read and agreed to the published version of the manuscript.

**Funding:** This study was partially supported by the Ministry of Science and Technology (Taiwan) research project (Grant Number MOST 111-2121-M-027-001) and by the National Science and Technology Council (Taiwan) research projects (Grant Numbers NSTC 112-2121-M-027-001 and NSTC 114-2121-M-027-001).

**Institutional Review Board Statement:** Not applicable.

**Informed Consent Statement:** Not applicable.

**Data Availability Statement:** The raw data used in this study are not publicly available due to restrictions imposed by the data owner or source. Consequently, these data cannot be disseminated or shared as part of this publication. Interested researchers may request access directly from the data owner or source, subject to their terms and conditions. The authors confirm that they do not have the right to distribute the raw data employed in this study. The derived GeoTIFF figures of erosion factors (Rm, Km, LS, LST) and modeled outputs from RUSLE and USPED, together with R scripts for rainfall event segmentation and erosivity (Rm) calculation, have been deposited in Figshare and are openly available at https://doi.org/10.6084/m9.figshare.30252529.

**Acknowledgments:** This article is partially derived from the master's thesis of Yu-Chieh Huang, submitted to the National Taipei University of Technology in partial fulfillment of the requirements for the Master of Science degree [51]. Part of this article builds on a preprint version we deposited in SSRN, which has since been expanded with additional analyses, corrected methodologies, new figures, and Supplementary Materials. We would like to express our sincere gratitude to Hao-Jun Sun, Yi-Jia Jiang, and Chiao-Shin Huang for their invaluable contributions during field excursions, particularly in verifying soil erosion and conducting in situ permeability tests. Furthermore, the authors acknowledge the use of ChatGPT 5 to enhance the readability and language of the manuscript. All content generated by the tool was subsequently reviewed and edited as needed, and the authors take full responsibility for the final version of the publication. The authors gratefully acknowledge the use of purchased satellite imagery obtained through the Center for Space and Remote Sensing Research (CSRSR), National Central University (NCU): SPOT-7 imagery © Airbus DS (2022), and Pléiades imagery © CNES (2022), Distribution Airbus DS. The imagery was sublicensed for academic research use from Airbus DS via CSRSR, NCU.

**Conflicts of Interest:** The authors declare no conflicts of interest.

## References

1. Pimentel, D.; Harvey, C.; Resosudarmo, P.; Sinclair, K.; Kurz, D.; McNair, M.; Crist, S.; Shpritz, L.; Fitton, L.; Saffouri, R.; et al. Environmental and economic costs of soil erosion and conservation benefits. *Science* **1995**, *267*, 1117–1123. [CrossRef] [PubMed]
2. Montgomery, D.R. Soil erosion and agricultural sustainability. *Proc. Natl. Acad. Sci. USA* **2007**, *104*, 13268–13272. [CrossRef] [PubMed]
3. Borrelli, P.; Alewell, C.; Alvarez, P.; Anache, J.A.A.; Baartman, J.; Ballabio, C.; Bezak, N.; Biddoccu, M.; Cerdà, A.; Chalise, D.; et al. Soil erosion modelling: A global review and statistical analysis. *Sci. Total Environ.* **2021**, *780*, 146494. [CrossRef] [PubMed]
4. Panagos, P.; Borrelli, P.; Poesen, J.; Ballabio, C.; Lugato, E.; Meusburger, K.; Montanarella, L.; Alewell, C. The new assessment of soil loss by water erosion in Europe. *Environ. Sci. Policy* **2015**, *54*, 438–447. [CrossRef]

5. Li, K.; Zhou, Z.; Wang, Z.; Zeng, Y.; Hubacek, K.; Borrelli, P.; Li, C.; Zhang, W.; Deng, Z.; Wang, J.; et al. Opportunity for mitigating soil loss by water erosion in cropland through crop switching and improved management in China. *Geogr. Sustain.* **2025**, *6*, 100262. [CrossRef]
6. de Faria Godoi, R.; Rodrigues, D.B.B.; Borrelli, P.; Oliveira, P.T.S. High-resolution soil erodibility map of Brazil. *Sci. Total Environ.* **2021**, *781*, 146673. [CrossRef]
7. Borrelli, P.; Robinson, D.A.; Panagos, P.; Lugato, E.; Yang, J.E.; Alewell, C.; Wuepper, D.; Montanarella, L.; Ballabio, C. Land use and climate change impacts on global soil erosion by water (2015–2070). *Proc. Natl. Acad. Sci. USA* **2020**, *117*, 21994–22001. [CrossRef]
8. Liu, Y.-H.; Li, D.-H.; Chen, W.; Lin, B.-S.; Seeboonruang, U.; Tsai, F. Soil erosion modeling and comparison using slope units and grid cells in Shihmen reservoir watershed in Northern Taiwan. *Water* **2018**, *10*, 1387. [CrossRef]
9. Tsai, F.; Lai, J.-S.; Nguyen, K.A.; Chen, W. Determining cover management factor with remote sensing and spatial analysis for improving long-term soil loss estimation in watersheds. *ISPRS Int. J. Geo-Inf.* **2021**, *10*, 19. [CrossRef]
10. Chen, W.; Li, D.-H.; Yang, K.-J.; Tsai, F.; Seeboonruang, U. Identifying and comparing relatively high soil erosion sites with four DEMs. *Ecol. Eng.* **2018**, *120*, 449–463. [CrossRef]
11. Lo, T.-W. Rainfall Factor and Mean Value Estimation of Shihmen Reservoir Sub-Watersheds. Master's Thesis, National Taipei University of Technology, Taipei, Taiwan, 2020. (In Chinese)
12. Liu, Y.-H. Analysis of Soil Erosion of Shihmen Reservoir Watershed in Taiwan and Lam Phra Ploeng Basin in Thailand. Master's Thesis, National Taipei University of Technology, Taipei, Taiwan, 2019. (In Chinese)
13. Liou, Y.-A.; Nguyen, Q.-V.; Hoang, D.-V.; Tran, D.-P. Prediction of soil erosion and sediment transport in a mountainous basin of Taiwan. *Prog. Earth Planet. Sci.* **2022**, *9*, 52. [CrossRef]
14. Chen, W.; Nguyen, K.A.; Huang, Y.-C. Soil erosion in Taiwan. *Agriculture* **2023**, *13*, 1945. [CrossRef]
15. Majhi, A.; Shaw, R.; Mallick, K.; Patel, P.P. Towards improved USLE-based soil erosion modelling in India: A review of prevalent pitfalls and implementation of exemplar methods. *Earth-Sci. Rev.* **2021**, *221*, 103786. [CrossRef]
16. Chen, W.; Huang, Y.-C.; Lebar, K.; Bezak, N. A systematic review of the incorrect use of an empirical equation for the estimation of the rainfall erosivity around the globe. *Earth-Sci. Rev.* **2023**, *238*, 104339. [CrossRef]
17. Gouma, M.; van Wijngaarden, G.J.; Soetens, S. Assessing the effects of geomorphological processes on archaeological densities: A GIS case study on Zakynthos Island, Greece. *J. Archaeol. Sci.* **2011**, *38*, 2714–2725. [CrossRef]
18. Mitasova, H.; Barton, M.; Ullah, I.; Hofierka, J.; Harmon, R.S. GIS-based soil erosion modeling. In *Treatise on Geomorphology: Volume 1–14*; Elsevier: Amsterdam, The Netherlands, 2013; pp. 228–258.
19. Renard, K.G.; Foster, G.R.; Weesies, G.A.; McCool, D.K.; Yoder, D.C. Predicting Soil Erosion by Water: A Guide to Conservation Planning with the Revised Universal Soil Loss Equation (RUSLE). In *Agricultural Handbook*; United States Department of Agriculture: Washington, DC, USA, 1997; p. 703.
20. Mitasova, H.; Hofierka, J.; Zlocha, M.; Iverson, L.R. Modelling topographic potential for erosion and deposition using GIS. *Int. J. Geogr. Inf. Syst.* **1996**, *10*, 629–641. [CrossRef]
21. Mitasova, H.; Brown, W.M.; Johnston, D.; Mitas, L. GIS tools for erosion/deposition modelling and multidimensional visualization. In *Part III: Process Based Erosion Simulation. Geographic Modelling and Systems Laboratory*; University of Illinois Urbana-Champaign: Champaign, IL, USA, 1996.
22. Chen, W.; Wang, W.-H.; Nguyen, K.A. Soil erosion and deposition in a Taiwanese watershed using USPED. *Sustainability* **2022**, *14*, 3397. [CrossRef]
23. Wu, C.-T.; Lin, E.-S. Geology and Landscapes of Kinmen. In *Kinmen Academic Series*; Dow Tien Publishing House: New York, NY, USA, 1998; Volume 11. (In Chinese)
24. Chen, C.-H. Kinmen Island Geological Survey and Research. *Geophysics* **1986**, *7*, 14–25. (In Chinese)
25. Lin, C.-C.; Chen, P.-Y.; Lin, F.-Y. *Report on the Geology and Mineral Deposits in Quemoy Islands, Fuchien*; MRSO: Hollywood, FL, USA, 1970. (In Chinese)
26. Chen, P.-Y. Studies of the Quaternary Clays and Clay Minerals from Chinmen Island, Fukien. *Taiwan Min. Ind.* **1965**, *2–4*, 66–92. (In Chinese)
27. Halperin, M.H. *The 1958 Taiwan Straits Crisis: A Documented History*; Rand Corporation: Santa Monica, CA, USA, 1996.
28. Wan, T.-L.A. Islands, geopolitics and language ideologies: Sociolinguistic differentiation between Taiwanese and Kinmenese Hokkien. *Lang. Commun.* **2022**, *83*, 36–48. [CrossRef]
29. Tsai, C.-W. A Strategy Study of Tourism Development for Kinmen after the Implementation of Individual Visit of Mini Three Links. *J. Natl. Quemoy Univ.* **2012**, *2*, 235–254. (In Chinese)
30. Kuo, C.-M.; Chen, Z.-S. *The Project of Investigation of Soil and Vegetation Analysis in Kinmen National Park*; Kinmen National Park Headquarters, Construction and Planning Agency, Ministry of the Interior: Kinmen, Taiwan, 2002.

31. Chen, C.-T. *Introduction to Kinmen Studies*; National Quemoy University: Kinmen, Taiwan, 2022. (In Chinese)
32. Lal, R.; Bouma, J.; Brevik, E.; Dawson, L.; Field, D.J.; Glaser, B.; Hatano, R.; Hartemink, A.E.; Kosaki, T.; Lascelles, B.; et al. Soils and sustainable development goals of the United Nations: An International Union of Soil Sciences perspective. *Geoderma Reg.* **2021**, *25*, e00398. [CrossRef]
33. Borrelli, P.; Märker, M.; Schütt, B. Modelling post-tree-harvesting soil erosion and sediment deposition potential in the Turano River Basin (Italian Central Apennine). *Land Degrad. Dev.* **2015**, *26*, 356–366. [CrossRef]
34. Harmon, B.A.; Mitasova, H.; Petrasova, A.; Petras, V. r.sim.terrain 1.0: A landscape evolution model with dynamic hydrology. *Geosci. Model Dev.* **2019**, *12*, 2837–2854. [CrossRef]
35. Mitasova, H.; Mitas, L. Multiscale soil erosion simulations for land use management. In *Landscape Erosion and Evolution Modeling*; Springer: Berlin/Heidelberg, Germany, 2001; pp. 321–347.
36. Lin, W.; Lee, C.-Y.; Yang, H.-C.; Chen, C.-H. *Geological Map of Taiwan Scale 1:50,000 in Kinmen Area*; Central Geological Survey, Ministry of Economic Affairs: New Taipei City, Taiwan, 2011. (In Chinese)
37. Lin, L.-L.; Chang, S.-T. Study of Soil Erodibility Factor Estimation Equations. *J. Chin. Soil Water Conserv.* **2008**, *39*, 355–366. (In Chinese)
38. Desmet, P.J.J.; Govers, G. A GIS procedure for automatically calculating the USLE LS factor on topographically complex landscape units. *J. Soil Water Conserv.* **1996**, *51*, 427–433. [CrossRef]
39. Chen, W.; Nguyen, K.A. The New Island-Wide LS Factors of Taiwan, with Comparison with EU Nations. *Sustainability* **2022**, *14*, 3059. [CrossRef]
40. Wang, W.-H. Analyzing Soil Erosion and Deposition Using USPED in the Shihmen Reservoir Watershed. Master's Thesis, National Taipei University of Technology, Taipei, Taiwan, 2022. (In Chinese)
41. Central Weather Administration. Kinmen Weather Station: Station Characteristics; Central Weather Administration: Taipei, Taiwan. Available online: https://etour.cwa.gov.tw/station_inpage.php?stn_id=17 (accessed on 16 August 2024). (In Chinese)
42. Fang, G.Q.; Ruan, F.S.; Wu, X.H.; Guo, Z.M. Main features of soil erodibility in Fujian province. *Soil Water Conserv. Fujian* **1997**, *2*, 19–23.
43. Wang, B.; Zheng, F.; Guan, Y. Improved USLE-K factor prediction: A case study on water erosion areas in China. *Int. Soil Water Conserv. Res.* **2016**, *4*, 168–176. [CrossRef]
44. Wang, X.Q.; Zeng, S.J.; Chen, X.G.; Lin, J.L.; Chen, S.M. Change analysis on soil erosion of Fujian Province from 1990 to 2015. *Int. Arch.* Photogramm. Remote Sens. Spat. Inf. Sci. **2017**, *42*, 1381–1386. [CrossRef]
45. Chen, S.; Zha, X. Effects of the ENSO on rainfall erosivity in the Fujian Province of southeast China. *Sci. Total Environ.* **2018**, *621*, 1378–1388. [CrossRef]
46. Ye, H.; Hua, W.; Huang, C.; Huang, T.; Wang, H.; Yuan, X.; Yang, H. Quantitative evaluation of soil erosion in the upper Minjiang River Basin of China based on integration of geospatial technologies using RUSLE. *Pol. J. Environ. Stud.* **2020**, *29*, 3419–3429. [CrossRef]
47. Li, C.; Lin, L.; Hao, Z.; Post, C.J.; Chen, Z.; Liu, J.; Yu, K. Developing a USLE cover and management factor (C) for forested regions of southern China. *Front. Earth Sci.* **2020**, *14*, 660–672. [CrossRef]
48. Gao, J.; Shi, C.; Yang, J.; Yue, H.; Liu, Y.; Chen, B. Analysis of spatiotemporal heterogeneity and influencing factors of soil erosion in a typical erosion zone of the southern red soil region, China. *Ecol. Indic.* **2023**, *154*, 110590. [CrossRef]
49. Xie, X.; Wang, X.; Wang, Z.; Lin, H.; Xie, H.; Shi, Z.; Liu, X. Influence of landscape pattern evolution on soil conservation in a red soil hilly watershed of Southern China. *Sustainability* **2023**, *15*, 1612. [CrossRef]
50. Xu, J.; Jiang, X.; Liu, H.; Yin, F.; Xu, Y.; Gao, C. Monitoring of spatiotemporal changes in ecosystem service functions and analysis of influencing factors in Pingtan Island. *Ecol. Indic.* **2024**, *158*, 111590. [CrossRef]
51. Huang, Y.-C. Soil Erosion and Deposition of Kinmen. Master's Thesis, National Taipei University of Technology, Taipei, Taiwan, 2023. (In Chinese)

sustainability

MDPI

Article

# Hydrochemical Characterisation and Assessment of Groundwater Suitability for Drinking and Irrigation Purposes in Sângeorz-Băi Area, Bistrița-Năsăud County (Romania)

Ramona Bălc [1], Carmen-Andreea Roba [1,*], Mircea Moldovan [1], Octavia Raluca Zglobiu [2], Gheorghe Roșian [1] and Lidia Vasilian [1]

[1] Faculty of Environmental Science and Engineering, Babeș-Bolyai University, Fântânele 30, 400084 Cluj-Napoca, Romania; ramona.balc@ubbcluj.ro (R.B.); mircea.moldovan@ubbcluj.ro (M.M.); gheorghe.rosian@ubbcluj.ro (G.R.); lvasilian2@gmail.com (L.V.)

[2] Department of Foreign Languages for Specific Purposes, Faculty of Letters, Babeș-Bolyai University, Horea 31, 400202 Cluj-Napoca, Romania; octavia.zglobiu@ubbcluj.ro

* Correspondence: carmen.roba@ubbcluj.ro

**Abstract**

Groundwater quality is a key factor and a critical determinant of public health, agriculture, and socio-economic development, particularly in regions where private wells and mineral springs constitute the primary water sources. This study presents an integrated hydrochemical, radiological, and toxicological assessment of groundwater in the Sângeorz-Băi area, Romania, a spa region where mineral waters hold both therapeutic and economic significance. Samples from mineral springs, the municipal supply system, and private wells were analyzed to evaluate compliance with national and international standards and to assess their suitability for drinking, therapeutic, and agricultural purposes. The results reveal distinct hydrochemical contrasts between sources. Mineral springs are characterized by elevated salinity, hardness, and Na–$HCO_3$ facies, whereas the municipal network and private wells are dominated by Ca–$HCO_3$ facies. More than half of the private wells exceeded permissible limits for $NO_3^-$, $NO_2^-$, $NH_4^+$, Pb, and Fe, with one well posing a significant nitrite-related health risk. Trace metal analysis indicated localized enrichment in Cu, Fe, and Pb. Radon and radium activities generally complied with regulations, although radium occasionally exceeded the more stringent WHO guidelines. Seasonal variation was minimal, reflecting stable groundwater chemistry. Health risk and irrigation assessments suggest that municipal supply water is largely safe for consumption, while private wells require targeted monitoring and mitigation. Despite elevated $Na^+$ and $Cl^-$, mineral springs retain therapeutic value under controlled use. This study provides a replicable framework for groundwater quality assessment in spa regions and offers critical insights for public health protection, sustainable tourism, and agricultural resilience.

**Keywords:** mineral springs; drinking water; geochemical facies; heavy metals; radioactivity; Romania

## 1. Introduction

As the global population continues to grow, the demand for drinking water has also been increasing, particularly over the past few decades. This rising demand for water, both for domestic and agricultural purposes, has led to the overexploitation of groundwater resources, especially in arid and warm climates. Numerous studies have examined groundwater quality [1–12]. Continuous monitoring of groundwater is crucial, as it helps identify potential toxic elements, their sources, mobility, and impact on both human health and environmental quality. The chemical composition of groundwater is closely linked to various factors, including meteorological conditions, geographical features of the area, geological characteristics of the aquifer, and the contact time between water and rock. Hydrochemical investigation of aquifers provides valuable data that local authorities can use for managing groundwater resources [13] or implementing remediation strategies [14].

The spatial and temporal distribution of water resources in Romania are primarily influenced by the country's geographical location in the temperate-continental climate zone and by the presence of the Carpathian Mountain range. Over the past few decades, Romania has experienced a growing demand for water across various sectors, including drinking, industry, agriculture, fish farming and tourism. Water consumption increased from 14.4 billion $m^3$ in 1975 to 46 billion $m^3$ between 2010 and 2015 [15]. Groundwater resources are estimated at 11 billion $m^3$ per year, with approximately 80% of these resources located in hilly and plateau regions. The Carpathian and Sub-Carpathian areas are home to significant mineral water resources, which are linked to volcanic activity and the presence of salt, oil, gas, and coal deposits. More than 2000 springs are found in these areas, the majority of which are of chloro-sodic, sulphurous-sulphate and carbonated types [15].

Groundwater represents a vital component of the freshwater system and is highly vulnerable to chemical contamination from both natural and anthropogenic sources. This study demonstrates the generally good quality of several groundwater sources in an historic spa resort area of Romania, underscoring the need to preserve their integrity while also recognizing the economic potential of mineral springs for bottling, commercial distribution, and therapeutic applications. These resources constitute a valuable asset that can contribute to the sustainable development of the region. The primary objectives of this study were: (1) to characterize the hydrochemical properties of groundwater in Sângeorz-Băi; (2) to assess its suitability for drinking and irrigation; and (3) to evaluate the temporal stability of its chemical composition.

## 2. Materials and Methods

### 2.1. Study Area

Sângeorz-Băi is located in Bistrița-Năsăud County, along the Someșul Mare River in the southern Rodna Mountains, approximately 56 km from the city of Bistrița. The local economy is primarily based on tourism, forestry, and agriculture. Geologically, the area belongs to a subvolcanic district in the inner Eastern Carpathians (northern Romania) [16]. The study site is situated within the volcanic segment of Țibleș–Rodna–Bîrgău, where rocks have been dated between 8.0 and 11.4 Ma [17].

This region is characterized by magmatic intrusions into diverse lithologies, including the crystalline units of the Rodna Mountains—dominated by garnet-bearing micaschists, paragneisses, and amphibolites—as well as the sedimentary successions of the Bârgău Mountains, consisting of conglomerates, limestones, marls, sandstones, and argillaceous marls [18,19]. The igneous suite is weakly evolved and comprises two main groups: (1) basic rocks such as amphibole microgabbro and amphibole- and clinopyroxene-bearing basaltic andesite, and (2) intermediate rocks including amphibole andesite, plagioclase andesite, and leucocratic andesite [20]. According to established classifications [20,21], most igneous rocks in the Rodna–Bârgău area belong to the calc-alkaline or low-iron subalkaline series. Contacts between magmatic intrusions and sedimentary host rocks are typically marked by hornfels zones and, less frequently, by breccias [17]. Intrusive bodies include laccoliths, sills, stocks, and dykes [18,19,22], which are interpreted as remnants of magma chambers that once sustained subaerial volcanic activity [23].

The climate of the region is characterized by a mean annual temperature of approximately 7.5 °C. Prevailing winds predominantly originate from the northwest, west, and southwest. In Bistrița-Năsăud County, annual precipitation averages 650–700 mm in the valleys and around 800 mm on the hilltops, with a maximum between May and August and a minimum during winter [24].

The mineral waters of Sângeorz-Băi have been renowned for their therapeutic properties since the 17th century. The first documented mention of these waters dates back to 1777, by Mattyus cited in [25], though they were likely known even earlier. The earliest physico-chemical analyses were conducted in the late 19th century [26,27]. In 1858, a spa resort was established in the area, transforming it into a tourist destination. By 1879, the mineral waters began to be commercially exploited and were bottled under the brand name "Hebe"—a bicarbonate-chloro-sodic water recommended for treating digestive, liver, and reproductive system disorders. Currently, the mineral springs are no longer in use and are in an advanced state of degradation.

In Sângeorz-Băi locality, groundwater serves as the primary source of drinking water. Residents have access to potable water through a local distribution network and private wells. The local distribution system is supplied by 14 boreholes, ranging in depth from 8 to 17 m, with flow rates varying between 7 and 25.2 $m^3/h$. The daily water demand in Sângeorz-Băi ranges from 1429.1 to 1728.9 $m^3/day$, with an average consumption of approximately 1571.6 $m^3/day$.

### 2.2. Sampling and Analytical Methods

To assess the quality of groundwater in the Sângeorz-Băi area, a total of 60 samples were collected: 16 from mineral springs, 14 from local water supply network, and 30 from private wells (Figure 1). To evaluate the chemical stability of the groundwater, two sets of samples were taken—one in autumn (October 2017) and another in spring (May 2018). Sampling was conducted in accordance with standard procedure outlined by national and international protocols [28,29].

The following physico-chemical parameters were analyzed: pH, redox potential (Eh—mV), electrical conductivity (EC—μS/cm), total dissolved salts (TDS—mg/L), and salinity (‰). The measurements were taken in situ using a pre-calibrated portable multiparameter device (WTW Multi 350i, Weilheim, Germany).

The samples were filtered using 0.45 μm nylon syringe filters and transferred into clean polyethylene vials, which had been thoroughly pre-rinsed before use. Samples intended for cation and metal analysis were acidified to approximately pH 2 using HCl and $HNO_3$, respectively, to prevent precipitation or adsorption during storage. All samples were stored in dark, refrigerated conditions at 4 °C. Prior to ion analysis, the samples were diluted with ultrapure water (18.2 MΩ·cm) to an electrical conductivity of approximately 100 μS/cm, in order to protect the chromatographic column and prevent detector oversaturation.

The major dissolved ions ($F^-$, $Cl^-$, $Br^-$, $NO_2^-$, $NO_3^-$, $PO_4^{3-}$, $SO_4^{2-}$, $Na^+$, $K^+$, $NH_4^+$, $Mg^{2+}$, $Ca^{2+}$) were analyzed using an ion chromatography system (Dionex 1500 IC, Sunnyvale, CA, USA) equipped with IonPac AS23/CS12A columns (4 × 250 mm), IonPac A23G/CG12A precolumns (4 × 50 mm) and a self-regenerating suppressor (ASRS/CSRS ultra II, 4 mm (Dionex, Sunnyvale, CA, USA) coupled to a conductivity detector. The eluents used were 4.5 mM $Na_2CO_3$/0.8 mM $NaHCO_3$ for anionis and 20 mM methanesulfonic acid for cations, with a flow rate of 1 mL/min. The analysis of $CO_3^{2-}$ and $HCO_3^-$ was performed using a gravimetric method, involving titration with 0.1 M HCl in the presence of phenolphthalein and methyl orange as indicators.

The total metal content (Ni, Cd, Cr, Pb, Zn, Cu, Fe, Mn) was determined by atomic absorption spectrometry (AAS), using a ZEEnit 700 system (Analytik Jena, Jena, Germany) equipped with an air—acetylene burner, a graphite furnace, and a specific hollow—cathode lamp for each metal.

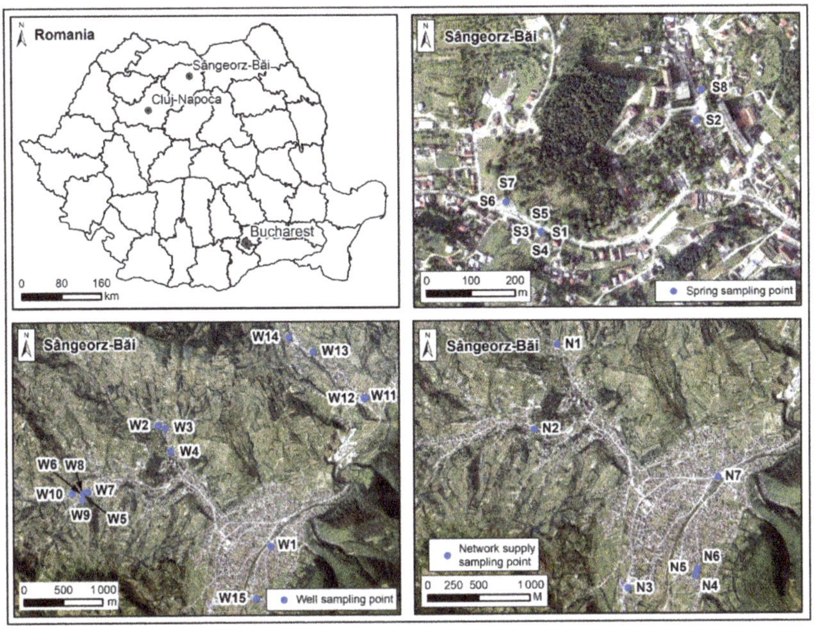


**Figure 1.** Location of the sampling points in the study area.

Radon and radium concentration in water were measured using a LUK-3A device specifically adapted for liquid samples (Jirí Plch—SMM (Prague, Czech Republic)). Radon measurements were performed with a LUK-VR system equipped with Lucas cells and a specialized apparatus for dissolved gas extraction [30]. Water samples were analyzed in the laboratory after reaching a temperature of 20 °C, employing the Lucas cell scintillation method as described by [31]. To determine radium concentration, water samples were stored in the laboratory for 30 days to allow for secular equilibrium between radium and its decay product radon, ensuring equal concentrations of both isotopes. Following this period, radium concentration was measured using the same Lucas cell scintillation technique [31,32].

### 2.3. Water Suitability for Drinking Usage

The potential health risk associated with the ingestion of chemical compounds (dissolved ions and metals) through drinking water was assessed by calculating the estimated daily intake (EDI), expressed in mg/kg body weigh/day [33,34] using the following formula:

$$\mathrm{EDI} = \frac{C \cdot IR}{BW}, \tag{1}$$

where C—element concentration in water (mg/L); IR—water ingestion rate (2 L/person/day); and BW—average adult body weight (60 kg for an adult person in Europe [34]).

To assess the potential health impact on consumers due to radon and radium intake through drinking water, the annual effective dose (Sv) resulting from water ingestion was estimated, as follows:

$$E = K \cdot C \cdot IR \cdot t, \tag{2}$$

where E—effective dose from $^{222}Rn$ ($^{226}Ra$) ingestion (Sv); K—conversion factor of $^{222}Rn$ ($^{226}Ra$) ingesting dose (Sv/Bq); C—concentration of $^{222}Rn$ ($^{226}Ra$) in water (Bq/L); IR—water ingestion rate for adults (1 L/person/day); t—duration of consumption (365 days) [35,36].

The dose conversion factors for adults used in the assessment of radon exposure were those recommended by the United Nations Scientific Committee on the Effects of Atomic Radiation [35], at $1.0 \times 10^{-8}$ Sv/Bq, and by the National Council on Radiation Protection and Measurements [37], at $0.35 \times 10^{-8}$ Sv/Bq. The committed effective dose from $^{226}Ra$ ingestion was calculated by multiplying the annual intake activity by the corresponding dose conversion factors: $2.8 \times 10^{-7}$ Sv/Bq [36] or $2.2 \times 10^{-7}$ Sv/Bq [38].

### 2.4. Water Suitability for Agricultural Usage

Elevated concentration of $Na^+$, $Cl^-$, $Ca^{2+}$, $Mg^{2+}$, and $HCO_3^-$ in irrigation waters can adversely affect soil structure and permeability, potentially leading to significant stress on crops. To assess the suitability of water from private wells and the local supply network for irrigation purposes, the following indices were calculated: Sodium Adsorption Ratio (SAR) [39], Sodium Percentage (%Na) [40], Kelly's Ratio (KR) [41], Residual Sodium Carbonate (RSC) [42], Magnesium Adsorption Ratio (MAR) [43], and Permeability Index (PI) [44]. All ion concentrations were expressed in milliequivalents per liter (meq/L):

$$SAR = \frac{Na^+}{\sqrt{\frac{Ca^{2+}+Mg^{2+}}{2}}}, \quad (3)$$

$$\%Na = \frac{Na^+ + K^+}{Ca^{2+} + Mg^{2+} + Na^+ + K^+} \cdot 100, \quad (4)$$

$$KR = \frac{Na^+}{Ca^{2+} + Mg^{2+}}, \quad (5)$$

$$RCS = \left(CO_3^{2-} + HCO_3^-\right)\left(Ca^{2+} + Mg^{2+}\right), \quad (6)$$

$$MAR = \frac{Mg^{2+}}{Ca^{2+} + Mg^{2+}} \cdot 100, \quad (7)$$

$$PI = \frac{Na^+ + \sqrt{HCO_3^-}}{Ca^{2+} + Mg^{2+} + Na^+} \cdot 100. \quad (8)$$

The SAR, %Na and KR indices are used to evaluate the potential hazard posed by excessive $Na^+$ in irrigation water. Sodium ions can be adsorbed by soil clay particles, displacing $Ca^{2+}$ and $Mg^{2+}$ ions, which leads to reduced soil permeability and impaired plant growth [45,46]. Additionally, $Na^+$ can react with $CO_3^{2-}$ to form alkaline soils or with $Cl^-$ to form saline soils—both conditions having a significant adverse effect on crop productivity [47].

RCS estimates the impact of elevated carbonate concentration in irrigation water, particularly the removal of $Ca^{2+}$ and $Mg^{2+}$ from the soil solution, which can significantly alter soil physical properties [45]. When the combined concentration of $Ca^{2+}$ and $Mg^{2+}$ exceeds that of $CO_3^{2-}$, $Na_2CO_3$ levels tend to decrease. Conversely, an excess of $CO_3^{2-}$ and $HCO_3^-$ can lead to the precipitation of $Ca^{2+}$ and $Mg^{2+}$ in the soil, deteriorating soil structure and enhancing the mobilization of $Na^+$ [48]. High RCS values are indicative of increased sodium adsorption in the soil, which can negatively impact soil quality [48]. MAR characterizes the balance between $Mg^{2+}$ and $Ca^{2+}$ in groundwater. Elevated levels of

$Mg^{2+}$ in irrigation water can disrupt soil chemistry and structure, ultimately reducing crop production [49].

PI is a useful index used to assess the suitability of irrigation water based on the concentrations of $Na^{+}$, $Ca^{2+}$, $Mg^{2+}$, and $HCO_3^{-}$, as these ions influence soil permeability and water infiltration capacity.

## 3. Results and Discussions

Table 1 presents a synthetic statistic of the analyzed parameters, comparing them to the values recommended by the national [50–52] and international [53] legislation.

### 3.1. Physico-Chemical Parameters

The pH of the water samples ranged from slightly acidic to neutral (6.5–7.1) in mineral spring, and from slightly acidic to slightly alkaline in local supply networks (6.8–7.5) and private wells (6.5–7.6) [54]. In all cases, the pH values fell within the acceptable limits established by national regulations (6.5–9.5, [50]) and international guidelines (6.5–8.5, [53,55]) (Table 1). Redox potential values were generally negative and showed an inverse correlation with pH, consistent with typical groundwater chemistry behaviour.

The pH values we observed (6.5–7.6 across all water types) are consistent with the typical range for Carpathian springs and shallow wells, which usually vary between slightly acidic and slightly alkaline, depending on carbonate buffering capacity and local lithology [56,57]. Recent studies in Apuseni Mountains (Romania) and Low Tatra (Slovakia) reported groundwater pH mostly between 6.7 and 7.5, supporting the view that our samples fit well into the regional pattern [12,56,58].

**Table 1.** Summary of the analyzed parameters and the permissible limits for drinking water as specified by Romanian regulation and WHO guidelines (the bold values represent the permissible limits exceeding).

| Parameter | Mineral Springs | Local Network | Private Wells | Romanian Standard (1,2) | WHO Standard (3) |
|---|---|---|---|---|---|
| pH | 6.5–7.1 (6.8) | 6.8–7.5 (7.1) | 6.5–7.6 (7.2) | 6.5–9.5 | 6.5–8.5 |
| Eh (mV) | −24.9–11.8 (−6.5) | −47.7–7.9 (−24.3) | −54.3–7.5 (−29.9) | - | - |
| EC (μS/cm) | 5280–7720 (6773) | 141.8–447.1 (205.3) | 98.6–731 (365.4) | 2500 | - |
| TDS (mg/L) | 3379–5018 (4369) | 91–286 (129.4) | 63–468 (233.8) | - | 1000 |
| Salinity (‰) | 2.9–4.3 (3.7) | 0.0–0.1 (0.01) | 0.0–0.3 (0.1) | - | - |
| TH (mg/L of $CaCO_3$) | 470.7–979.0 (709.5) | 51.5–243.5 (97.2) | 59.1–308.5 (143.4) | - | 1.5 |
| $Cl^{-}$ (mg/L) | 1100.8–2160.1 (1773.3) | 9.2–34.2 (14.1) | 6.8–47.9 (26.8) | 250 | 250 |
| $F^{-}$ (mg/L) | ND | 0.02–0.10 (0.04) | 0.04–0.07 (0.02) | 1.2 | 1.5 |
| $NO_2^{-}$ (mg/L) | ND | 0.01–0.02 (0.01) | 0.1–**2.0 (0.7)** | 0.5 | 0.5 |
| $NO_3^{-}$ (mg/L) | ND | 3.2–16.5 (6.4) | 2.2–**71.3** (32.1) | 50 | 50 |
| $SO_4^{2-}$ (mg/L) | 19.7–25.8 (22.3) | 15.8–85.9 (30.1) | 14.8–150.1 (41.2) | 250 | 250 |
| $HCO_3^{-}$ (mg/L) | 2453.1–3297.2 (2892.2) | 81.4–162.0 (108.6) | 101.2–330.1 (178.4) | - | - |
| $Na^{+}$ (mg/L) | 1571.5–2207.1 (1926.7) | 7.0–36.2 (12.8) | 7.4–37.3 (21.3) | 200 | 200 |
| $K^{+}$ (mg/L) | 60.1–180.9 (119.1) | 0.3–6.6 (4.6) | 0.7–41.7 (14.5) | - | - |
| $NH_4^{+}$ (mg/L) | ND | ND | 0.1–**0.7** (0.3) | 0.5 | - |
| $Mg^{2+}$ (mg/L) | 61.8–100.6 (76.7) | 3.0–19.6 (7.8) | 4.9–32.0 (10.7) | - | 50 |
| $Ca^{2+}$ (mg/L) | 46.4–265.1 (144.9) | 15.7–73.0 (26.0) | 14.7–**95.2** (39.9) | - | 75 |
| Fe (mg/L) | 1.1–1.4 (1.2) | 0.07–0.14 (0.09) | 0.05–**0.68** (0.11) | 0.2 | 2 |
| Cu (mg/L) | ND | **ND–1.3 (0.2)** | ND | 0.1 | 2 |
| Zn(μg/L) | 2.2–28.4 (13.8) | 17.4–304.1 (63.1) | 6.9–1661.1 (13.4) | 5000 | 3000 |
| Mn (μg/L) | ND | ND | ND | 50 | 400 |
| Ni (μg/L) | 3.1–18.2 (11.7) | 5.6–17.4 (11.7) | 5.0–19.6 (14.2) | 20 | 70 |
| Cr (μg/L) | ND | ND | ND | 50 | 50 |
| Pb (μg/L) | ND | **ND**–3.5 (0.2) | 4.1–**23.1 (13.9)** | 10 | 10 |
| Cd (μg/L) | ND | ND–0.8 (0.3) | 0.9–3.8 (2.1) | 5 | 3 |
| $^{222}Rn$ (Bq/L) | 9.6–24.8 (16.6) | no data | 5.3–18.8 (11.1) | 100 (5) | 100 (3)/11.1 (4) |
| $^{226}Ra$ (Bq/L) | 0.08–0.2 (0.18) | no data | 0.07–0.25 (0.14) | | 0.1 (3)/0.185 (4) |

(1) Maximum permissible limit for drinking water [50]; (2) threshold level for groundwater bodies used as drinking water from alluvial cone Rodnei Mountains (ROSO15) [51]; (3) WHO standard [53,59]; (4) EPA standard [60]; (5) Law 301/2015 [61].

Mineral springs exhibited significantly higher values of EC, TDS and salinity compared to water from the local distribution network and private wells (Table 1), reflecting a higher concentration of inorganic salts—primarily $Ca^{2+}$, $Mg^{2+}$, $K^+$, $Na^+$, $HCO_3^-$, $Cl^-$, and $SO_4^{2-}$—along with minor amounts of organic matter. In contrast, EC and TDS levels in local network and private wells water were within recommended safety limits (Table 1). According to TDS classification, water from the local supply and private wells falls within the freshwater category (TDS < 1000 mg/L), whereas the mineral spring water is classified as saline (3000 < TDS < 10,000 mg/L) [49]. The higher electrical conductivity (EC), total dissolved solids (TDS), and salinity in mineral spring waters compared to private wells and supply networks reflect intense water–rock interaction and longer residence times, a common feature of mineralized waters in Carpathian and Central European contexts [62]. Classification of mineral springs as saline (TDS > 3000 mg/L) aligns with European mineral-water surveys [62], which frequently categorize commercialized springs in this range [63], while private wells and network waters fit the "freshwater" class (TDS < 1000 mg/L).

Total hardness (TH) was calculated based on the concentration of calcium and magnesium. The mineral springs were classified as very hard waters, with TH values exceeding 180 mg $CaCO_3$/L [54,55,64]. In contrast, most samples from the local distribution network were moderately hard (60–120 mg $CaCO_3$/L), while water from private wells was generally hard (120–180 mg $CaCO_3$/L). Total hardness (TH) patterns also correspond to published values: Carpathian mineral springs are often "very hard" (>180 mg $CaCO_3$/L) owing to carbonate lithologies, whereas community supplies and wells in rural areas typically fall into "moderately hard to hard" categories, depending on aquifer depth and limestone/dolomite contributions [56,57]. Our results therefore mirror the broader regional hydrogeochemical trends and confirm the strong lithological control on hardness in the Carpathians.

The physico-chemical parameters remained relatively stable over time (Figure 2). The EC, TDS and salinity showed no significant seasonal variation. However, pH and TH were slightly lower during the spring sampling campaign (May 2018). The lack of strong seasonal variation in EC and TDS is also in agreement with hydrogeochemical monitoring in aquifers of the Carpathians, where buffering by carbonate reservoirs stabilizes ionic composition across seasons, with only minor springtime dilution effects observed in pH and hardness [56].

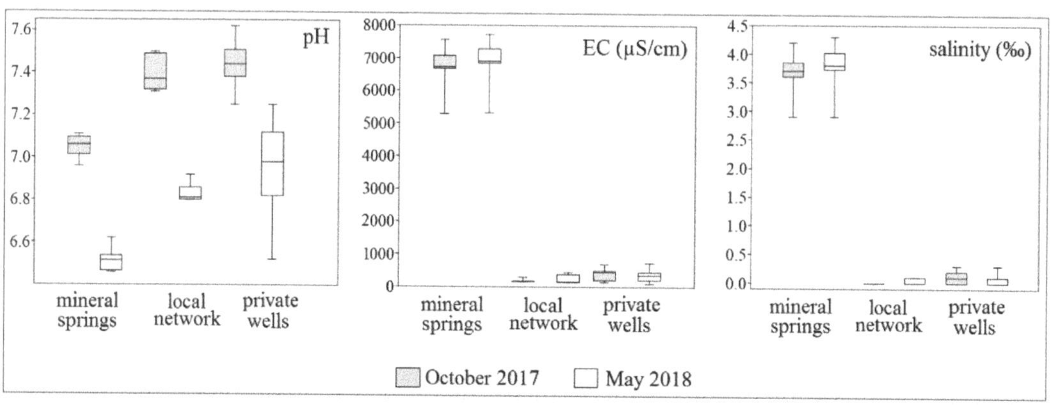


**Figure 2.** Seasonal fluctuation of pH, electrical conductivity and salinity in the investigated water sources.

### 3.2. Major Ions Content

In groundwater from private wells and the local distribution system, the dominant cations and anions follow the order $Ca^{2+} > Na^{+} > K^{+}, Mg^{2+}$ and $HCO_3^{-} > SO_4^{2-} > NO_3^{-}$, $Cl^{-} > NO_2^{-}, F^{-}$. In contrast, the mineral springs exhibit a different ionic composition, characterized by $Na^{+} > Ca^{2+} > K^{+} > Mg^{2+}$ and $HCO_3^{-} > Cl^{-} > SO_4^{2-}$ (Table 1). The concentration of major ions in all samples from the local supply network were within the permissible limits established by national drinking water standards.

However, 7%, 20% and 27% of private wells exceeded national limits for $NH_4^{+}$, $NO_2^{-}$ and $NO_3^{-}$. The occurrence of nitrogenous compounds ($NO_3^{-}$, $NO_2^{-}$, $NH_4^{+}$) in groundwater may be attributed to both natural processes—such as those governed by the nitrogen cycle—and anthropogenic sources, including leachate from septic systems, livestock waste, and agricultural runoff. Among these, diffuse agricultural pollution and the erosion of nitrogen-rich natural deposits are considered the primary contributors to nitrate contamination of groundwater [65,66]. Our findings that 7–27% of private wells exceeded national limits for $NH_4^{+}$, $NO_2^{-}$, and $NO_3^{-}$ are consistent with European-scale evidence that agricultural diffuse pollution and inadequate sanitation are leading sources of nitrogenous compounds in groundwater. The European Environment Agency [67] reports that roughly 14% of monitoring stations across the EU exceed the 50 mg/L nitrate threshold, with rural wells in Eastern and Central Europe disproportionately affected.

In one of the wells, the calcium concentration exceeded the maximum recommended value set by the World Health Organization [53]. Elevated calcium levels in drinking water may lead to gastrointestinal discomfort and are generally undesirable for domestic use due to their potential to cause encrustation and scaling in household appliances [68]. Calcium concentrations exceeding WHO guidance in one well highlight localized geogenic influences. Elevated $Ca^{2+}$ has also been documented in Apuseni karst aquifers [56] and in bottled European mineral waters [62], indicating that such anomalies are not uncommon in carbonate-dominated settings.

According to national standards for the exploitation and marketing of natural mineral waters [52], all investigated springs in the study area are classified as chlorinated waters ($Cl^{-} > 200$ mg/L), hydrogen carbonate waters ($HCO_3^{-} > 600$ mg/L), sodic waters ($Na^{+} > 200$ mg/L), and magnesium-rich waters ($Mg^{2+} > 50$ mg/L). Four springs also meet the criteria for calcium-rich waters ($Ca^{2+} > 150$ mg/L). These mineral springs hold therapeutic potential for internal use (drinking, aerosols, inhalations), parenteral administration (injections), and external treatments (bathing), but only under the strict supervision of qualified medical personnel. The results are relatively similar with other studies focused on hydrochemistry of mineral springs in EU [62]. The study performed on a total of 692 mineral waters from 13 countries showed that in terms of the ionic content, the dominant water type was bicarbonate, followed by fluoride, magnesium and calcium, and then sulphate and acidic had a lower prevalence [62].

Saline springs in the area are traditionally prescribed for digestive disorders (e.g., chronic gastritis), chronic bronchitis, chronic rhinitis, and for external therapies targeting musculoskeletal conditions, including inflammatory and rheumatic diseases [69–71]. Hydrogen carbonate-rich springs in the Sângeorz-Băi region may also be exploited economically through bottling and commercial distribution or used therapeutically in internal treatments for digestive and hepato-biliary disorders, as well as in external therapies such as inhalation and pulverization for respiratory conditions [69–71].

To further illustrate the hydrochemical type of the analyzed waters, a Piper diagram was constructed based on the major ions' concentrations (meq/L) (Figure 3).

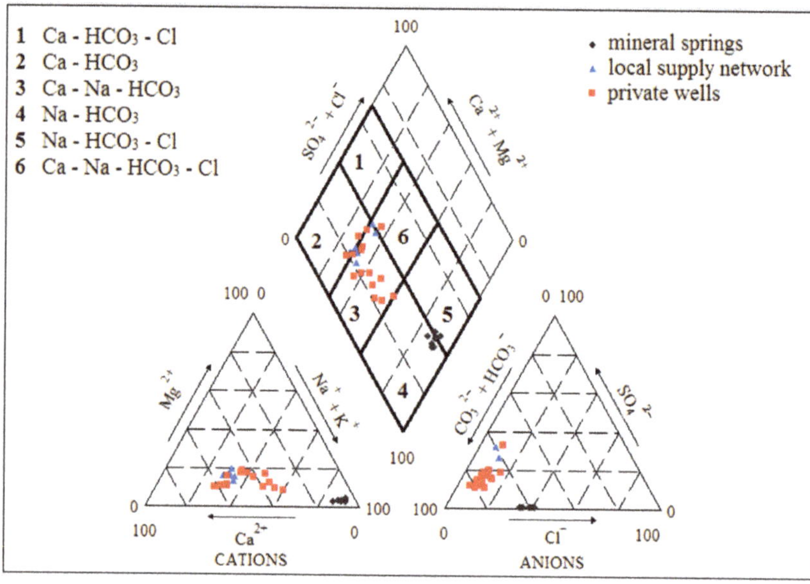


**Figure 3.** Piper diagram illustrating the hydrochemical facies of the analyzed water samples.

The investigated mineral springs are predominantly of the Na–$HCO_3$ and Na–$HCO_3$–Cl type, whereas most water samples from the local supply network and private wells correspond to the Ca–Na–$HCO_3$ type. This classification suggests that some private wells share a similar hydrochemical composition with the local supply network, potentially indicating a common aquifer or aquifers with comparable hydrogeochemical characteristics. The major ion composition of our local supply and private wells (Ca–Na–$HCO_3$ type) reflects typical shallow aquifers in the Carpathians, where carbonate dissolution governs the dominance of $Ca^{2+}$ and $HCO_3^-$, with secondary inputs from $Na^+$ and $SO_4^{2-}$ [56]. In contrast, the Na–$HCO_3$ and Na–$HCO_3$–Cl facies observed in our mineral springs correspond to hydrochemical signatures reported in highly mineralized or saline Carpathian springs, particularly in Transylvania and northern Romania, where evaporitic or deep-seated sources contribute to enrichment in $Na^+$ and $Cl^-$ [62].

Gibbs diagrams are commonly employed to evaluate the relationship between water chemistry and the lithological features of aquifers [72]. These diagrams help infer the dominant mechanisms influencing groundwater chemistry: rock-water interaction (rock dominance), evaporation, and precipitation [45,72]. In this study, Gibbs diagrams were constructed using two ratios: TDS versus (Na + K)/(Na + K + Ca), and TDS versus Cl/(Cl + $HCO_3$), where TDS is expressed in mg/L and ion concentrations in meq/L (Figure 4). The plotted diagrams (Figure 4), indicate that the hydrochemical evolution of groundwater from the local supply network and private wells is primarily controlled by rock weathering and water-rock interactions. In contrast, the chemical composition of the mineral springs appears to be influenced predominantly by evaporation processes. Notably, in the TDS versus (Na + K)/(Na + K + Ca) diagram, several mineral spring samples fall outside the defined fields, suggesting that anthropogenic activities may be affecting the concentrations of these ions.

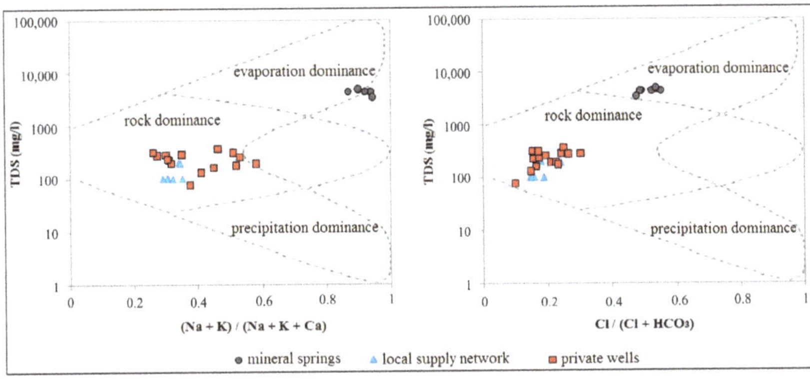


**Figure 4.** Gibbs diagram showing the relationship between total dissolved solids and the weight ratio of major ions in the analyzed water samples.

### 3.3. Total Content of Metals

Water pollution by heavy metals is an increasingly significant global concern [73]. These elements are widely distributed in the environment, affecting water, air and soil, primarily due to anthropogenic activities such as industrial processes, agriculture, mining, and vehicular traffic. However, it is also essential to consider natural sources, including magmatic, metamorphic, and sedimentary rocks, which can contribute to background levels of these metals [74].

The concentrations of metals in the local water supply network were generally within permissible limits, except for Cu in a single sample (Table 1). The presence of higher Cu levels in the same sample across both sampling campaigns suggests a localized point source of contamination, potentially due to aging copper piping.

In contrast, several private wells exhibited concentrations of Fe and Pb that exceeded regulatory thresholds. Lead was detected in four wells, with three surpassing the acceptable safety limits. Lead is classified by IARC as possibly carcinogenic to humans. Chronic ingestion of high amounts of lead can adversely affect the kidneys, liver, spleen, and lungs. These wells should be subject to continuous monitoring to determine whether an accidental pollution source is present. In the absence of corrective measures, consumption of water from these sources should be either discontinued or significantly restricted to protect public health. The exceedance of Pb and $NO_2^-$ limits in several wells has significant implications for public health and groundwater management. According to national legislation, the authorities must ensure access to water sources of adequate quality, which does not pose a risk to consumers health, to constantly monitor these water sources, and to provide free access to data related to monitors quality parameters. If locals want to use private wells as water sources, other than those monitored by the local authorities, they must be aware that they are assuming high risks. In this situation, the authorities have a major role, namely, to disseminate information on the water quality from the local distribution network and the negative impact on human health associated with the consumption of water of inadequate quality [75–79]. Academic community should also address this issue, by organizing different citizen science actions, together with local authorities. These actions could be focused on evaluating the water quality from private wells by performing specific analyses, disseminating the results, and raising awareness among locals related to the importance of water quality monitoring. Authorities can provide guidance related to possible solutions to improve water quality for private wells. Some of these solutions may require targeted interventions, such as point-of-use treatment (e.g., activated carbon filters for Pb, ion exchange or reverse osmosis for nitrate/nitrite) or may include stricter control of potential contamination sources (e.g., agricultural fertilizer usage).

The concentration of Cu, Pb, Cd, Cr, Ni and Mn in the analyzed mineral springs were all within the maximum limits established by national legislation for the exploitation and commercialization of natural mineral waters (1000 µg/L for Cu, 10 µg/L for Pb, 3 µg/L for Cd, 50 µg/L for Ni, 20 µg/L and 500 µg/L for Mn) [52].

The results indicate that the waters comply with safety standards for bottling as mineral water and for use in therapeutic treatment regimens. Iron concentrations were consistently below 10 mg/L, which does not satisfy the threshold for classification as ferruginous water, according to criteria established in previous studies [69–71]. No significant seasonal variations were observed in the concentrations of the analyzed metals. Slightly higher levels of Cu, Ni, Cd, and Fe were recorded during winter, whereas Zn concentrations showed a modest increase in spring.

To emphasize the trends and the relationships between the general physico-chemical parameters and the specific parameters (dissolved ions, heavy metals) and to identify the directions (principal components) along which the data variation is maximal, principal component analysis (PCA) was used (soft Past14.5) (Figure 5), which explained 68.3% of total variance by the first two principal components (PCs): PC1 represents 59.4% of the total variance in the water quality datasets and the PC2 represents 8.9%. The three clusters separated by the PCA consist in: (1) I1–I8; (2) F1, F2, F4, F5, F8–F11; (3) F3, F6, F7, F12–F15, R1–R7. Based on their distribution on PCA, the hydrochemical characteristics of cluster 1 are quite different from those of the other two clusters. The principal components did not indicate significant loadings between the analyzed parameters, the loading coefficients being lower than 0.6. In the present study, the higher loadings for PC2 had positive values for Cd (0.58), $NO_2^-$ (0.37), $NO_3^-$ (0.34), $SO_4^{2-}$ (0.33), Ni (0.3) and negative for $F^-$ (−0.33). For PC1 the higher loadings were positive in the case of CE, TDS, salinity, alkalinity, hardness, $K^+$, $Mg^{2+}$, $HCO_3^-$, Fe (0.26), $Ca^{2+}$ (0.24), ORP (0.23) while a negative loading was observed for pH (−0.23). As it is shown in Figure 5, these variables are very closed to each other's, reflecting a strong correlation among them and a greater contribution to PC1, comparing to other variables. In the case of PC2, the parameters had a higher dispersion, which represents a weaker correlation among them. The PCA indicated that dissolved ions had a higher contribution to the hydrochemical characteristics of the analyzed water sources comparing to heavy metals (with the exception of Cd).

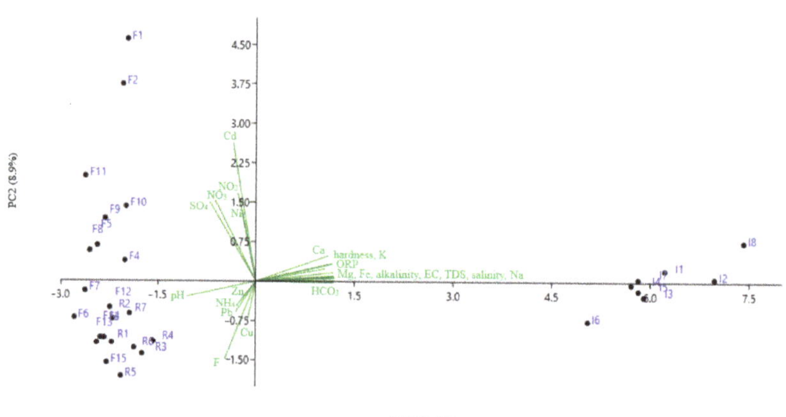


**Figure 5.** Biplot of the first two principal components (PC1 and PC2, explaining 59.4% and 8.9% of the variance, respectively) extracted by the correlation matrix.

### 3.4. Radon and Radium Activity

Radon activity concentrations in mineral springs ranged from 9.6 Bq/L to 24.8 Bq/L, with an arithmetic mean of 16.6 Bq/L. In private wells, concentrations varied between 5.3 Bq/L and 18.8 Bq/L, with an average value of 11.1 Bq/L. The overall mean radon concentration across all samples was 13.07 Bq/L, which is slightly higher than the national average of 12.3 Bq/L reported in over 2000 samples collected across Romania [31,80]. These values are, however, consistent with those reported in previous studies, which indicate typical radon concentrations in groundwater ranging from 10 Bq/L to 50 Bq/L [81,82]. Earlier research also supports the observation of higher radioactivity levels in these waters compared to other Romanian water sources [83]. The safe level recommended by [60] for radon in drinking water is 11.1 Bq/L, while WHO recommends a maximum permitted limit of 100 Bq/L [53,59], which aligns with the current Romanian national legislation [61].

Radium activity was also found to be slightly higher in mineral springs, with values ranging from 0.08 Bq/L to 0.28 Bq/L and an average of 0.18 Bq/L, compared to 0.07 Bq/L to 0.25 Bq/L (mean 0.148 Bq/L) in private wells. In 29% of the analyzed samples, radium activity exceeded the EPA's recommended limit of 0.185 Bq/L [60]. When compared to the more stringent WHO guideline of 0.1 Bq/L [59] 83% of the samples surpassed this threshold, indicating a widespread exceedance of international safety standards for radium content in drinking water.

Considering that the activity concentrations of the investigated radioisotopes fall within the limits established by the national regulations and guidelines for radon [59], as well as the standards for radium [60], their presence in the analyzed water sources is not expected to pose a significant risk to human health or to the environment.

### 3.5. Water Suitability for Drinking Usage

The analysis showed that 53% of private well water samples exceeded permissible limits for several chemical parameters, including $NO_2^-$, $NO_3^-$, $NH_4^+$, $Ca^{2+}$, Pb, and Fe. By contrast, only one sample from the municipal distribution system exhibited an elevated Cu concentration, exceeding the recommended threshold. To evaluate the potential health risks associated with ingestion of these constituents through drinking water, estimated daily intakes were calculated and compared with guideline values established by relevant health authorities (Table 2).

As expected, the intake of mineral nutrients through water ingestion was substantially higher for mineral spring waters. The estimated daily intake of $Na^+$ and $Cl^-$ exceeded the adequate intake levels recommended for human consumption (Table 2). Consequently, such waters are unsuitable for routine drinking. Instead, they are more appropriate for therapeutic applications -such as the treatment of digestive disorders, chronic bronchitis, and chronic rhinitis -administered in specific doses over short periods (typically 7–10 days) under the supervision of qualified medical personnel. Mineral waters with elevated salinity are also commonly used in external therapies, particularly for musculoskeletal disorders. However, prolonged and continuous consumption of drinking water with high $Na^+$ and $Cl^-$ concentrations may pose health risks, notably by contributing to the onset or progression of cardiovascular conditions.

In one of the analyzed private wells, the EDI of $NO_2^-$ exceeded the acceptable daily intake (ADI) established by the European Commission's Scientific Committee on Food (SCF) [84,85] (Table 2). This finding indicates that continuous consumption of water from this source poses a significant health risk. Nitrites are classified by the International Agency for Research on Cancer (IARC) as Group 2A agents, meaning they are probably carcinogenic to humans. Furthermore, epidemiological studies have linked nitrite exposure to adverse health outcomes such as methemoglobinemia (commonly known as "blue baby syndrome"), gastric cancer, thyroid disfunction, and diabetes [68]. Consequently, the use of water from this well for drinking purposes should be discontinued. In the remaining private wells where the concentrations of $NO_3^-$, $NH_4^+$, $Ca^{2+}$, Pb and Fe exceeded the maximum allowable limits, the corresponding EDIs remained below health-based guideline values. Nonetheless, it is recommended that these sources be used with caution and that their use for drinking purposes be reduced, particularly for vulnerable populations such as infants and individuals with pre-existing health conditions.

**Table 2.** Groundwater suitability for drinking based on the concentrations of major ions and trace metals (the bold values represent the exceeding of the recommended values).

| Parameter | EDI (μg/kg bw/day) | | | Recommendation (μg/kg bw/day) |
|---|---|---|---|---|
| | Mineral Springs | Local Supply System | Private Wells | |
| $F^-$ | ND | 0.67–3.33 (1.38) | 1.5–2.33 (2.1) | 56.66 [(1)] |
| $NO_2^-$ | ND | 0.3–0.5 (0.3) | 4.0–**68.1** (23.8) | 60 [(3)] /70 [(4)] |
| Fe | 37.9–42.8 (40.85) | 2.56–4.13 (3.02) | 2.18–12.9 (3.63) | 800 [(9)] |
| Zn | 0.26–0.79 (0.46) | 0.50–9.06 (2.10) | 0.26–50.98 (4.46) | 300–1000 [(9)] |
| Ni | 0.18–0.60 (0.39) | 0.21–0.52 (0.39) | 0.22–0.63 (0.48) | 11 [(10)] /2.8 [(11)] |
| Pb | ND | 0.01–0.12 (0.02) | 0.24–0.71 (0.47) | 3.57 [(12)] |
| Cd | ND | 0.01–0.03 (0.01) | 0.02–0.12 (0.06) | 0.83 [(13)] |
| | EDI (mg/kg bw/day) | | | Recommendation (mg/kg bw/day) |
| $Cl^-$ | **46.36–67.84 (59.11)** | 0.34–0.79 (0.47) | 0.23–1.48 (0.89) | 38.3–41.7 [(2)] |
| $NO_3^-$ | ND | 0.13–0.46 (0.21) | 0.07–2.33 (1.07) | 3.7 [(4)] |
| $Na^+$ | **52.76–71.74 (64.22)** | 0.29–0.81 (0.43) | 0.28–1.09 (0.71) | 25 [(5)] |
| $K^+$ | 2.59–5.29 (3.97) | 0.08–0.19 (0.11) | 0.09–1.22 (0.45) | 58.3 [(6)] |
| $Mg^{2+}$ | 2.16–3.19 (2.56) | 0.17–0.46 (0.26) | 0.18–0.88 (0.36) | 5.8 [(7)] |
| $Ca^{2+}$ | 2.67–5.82 (5.25) | 0.54–1.51 (0.87) | 0.50–2.48 (1.33) | 15.83 [(8)] |
| Cu | ND | 0.00007–0.04 (0.006) | ND | 0.5 [(9)] |

[(1)] Based on the Adequate Intake (AI) of 3.4 mg/day for an adult male [86]; [(2)] based on the adequate intake of 2300–2500 mg/day for all adults [87]; [(3)] acceptable daily intake (ADI) recommended by the European Commission's Scientific Committee on Food (SCF) [84,85]; [(4)] acceptable daily intake (ADI) recommended by the Joint Expert Committee of the Food and Agriculture (JEFCA) of the United Nations/World Health Organization (WHO) [88,89]; [(5)] based on the adequate intake of 1500 mg/day was set for all adults [86]; [(6)] based on the Adequate Intake (AI) of 3500 mg/day for an adult male [90]; [(7)] based on the Adequate Intake (AI) of 350 mg/day for adult male [90]; [(8)] based on the Population Reference Intake (PRI) of 950 mg/day for an adult male [90]; [(9)] PMTDI (Provisional Maximum Tolerable Daily Intake) [91]; [(10)] TDI (Tolerable Daily Intake) [92]; [(11)] TDI [93]; [(12)] TDI [94]; [(13)] provisional tolerable daily intake based on the provisional tolerable monthly intake of 25 μg/kg by [91].

Water quality parameters from the local distribution network, as well as from 47% of the sampled private wells, were within the regulatory limits. The EDIs for all analyzed substances in these sources were also below recommended thresholds, indicating that these water sources are safely for human consumption. Ensuring access to high-quality drinking water is particularly important in regions such Sângeorz-Băi, where tourism constitutes a primary economic activity. The availability of safe and reliable drinking water is essential for both public health and sustainable development of the local tourism industry.

For all analyzed water samples, the committed effective dose resulting from the ingestion of $^{222}Rn$ (Table 3), remained below the recommended annual limit of 1 mSv/year [59]. This value represents less than half of the average global exposure to natural background radiation, which is approximately 2.4 mSv/year. Regarding $^{226}Ra$, the results indicate that the daily ingestion 1 L of water from the studied sources does not pose a radiological health risk to consumers. The mean committed effective dose from $^{226}Ra$ ingestion (Table 3) was significantly lower than the WHO recommended annual internal dose limit of 0.1 mSv/year [59]. Overall, the ingestion-related doses from both radionuclides were negligible when compared to other natural sources of radiation exposure, such as indoor radon inhalation or the ingestion of naturally occurring radionuclides like $^{40}K$.

**Table 3.** Estimated committed effective dose (CED) for adults resulting from the ingestion of drinking water containing radon ($^{222}Rn$) and radium ($^{226}Ra$).

| Source | Average Value of Estimated Committed Effective Dose (mSv/year) | |
|---|---|---|
| | **$^{222}$Rn** | **$^{226}$Ra** |
| Mineral springs | 0.035–0.089 (0.060) [(1)]<br>0.012–0.031 (0.021) [(2)] | 0.009–0.022 (0.020) [(3)]<br>0.007–0.017 (0.015) [(4)] |
| Private wells | 0.021–0.073 (0.043) [(1)]<br>0.007–0.025 (0.015) [(2)] | 0.007–0.025 (0.014) [(3)]<br>0.006–0.019 (0.011) [(4)] |
| Recommendations [(5)] | 1 | 0.1 |

[(1)] Based on conversion factor recommended by UNSCEAR [35]; [(2)] based on conversion factor recommended by IAEA [36]; [(3)] based on conversion factor recommended by IAEA [36]; [(4)] based on conversion factor recommended by WHO [38]. [(5)] based on WHO recommendation [59].

### 3.6. *Water Suitability for Agriculture Usage*

In this study, the suitability of water from private wells and the local supply system for irrigation was evaluated, while mineral springs were excluded due to their high concentrations of dissolved ions. A summary of the results is presented in Table 4. Although most samples were classified as moderately hard or hard waters, the concentrations of $Ca^{2+}$, $Mg^{2+}$, $Na^{+}$, $K^{+}$, $Cl^{-}$, and $HCO_3^{-}$ did not present significant risks to crops. Based on the calculated indices, the majority of sources were classified as having excellent, good, or acceptable quality for irrigation (Table 4).

**Table 4.** Groundwater suitability for irrigation (the suitability class is marked in bold, according to the values obtained for each quality parameter).

| Parameter | Local Supply System | Private Wells | Suitability |
|---|---|---|---|
| **EC (µS/cm)** | 141.8–447.1 (205.3) | 98.6–731.0 (365.4) | **excellent** (<250), **good** (250–750), permissible (750–2000), doubtful (2000–3000), not suitable (>3000) [95,96] |
| **TDS (mg/L)** | 91–286 (129.4) | 63–468 (233.8) | **desirable** (<500), permissible (50–1000), useful for irrigation (1000–3000), unfit for irrigation (>3000) [95,96] |
| **TH (mg/L of $CaCO_3$)** | 51.5–243.5 (97.2) | 59.1–308.5 (143.4) | **soft** (<75), **moderately hard** (75–150), **hard** (150–300), very hard (>300); or: soft (<60), moderately hard (60–120), hard (121–180), very hard (>180) [95,96] |
| **SAR** | 0.4–0.8 (0.5) | 0.5–1.1 (0.8) | **excellent** (<10), good (10–18), doubtful (18–26), not safe (>26) [39,45,96] |
| **%Na** | 22–27 (24) | 22–49 (31) | excellent (<20), **good** (20–40), permissible (40–60), doubtful (60–80), not safe (>80) [40,96] |
| **KR** | 0.23–0.33 (0.27) | 0.18–0.65 (0.35) | **suitable** (<1), unsuitable (>1) [41,95,96] |
| **RSC** | −1.07–0.45 (−0.17) | −1.49–1.26 (0.05) | **suitable** (<1.25), marginally suitable (1.25–2.5), not suitable (>2.5) [96,97] |
| **MAR** | 29–39 (33) | 22–45 (32) | **suitable** (<50), not suitable (>50) [43,96] |
| **PI** | 59–104 (81) | 46–107 (75) | **good** (<80), moderate (80–100), poor (10–20) [95,96] |

## 4. Conclusions

This study provides a comprehensive, multidisciplinary assessment of groundwater quality in the Sângeorz-Băi area, a former emblematic spa region of Romania. Private wells, the municipal supply, and mineral springs remain central to drinking water provision, therapeutic practices, and socio-economic development. By integrating hydrochemical, radiological, toxicological, and irrigation suitability analyses, the study yields conclusions of direct relevance to public health, water management, and sustainable regional development.

Groundwater from the municipal network generally complies with regulatory standards. In contrast, more than half of the private wells exceeded permissible limits for $NO_3^{-}$,

$NO_2^-$, $NH_4^+$, Pb, and Fe. One well showed acute nitrite contamination, representing an immediate health risk that requires discontinuation of use. Elevated Pb and Fe levels highlight the vulnerability of aquifers to diffuse agricultural inputs and infrastructure deficiencies. Mineral springs complied with regulatory limits for trace metals and radionuclides, but their high $Na^+$ and $Cl^-$ concentrations make them unsuitable for continuous consumption. They remain valuable for short-term therapeutic use under medical supervision and retain potential for economic exploitation through bottling. Radiological analyses indicated radon activities below international safety thresholds. Radium occasionally exceeded the stricter WHO guideline, requiring continued monitoring but not indicating immediate health risks.

The findings underscore the dual role of groundwater in the region: a critical source of safe drinking water under regulated conditions, but also a vector of contamination where private wells remain unmonitored. Effective protection requires systematic water quality monitoring, stricter regulation of agricultural practices, infrastructure improvements, and increased community-level risk awareness.

A key limitation of this study lies in the restricted temporal coverage of sampling, which was conducted only twice (autumn 2017 and spring 2018). Although these campaigns provide valuable insights under contrasting seasonal conditions, they do not capture the full extent of seasonal or inter-annual variability in groundwater quality. Long-term monitoring is therefore required to achieve a more comprehensive characterization of temporal dynamics.

In conclusion, this study advances the understanding of groundwater quality in a complex hydrogeological and socio-economic setting. It also provides a replicable framework for groundwater management in spa and rural regions facing similar challenges at the intersection of environment, public health, and sustainable development.

**Author Contributions:** Conceptualization, R.B. and C.-A.R.; methodology, R.B., C.-A.R. and M.M.; software, R.B., C.-A.R., M.M. and G.R.; validation, R.B., C.-A.R. and M.M.; formal analysis, R.B., C.-A.R., M.M. and L.V.; investigation, R.B., C.-A.R., M.M. and L.V.; data curation, R.B. and C.-A.R.; writing—R.B., C.-A.R., M.M. and O.R.Z.; visualization, R.B., C.-A.R., M.M. and G.R. All authors have read and agreed to the published version of the manuscript.

**Funding:** This research received no external funding.

**Institutional Review Board Statement:** Not applicable.

**Informed Consent Statement:** Not applicable.

**Data Availability Statement:** The raw data supporting the conclusions of this article will be made available by the authors on request.

**Acknowledgments:** We thank the anonymous reviewers for their comments and suggestions that considerably helped improve the previous version of our manuscript.

**Conflicts of Interest:** The authors declare that the research was conducted in the absence of any commercial or financial relationships that could be construed as a potential conflict of interest.

## References


1. Bai, L.; Jiang, L.; Zhao, Y.; Li, Z.; Cao, G.; Zhao, C.; Liu, R.; Wang, H. Quantifying the influence of long-term overexploitation on deep groundwater resources across Cangzhou in the North China Plain using InSAR measurements. *J. Hydrol.* **2022**, *605*, 127368. [CrossRef]
2. Taucare, M.; Viguier, B.; Figueroa, R.; Daniele, L. The alarming state of Central Chile's groundwater resources: A paradigmatic case of a lasting overexploitation. *Sci. Total Environ.* **2024**, *906*, 167723. [CrossRef] [PubMed]
3. El Alaoui, A.; Haidara, I.; Bouya, N.; Moussaid, B.; Faqeih, K.Y.; Alamri, S.M.; Alamery, E.R.; AlAmri, A.R.; Moussaid, Y.; Haddou, M.A. Sustainable Groundwater Management in the Coastal Aquifer of the Témara Plain, Morocco: A GIS-Based Hydrochemical and Pollution Risk Assessment. *Sustainability* **2025**, *17*, 5392. [CrossRef]

4. Guettaia, S.; Boudjema, A.; Derdour, A.; Laoufi, A.; Almohamad, H.; Al-Mutiry, M.; Abdo, H.G. Hydrochemical Characterization and Predictive Modeling of Groundwater Quality in Karst Aquifers Under Semi-Arid Climate: A Case Study of Ghar Boumaaza, Algeria. *Sustainability* **2025**, *17*, 6883. [CrossRef]
5. Halecki, W.; Kalarus, K.; Kowalczyk, A.; Garbowski, T.; Chudziak, J.; Grabowska-Polanowska, B. Reducing Water Resource Pressure and Determining Gross Nitrogen Balance of Agricultural Land in the European Union. *Appl. Sci.* **2025**, *15*, 9216. [CrossRef]
6. Karandish, F.; Liu, S.; de Graaf, I. Global groundwater sustainability: A critical review of strategies and future pathways. *J. Hydrol.* **2025**, *657*, 133060. [CrossRef]
7. Lu, T.; Luo, P.; Wang, J.; Lu, Y.; Huo, A.; Liu, L. Soil salinity accumulation and groundwater degradation due to overexploitation over recent 40-year period in Yaoba Oasis, China. *Soil Tillage Res.* **2025**, *248*, 106398. [CrossRef]
8. Minea, I.; Chelariu, O.E.; Boicu, D.; Iosub, M.; Mărgărint, M.C. Assessing the rural social vulnerability associated with groundwater resources in Eastern Romania. *Environ. Sustain. Indic.* **2025**, *28*, 100910. [CrossRef]
9. Tazhiyev, S.; Murtazin, Y.; Rakhimova, V.; Rakhmetov, I.; Adenova, D.; Koshpanova, K.; Sotnikov, Y.; Abdizhalel, M.; Akylbayeva, A.; Yerezhep, D. Applied Hydrogeological Assessment and GIS-Based Modeling of Transboundary Aquifers in the Shu River Basin. *Water* **2025**, *17*, 2476. [CrossRef]
10. Zhao, L.; Geng, B.; Zhao, M.; Li, B.; Miao, Q.; Liu, S.; Zhao, Z.; Wang, H.; Li, Y.; Jin, W.; et al. Evaluation of Groundwater Quality and Health Risk Assessment During the Dry Season in the Xin'an River Basin, China. *Water* **2025**, *17*, 2412. [CrossRef]
11. Roba, C.; Bălc, R.; Creța, F.; Andreica, D.; Pădurean, A.; Pogăcean, P.; Chertes, P.; Moldovan, F.; Mocan, B.; Roșu, C. Assessment of groundwater quality in NW of Romania and its suitability for drinking and agricultural purposes. *Environ. Eng. Manag. J.* **2021**, *20*, 435–447. [CrossRef]
12. Petruța, N.-L.; Sur, I.M.; Rusu, T.A.; Gabor, T.; Rusu, T. Integrated Assessment of Groundwater Vulnerability and Drinking Water Quality in Rural Wells: Case Study from Ceanu Mare Commune, Northern Transylvanian Basin, Romania. *Sustainability* **2025**, *17*, 6530. [CrossRef]
13. Islam, R.; Kushwah, V.K.; Gupta, N.; Kumar, A.; Goyal, R.; Berwal, P.; Alfaisal, F.M.; Majdi, A.; Al-sareji, O.J.; Alsubih, M. Integrated evaluation of groundwater hydrochemistry using multivariate statistics and irrigation-based water quality indices. *Sci. Rep.* **2025**, *15*, 24923. [CrossRef] [PubMed]
14. Qin, B.; Zhu, Q.; Yang, Y.; Liang, C.; Fu, Y.; Meng, F.; Shi, Y.; Zhang, F.; Ren, L. Embedded slow-release materials for the remediation of organic pollutants in groundwater: A review. *Desalination Water Treatement* **2025**, *322*, 101108. [CrossRef]
15. Gâștescu, P. Water resources from Romania. Potential, quality, teritorial distribution, management. In *Water Resources from Romania. Vulnerability to the Pressure of Man's Activities*; Gâștescu, P., Zăvoianu, I., Driga, B., Ciupitu, D., Drăgoi, I.J., Eds.; Transversal Press: Târgoviște, Romania, 2010; pp. 10–30.
16. Săndulescu, M. *Geotectonics of Romania*; Editura Tehnică: Bucuresti, Romania, 1984; p. 334. (In Romanian)
17. Pécskay, Z.; Seghedi, I.; Kovacs, M.; Szakács, A.; Fülöp, A. Geochronology of the Neogene calc-alkaline intrusive magmatism in the "Subvolcanic Zone" of the Eastern Carpathians (Romania). *Geol. Carpathica* **2009**, *60*, 181–190. [CrossRef]
18. Niţoi, E.; Munteanu, M.; Marincea, Ş.; Paraschivoiu, V. Magma–enclave interactions in the East Carpathian Subvolcanic Zone, Romania: Petrogenetic implications. *J. Volcanol. Geotherm. Res.* **2002**, *118*, 229–259. [CrossRef]
19. Papp, D.C.; Ureche, I.; Dallai, L.; Seghedi, I.; Downes, H. Petrogenesis of convergent-margin calc-alkaline rocks and the significance of the low oxygen isotope ratios: The Rodna—Bârgău Neogene subvolcanic area (East Carpathian). *Geol. Carpathica* **2005**, *56*, 77–90.
20. Fedele, L.; Seghedi, I.; Chung, S.-L.; Laiena, F.; Lin, T.-H.; Morra, V.; Lustrino, M. Post-collisional magmatism in the Late Miocene Rodna-Bîrgău district (East Carpathians, Romania): Geochemical constraints and petrogenetic models. *Lithos* **2016**, *266–267*, 367–382. [CrossRef]
21. Miyashiro, A. Volcanic rock series in island arcs and active continental margins. *Am. J. Sci.* **1974**, *274*, 321–355. [CrossRef]
22. Peltz, S.; Vasiliu, C.; Udrescu, C. Petrology of igneous rocks from the Neogene East Carpathian subvolcanic zone. *Anu. Institutului Geol. Geofiz.* **1972**, *39*, 177–218. (In Romanian)
23. Seghedi, I.; Downes, H. Geochemistry and tectonic development of Cenozoic magmatism in the Carpathian–Pannonian region. *Gondwana Res.* **2011**, *20*, 655–672. [CrossRef]
24. Vasiliță-Crăciun, I.-C.; Gavra, C.-I. The role of viticultural landscape and microregional differentiation. Comparative study: Alba and Bistrița-Năsăud Counties. *Rom. Rev. Reg. Stud.* **2016**, *XII*, 79–90.
25. Moisilu, I. *The Alkaline-Muriatic Baths of Sângeorgiul Românesc Near Năseud, Bistriță-Năseud District in Transilvania*; Tipo-Litografiei și Librăriei, N.D. Miloșescu: Târgu-Jiu, Romania, 1897. (In Romanian)
26. Bélteki, S. *Conspectus Systematics Practicae Aquarum Mineralium in Transylvaniae Principatu*; Wimmer: Vienna, Austria, 1818.
27. Pataki, S. *Descriptio Physico-Chemica Aquar*; M. P. Transsylvan.: Pestini, Romania, 1820.
28. *ISO 5667-2/1998*; Water Quality. Sampling. Part 1: General Guidance on Sampling Techniques. International Organization for Standardization (ISO): Geneva, Switzerland, 1998. (In Romanian)

29. *ISO 5667-3/1998*; Water Quality. Sampling. Part 1: General Guidance on Sample Preservation and Handling. Internation Organization for Standardization (ISO): Geneva, Switzerland, 1998. (In Romanian)
30. Plch, J. Radon Detector LUK 3C. In *Manual for Operating LUK 3C*; Jiri PlchMEng SMM: Prague, Czech, 2002.
31. Cosma, C.; Moldovan, M.; Dicu, T.; Kovacs, T. Radon in water from Transylvania (Romania). *Radiat. Meas.* **2008**, *43*, 1423–142 [CrossRef]
32. Moldovan, M.; Cosma, C.; Encian, I.; Dicu, T. Radium-226 concentration in Romanian bottled mineral waters. *J. Radioanal. Nuc Chem.* **2009**, *279*, 487–491. [CrossRef]
33. Li, Z.; Ma, Z.; Kuijp, T.J.; Yuan, Z.; Huang, L. A review of soil heavy metal pollution from mines in China: Pollution and healt risk assessment. *Sci. Total Environ.* **2014**, *468–469*, 843–853. [CrossRef] [PubMed]
34. *CAC/GL 3-1989*; Guidelines for the Simple Evaluation of Dietary Exposure to Food Additives. International Food Standard Revision 2014. Codex Alimentarius: Rome, Italy, 1989.
35. Report to the General Assembly, Volume I: Sources. In *Sources and Effects of Ionizing Radiation*; UNSCEAR 2000; United Natior Scientific Committee on the Effects of Atomic Radiation: Vienna, Austria, 2000.
36. IAEA (International Atomic Energy Agency). *International Basic Safety Standards for Protection Against Ionizing Radiation and for th Safety of Radiation Sources*; IAEA (International Atomic Energy Agency): Vienna, Austria, 1995; No. 115.
37. NRCP. *Measurement of Radon and Radon Daughters in Air*; Report No: 97; National Council on Radiation Protection and Measure ments: Bethesda, MD, USA, 1988.
38. WHO. Health Criteria and Other Supporting Information, Addendum to Vol. 2. In *Guidelines for Drinking-Water Quality*, 2nd ed World Health Organization: Geneva, Switzerland, 1998.
39. Richards, L.A. Diagnosis and improvement of saline alkali soils: Agriculture. In *Handbook 60*; Department of Agricultur Washington, DC, USA, 1954; Volume 160.
40. Wilcox, L.V. *Classification and Use of Irrigation Waters*; U.S. Department of Agriculture: Washington, DC, USA, 1955.
41. Kelley, W.P.; Brown, S.M.; Liebig, G.F. Chemical effects of saline irrigation water on soils. *Soil Sci.* **1940**, *49*, 95–107. [CrossRef]
42. Eaton, F.M. Significance of carbonate in irrigation water. *Soil Sci.* **1950**, *69*, 123–133. [CrossRef]
43. Paliwal, K.V. *Irrigation with Saline Water*; Monogram No. 2 (New Series); IARI: New Delhi, India, 1972.
44. Doneen, L.D. *Notes on Water Quality in Agriculture Published as a Water Science and Engineering Paper 4001*; Department of Wate Science and Engineering, University of California: Berkeley, CA, USA, 1964.
45. Song, T.; Yang, F.; Chen, Y.; Du, S. Hydrogeochemical evolution and risk assessment of human health in a riverbank filtration site Northeastern China. *Hum. Ecol. Risk Assess.* **2017**, *23*, 705–726. [CrossRef]
46. Jain, C.K.; Vaid, U. Assessment of groundwater quality for drinking and irrigation purposes using hydrochemical studies ir Nalbari district of Assam, India. *Environ. Earth Sci.* **2018**, *77*, 254. [CrossRef]
47. Todd, D.K. *Groundwater Hydrology*, 2nd ed.; Wiley: New York, NY, USA, 1980.
48. Rawat, K.S.; Singh, S.K.; Gautam, S.K. Assessment of groundwater quality for irrigation use: A peninsular case study. *Appl. Wate Sci.* **2018**, *8*, 233. [CrossRef]
49. He, X.; Wu, J.; He, S. Hydrochemical characteristics and quality evaluation of groundwater in terms of health risks in Luohe aquifer in Wuqi County of the Chinese Loess Plateau, northwest China. *Hum. Ecol. Risk Assess.* **2019**, *25*, 32–51. [CrossRef]
50. *Law No. 458 of 8 July 2002 on the Quality of Drinking Water*; The Official Gazette: Bucharest, Romania, 12 December 2011. No. 875 (In Romanian)
51. *Order No. 621/2014 on the Approval of Threshold Values for Groundwater in Romania*; The Official Gazette: Bucharest, Romania, 18 July 2014. Part I, No. 535. (In Romanian)
52. *The Technical Norms for the Exploitation and Marketing of Natural Mineral Waters*; The Official Gazette: Bucharest, Romania, 1 September 2005. Government Decision No. 1020. (In Romanian)
53. WHO. *Guidelines for Drinking Water Quality*, 4th ed.; Incorporating the First Addendum; World Health Organization: Geneva Switzerland, 2017.
54. Bălc, R.; Roba, C.; Moldovan, M.; Vasilian, L. Quality assessment of mineral and underground water from an old spa resort from Romania (Sângeorz-Băi locality). In Proceedings of the XXI International Congress of Carpathian Balkan Geological Association (CBGA), Salzburg, Austria, 10–13 September 2018.
55. WHO. Hardness in Drinking-water Background document for development. In *WHO Guidelines for Drinking-Water Quality*; World Health Organization: Geneva, Switzerland, 2011.
56. Hoaghia, M.-A.; Moldovan, A.; Kovacs, E.; Mirea, I.C.; Kenesz, M.; Brad, T.; Cadar, O.; Micle, V.; Levei, E.A.; Moldovan, O.T Water Quality and Hydrogeochemical Characteristics of Some Karst Water Sources in Apuseni Mountains, Romania. *Water* **2021** *13*, 857. [CrossRef]
57. Moldovan, A.; Török, A.I.; Mirea, I.C.; Micle, V.; Moldovan, O.T.; Levei, E.A. Health Risk Assessment in Southern Carpathians Small Rural Communities Using Karst Springs as a Drinking Water Source. *Int. J. Environ. Res. Public Health* **2022**, *19*, 234 [CrossRef]

58. Backman, B.; Bodiš, D.; Lahermo, P.; Rapant, S.; Tarvainen, T. Application of a groundwater contamination index in Finland and Slovakia. *Environ. Geol.* **1998**, *36*, 55–64. [CrossRef]
59. WHO. *Guidelines for Drinking Water Quality*, 3rd ed.; Incorporating the First and Second Addenda; Recommendation, NCW classifications WA675; World Health Organization: Geneva, Switzerland, 2008; Volume 1.
60. EPA (US Environmental Protection Agency). *National Primary Drinking Water Regulations*; Federal Register, No. 211; Radon-222; Proposed Rules; EPA (US Environmental Protection Agency): Washington, DC, USA, 1999; Volume 64.
61. *Law/301/2015*; Commission Recommendation of 27 November 2015 on the Establishment of Health Protection Requirements for the Population Regarding Radioactive Substances in Drinking Water. MO nr. 904/7.12.2015; European Commission: Brussels, Belgium, 2015.
62. Bodor, K.; Tokos, B.; Bodor, Z.; Keresztesi, A.; László, S.; Garbacea, G.; Szép, R. Hydro-geochemical characterization of the main European mineral water brands. *J. Food Compos. Anal.* **2023**, *122*, 105438. [CrossRef]
63. Chau, N.D.; Tomaszewska, B. 1-Mineral and bottled water as natural beverages. *Bottled Packag. Water* **2019**, *4*, 1–38.
64. McGowan, W. *Water Processing: Residential, Commercial, Light-Industrial*, 3rd ed.; Water Quality Association: Lisle, IL, USA, 2000.
65. Elzain, H.E.; Chung, S.Y.; Senapathi, V.; Sekar, S.; Lee, S.Y.; Roy, P.D.; Hassan, A.; Sabarathinam, C. Comparative study of machine learning models for evaluating groundwater vulnerability to nitrate contamination. *Ecotoxicol. Environ. Saf.* **2022**, *229*, 113061. [CrossRef]
66. Boschetti, T.; Mohamed, Y.; Hadji, R.; Barbieri, M.; Gentilucci, M.; Rossi, M.; Khalil, R.; Khan, S.D.; Asghar, B.; Al-Omran, A.; et al. Using principal component analysis to distinguish sources of radioactivity and nitrates contamination in Southern Tunisian groundwater samples. *J. Geochem. Explor.* **2025**, *271*, 107670. [CrossRef]
67. EEA (European Environmental Agency). Nitrate in Groundwater in Europe. 2024. Available online: https://www.eea.europa.eu/en/analysis/indicators/nitrate-in-groundwater-8th-eap (accessed on 16 August 2025).
68. Krishna Kumar, S.; Logeshkumaran, A.; Magesh, N.S.; Godson, P.S.; Chandrasekar, N. Hydro-geochemistry and application of water quality index (WQI) for groundwater quality assessment, Anna Nagar, part of Chennai City, Tamil Nadu, India. *Appl. Water Sci.* **2015**, *5*, 335–343. [CrossRef]
69. Berlescu, E.; Chioreanu, T.; Cociasu, E.; Dene, A.; Dinculescu, T.; Ștefănescu, S.; Tătăranu, I. *Guide for Referrals to the Balneo-Climatic Cure*; Medical Publishing House: Bucharest, Romania, 1965; pp. 28–41, 172–173. (In Romanian)
70. Feru, A. *Natural Mineral Water Guide*; Novis SRL Publishing House: Cluj Napoca, Romania, 2012. (In Romanian)
71. Munteanu, C. *Therapeutic Mineral Waters*; Editura Balneară: Bucharest, Romania, 2013.
72. Gibbs, R.J. Mechanisms controlling world water chemistry. *Science* **1970**, *170*, 1088–1090. [CrossRef]
73. Da Costa, L.S.; Teixeira, D.S.; Bovendorp, R.S. Heavy metal accumulation in non-human primates: A global systematic review. *Environ. Pollut.* **2025**, *385*, 127042. [CrossRef]
74. Siegel, F.R. *Environmental Geochemistry of Potentially Toxic Metals*; Springer: Berlin, Germany, 2002.
75. Farzana, F.; Roy, T.K.; Hossain, S.A.; Mazrin, M.; Islam, S.; Mahiddin, N.A.; Jayoti, J.R.; Ghosh, R.; Al Bakky, A.; Ismail, Z.; et al. Assessment of groundwater quality and potential health risks related to heavy metals in a peri-urban area of a developing country. *Sci. Rep.* **2025**, *15*, 27970. [CrossRef]
76. Aslani, R.; Esmaeili, S.; Akbari, M.E.; Aghaee, E.M.; Sadighara, P.; Nazmara, S.; Mahmoudi, B. Determination of heavy metals, nitrate and nitrite in mineral and drinking bottled water in Tehran, Iran: A health risk assessment by Monte-Carlo simulation method. *Heliyon* **2024**, *10*, e40714. [CrossRef]
77. Salami, I.R.S.; Thufailah, N.A.; Fahimah, N.; Roosmini, D. Health risk assessment of physicochemical and heavy metals exposures of the usage of shallow groundwater located at the proximity to Citarum River, Indonesia. *Case Stud. Chem. Environ. Eng.* **2025**, *11*, 101153. [CrossRef]
78. Abdipour, H.; Azari, A.; Kamani, H.; Pirasteh, K.; Mostafapour, F.K.; Rayegnnakhost, S. Human health risk assessment for fluoride and nitrate contamination in drinking water of municipal and rural areas of Zahedan, Iran. *Appl. Water Sci.* **2025**, *15*, 47. [CrossRef]
79. Moradnia, M.; Attar, H.M.; Hajizadeh, Y.; Lundh, T.; Salari, M.; Darvishmotevalli, M. Assessing the carcinogenic and non-carcinogenic health risks of metals in the drinking water of Isfaha, Iran. *Sci. Seport* **2024**, *14*, 5029. [CrossRef]
80. Moldovan, M.; Benea, V.; Niță, D.C.; Papp, B.; Burghele, B.D.; Bican-Brișan, N.; Cosma, C. Radon and radium concentration inwater from North-West of Romania and the estimated doses. *Radiat. Prot. Dosim.* **2014**, *162*, 96–100. [CrossRef]
81. Moldovan, M.; Niță, D.C.; Cucoș-Dinu, A.; Dicu, T.; Bican-Brișan, N.; Cosma, C. Radon concentration in drinking water and supplementary exposure in Băița Ștei mining area, Bihor County (Romania). *Radiat. Prot. Dosim.* **2013**, *158*, 447–452. [CrossRef]
82. Niță, D.C.; Moldovan, M.; Sferle, T.; Ona, V.D.; Burghele, B.D. Radon concentrations in water and indoor air in north–west regions of Romania. *Rom. Journ. Phys.* **2013**, *58*, 196–201.
83. Chintăuan, I.; Rusu, I. Considerations regarding the history of the knowledge and use of carbonated water from northeastern Transylvania (Bistrița-Năsăud County). *Hist. Pap.-Bistrița Mus.* **1989**, *6*, 238–275. (In Romanian)

84. EC. *Opinion on Nitrate and Nitrite*; Reports of the Scientific Committee for Food (SCF); 26th Series: 21–28; European Commission: Brussels, Belgium, 1992; pp. 21–28.
85. EC. *Opinion on Nitrate and Nitrite*; Reports of the Scientific Committee for Food (SCF); Report No: 197 38th Series; European Commission: Brussels, Belgium, 1997; pp. 1–33.
86. EFSA NDA Panel (EFSA Panel on Dietetic Products, Nutrition and Allergies); Turck, D.; Bresson, J.L.; Burlingame, B.; Dean, T.; Fairweather-Tait, S.; Heinonen, M.; Hirsch-Ernst, K.I.; Mangelsdorf, I.; McArdle, H.J.; et al. Scientific opinion on Dietary Reference Values for sodium. *EFSA J.* **2011**, *9*, 2246. [CrossRef]
87. EFSA NDA Panel (EFSA Panel on Nutrition, Novel Foods and Food Allergens); Turck, D.; Castenmiller, J.; de Henauw, S.; Hirsch-Ernst, K.I.; Kearney, J.; Maciuk, A.; Mangelsdorf, I.; McArdle, H.J.; Pelaez, C.; et al. Scientific opinion on Dietary Reference Values for chloride. *EFSA J.* **2019**, *17*, 5779. [CrossRef]
88. EFSA. Nitrate in vegetables. Scientific opinion of the panel on contaminants in the food chain. *EFSA J.* **2008**, *689*, 1–79.
89. FAO-WHO. *Nitrate (and Potential Endogenous Formation of N-Nitroso Compounds)*; WHO Food Additive Series 50; World Health Organisation: Geneva, Switzerland, 2003.
90. EFSA. Overview on Dietary Reference Values for the EU population as derived by the EFSA Panel on Dietetic Products, Nutrition and Allergies (NDA). In *Summary of Dietary Reference Values*; Version 4 (September 2017); EFSA: Parma, Italy, 2017; p. 15.
91. FAO-WHO. Codex alimentarius commission. In Proceedings of the Codex Committee on Contaminants in Foods 12th Session, Utrecht, The Netherlands, 12–16 March 2018.
92. WHO. *Nickel in Drinking Water—Background Document for Development of WHO Guidelines for Drinking-Water Quality*; WHO: Geneva, Switzerland, 2007.
93. EFSA. Scientific Opinion on the risks to public health related to the presence of nickel in food and drinking water. *EFSA J.* **2015**, *13*, 4002. [CrossRef]
94. FAO-WHO. Food additives and contaminants. In Proceedings of the Codex Alimentarius Commission, Twenty-fourth Session, Geneva, Switzerland, 2–7 July 2001.
95. Sharma, C.; Mahajan, A.; Garg, U.K. Fluoride and nitrate in groundwater of south-western Punjab, India—Occurrence, distribution and statistical analysis. *Desalination Water Treat.* **2016**, *57*, 3928–3939. [CrossRef]
96. Singh, M.C. Groundwater pollution, causes, assessment methods and remedies for mitigation: A special attention to Indian Punjab. In *Contaminants in Agriculture and Environment: Health Risks and Remediation*; Kumar, V., Kumar, R., Singh, J., Kumar, P., Eds.; Agro. Environ. Media: Haridwar, India, 2019; Volume I, pp. 148–172.
97. Lloyd, J.W.; Heathcoat, J.A. *Natural Inorganic Chemistry in Relation to Groundwater*; Clarendon Press: Oxford, UK, 1985.

sustainability

MDPI

*Article*

# Characterizing the Green Watershed Index (GWI) in the Razey Watershed, Meshginshahr County, NW Iran

Akbar Irani [1], Roghayeh Jahdi [2,*], Zeinab Hazbavi [3], Raoof Mostafazadeh [3] and Abazar Esmali Ouri [3]

1 Department of Rangeland and Watershed Management, Faculty of Natural Resources, University of Mohaghegh Ardabili, Ardabil 56199-11367, Iran; akbar.irani0102@gmail.com
2 Department of Forest Science and Engineering, Faculty of Natural Resources, University of Mohaghegh Ardabili, Ardabil 56199-11367, Iran
3 Department of Rangeland and Watershed Management, Faculty of Natural Resources, Water Management Research Center, University of Mohaghegh Ardabili, Ardabil 56199-11367, Iran; z.hazbavi@uma.ac.ir (Z.H.); raoofmostafazadeh@uma.ac.ir (R.M.); esmaliouri@uma.ac.ir (A.E.O.)
* Correspondence: roghayeh.jahdi@uma.ac.ir

**Abstract**

This paper presents the Green Watershed Index (GWI) methodology, focusing on the 17 sustainability indicators selected in the Razey watershed, NW Iran. Field surveys and data collection have provided the possibility of field inspection and measurement of the present condition of the watershed and the indicators taken. Based on the degree of compliance with the required process, each indicator was scored from 0 to 10 and classified into three categories: unsustainable, semi-sustainable, and sustainable. Using the Entropy method to assign weight to each indicator and formulating a proportional mathematical relationship, the GWI score for each sub-watershed was derived. Spatial changes regarding the selected indicators and, consequently, the GWI were detected in the study area. Development of water infrastructure, particularly in the upstream sub-watersheds, plays a great role in increasing the GWI score. The highest weight is related to environmental productivity (0.26), and the five indicators of water footprint, knowledge management and information quality system, landscape attractiveness, waste recycling, and corruption control have approximately zero weight due to their monotonous spatial distribution throughout sub-watersheds. Only sub-watershed R1 has the highest score (5.13), indicating a semi-sustainable condition. The rest of the sub-watersheds have unsustainable conditions (score below 5). Concerning the GWI, the watershed is facing a critical situation, necessitating the implementation of management and conservation strategies that align with the sustainability level of each sub-watershed.

**Keywords:** green watershed; monitoring indicators; sustainability; water resources

## 1. Introduction

Environmental degradation in Iran's watersheds has escalated to a critical point, endangering public health and ecological stability [1–3]. Iran, located in the Middle East, with an approximate population of 92 million, is a fast-growing developing country https://www.worldometers.info/world-population/iran-population/ (accessed on 20 May 2024). Iran's pressing environmental issues, including diminishing water resources, severe air pollution, and the overuse of natural resources, emphasize adapting short-term and long-term green strategies [4–6]. This article outlines critical factors for adopting green leadership in Iran, including engaging stakeholders, developing training initiatives, and aligning environmental policies. Additionally, Iran encounters significant challenges like financial limitations and an underdeveloped regulatory system. To overcome these hurdles, applying practical solutions aimed at promoting sustainable development and calling for more in-depth research to address Iran's distinct environmental sustainability and green development conditions is essential [2,7,8].

Sustainable and healthy watersheds offer numerous ecosystem services that provide for both humans and the environment [9–11]. To achieve watershed sustainability, it should be ensured that a watershed can withstand natural disturbances, is self-sustainable, has no effects on the surrounding environment, and is useful for communities with ecological value. Advancing watershed management in the 1990s presented a holistic view in which the community and watershed sustainability are interdependent, and they are mutually beneficial [12–14]. The achievement of watershed sustainability necessitates the development of multidimensional indices to monitor and evaluate resource management [15,16]. In the previous studies on watershed sustainability measurements, numerous watershed sustainability indices (e.g., the Water Exploitation Index [17], European Green City Index [18], West Java Water Sustainability Index (WJWS) [19], Watershed Sustainability Index (WSI) [20], and Environmental Performance Index (EPI) [21]) were developed to estimate the watershed sustainability in terms of a specific aspect [10,22,23]. However, there is now a need to develop tools that enable the monitoring of watershed sustainability through different aspects of water, soil, and environmental resource management, emphasizing the green watershed development.

Green development and management in different ecosystems are closely linked to sustainable development [18,24–26]. Since the introduction of the green development concept, researchers have engaged in extensive discussions regarding its implications in diverse fields and have made noteworthy advancements in defining and applying related green initiatives [27,28]. Nonetheless, current researchers have focused less on developing a temporally and spatially dynamic and multidimensional index at the watershed scale. For instance, Hall and Kerr [29] introduced the idea of a "green index," organizing it into three levels, with the first level containing two key categories—the green state and green policy, which together include 256 specific indicators. Cobb et al. [30] also developed the Genuine Progress Indicator (GPI), a sustainability-oriented measure that encompasses social, economic, and environmental components. Later, Lawn [31,32] argued that, in addition to the GPI and the Index of Sustainable Economic Welfare (ISEW), alternative indicators should be developed to encourage broader acceptance and application. Furthermore, Wang et al. [33] evaluated the green development levels of nine cities in China using a multi-level index system covering living environment, ecological efficiency, economic growth, pollutant treatment, and innovation potential. Results reveal significant differences among cities: Shenzhen, Zhuhai, and Guangzhou lead in green development, particularly in ecological efficiency and economic optimization. Their study recommends targeted policies to improve energy and industrial structures, public transport, and forestry carbon sinks to promote a low-carbon, circular development model in the PRD. Recently, the Green Watershed Index (GWI) was developed by Caro and Bladé [26] to analyze natural resource management in rural areas. Applied in various rural regions of Colombia, the method revealed both challenges in implementation and opportunities to enhance water resource management at the watershed level. Their findings revealed that the GWI can act as a crucial foundation for future in-depth studies, offering a new perspective on green watershed assessment. Their methodology is adaptable to other key ecological areas. Therefore, it is inferred from the literature that research aimed at green assessment is essential, and whether a watershed is green, which indicators ought to be utilized, and how to define the level of greenness in a watershed are intricate topics that merit further investigation. There is room for improvement.

Quantitative assessment of watershed status in terms of green representative indicators provides valuable knowledge for conserving and enhancing ecosystem services, especially for Iranian watersheds where immediate managerial solutions are needed. As mentioned earlier, today, climate change, pollution control, overexploitation of natural resources, loss of biodiversity, and water resources scarcity are important public concerns worldwide. In addition, various related crises, destructions, and hazards that affect human health or the environment, along with the persistence of poverty, have reinforced concerns about technological evolution and economic development, while concurrently questioning the ability of public policies to address such problems in a timely manner, especially in developing countries [34]. Therefore, determining indicators for exploring the impact of environmental, economic, and social factors on the green watershed status is essential in management frameworks. Various standards have been proposed for choosing suitable indicators to assess watershed sustainability and associated issues, such as sensitivity to environmental changes, being accessible and cost-effective in nature, and having the capacity to offer predictive insights into watershed functioning [22,23,35,36]. Specifically, in line with literature reviews [27,28,31,32], a selection of diagnostic indicators on the substantive and formal criteria should be completed. The substantive criterion means prioritizing indicators that accurately represent and explain the green status of the study regional condition. Furthermore, in terms of the formal criterion, it is important to ensure that the indicators are quantitative, complete, and continuous, while eliminating variables that could compromise the comparability of the analyzed objects.

Razey watershed, located in Ardabil Province (northwest Iran), is among the few watersheds in the region where measures for soil and water conservation, both structural and biological, have been implemented [37]. Previous studies have primarily examined the impacts of land use changes—such as deforestation and expansion of residential areas—on soil erosion and flood hazards within the watershed [3,37]. However, there is a significant lack of studies focused on the conceptualization and assessment of the green status of the watershed aimed at evaluating the sustainability of various ecosystems in Ardabil Province. This study was, therefore, designed to (1) apply the GWI to offer actionable insights for policymakers in developing a balanced and comprehensive green development roadmap in the Razey watershed, especially in light of significant land use change, soil degradation, and sedimentation; (2) determine the relative weight of key indicators influencing the watershed's green status; and (3) conduct spatial zoning of both individual indicators and the overall GWI at the sub-watershed level. The present research supports the broader goals of previous watershed management efforts in the Razey watershed, while introducing an advanced framework to assist decision-makers in optimizing the management of the province's natural and anthropogenic resources.

This research serves as an initial step toward more comprehensive investigations, providing a new lens for policymaking in watershed management and introducing an innovative approach to sustainability assessment. Notably, evaluating sustainability at the watershed scale—especially through green development indicators—has become an increasingly important global concern. Therefore, it is essential for watershed managers to implement a context-specific, structured framework that effectively captures the dynamic relationship between anthropogenic pressures and ecological integrity. Accordingly, this study can be viewed as a complementary advancement of Caro and Bladé's work [26], offering new insights derived from a case with distinct climatic, ecological, economic, social, and policy-related features. Based on an extensive review of the existing literature, this study is among the first to tailor and apply the GWI framework to such conditions. The results enhance our understanding of watershed sustainability and establish a foundation for future initiatives aimed at reducing watershed degradation and encouraging equitable development. Although the GWI framework developed by Caro and Bladé [26] presents a valuable tool for sustainability assessment, several challenges remain in terms of its customization, modeling, and quantification. In particular, the selection of appropriate evaluation indicators that account for varying local and regional climatic and ecological conditions requires further exploration and discussion. In this context, the current research marks the first attempt to adapt the GWI framework to a case study in Iran, thereby extending its applicability and relevance.

## 2. Materials and Methods

### *2.1. Description of the Study Area*

The Razey watershed (18,194 ha) is located in Meshginshahr, Ardabil (NW Iran; Figure 1a,b). In terms of topography, the minimum and maximum elevations are 1151.75 and 2408.33 m amsl, respectively, and its average slope is 13.3%. The mean annual rainfall is 315.6 mm. In terms of the mean temperature, there is a minimum of 3.3 °C and a maximum of 13.9 °C. According to the Emberger climatic classification, the region is placed in the highlands, and based on the Ambrothermic curve method, a wide area is classified as a semi-arid ultra-cold region. This watershed is divided into 17 sub-watersheds (Figure 1c; Table 1).

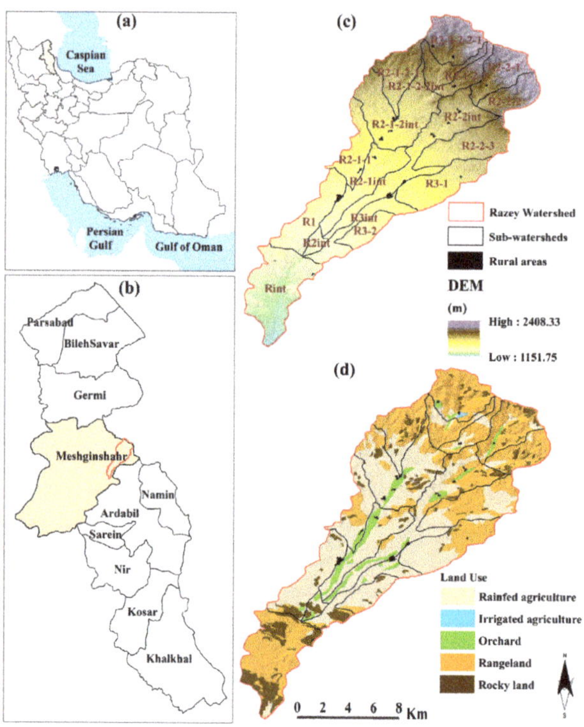


**Figure 1.** Location of Razey watershed in Ardabil Province (**a**), Meshginshahr county (**b**), digital elevation model map with sub-watersheds boundary (**c**), and land use map (**d**).

**Table 1.** Sub-watershed characteristics in the study area.

| Sub-Watershed | Area (ha) | Area (%) | Land Use (ha) | | | | | | Mean Elevation (M) |
|---|---|---|---|---|---|---|---|---|---|
| | | | Residential | Rainfed Agriculture | Orchard | Rangeland | Rocky land | Irrigated Agriculture | |
| R1 | 1146.3 | 6.30 | 10.30 | 944.80 | 17.00 | 37.40 | 136.80 | 0.00 | 2129.44 |
| R2-1-1 | 842.5 | 4.63 | 3.50 | 546.30 | 55.20 | 199.70 | 37.90 | 0.00 | 1807.92 |
| R2-1-2-1 | 909.4 | 5.00 | 0.70 | 197.60 | 2.30 | 611.50 | 97.40 | 0.00 | 1916.61 |
| R2-1-2-2-1 | 1121.2 | 6.16 | 4.30 | 51.20 | 36.90 | 945.70 | 63.60 | 19.50 | 1972.65 |
| R2-1-2-2-int | 582.8 | 3.20 | 0.00 | 127.10 | 12.70 | 439.80 | 3.20 | 0.00 | 2100.93 |
| R2-1-2-3 | 489.9 | 2.69 | 2.30 | 139.90 | 6.70 | 336.90 | 4.00 | 0.00 | 1663.82 |
| R2-1-2-int | 2098.6 | 11.53 | 12.20 | 1318.80 | 110.10 | 513.90 | 143.20 | 0.00 | 2033.07 |
| R2-1-int | 1129 | 6.21 | 10.10 | 730.20 | 248.90 | 110.00 | 29.80 | 0.00 | 1628.60 |
| R2-2-1 | 944.2 | 5.19 | 0.00 | 69.90 | 43.70 | 713.80 | 116.70 | 0.00 | 1544.93 |
| R2-2-2 | 742.9 | 4.08 | 0.00 | 59.60 | 0.00 | 592.10 | 91.20 | 0.00 | 1714.98 |
| R2-2-3 | 1313.2 | 7.22 | 0.00 | 570.10 | 0.00 | 724.80 | 18.20 | 0.00 | 1479.09 |
| R2-2-int | 2061.8 | 11.33 | 18.90 | 1031.70 | 223.70 | 683.60 | 104.20 | 0.00 | 1579.92 |
| R2-int | 178.6 | 0.98 | 0.00 | 74.90 | 37.50 | 0.00 | 66.20 | 0.00 | 1402.51 |
| R3-1 | 1269 | 6.97 | 0.60 | 749.30 | 0.60 | 492.30 | 26.20 | 0.00 | 1415.51 |
| R3-2 | 489.4 | 2.69 | 0.00 | 398.70 | 0.00 | 53.30 | 37.30 | 0.00 | 1418.93 |
| R3-int | 768 | 4.22 | 6.00 | 538.70 | 43.10 | 62.50 | 117.60 | 0.00 | 1337.67 |
| R-int | 2107.25 | 11.58 | 2.30 | 57.40 | 0.60 | 1390.30 | 657.10 | 0.00 | 1288.62 |
| Total | 18,194.05 | 100 | 71.20 | 7606.20 | 839.00 | 7907.60 | 1750.60 | 19.50 | 1650.37 |

This watershed has relatively diverse land uses such as rangeland, rainfed agriculture, rocky land, orchard, residential areas, and irrigated agriculture (Figure 1d). The main activity of the local residents is agriculture and animal husbandry. Agricultural lands of the domain include rainfed, irrigated, and orchard lands. In agriculture, wheat cultivation, and orchards, walnuts have the highest net income in the area.

### 2.2. Setting the Sustainability Indicators

To monitor ecosystem sustainability, there are many indicators within the Sustainable Development Goals (SDGs) in urban and rural areas. These indicators are designed to monitor the natural resources supplied by land, air, water, and biodiversity [38]. To select the most suitable indicators for monitoring and evaluation of natural resources, their inputs, processes, outputs, outcomes, impacts, and exogenous factors should have been identified [39]. Here, the research design was grounded in the methodology originally developed by Caro and Bladé [26]. Following this approach, the Green Watershed Index (GWI) framework was implemented to evaluate the environmental conditions of the Razey watershed. While most input indicators were chosen in alignment with this framework, certain ones—such as the *water retention index (WRI)*, *Simpson's biodiversity index (BD)*, *environmental productivity (EP)*, and *recycling (R)* indicators—were selected differently due to data limitations and specific regional constraints. This study made use of the most comprehensive and accessible datasets and maps available. Notably, the selected indicators are widely recognized and valued by experts in hydrology, geomorphology, sociology, and economics for their capacity to analyze diverse and complex watershed dynamics, ranging from physical and hydrological to social, economic, and ecological aspects and supporting informed improvements. To support this analysis, a GIS database was developed, incorporating the necessary data layers for each sub-watershed, as defined by the General Office of Natural Resources and Watershed Management in Ardabil Province.

Accordingly, in this research, we selected 17 sustainability indicators to generate the GWI in the study area. The first eight indicators were widely used in assessing different aspects of hydrologic sustainability [14,37,40–43]. These indicators highlighted the specific importance and role of hydrology dynamics on the different aspects of watershed sustainability.

For *infrastructure maintenance (IM)* calculation, the gray and blue infrastructure play an influential role in the development of the Razey watershed. Rivers, water wells, and springs were examined as blue infrastructures, and fuel supply (gasoline, kerosene, and gas), access to drinking water, health, schools, and public transportation roads were examined as grey infrastructure.

Three influential sub-indicators, including water production, water quality, and sewage system coverage, were considered in calculating *effectiveness in public water management (EPWM)*. According to Dalavi et al. [44], the Coutagine model [45] demonstrated the highest accuracy among the four empirical models examined for runoff estimation. This model was applied to estimate water production volume, which is a basic input for the calculation of the *EPWM*, *water vulnerability (WV)*, *water stress (WS)*, and *water self-sufficiency (WSS)* indicators. Coutagine [45] analyzed multiple catchments across various countries and developed a formula that links temperature, precipitation, and runoff.

Using the results obtained from the chemical analysis of four wells drilled in the groundwater aquifer and four wells drilled in rock units, the water quality of the Razey watershed was evaluated. The four variables used in this regard included pH, electrical conductivity (EC), total dissolved solids (TDS), and the sodium absorption ratio (SAR). TDS is very effective in creating the taste of drinking water. So, in the range of less than 500 mg/L, it is very good, between 500 and 1000 is desirable, between 1000 and 1500 is permissible for drinking, and above 1500 is not permissible. EC is the ability to pass electric current in water, and a TDS value less than 1000 mg/L is suitable for drinking. According to WHO standards, the standard pH value for groundwater varies between 6.5 and 8.5. An SAR value up to 6 is very good, from 6 to 10 is acceptable, and more than 10 is suitable for agriculture.

*Water vulnerability (WV)* for each sub-watershed was evaluated based on the ratio of water production volume to its population. The *water retention index (WRI)*, which is also considered one of the ecosystem services, was calculated using InVEST 3.16.1 software and land use data [46].

*Water stress (WS)* and *water self-sufficiency (WSS)* indicators were also calculated based on the water production volume of each sub-watershed and population and by considering an annual per capita consumption of 1500 m$^3$. In addition, the *water footprint (WF)* is a multidimensional indicator that expresses the direct and indirect consumption and production of water and the induced pollution [42].

To calculate the *water saving and efficient use programs (WSE)*, the implemented cultural promotion activities were taken into account. For example, in Razey City—located within the sub-watersheds R1 and R2-1—these efforts involve not only advertising but also educational messages displayed on city walls and the organization of workshops by the Meshginshahr County water departments. Additionally, incentive and enforcement strategies are applied, such as modifying water consumption and controlling water pressure using specialized equipment. In comparison, surrounding villages mainly utilize wall writings to foster a culture of water conservation and encourage responsible water use.

Malvandi et al. [47] and Alaei et al. [36] used *Simpson's biodiversity index (BD)* to study the habitat fragmentation and quality, which are a prerequisite to assessing the environmental sustainability. Indeed, this index quantifies the landscape composition. The plant vegetation and green spaces are a critical component that sustains the watershed performance in the long term [28,33]. Three activities of afforestation, pit seeding, and seeding, which were conducted sparsely in the Razey watershed, were investigated with the *environmental productivity (EP)* indicator. In addition, *waste recycling (R)* is one of the important green indicators considering sustainability, especially in urban ecosystems [48], which was also investigated in our study.

Three indicators, *landscape attractiveness (LA)*, *human capital (HC)*, and *population participation (PP)*, were calculated based on the demographic information in terms of social sustainability. To examine the desire and sense of belonging of people to their place of residence (i.e., LA), population changes were compared over 10 years based on the available data (2006 to 2016). Then, the HC was evaluated by considering the formal and informal staff in the related organizations to sustainability issues in the Razey watershed. Finally, the PP was assessed based on the participation level of the local communities in gatherings and educational workshops, which are mostly held in local mosques. In terms of the *adaptation to climate change (ACC)* indicator, the design and implementation of stone-mortar and gabion (stone nets) structures and canalization in residential areas, as well as the improvement of old wells and aquifers, were examined as factors affecting the water supply and consumption system, as well as protection against floods [16,49,50].

The last two indicators of the *knowledge management and information quality system (KM) and corruption control (CC)* are among the significant indicators for assessing the green sustainability of organizations. Based on field visits to various departments in the Razey watershed, it can be stated that there is a spatial difference in the application of the integrated virtual system in the departments of natural resources and the environment for responding to letters and inquiries from departments, the public, and reviewing natural resources and watershed management plans for transfer to the private sector online (which has reduced the need for in-person visits and administrative bureaucracy). Finally, to improve environmental sustainability performance, it is crucial to engage citizens effectively and create transparent governments that help combat corruption in environmental and natural resource issues [51]. In this regard, the online finding of the Transparency International organization (https://www.transparency.org.uk/) was considered. It is a global organization that works in more than 100 countries to end the injustice of corruption. The association was founded in 1993 by former World Bank employees. According to the statistics provided, the global score of the Corruption Perceptions Index has remained stable over the past decade (statistics up to 2022) and is equal to 43. Over two-thirds of the world's countries have a rank below 50, and 26 countries have been assigned the minimum score.

The above-mentioned indicators, along with their abbreviations, is presented in Table 2. Each indicator receives a score ranging from 0 (worst) to 10 (best), based on the degree of conformity of the required process. Then, by weighting each of the indicators and formulating a proportional mathematical relationship, the score of the GWI of each sub-watershed was obtained. A sub-watershed that belongs to a developed region is expected to maintain its management indicators within a very sustainable framework, and a sub-watershed that belongs to a developing region is expected to have lower levels of sustainability. Therefore, following a thorough assessment of each indicator and the sub-indicators involved in their evaluation and implementation, 17 indicators were selected that are suitable for describing a watershed [26].

**Table 2.** Monitoring indicators for Green Watershed Index (GWI) (adapted from [26] with some modifications).

| No. | Indicator * | Abbreviation | Source |
|---|---|---|---|
| 1 | *Infrastructure maintenance* | *IM* | Health Department of Razey, Education Department of Razey, General Office of Natural Resources and Watershed Management of Ardabil Province, Iran, and Field Visit |
| 2 | *Effectiveness in public water management* | *EPWM* | Coutagine formula [45], Soil and Water Laboratory, and General Office of Natural Resources and Watershed Management of Ardabil Province, Iran |
| 3 | *Water vulnerability* | *WV* | Coutagine formula [45] |
| 4 | *Water retention index* | *WRI* | INVEST software |
| 5 | *Water stress* | *WS* | Coutagine formula [45] |
| 6 | *Water footprint* | *WF* | Water and Wastewater Department of Meshginshahr and Razey |
| 7 | *Water self-sufficiency* | *WSS* | Coutagine formula [45] |
| 8 | *Water saving and efficient use programs* | *WSE* | Water and Wastewater Department of Meshginshahr and Razey |
| 9 | *Simpson's biodiversity index* | *BD* | Fragstat 3.16.1 software |
| 10 | *Environmental productivity* | *EP* | General Office of Natural Resources and Watershed Management of Ardabil Province, Iran |
| 11 | *Recycling* | *R* | Municipality and Village Councils of Razey |
| 12 | *Landscape attractiveness* | *LA* | Statistical Center of Iran |
| 13 | *Human Capital* | *HC* | Water and Wastewater Department of Meshginshahr and Razey |
| 14 | *Population participation* | *PP* | General Office of Natural Resources and Watershed Management of Ardabil Province, Iran |
| 15 | *Adaptation to climate change* | *ACC* | General Office of Natural Resources and Watershed Management of Ardabil Province, Iran, and field visit |
| 16 | *Knowledge management and information quality system* | *KM* | General Office of Natural Resources and Watershed Management of Ardabil Province, Iran |
| 17 | *Corruption control* | *CC* | Transparency International Association |

* The names of indicators are shown in *italic* in the tables and context for better understanding.

All indicators were calculated based on defined methods and existing standards. Some of the information needed to calculate the indicators was obtained from government organizations and institutions. For example, information related to population statistics was collected through the statistics provided by the Statistical Center of Iran. Furthermore, field visits and data collection have provided the possibility of field surveys and measurements of the current state of the study area and the set of measures taken, including biological and hydraulic activities. The justification of the eligibility of each indicator was completed according to the extent of its compliance with the purpose of the GWI.

### 2.3. Scoring of the Sustainability Indicators

Considering the scores presented in Table 3, the indicators are classified into three categories: unsustainable, semi-sustainable, and sustainable (Table 3).

Table 3. The general evaluation criteria of the selected indicators (adapted from [26]).

| Score | Category |
|---|---|
| 0–5 | Unsustainable |
| 6–8 | Semi-sustainable |
| 9–10 | Sustainable |

To describe the performance of the indicators, six levels of conformity were established for each indicator (Table 4). In this context, a score in the range of 0 to 10 was assigned, based on the effectiveness and utilization of resources for delivering the service. Furthermore, this scale from 0 to 10 was preferred as a basis for ranking, where a green representative criterion of 0 specifies the worst possible green state and 10 represents the best possible green state. Each score highlights the importance of strategic planning and implementing processes to fulfill the fundamental requirements of the watershed in optimal conditions.

Table 4. Specific evaluation of green management criteria (adapted from [26] with some modifications).

| Score | How to Evaluate |
|---|---|
| 0 | Designated and desired criteria for green watershed management have not been met. |
| 2 | Activities for achieving green watershed management have been started. |
| 4 | Operational plans for achieving green watershed management have been completed. |
| 6 | Approaches to the development of executive measures for achieving green watershed management have been initiated. |
| 8 | Measures for achieving green watershed management have been partially implemented. |
| 10 | Fully compliant with green management criteria, and the intended measures have been fully implemented. |

### 2.4. Weighting the Sustainability Indicators

The indicators that are the most important in natural ecosystems and for ensuring sustainability have the most impact on green watershed management. The Entropy method was applied to define this importance and evaluate the weight of each indicator. The Entropy is expressed by the specific probability distribution of Pi with Equation (1):

$$\mathrm{Ei} = \mathrm{s}(\mathrm{p1}, \mathrm{p2}, \ldots, \mathrm{pn}) = -\mathrm{K} = \sum_{\mathrm{i}=1}^{\mathrm{n}} [\mathrm{Pi} - \mathrm{LnPi}] \quad (\mathrm{i} = 1, 2, 3, \ldots, \mathrm{m}) \tag{1}$$

To calculate the importance coefficient, the following four steps were taken:

(a) The available information was normalized. In this way, the values of each indicator were divided by the sum of the values of that indicator throughout all sub-watersheds.

(b) Equation (1) was used to calculate Ei.

(c) The degree of dispersion was obtained from Equation (2):

$$\mathrm{Dj} = 1 - \mathrm{Ej} \tag{2}$$

(d) By dividing the values of Dj by their sum, the weight of the indicators was obtained.

### 2.5. Calculation of Green Watershed Index (GWI)

The *GWI* was calculated from the weighted combination of 17 indicators as explained in Equation (3) [26].

$$GWI = \frac{Pi * (IM + EPWM + VH + \ldots + ACC + KM + CC)}{17} \tag{3}$$

Here, each indicator was scored with a value between 0 and 10, and a distinct weight coefficient *P* was placed for each of the indicators based on the study area. The abbreviated form of the previous relation for better explanation and understanding is presented in Equation (4):

$$GWI = \frac{\sum_{i=1}^{n} Fi * Pi}{\Sigma Pi} \tag{4}$$

In this equation, the *GWI* represents the Green Watershed Index, *n* denotes the total count of management indicators, *Fi* indicates the numerical outcome of management indicator *i*, and *P* signifies the weighting coefficient assigned to each indicator based on the study area. It should be noted that the full title of each of the indicators, along with their abbreviations, is presented in Table 2.

## 3. Results

The findings of the water-related indicators chosen for the Green Watershed Index (GWI) in the study area are shown in Table 5. Each sub-watershed has its indicator scores that are used to evaluate its sustainability category. Based on the zoning of the unsustainable, semi-sustainable, and sustainable categories in the study area, the spatial changes in the GWI are confirmed (Figure 2). Based on Table 5, the minimum (0.25) and maximum (5.13) values of the GWI with the standard deviation of 1.42 were obtained, which indicates relatively moderate spatial variability. Furthermore, in Figure 3, a sub-watershed weight radar chart is used to display the results of the GWI with a range of 0 to 6, where 6 is the most important in sustainability development (Figure 3).

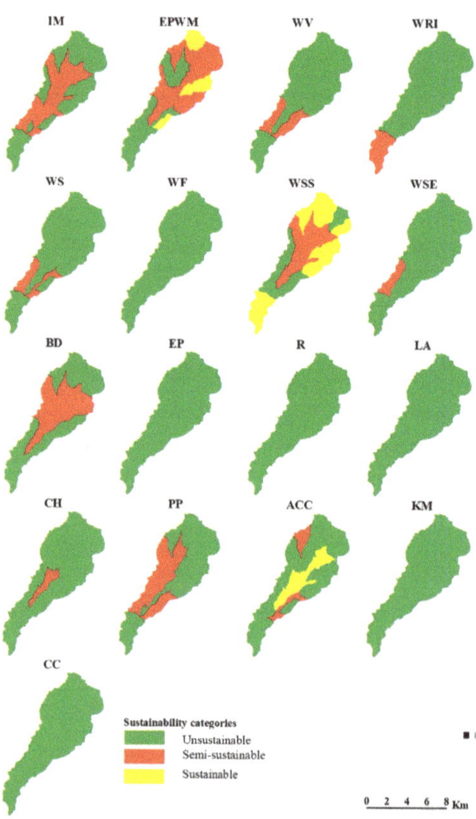


**Figure 2.** Zoning of sustainability categories of the 17 indicators selected for the Green Watershed Index (GWI) in the Razey Watershed. The full title of each indicator, along with its abbreviations, is presented in Table 2.

**Table 5.** The results of calculating the Green Watershed Index (GWI) and the related indicators (Weighted values).

| Sub-Watershed | | Indicators * | | | | | | | | | | | | | | | | GWI | Category |
|---|---|---|---|---|---|---|---|---|---|---|---|---|---|---|---|---|---|---|---|
| | | *IM* | *EPWM* | *WV* | *WRI* | *WS* | *WF* | *WSS* | *WSE* | *BD* | *EP* | *R* | *LA* | *HC* | *PP* | *ACC* | *KM* | *CC* | | |
| | **Weight** | **0.030** | **0.006** | **0.109** | **0.005** | **0.109** | **0.000** | **0.015** | **0.150** | **0.004** | **0.262** | **0.000** | **0.000** | **0.191** | **0.098** | **0.022** | **0.000** | **0.000** | | |
| R1 | | 0.18 | 0.03 | 0.87 | 0.02 | 0.87 | 0.00 | 0.03 | 0.90 | 0.00 | 0.52 | 0.00 | 0.00 | 0.95 | 0.69 | 0.07 | 0.00 | 0.00 | 5.13 | Semi-sustainable |
| R2-1-1 | | 0.12 | 0.03 | 0.43 | 0.02 | 0.43 | 0.00 | 0.06 | 0.45 | 0.00 | 0.00 | 0.00 | 0.00 | 0.57 | 0.69 | 0.09 | 0.00 | 0.00 | 2.91 | |
| R2-1-2-1 | | 0.05 | 0.05 | 0.00 | 0.02 | 0.00 | 0.00 | 0.15 | 0.00 | 0.00 | 0.00 | 0.00 | 0.00 | 0.00 | 0.00 | 0.15 | 0.00 | 0.00 | 0.43 | |
| R2-1-2-2-1 | | 0.03 | 0.05 | 0.00 | 0.02 | 0.00 | 0.00 | 0.14 | 0.15 | 0.00 | 1.31 | 0.00 | 0.00 | 0.00 | 0.10 | 0.09 | 0.00 | 0.00 | 1.89 | |
| R2-1-2-2-int | | 0.06 | 0.05 | 0.00 | 0.02 | 0.00 | 0.00 | 0.15 | 0.00 | 0.00 | 0.00 | 0.00 | 0.00 | 0.00 | 0.10 | 0.11 | 0.00 | 0.00 | 0.49 | |
| R2-1-2-3 | | 0.01 | 0.05 | 0.00 | 0.02 | 0.00 | 0.00 | 0.15 | 0.00 | 0.00 | 0.79 | 0.00 | 0.00 | 0.00 | 0.00 | 0.04 | 0.00 | 0.00 | 1.07 | |
| R2-1-2-int | | 0.19 | 0.04 | 0.54 | 0.02 | 0.54 | 0.00 | 0.09 | 0.45 | 0.00 | 0.00 | 0.00 | 0.00 | 0.57 | 0.69 | 0.11 | 0.00 | 0.00 | 3.25 | |
| R2-1-int | | 0.24 | 0.04 | 0.22 | 0.02 | 0.22 | 0.00 | 0.11 | 0.60 | 0.00 | 0.00 | 0.00 | 0.00 | 1.53 | 0.69 | 0.22 | 0.00 | 0.00 | 3.87 | |
| R2-2-1 | | 0.03 | 0.05 | 0.54 | 0.02 | 0.54 | 0.00 | 0.08 | 0.00 | 0.00 | 0.00 | 0.00 | 0.00 | 0.00 | 0.00 | 0.09 | 0.00 | 0.00 | 1.35 | |
| R2-2-2 | | 0.03 | 0.05 | 0.00 | 0.02 | 0.00 | 0.00 | 0.15 | 0.00 | 0.00 | 0.00 | 0.00 | 0.00 | 0.00 | 0.00 | 0.00 | 0.00 | 0.00 | 0.25 | Unsustainable |
| R2-2-3 | | 0.07 | 0.05 | 0.00 | 0.02 | 0.00 | 0.00 | 0.15 | 0.00 | 0.00 | 0.00 | 0.00 | 0.00 | 0.00 | 0.00 | 0.13 | 0.00 | 0.00 | 0.43 | |
| R2-2-int | | 0.16 | 0.03 | 0.11 | 0.02 | 0.11 | 0.00 | 0.12 | 0.15 | 0.00 | 0.00 | 0.00 | 0.00 | 0.00 | 0.39 | 0.20 | 0.00 | 0.00 | 1.29 | |
| R2-int | | 0.13 | 0.03 | 0.54 | 0.01 | 0.54 | 0.00 | 0.06 | 0.00 | 0.00 | 0.00 | 0.00 | 0.00 | 0.00 | 0.00 | 0.09 | 0.00 | 0.00 | 1.41 | |
| R3-1 | | 0.13 | 0.05 | 0.00 | 0.02 | 0.00 | 0.00 | 0.15 | 0.00 | 0.00 | 0.00 | 0.00 | 0.00 | 0.00 | 0.29 | 0.04 | 0.00 | 0.00 | 0.69 | |
| R3-2 | | 0.07 | 0.05 | 0.65 | 0.01 | 0.65 | 0.00 | 0.05 | 0.00 | 0.00 | 0.00 | 0.00 | 0.00 | 0.00 | 0.69 | 0.09 | 0.00 | 0.00 | 2.25 | |
| R3-int | | 0.18 | 0.01 | 0.76 | 0.01 | 0.76 | 0.00 | 0.03 | 0.45 | 0.00 | 0.00 | 0.00 | 0.00 | 0.57 | 0.69 | 0.13 | 0.00 | 0.00 | 3.59 | |
| R-int | | 0.12 | 0.03 | 0.00 | 0.04 | 0.00 | 0.00 | 0.15 | 0.00 | 0.00 | 0.00 | 0.00 | 0.00 | 0.00 | 0.00 | 0.09 | 0.00 | 0.00 | 0.42 | |
| Total | | 0.11 | 0.04 | 0.27 | 0.02 | 0.27 | 0.00 | 0.11 | 0.15 | 0.00 | 0.44 | 0.00 | 0.00 | 0.84 | 0.29 | 0.10 | 0.00 | 0.00 | 2.65 | |

* The full title of indicators is presented in Table 2, and they are shown in *italic* in the tables and context for better understanding.

| Legend | |
|---|---|
| GWI Score | Category |
| 0–5 | Unsustainable |
| 6–8 | Semi-sustainable |
| 9–10 | Sustainable |

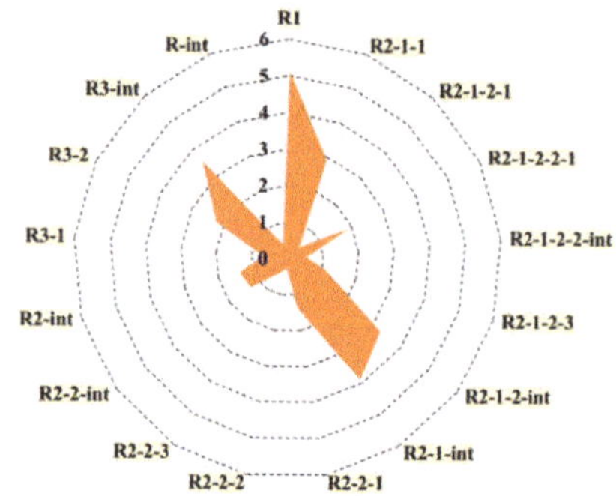


**Figure 3.** Green Watershed Index (GWI) results with a radar for different sub-watersheds.

According to the results of Entropy weighting (Table 5), the variability in indicator weights within the GWI assessment was validated, offering valuable guidance for policymakers in formulating effective management strategies. Among the indicators, *environmental productivity (EP), human capital (HC), and water-saving and efficient use programs (WSE)* were the most influential, with respective weights of 0.26, 0.19, and 0.15. In contrast, the monotonous spatial pattern of *recycling (R), landscape attractiveness (LA), knowledge management and information quality system (KM),* and *corruption control (CC)* throughout sub-watersheds leads to their least contribution to the GWI calculation.

In the study area, at first, the *infrastructure maintenance (IM)* indicator was calculated from the total scores of the two blue-green and gray infrastructure types. According to the regional condition, the blue-green infrastructure was related to the number of groundwater wells, rivers, and drinking fountains, while the gray infrastructure was related to the total number of gas stations, kerosene, gas, water piping, health, schools, and public transportation. The mean scores for blue-green and gray infrastructures were obtained as 0.64 and 0.24, respectively. The results showed that the mean and standard deviations of the IM score were 3.58 and 2.22, respectively. Furthermore, the highest score of IM was obtained in the sub-watershed R2-1-int located in Rahim Bigloi and Razey. In return, sub-watersheds R2-1-2-2-1, R2-2-1, and R2-2-2 had the lowest score. The mean values obtained from the three sub-indicators of water production, water quality, and sewage system coverage were 4.29, 2.18, and 0.26, respectively. The lowest score is related to sewage coverage and then the water quality. The total score of these three sub-indicators that led to the production of the *effectiveness in public water management (EPWM)* indicator varies between 2 and 8.5. The lowest score is related to the sub-watershed R3-int (located in Karamshahlo village). The three sub-watersheds R2-1-2-2-1, R2-2-3, and R3-2 have the best situation in terms of this indicator.

The indicators of *water vulnerability (WV), water stress (WS),* and *water self-sufficiency (WSS)* were calculated based on the amount of irrigation and the amount of water consumption in the sub-watersheds, using the Coutagine formula [45]. Therefore, these indicators have a direct relationship with the population. Regarding all three indicators, the sub-watershed R1 obtains lower values than the other sub-watersheds due to the larger population and excessive consumption. In addition, the *mean water retention index (WRI)* was equal to 3.8 and varied between 2 and 6.7 across the study area. The index shows particularly low values in almost all sub-watersheds (<4.6), resulting from a combination of typically sparse vegetation, low permeability, and limited, shallow presence of soil. The sub-watershed R-int recorded the highest WRI value at 6.7.

Regarding the *water footprint (WF)* indicator, given that the volume of water utilized for production is insignificant in comparison to the water used for drinking, all sub-watersheds were given a score of 8. Furthermore, in terms of the *water-saving and efficient use programs (WSE)* indicator, factors such as advertising, holding training workshops, incentive and punishment measures, actualization of bills, and use of equipment were investigated. High levels of WSE are associated with the central areas of the watershed, especially in Razey City and its nearby villages (i.e., sub-watershed R1 with a score of 6). In sub-watersheds located on the border of Iran and Azerbaijan (i.e., sub-watershed R2-1-2-2-1 with score 1), levels were low (<4). The mean WSE was equal to 1 in the study area.

In terms of *Simpson's biodiversity (BD)* indicator, the results obtained from the Fragstat software show that the mean value is 4.58 (varies between 2.8 to 6.4), which indicates a semi-sustainable condition in most of the sub-watersheds. The highest values of the Simpson index (>6) were observed in sub-watersheds R2-int and R2-2-int, and the lowest value (<3) was in the sub-watershed R2-1-2-2-1. In terms of the *environmental productivity (EP)* indicator, the mean score is 1.67, and most of the sub-watersheds have zero values, which shows the unsustainable conditions. The upstream sub-watersheds are in a better condition than other sub-watersheds, although the measures taken are insufficient and have little impact. Furthermore, in terms of the *waste recycling (R)* indicator, during the research carried out by the Razey Municipality and the Islamic Council of Villages, no practical action has been taken in the field of waste recycling, and the Razey watershed has an unsustainable condition in this regard.

In terms of the *landscape attractiveness (LA)* indicator, the score was measured based on the rate of population increase or decrease in the past decade in the study area. A slight increase in population was observed only in the sub-watershed R2-1-1 (1.79%), and the LA score was considered 1 for this sub-watershed. For other sub-watersheds, due to population decrease, a score of zero was considered for LA. Therefore, the average LA score for the entire study area was very low (<0.1).

Regarding the *human capital (HC)* indicator, the maximum score is related to the sub-watershed R2-1-int, which is higher than the mean of 4.4 for the entire watershed due to the participation of both the contracting and administrative parts in this sub-watershed. Findings of this study also indicated that the extent of the *population participation (PP)* indicator in watershed program planning and implementation phases, as well as participatory watershed management training (i.e., attending events and workshops), was higher in large, densely populated villages than in small villages. However, the average PP was relatively low (=3), with a range of variation between 0 and 7. Specifically, six sub-watersheds have the maximum PP, including R1, R2-1-1, R2-1-2-int, R2-1-int, R3-2, and R3-int. This shows that the extent of the *adaptation to climate change (ACC)* indicator was moderate in the study area. This indicator assesses the level of sub-watersheds' access to measures for reducing the effects of drought and floods. In the study area, the average value of ACC was about 4.7, and this indicator ranges between 0 and 10. The highest ACC score (10) was observed in the sub-watershed R2-1-int, followed by the sub-watershed R2-2-int (9).

Regarding the *knowledge management and information quality system (KM)* indicator, the existence of an integrated virtual system in the Natural Resources and Environment Departments to respond to requests without the need for in-person visits, and the lack of public access to maps, location of plans, and other important data and information by the public and experts in the study area, led to a score of 5 out of 10. To improve environmental performance, people's participation and transparent governance that help control corruption in the environment and natural resources issues are crucial [51]. Accordingly, *corruption control (CC)* and its ranking were obtained using information provided by Transparency International (https://www.transparency.org/), which ranked Iran at 25 out of 100, ranking 147 out of 180 countries. In other words, the CC value for Iran and, as a result, in the Razey watershed, is 2.5.

## 4. Discussion

Comparative analysis of land use type and elevation (Table 1), with GWI results, also showed that the sub-watershed R1, which is allocated the highest value of the GWI (5.13), had covered 37.4, 944.8, 136.8, 17, and 10.3 ha of rangeland, rainfed agriculture, rocky land, orchard, and residential areas, respectively. In addition, its mean elevation is about 2129.4 m, which is the maximum value among all sub-watersheds. Indeed, this watershed ranked third in terms of residential areas, rainfed agriculture, and rocky land. The sub-watershed R2-2-2 has the lowest value of the GWI (0.25), situated among other sub-watersheds with the lowest residential areas, rainfed agriculture, and orchard areas. The mean elevation of this sub-watershed is also obtained as 1714.98 m.

Different gray and blue infrastructures were adapted for assessing the sustainability and performance of various ecosystems [26,27,52]. The gray infrastructure refers to traditional, engineered structures like piped water systems, public transportation, and buildings. Furthermore, blue infrastructure focuses on water features, both natural (rivers, lakes, and wetlands) and engineered (drainage systems and canals). Finally, the green infrastructure encompasses natural land or plant-based features like trees, parks, and green roofs. In essence, blue-green infrastructure leverages natural and engineered systems to manage water and other environmental aspects, while gray infrastructure represents the traditional built environment [24,25,41,53].

In this study area, the mean and standard deviation of the *infrastructure maintenance (IM)* score were 3.58 and 2.22, respectively. The highest score was obtained in the sub-watershed R2-1-int located in Rahim Bigloi and Razey villages. This sub-watershed has more groundwater wells and rivers. It is also in very good condition in terms of gray infrastructure, so that in terms of gas station, kerosene, gas, water piping, health, and school conditions, it received the maximum score, and in terms of public transportation and asphalt road, it scored one out of two. In return, sub-watersheds R2-1-2-2-1, R2-2-1, and R2-2-2 had the lowest score. These results indicate the prioritization of sub-watersheds for planning and adopting suitable management and protection strategies for resiliency from environmental stresses. It can be concluded that the sub-watersheds with the highest IM scores are more resilient to various environmental stresses. In this regard, Rezaei Ghaleh et al. [54] reported the importance of green-blue infrastructure management in improving the people's health of Qazvin city (Iran) in dealing with climate change. There was an inverse relationship between the per capita green-blue infrastructure and the number of emergency calls regarding respiratory, cardiac, and stroke patients. Therefore, the watershed had more green-blue infrastructure and healthier people. Benedict and McMahon [24] also proposed that green infrastructures are linked to gray infrastructures, which consider a watershed as a unit at a specific time and scale. It is widely accepted that green infrastructure can effectively reduce combined sewer overflow incidents by increasing infiltration [25]. Water resulting from diminished runoff can be captured in the infrastructure and reintroduced into the watershed during dry periods, thereby supporting an ecological base flow to maintain ecosystem services [55].

Specifically, the *human capital (HC)* indicator with a score of 0.84 and *environmental productivity (EP)* with a score of 0.44 have the highest impact on the GWI. Among the 17 indicators studied, six indicators, including *infrastructure maintenance (IM)*, *effectiveness in public water management (EPWM)*, *water conservation (WC)*, *Simpson's biodiversity index (BD)*, *environmental productivity (EP)*, and *adaptation to climate change (ACC)*, are significantly affected by green infrastructure. Furthermore, five indicators, including *waste recycling (R)*, *human capital (HC), population participation (PP), knowledge management system (KM),* and *corruption control (CC),* are affected by gray infrastructure, and the rest are affected by both green and gray infrastructure.

The total score of the three sub-indicators (water production, water quality, and sewage system coverage) that led to the production of the *effectiveness in public water management (EPWM)* indicator varies between 2 and 8.5. The sub-watershed R3-int showed the lowest score, while sub-watersheds R2-1-2-2-1, R2-2-3, and R3-2 presented the highest scores. Maleki et al. [56] showed that, among the 31 investigated provinces, the water security index of Ardabil Province ranked 12th with a value of 0.101 in 2011 and 7th with a value of 0.119 in 2016. Despite the fact that a very slight improvement has been achieved during the period, the obtained average indicates the poor state of water resource management in the province and even in the whole country.

In terms of the *water vulnerability (WV), water stress (WS),* and *water self-sufficiency (WSS)* indicators, the least populated sub-watersheds and, as a result, the controlled water consumption or native "virgin" rangelands obtain higher values in the study area. Meanwhile, the sub-watershed R1 gains lower values than other sub-watersheds, due to the larger population and excessive water consumption. As a result, water planning, development, and management are needed to ensure sustainable supplies and meet growing demands for both humans and natural ecosystems [43] in the study area, especially in the sub-watershed R1 (located in Razey City).

The mean *water retention index (WRI)* in the study area reflects its unfavorable capacity in terms of water regulation and water quantity management in the soil. Vandecasteele et al. [40] indicated that the WRI is a composite measure that considers factors related to the potential for water retention in soil, plants, water bodies, and underlying aquifers, along with the effects of slope and artificially sealed surfaces. In the same context, Crossman et al. [57] stated that water regulation refers to how land cover influences the runoff and discharge of rivers. In Ardabil Province, Ghabelnezam et al. [46] prepared a water retention map for all watersheds of the province and reported the minimum and maximum water retention in residential and forest land uses, respectively. This study provides a dynamic assessment and warning of the risk of future degradation of water retention, which can be valuable and applicable to other regions.

The average daily consumption of fresh water in Razey City, with a population of 1581, is 602 $m^3$. On the other hand, the only industrial town in the watershed has an average consumption of 2004 $m^3$ over two months, resulting in a daily consumption of 33.4 $m^3$. According to a survey conducted throughout the watershed, it was found that the piped water is mainly used for domestic consumption and is not used for industrial production, agriculture, or animal husbandry except in exceptional and rare cases. Therefore, the only water-bearing industrial and production unit is Razey Industrial Town, which is small in comparison to the production of water in the watershed, so the ratio of water consumption in production is low for all watersheds, and the score of the *water footprint (WF)* for all sub-watersheds is 8.

The results of the *water-saving and efficient use programs (WSE)* index show that the study area is characterized by a very low mean value ($WSE = 1$). In Razey City and its nearby villages, the indices are better than in other sub-watersheds, and sub-watershed R1 has the maximum score. In remote areas and sub-watersheds located on the border of Iran and Azerbaijan (i.e., R2-1-2-2-1), due to the lack of necessary equipment and facilities, the score is naturally low, and paying attention to culture and low-cost advertising can help improve the situation.

Based on the results, *Simpson's biodiversity index (BD)* was significantly higher in upstream areas (i.e., sub-watersheds R2-int and R2-2-int) than downstream areas (i.e., sub-watersheds R2-1-2-2-1, R1, and R3-2). These spatial variations are related to the distribution of patches of land use. The Simpson's biodiversity index (BD) is particularly sensitive to the most dominant patches in a landscape. A homogeneity value of one indicates a highly diverse landscape, whereas values approaching zero reflect a decline in landscape diversity [58]. Malvandi et al. [47] also used diversity indices to assess water quality in eastern Iran. They found that diversity indices such as Simpson's diminished from upstream to downstream, which corresponded with a decline in water quality.

Results of *environmental productivity (EP)* show very low mean values, and most sub-watersheds have zero values, which shows the unstable situation. The lack of implementation of plant mulching and seeding in sub-watersheds R2-2-1, R2-2-2, and R2-1-2-1, which are located upstream, is quite evident in the instability of this watershed. In this regard, Wang et al. [33] and Cegielska et al. [28] stated that vegetation and green spaces provide economic and health advantages, support mitigating and adapting to climate change, help counteract the negative environmental and social impacts of development, and enhance both living standards and environmental quality.

Regarding *waste recycling (R)*, there are no records of practical actions in the study area, leading to a highly unstable condition in this aspect. However, efficient approaches for sustainably reducing waste facilitate sustainable resource management by enhancing the availability of clean water, decreasing the pressure on natural resources, recovering energy, and supporting agriculture [48].

For *landscape attractiveness (LA)*, the population growth rate in the sub-watersheds was compared. In the sub-watershed R2-1-1 only, there was a very slight increase in population, and in other sub-watersheds, a population decrease was observed. This watershed is unstable in this regard and needs basic and comprehensive management decisions.

*Human capital (HC)*, or the quality and knowledge of the workforce, is the main driver of innovation and economic growth [59]. So, if capital, raw materials, and other physical factors are available, it is the HC that shapes them. Regarding the repair and maintenance of water supply equipment, HC is the main factor in the Razey sub-watershed, which is managed by contracting and is only addressed and reviewed by administrative staff in the city of Razey itself. The maximum score is related to the sub-watershed R2-1-int, which is higher than the average of 4.4 of the entire watershed, due to the participation of the contracting and administrative parts in the area. Based on the field survey, it was found that several upstream villages are not covered by services at all. For this reason, sending specialized personnel to these areas, shaping a basis for the integrated development of all areas, can be provided.

*Population participation (PP)* in watershed management programs is an important strategy for making watershed programs successful and is considered a key element for effective long-term monitoring efforts [60]. The data in Table 5 show that the overall extent of PP in watershed development programs in the study area exhibited a low level, with a calculated mean PP equal to 3. This shows that the population contributed less to interventions implemented during watershed development programs. Surveys show that planning in sparsely populated areas such as Qasim-Kandi and Post-Kandi villages and proper cultural development increase the rate of participation.

In assessing the *adaptation to climate change (ACC)* indicator, the level of measures to reduce the effects of drought and flood carried out in the sub-watershed was investigated. Despite this, several previous studies, especially in arid and semi-arid regions [50,61], have shown that ACC has a direct relationship with rural livelihoods. Furthermore, creating new businesses that are less dependent on nature has been proposed as a new approach to ACC, reducing vulnerability and improving livelihoods in the region [49]. In the study area, measures to improve people's livelihood are not satisfactory. A series of flood protection measures have been carried out. Specifically, the construction of mortared stones and cobblestone structures, as well as the construction and modification of canals, water supply, production systems, and flood protection systems, were investigated. The maximum score of 10 compared to the mean score of 4.65 indicates a large difference in investment and management outlook between sub-watersheds. Based on the research conducted, political tastes play an important role in the promotion or decline of this index.

Finally, the values of *knowledge management and information quality system (KM)* and *corruption control (CC)* indicators were, respectively, 5 and 2.5 for all sub-watersheds. Both indicators are in completely unstable conditions. However, taking a knowledge-based view of the watershed management programs can provide more successful management practices. Furthermore, low corruption control, with the value of 2.5, is a driver of inefficient water governance [62] and can be a barrier to sustainable watershed management strategies. Widespread corruption in governance not only leads to inefficiency and reduced quality of public services but also negatively impacts social capital, erodes public trust, and ultimately weakens citizen participation by diminishing confidence in governing institutions. Moeinifar et al. [63] created a model for anti-corruption policy in Iran and discovered that administrative corruption, lack of transparency, and exclusion of public participation had fostered an unstable and inefficient management system in terms of a causal perspective. Corruption both drives and stems from the weaknesses of formal institutions. The findings of Taghva et al. [64] revealed that widespread legal violations are often facilitated by the corrupt actions of current and former officials. In response, many farmers have internalized these corrupt practices as a means to gain access to water, exacerbating social conflict and obstructing effective planning and water governance. This persistent dysfunction continues to degrade the watershed's ecological integrity [62].

The Razey watershed in Ardabil Province is a vital area for soil and water conservation initiatives, particularly due to its limited size and increasing pressures from land use changes. Our focus on utilizing the Green Watershed Index (GWI) framework to evaluate ecosystem sustainability, especially at the sub-watershed level, addresses a significant gap in current research. While previous studies [3,37] have mainly explored how land use changes—including deforestation and urban growth—affect soil erosion and flood risks, this approach to employing a sustainability assessment like the GWI offers a more holistic view of the watershed's ecological condition and its capacity to maintain resilience.

To improve the effectiveness of green management strategies, establishing regional priorities is crucial for making informed decisions regarding resource allocation. By ranking different sub-watersheds based on the severity of environmental issues, managers gain valuable insights that help streamline interventions. This approach not only benefits watershed managers by enabling more efficient and cost-effective solutions but also aids government agencies in developing targeted regional development plans. Prioritization results can guide budgetary agencies in distributing funds among departments involved in watershed management and environmental protection. Similarly, research funding organizations can use these rankings to identify sub-watersheds that require more or less research attention. Prioritizing sub-watersheds from various environmental perspectives supports smarter, more strategic green management and is a widely recognized practice in related studies.

## 5. Conclusions

This study established an integrated Green Watershed Index (GWI) using 17 indicators, which were calculated based on water resources, biodiversity, environmental sustainability, and social well-being data that were collected from field survey, remote sensing (RS), and a geographical information system (GIS) in the Razey watershed in a cold semi-arid region of Iran. The spatial dynamics of the calculated indicators were also well analyzed. It was found that the minimum and maximum values of the GWIs were equal to 0.25 and 5.13. Furthermore, the standard deviation is equal to 1.42, which indicates relatively moderate spatial variability. In conclusion, GWI values, along with its assessment indicators and criteria, vary significantly across different locations and over time. As such, continuous and systematic monitoring of the watershed's green status is essential to ensure accurate evaluation and informed decision-making. Given the innovative nature of this research and the practical relevance of the selected indicators, further studies are recommended to calculate the GWI for additional watersheds across Iran and other countries. Such investigations could contribute to the creation of a comprehensive GWI atlas, allow for testing various methods of indicator integration and weighting for sensitivity analysis, assess the uncertainty associated with GWI results, and support the development of detailed management guidelines. Finally, using remotely sensed data and products of Google Earth Engine could be very helpful and widely applicable to define more applicable indicators, especially for large regions, for monitoring spatio-temporal changes. In addition, it is highly encouraged to facilitate equipping the watershed for collecting high-resolution data associated with green representative indicators regarding the accuracy assessment of GWI results.

**Author Contributions:** Conceptualization, Z.H. and R.M.; methodology, Z.H.; software, A.I.; validation, R.J. and Z.H.; formal analysis, A.I.; investigation, R.J. and Z.H.; resources, R.J.; data curation, Z.H.; writing—original draft preparation, R.J.; writing—review and editing, Z.H., R.M. and A.E.O.; visualization, R.J.; supervision, R.J.; project administration, R.J. All authors have read and agreed to the published version of the manuscript.

**Funding:** This research received no external funding.

**Institutional Review Board Statement:** Not applicable.

**Informed Consent Statement:** Not applicable.

**Data Availability Statement:** The data are available on request.

**Acknowledgments:** This research was supported by the University of Mohaghegh Ardabili, Ardabil, Iran.

**Conflicts of Interest:** The authors declare no conflicts of interest.

## References

1. Zakeri, M.A.; Mirnia, S.K.; Moradi, H.R. Assessment of water security in the large watersheds of Iran. *Environ. Sci. Policy.* **2022**, *127*, 31–37. [CrossRef]
2. Afshar Bakeshlo, Z.; Omidvar, M.; Gigauri, I. Sustainable Path and Green Leadership: Navigating the Challenges of Environmental Sustainability in Iran. In *Navigating Corporate Social Responsibility Through Leadership and Sustainable Entrepreneurship*; Gigauri, I., A Khan, A., Eds.; IGI Global Scientific Publishing: Hershey, PA, USA, 2025; pp. 257–278. [CrossRef]
3. Abedini, M.; Sabouri, H.; Pasban, A. Flood Hazard Zoning and Its Relationship with Land Use Using the Analytic Network Process Model (Case Study: Razi Chay Watershed, Ardabil Province). *J. Sustain. Urban Reg. Dev. Stud.* **2025**, *6*, 68–84. Available online: https://www.srds.ir/article_214387_en.html (accessed on 26 May 2025). (In Persian)
4. Shamaee, S.H.; Yousefi, H.; Zahedi, R. Assessing urban development indicators for environmental sustainability. *Discov. Sustain* **2024**, *5*, 341. [CrossRef]
5. Amghani, M.S.; Sabouri, M.S.; Baghernejad, J.; Norozi, A. Factors affecting the livelihood sustainability of smallholder farmers in Iran. *Environ. Sustain. Indic.* **2025**, *26*, 100601. [CrossRef]
6. Tajbakhshian, M. Groundwater quality and hydrogeochemical challenges in the Sarakhs Plain, NE Iran: A call for sustainable management. *Environ. Geochem. Health.* **2025**, *47*, 75. [CrossRef] [PubMed]
7. Heshmati, S.; Hafezparast, M.; Mastali, F. Productivity and sustainability index evaluation for Harsin dam. *J. Rainwater Catchment Syst.* **2019**, *7*, 53–63. (In Persian)

8. Saeedi, I.; Mikaeili Tabrizi, A.R.; Bahremand, A.; Salmanmahiny, A. A soft systems methodology and interpretive structural modeling framework for green infrastructure development to control runoff in Tehran metropolis. *Nat. Resour. Model.* **2022**, *35*, e12339. [CrossRef]
9. He, D.; Hu, J.; Zhang, J. Assessment of sustainable development suitability in linear cultural heritage—A case of Beijing Great Wall Cultural Belt. *Land* **2023**, *12*, 1761. [CrossRef]
10. Núñez-Razo, I.; de Anda, J.; Barrios-Piña, H.; Olvera-Vargas, L.A.; García-Ruíz-García, M.; Hernández-Morales, S. Development of a watershed sustainability index for the Santiago River Basin, Mexico. *Sustainability* **2023**, *15*, 8428. [CrossRef]
11. Mocuta, D.; Stoian, E. Interdependence between sustainable development and human health. In *Scientific Papers. Series "Management, Economic Engineering in Agriculture and Rural Development"*; U.S.A.M.V.: Bucharest, Romania, 2013; Volume 13, pp. 241–246, ISSN 2284-7995.
12. Flotemersch, J.; Leibowitz, S.; Hill, R.; Stoddard, J.; Thoms, M.; Tharme, R. A watershed integrity definition and assessment approach to support strategic management of watersheds. *River Res. Appl.* **2016**, *32*, 1654–1671. [CrossRef]
13. Naiman, R.J. New Perspectives for Watershed Management: Balancing Long-Term Sustainability with Cumulative Environmental Change. In *Watershed Management*; Naiman, R.J., Ed.; Springer: New York, NY, USA, 1992. [CrossRef]
14. Kateb, F.; Ouri, A.E.; Mostafazadeh, R.; Sharari, M.; Hazbavi, Z. Index-based assessment of stakeholder perception for participatory sustainable soil and water resources in a mountainous landscape. *Eval. Program Plan.* **2025**, *112*, 102643. [CrossRef]
15. Jarzebski, M.P.; Karthe, D.; Chapagain, S.K.; Setiawati, M.D.; Wadumestrige Dona, C.G.; Pu, J.; Fukushi, K. Comparative Analysis of Water Sustainability Indices: A Systematic Review. *Water* **2024**, *16*, 961. [CrossRef]
16. Nugroho, H.Y.S.H.; Purwanto, P.; Nandini, R.; Adi, R.N.; Savitri, E.; Putra, P.B.; Indrawati, D.R.; Nada, F.M.H.; Nugroho, A.W. Building Adaptive and Resilient Community to Climate Change Through Micro Watershed Management Strategies to Support the Achievement of Village SDGs. In *Climate Change: Conflict and Resilience in the Age of Anthropocene*; Pal, S.C., Chatterjee, U., Saha, A., Ruidas, D., Eds.; Advances in Global Change Research; Springer: Cham, Switzerland, 2025; Volume 80. [CrossRef]
17. Eurostat. Water Exploitation Index in Europe, 2016, 310. Available online: http://maps.eea.europa.eu/EEAViewer/?appid=eaf7bb088ba-548368c6e102c4a876c13 (accessed on 1 June 2023).
18. Dening, S. *European Green City Index*; Siemens AG: Munich, Germany, 2009.
19. Juwana, I.; Perera, B.J.; Muttil, N. A water sustainability index for West Java. Part 1: Developing the conceptual framework. *Water Sci. Technol.* **2010**, *62*, 1629–1640. [CrossRef]
20. Chaves, H.M.L.; Alipaz, S. An Integrated Indicator Based on Basin Hydrology, Environment, Life, and Policy: The Watershed Sustainability Index. *Water Resour. Manage* **2007**, *21*, 883–895. [CrossRef]
21. Emerson, J.; Esty, D.C.; Levy, M.A.; Kim, C.H.; Mara, V.; de Sherbinin, A.; Srebotnjak, T. *2010 Environmental Performance Index*; Yale Center for Environmental Law and Policy: New Haven, CT, USA, 2010. Available online: http://epi.yale.edu/ (accessed on 26 May 2025).
22. Mititelu-Ionuş, O. Watershed Sustainability Index Development and Application: Case Study of the Motru River in Romania. *Pol. J. Environ. Stud.* **2017**, *26*, 2095–2105. [CrossRef]
23. Rendrarpoetri, B.L.; Rustiadi, E.; Fauzi, A.; Pravitasari, A.E. Sustainability Assessment of the Upstream Bengawan Solo Watershed in Wonogiri Regency, Central Java Province, Indonesia. *Sustainability* **2024**, *16*, 1982. [CrossRef]
24. Benedict, M.A.; McMahon, E.T. Green infrastructure: Smart Conservation for the 21st Century. *Renew. Resour. J.* **2002**, *20*, 12–17. Available online: http://www.sprawlwatch.org/greeninfrastructure.pdf (accessed on 2 June 2023).
25. Buckley, M. *Economic Benefits of Green Infrastructure: Chesapeake Bay Region*; ECONorthwest: Portland, OR, USA, 2011.
26. Caro, C.A.; Bladé, E. Water Resources Management: Green Watershed Index (GWI). In Proceedings of the IOP Conference Series: Earth and Environmental Science, Volume 690, 2020 International Symposium on Water, Ecology and Environment, Beijing, China, 6–8 December 2020.
27. Abdel-Basset, M.; Gamal, A.; Chakrabortty, R.K.; Ryan, M.; El-Saber, N. A Comprehensive Framework for Evaluating Sustainable Green Building Indicators under an Uncertain Environment. *Sustainability* **2021**, *13*, 6243. [CrossRef]
28. Cegielska, K.; Kukulska-Kozieł, A.; Hernik, J. Green Neighbourhood Sustainability Index–A measure of the balance between anthropogenic pressure and ecological relevance. *Ecol. Indic.* **2024**, *160*, 111815. [CrossRef]
29. Hall, B.; Kerr, M.L. *1991–1992 Green Index: A State-by-State Guide to the Nation's Environmental Health*, 1st ed.; Island Press: Washington, DC, USA, 1991; 170p.
30. Covv, C.; Halstead, T.; Rowe, J. If the GDP is Up, why is America Down? *Atl. Mon.* **1995**, *276*, 5978.
31. Lawn, P.A. A theoretical foundation to support the Index of Sustainable Economic Welfare (ISEW), Genuine Progress Indicator (GPI), and other related indexes. *Ecol. Econ.* **2003**, *44*, 105–118. [CrossRef]
32. Lawn, P.A. An assessment of the valuation methods used to calculate the index of sustainable economic welfare (ISEW), genuine progress indicator (GPI), and sustainable net benefit index (SNBI). *Environ. Dev. Sustain.* **2005**, *7*, 185–208. [CrossRef]
33. Wang, M.X.; Zhao, H.H.; Cui, J.X.; Fan, D.; Lv, B.; Wang, G.; Li, Z.H.; Zhou, G.J. Evaluating green development level of nine cities within the Pearl River Delta, China. *J. Clean. Prod.* **2018**, *174*, 315–323. [CrossRef]

34. Ahmadov, E. Water resources management to achieve sustainable development in Azerbaijan. *Sustain. Futures* **2020**, *2*, 10003C [CrossRef]
35. Hazbavi, Z.; Sadeghi, S.H.; Gholamalifard, M.; Davudirad, A.A. Watershed health assessment using the pressure–state–response (PSR) framework. *Land Degrad. Dev.* **2020**, *31*, 3–19. [CrossRef]
36. Alaei, N.; Mostafazadeh, R.; Esmali Ouri, A.; Hazbavi, Z.; Sharari, M.; Huang, G. Spatial comparative analysis of landscape fragmentation metrics in a watershed with diverse land uses in Iran. *Sustainability* **2022**, *14*, 14876. [CrossRef]
37. Azizi, E.; Nikoo, M.R.; Mostafazadeh, R.; Hazbavi, Z. Flood vulnerability analysis using different aggregation frameworks across watersheds of Ardabil province, northwestern Iran. *Int. J. Disaster Risk Reduct.* **2023**, *91*, 103680. [CrossRef]
38. Dickens, C.; McCartney, M.; Tickner, D.; Harrison, I.J.; Pacheco, P.; Ndhlovu, B. Evaluating the global state of ecosystems and natural resources: Within and beyond the SDGs. *Sustainability* **2020**, *12*, 7381. [CrossRef]
39. Ali, A. Monitoring, Evaluation and Learning of Natural Resources Management Projects, Part II. Project "Strengthening Natural Resources Management Capacities to Revitalize Agriculture in Fragile Contexts". E-nugget. 2020. Available online: https://www.fao.org/3/cb2139en/cb2139en.pdf (accessed on 12 May 2023).
40. Vandecasteele, I.; Mari Rivero, I.; Baranzelli, C.; Becker, W.; Dreoni, I.; Lavalle, C.; Batelaan, O. The Water Retention Index: Using land use planning to manage water resources in Europe. *Sustain. Dev.* **2018**, *26*, 122–131. [CrossRef]
41. Alves, A.; Gersonius, B.; Kapelan, Z.; Vojinovic, Z.; Sanchez, A. Assessing the Co-Benefits of green-blue-grey infrastructure for sustainable urban flood risk management. *J. Environ. Manag.* **2019**, *239*, 244–254. [CrossRef]
42. Hogeboom, R.J. The water footprint concept and water's grand environmental challenges. *One Earth* **2020**, *2*, 218–222. [CrossRef]
43. Loucks, D.P.; van Beek, E. Water Resources Planning and Management: An Overview. In *Water Resource Systems Planning and Management*; Springer: Cham, Switzerland, 2017. [CrossRef]
44. Dalavi, P.; Bhakar, S.R.; Rajput, J.; Gaddikeri, V.; Tiwari, R.K.; Shukla, A.; Vishwakarma, D.K. Modeling runoff in Bhima River catchment, India: A comparison of artificial neural networks and empirical models. *Water Pract. Technol.* **2024**, *19*, 2595–2612 [CrossRef]
45. Coutagine, J. A new method of determining the runoff coefficient of small catchments. *J. Hydrol. Hydromech.* **1967**, *5*, 94–108.
46. Ghabelnezam, E.; Mostafazadeh, R.; Esmali Ouri, A.; Hazbavi, Z. The Importance of Watershed Ecosystem Services with Emphasis on Runoff yield and Erosion Control. *Hum. Environ.* **2022**, *62*, 137–155. Available online: https://sanad.iau.ir/en/Article/847705 (accessed on 20 May 2023). (In Persian)
47. Malvandi, H.; Moghanizade, R.; Abdoli, A. The use of biological indices and diversity indices to evaluate water quality of rivers in Mashhad, Iran. *Biologia* **2021**, *76*, 959–971. [CrossRef]
48. Silva, J.A. Wastewater Treatment and Reuse for Sustainable Water Resources Management: A Systematic Literature Review. *Sustainability* **2023**, *15*, 10940. [CrossRef]
49. Tohidimoghadam, A.; PourSaeed, A.; Bijani, M.; Eshraghi Samani, R. Rural sustainable livelihood resilience to climate change: A strategic analysis. *Environ. Sustain. Indic.* **2023**, *20*, 100292. [CrossRef]
50. Ullah, A.; Bavorova, M.; Ahmad Shah, A.; Kandel, G.P. Climate change and rural livelihoods: The potential of extension programs for sustainable development. *Sustain. Dev.* **2024**, *32*, 4992–5004. [CrossRef]
51. Rodríguez-Martínez, C.C.; García-Sánchez, I.M.; Vicente-Galindo, P.; Galindo-Villardón, P. Exploring relationships between environmental performance, e-government and corruption: A multivariate perspective. *Sustainability* **2019**, *11*, 6497. [CrossRef]
52. Lu, P.; Sun, Y.; Steffen, N. Scenario-based performance assessment of green-grey-blue infrastructure for flood-resilient spatial solution: A case study of Pazhou, Guangzhou, greater Bay area. *Landsc. Urban Plan.* **2023**, *238*, 104804. [CrossRef]
53. Ncube, S.; Arthur, S. Influence of Blue-Green and Grey Infrastructure Combinations on Natural and Human-Derived Capital in Urban Drainage Planning. *Sustainability* **2021**, *13*, 2571. [CrossRef]
54. Rezaei Ghaleh, M.; Hagh Parast, F.; Maleki, A. Investigating the correlation between blue-green infrastructure and reduction of heat-related health effect under climate change (Case study: Qazvin City). *Bagh-e Nazar.* **2022**, *19*, 69–84. (In Persian) [CrossRef]
55. Risch, E.; Gasperi, J.; Gromaire, M.C.; Chebbo, G.; Azimi, S.; Rocher, V.; Roux, P.; Rosenbaum, R.K.; Sinfort, C. Impacts from urban water systems on receiving waters–how to account for severe wet-weather events in LCA? *Water Res.* **2018**, *128*, 412–423. [CrossRef] [PubMed]
56. Maleki, N.; Shakeri Bostanabad, R.; Salehi komroudi, M.; Seiedabadi, S. Investigating the status of the combined water security index of Iranian Provinces in the Period of 2012-2017: Application of Multi-Criteria Analysis Methods. *J. Water Sustain. Dev.* **2021**, *8*, 21–32. (In Persian) [CrossRef]
57. Crossman, N.D.; Nedkov, S.; Brander, L. Discussion Paper 7: Water Flow Regulation for Mitigating River and Coastal Flooding. Paper Submitted to the Expert Meeting on Advancing the Measurement of Ecosystem Services for Ecosystem Accounting, New York, 2019 and Subsequently Revised. Version of 1 April 2019. Available online: https://seea.un.org/events/expert-meeting-advancing-measurement-ecosystem-services-ecosystem-accounting (accessed on 14 April 2023).
58. McGarigal, K.; Cushman, S.A.; Ene, E. *Spatial Pattern Analysis Program for Categorical and Continuous Maps, FRAGSTATS v4*; University of Massachusetts: Amherst, MA, USA, 2012.

9. Brodny, J.; Tutak, M. A multi-criteria measurement and assessment of human capital development in EU-27 countries: A 10-year perspective. *J. Open Innov. Technol. Mark. Complex.* **2024**, *10*, 100394. [CrossRef]
0. Zuniga-Teran, A.A.; Fisher, L.A.; Meixner, T.; Le Tourneau, F.-M.; Postillion, F. Stakeholder participation, indicators, assessment, and decision-making: Applying adaptive management at the watershed scale. *Environ. Monit. Assess.* **2022**, *194*, 156. [CrossRef]
1. Piya, L.; Maharjan, K.L.; Joshi, N.P. Climate Change and Rural Livelihoods in Developing Countries. In *Socio-Economic Issues of Climate Change*; Springer: Singapore, 2019. [CrossRef]
2. Lopez Porras, G.; Stringer, L.C.; Quinn, C.H. Corruption and conflicts as barriers to adaptive governance: Water governance in dryland systems in the Rio del Carmen watershed. *Sci. Total Environ.* **2019**, *10*, 519–530. [CrossRef]
3. Moeinifar, M.; Givarian, H.; Rabiei Mondajin, M.R.; Momeni, M. Indicators and Components for Developing a Model of Anti-Corruption Policy in Iran (With an Electronic Government Approach). *Int. J. Innov. Manag. Organ. Behav.* **2023**, *3*, 77–83. [CrossRef]
4. Taghva, M.R.; Masnavi, H.; Taghavifard, M.T.; Zarandi, S. E-Governance Development Model Towards Anti-Corruption in Iran. *Public Manag. Res.* **2023**, *16*, 163–190. (In Persian) [CrossRef]

*sustainability*

MDPI

*Article*

# Sedimentation Pattern as a Response to Hydrodynamics in a Near-Symmetric River Confluence

**João Nuno Fernandes [1,*] and Leila Alizadeh [2]**

1 National Laboratory for Civil Engineering, 1700-066 Lisbon, Portugal
2 Civil Engineering Research and Innovation for Sustainability, Instituto Superior Técnico, Lisbon University, 1049-001 Lisbon, Portugal
* Correspondence: jnfernandes@lnec.pt

**Abstract:** River confluences are dynamic zones where hydrodynamic interactions between tributary flows—varying in velocity, direction, and sediment concentration—can significantly alter hydro morphology. These changes feature substantial consequences for the stability of riverbanks, nearby hydraulic structures, and the surrounding environment. This paper investigates flow mechanisms and sediment dynamics in a symmetric 50° confluence through laboratory experiments on a scaled physical model of a real confluence located on Madeira Island, Portugal. Acoustic Doppler velocity measurements were used to analyze the hydrodynamic characteristics, while bathymetry was surveyed using an RGB sensor and the Structure from Motion technique. Sedimentation patterns were correlated with key flow zones within the confluence. This study highlights how variations in discharge and momentum ratios influence sediment distribution and morphology, potentially destabilizing riverbanks and contributing to sediment deposition and erosion patterns. Understanding these mechanisms is critical for improving the sustainable management of water resources and minimizing anthropogenic impacts on fluvial systems. The findings provide valuable insights for enhancing river resilience, protecting natural watercourses, and supporting sustainable development by promoting informed planning of hydraulic structures and sediment management strategies.

**Keywords:** river confluences; tributaries; mixing layers; hydromorphology; flow structures

## 1. Introduction

The interaction of two streams at fluvial confluences leads to significant transfers of mass and momentum, establishing these nodes as critical elements of the fluvial corridor. The interactions can profoundly influence downstream conditions by introducing water, sediment, wood, or ice into the main river system (cf. [1]). Differences in velocity, flow direction, and sediment concentration between tributaries create complex hydrodynamic and morphodynamic processes in the downstream reach. Due to the inherent complexity of these processes, scientific investigations into river channel confluences have evolved gradually over time [2].

The flow mechanisms in fluvial confluences are primarily governed by the mixing of tributary flows. Mosley [3] distinguished between symmetrical confluences, where tributaries merge to form a new downstream channel, and asymmetrical confluence, where a tributary joins a main channel laterally. In symmetrical confluences, the main flow mechanisms, as illustrated in Figure 1, include flow stagnation immediately downstream of the junction, deflection of the tributary flow, flow separation zones, shear layers, maximum velocity regions, gradual flow recovery, and secondary currents (see, for example, [3,4]).



https://doi.org/10.3390/su17093790

These mechanisms play a critical role in determining the spatial and temporal distribution of flow velocities and sediment transport, influencing the long-term morphology and stability of the confluence zone.

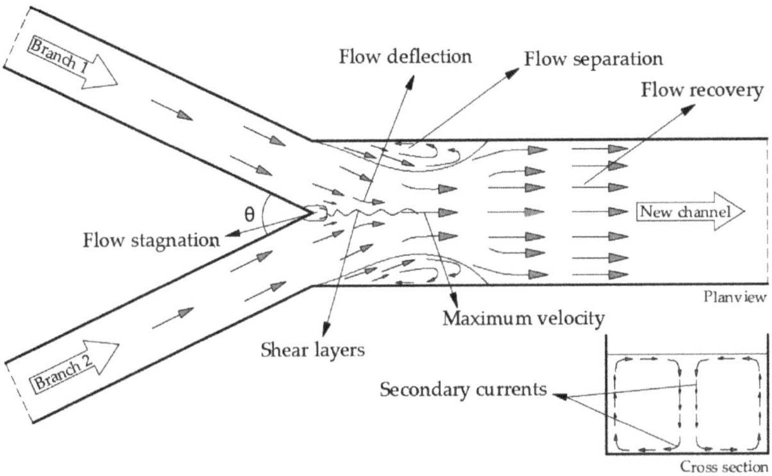


**Figure 1.** Main flow mechanisms in symmetrical confluences (adapted from [5]).

The geometry of a confluence significantly influences flow structures, with factors such as confluence angle (θ), momentum and discharge ratios, and symmetry playing key roles [5]. Increased confluence angles have been linked to enhanced flow stagnation and separation [6]. While most flow studies use in situ or laboratory methods, Bradbrook et al. [7] employed numerical approaches to show the effects of confluence angles and tributary turbulence on downstream flow. Similarly, Nazari-Sharabian et al. [8] developed numerical guidelines for managing flooding at a submerged drainage confluence. Confluences with bed discordance distort mixing layers as deeper channel flows laterally entrain into the separation zones [9]. Downstream of such confluences, secondary cells form between jet flows and inner banks.

Understanding the influence of flow dynamics on sedimentation is essential for effective river management, as erosion affects riverbank stability, vegetation, and sediment deposition [10]. While clear water flow mechanisms are well-studied, sediment transport processes at confluences remain less explored [11]. Guillén-Ludeña et al. [12,13] studied hydro-morphological processes in confluences where narrow, steep tributaries join wide, low-gradient main channels. They observed key discordant features, such as a so-called avalanche face at the tributary mouth and a bank-attached bar in the post-confluence zone, highlighting areas of significant sediment deposition and erosion. In mountain river confluences, junction angle and discharge ratio were identified as critical factors influencing hydrodynamics and morphology. High sediment loads from tributaries contribute to bed discordance, affecting flow, habitat, flood risk, and sediment management.

For steep tributaries with high sediment loads, Leite Ribeiro et al. [14] identified significant bed discordance due to contrasting flow regimes: tributary flows occupy upper layers, while the main channel dominates the lower layers, shaping downstream sedimentation patterns.

The impact of shear layer distortion on turbulent flow structures in asymmetric channel confluences was demonstrated by Yuan et al. [15], who found stronger helical cells formed when tributary flow rates exceeded those of the main channel. Key turbulence characteristics—such as Reynolds shear stress, maximum turbulent kinetic energy, and ejection and sweep events—were concentrated at mid-water depths, closely tied to shear layer distortion. Higher discharge ratios increased velocity, turbulence, and shear stress, while reduced discharge amplified shear layer distortion. These insights are essential for understanding sediment transport, as turbulent structures and shear stresses drive sediment entrainment, transport, and deposition.

The influence of sediment discharge on the morphology of a movable bed in an asymmetric open channel confluence was investigated by Bombar and Cardoso [16]. The authors observed that larger sediment inputs, associated with higher stream powers, caused significant morphological changes for a given water discharge ratio. Similarly, Lewis and Rhoads [17] measured real-world hydrodynamics, linking flow mechanisms to scour hole formation. Higher discharges produced pronounced scour near the margins, while lower discharges shifted scour holes toward the channel center. Helical currents and shear layers emerged as key influencing factors.

A defining feature of confluence flow mechanisms is shear layer distortion, caused by tributary flow penetrating the main channel [15]. This behavior distinguishes confluences from other sheared flows, such as compound channels (e.g., [18]). Zhang et al. [19] explored hydrodynamics and bed morphology at the confluence of the sediment-laden Yellow River and the clear Fen River, revealing elevated water surfaces in stagnation zones and lower elevations at the junction corner. The size and location of critical regions—such as stagnation zones, acceleration areas, shear layers, and flow recovery zones—were primarily shaped by momentum transfer ratios.

Nazari-Giglou et al. [20] examined the flow and geometrical conditions leading to sediment motion in channel confluences, emphasizing the significance of flow velocity, confluence angle, discharge and width ratios, and bed material properties. These factors play a vital role in predicting sediment transport and deposition patterns, which are fundamental for river engineering, habitat restoration, and sediment management practices.

This highlights the importance of sediment transport studies in confluences, as variations in sediment load and flow conditions can lead to substantial changes in channel morphology, affecting floodplain stability and infrastructure. Understanding these patterns is crucial, as the formation and evolution of scour holes can influence sediment routing, habitat structures, and the long-term stability of confluence zones.

This paper investigates the three-dimensional flow structure and turbulence field in a symmetric 50° confluence, aiming to establish links between these hydrodynamic features and observed riverbed hydromorphology. A non-distorted 1:60 scale physical model of the confluence was developed for experimental purposes. Following Alizadeh and Fernandes [21], controlled experiments were conducted to analyze the effects of tributary momentum and width ratios on hydrodynamics and sediment routing. Velocity fields for five flow cases with varying tributary momentum and width ratios were measured using acoustic Doppler velocimetry. These high-frequency measurements allowed the characterization of the turbulent statistics. Additionally, sedimentation patterns downstream of the confluence were characterized through point measurements, infrared imaging, and photogrammetry. These findings are critical for managing sediment transport in natural and engineered river systems, supporting aquatic habitat stability, and improving sediment management strategies.

The scientific gaps addressed in this study primarily relate to the limited understanding of sediment transport dynamics and hydrodynamic interactions in symmetric river confluences. While previous research has extensively investigated flow structures and turbulence in natural and experimental settings, sediment transport mechanisms remain insufficiently explored, particularly in controlled laboratory environments. Many studies have focused on clear-water hydrodynamics, often neglecting the complex interplay between flow structures, sediment entrainment, and deposition. Additionally, existing research has predominantly examined asymmetric confluences, leaving a gap in knowledge regarding sedimentation patterns in (near-)symmetric confluences, where tributary momentum and discharge ratios play a crucial role in shaping riverbed morphology. The absence of high-resolution experimental studies has further constrained the development of analytical models for sediment transport in engineered and natural river systems.

Using acoustic Doppler velocimetry, photogrammetry, and RGB sensor-based bathymetric surveys, this study provides high-resolution data on sedimentation patterns and hydrodynamic interactions. The findings reveal the impact of tributary momentum and discharge ratios on sediment deposition, scour formation, and shear layer development. These insights enhance the understanding of sediment transport processes in confluences, offering valuable guidance for sustainable river management, flood risk mitigation, and hydraulic structures management. This study also contributes to refining numerical models by providing empirical data that can improve the accuracy of sediment transport simulations in river confluences.

## 2. Materials and Methods

### 2.1. Site and Scaled Model

The experiments were conducted in a model of a symmetric confluence located in Funchal, Madeira Island, Portugal. João Gomes and Santa Luzia Creeks are the two tributaries that join in a confluence at coordinates 32°38′49.7″ N and 16°54′17.9″ W and flow directly to the Atlantic Ocean, as presented in Figure 2a,c.

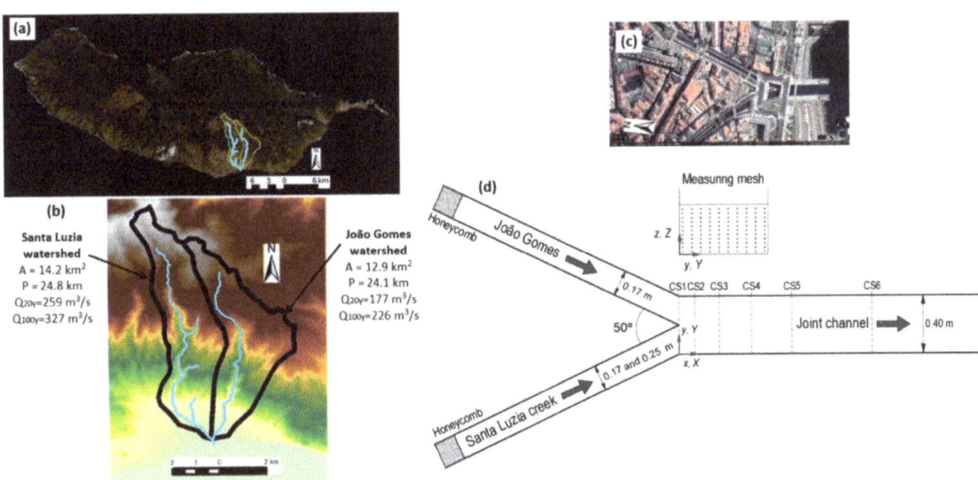


**Figure 2.** (**a**) Location of the watersheds on Madeira Island; (**b**) hipsometry and watershed characteristics; (**c**) Google aerial view; (**d**) schematic representation of the confluence flume.

The watersheds of these creeks are presented in the hypsometric map in Figure 2b. They have a similar area and shape, and as the soil characteristics are also approximately the same, previous studies showed that the sediment transport due to a 100-year return period flood are similar for both tributaries. Figure 2b also features the corresponding flood discharge flows for return periods of 20 and 100 years. The land use is predominantly forest and urbanization. The final reach comprises a confluence that was constructed in concrete as a regularized channel.

In order to study the hydrodynamic characteristics and sediment transport in this confluence, a 1:60 scale model was constructed at the National Laboratory for Civil Engineering (LNEC) in Lisbon, Portugal. For this open-channel model, Froude similarity was used due to the highly turbulent phenomena and the fact that energy dissipation depends mainly on turbulent shear stress terms [22].

In the flume, João Gomes and Santa Luzia creeks were designated as tributaries 1 and 2, respectively, with widths (b1 and b2) of 0.17 m and 0.25 m and heights of 0.25 m each. The joint channel is a concrete structure with a width (b3) of 0.40 m and a height of 0.15 m. Additional experiments were conducted by reducing the width of Tributary 2 to 0.17 m to analyze its influence during the experimental campaign, achieving a fully symmetrical configuration in this case. Both tributaries have a slope of 0.028 m/m, while the joint channel is horizontal. A schematic representation of the experimental recirculating hydraulic circuit is shown in Figure 2d.

The streamwise flow alignment was provided by a 3 cm diameter and 30 cm long circular honeycomb screen located upstream in each tributary. This screen improved the approach flow and the transition between the constant level reservoirs (not represented) and the tributary channels. In the downstream section, a flap tailgate imposes the subcritical flow depth.

### 2.2. Control Variables and Parameters

This experimental study aimed to characterize how channel width and momentum ratio influence flow structure and sediment transport in confluences. To achieve this goal, five flow cases, labeled FC1 to FC5, were conducted. Table 1 summarizes the key parameters of these experiments. Each flow case features a constant discharge in each tributary, denoted as $Q_1$ and $Q_2$. Additionally, the corresponding momentum transfer, calculated as $M = \rho\, QU$ (where $\rho$ is the volumetric mass density and $U$ is the cross-sectional averaged velocity), is provided.

**Table 1.** Characteristics of the flow cases (FCs).

| Flow Case | $b_1$ (m) | $b_{2.5}$ (m) | $Q_1$ (l s$^{-1}$) | $Q_2$ (l s$^{-1}$) | $Q_2/Q_1$ | $M_1$ (kg m s$^{-2}$) | $M_2$ (kg m s$^{-2}$) | $M_1/M_2$ | $Fr_1$ | $Fr_2$ | $Re_1$ ($\times 10^4$) | $Re_2$ ($\times 10^4$) | Sediment Discharge (l/s) | |
|---|---|---|---|---|---|---|---|---|---|---|---|---|---|---|
| | | | | | | | | | | | | | 1 | 2 |
| FC1 | 0.17 | 0.25 | 7 | 3 | 0.4 | 2.86 | 0.33 | 8.6 | 0.60 | 0.14 | 7.94 | 2.69 | 0.007 | 0.011 |
| FC2 | 0.17 | 0.25 | 3 | 7 | 2.3 | 0.49 | 1.79 | 0.23 | 0.24 | 0.34 | 3.29 | 6.29 | 0.007 | 0.011 |
| FC3 | 0.17 | 0.25 | 5 | 5 | 1 | 1.35 | 0.90 | 1.5 | 0.39 | 0.23 | 5.44 | 4.46 | 0.007 | 0.011 |
| FC4 | 0.17 | 0.17 | 5 | 5 | 1 | 1.32 | 1.34 | 0.98 | 0.38 | 0.39 | 5.36 | 5.41 | 0.007 | 0.011 |
| FC5 | 0.17 | 0.17 | 3 | 7 | 2.3 | 0.48 | 2.62 | 0.18 | 0.23 | 0.54 | 3.22 | 7.56 | 0.007 | 0.011 |

The table also includes the Froude number ($Fr = U/\sqrt{gh}$, where $g$ is the acceleration due to gravity and $h$ is the flow depth) and the Reynolds number ($Re = 4UR/\nu$, where $R$ is the hydraulic radius and $\nu$ is the kinematic viscosity). These parameters are crucial for understanding the dynamics of flow in the experimental setup and provide insights into the hydraulic conditions governing sediment transport and flow characteristics at the confluence.

Figure 2d shows the Cartesian orthogonal coordinate system (the $x$, $y$, $z$ and $u$, $v$, $w$ symbols were used for position and velocities in longitudinal, transverse, and vertical directions, respectively). The origin is located on the right side, downstream of the confluence. When relevant, the positions $x$, $y$, and $z$ were nondimensionalized by the downstream channel width, $B_m$. These normalized variables were named $X = x/B_m$, $Y = y/B_m$, and $Z = z/B_m$. As only one flow case features complete "symmetric" conditions, this confluence may be referred as "nearly-symmetric".

The sedimentation was characterized in additional experiments for the five flow cases. The sediment load was fixed according to the 100 years return period flood for tributaries 1 and 2, respectively. These sediments were steadily placed in the upstream section of each tributary during 1 h (according to Froude similarity). At the beginning of the experiment, the joint channel downstream of the tributaries was empty and without any sediment. This setup was designed to simulate actual conditions observed on Madeira Island.

Using a similar Shields parameter, these sediments were characterized by a mean diameter $d_{50}$ = 0.43 mm (corresponding to 3.2 mm in the prototype) and a bulk mass density of 1250 $kg/m^3$.

*2.3. Equipment and Measuring Meshes*

The discharge of each tributary was measured by two dedicated electromagnetic flowmeters with a precision equal to 0.01 l/s.

Water depths were measured as the difference between the water surface and bottom elevations, as given by an ultrasonic probe (UNDK 20I6903/S35A from Baumer, Frauenfeld, Switzerland).

A Vectrino Accoustic Doppler Velocimeter from Nortek, Sandvika, Norway (with an acquisition frequency of 100 Hz) was used to measure the 3 velocity components. The measurement uncertainty in velocity is $\pm$0.5% of the measured value or $\pm$1 $mm.s^{-1}$.

Measurement cross sections (CS1 to CS6) are presented in Figure 2d. Each one of these 6 cross sections comprised a total of 80 measuring points (10 equidistant verticals of 8 points each).

Following the procedures and the setup defined in Goring and Nikora [23] and Fernande and Jónatas [24], in each point, 9000 samples (90 s times 100 samples per second) were collected, and the phase-space threshold despiking method was used to treat and filter the raw velocity data.

Two main techniques were used to characterize the sedimentation pattern. After the experiment, during the equilibrium phase, the flume was drained very slowly to preserve the final sedimentation pattern. Using this preserved bottom, a 3D reconstruction was performed with Structure from Motion (SfM) and an RGB sensor to obtain a detailed point cloud of the bathymetry.

The application VisualSFM proposed by Wu [25] was used for the 3D reconstruction using the SfM technique. This photogrammetric method is based on a set of multiple photographs collected from different angles.

An RGB sensor (MS Kinect from Microsoft, Seatle, WA, USA) was used to automatically obtain this point cloud. This sensor comprised a color camera and a depth sensor. The resolutions of the RGB and depth images were 1920 $\times$ 1080 px and 640 $\times$ 480 px, respectively. The frame rate was 30 fps, and the measurement distance range of the target objects in this study was between 170 and 320 cm. Point cloud data were treated using the Cloud Compare v.2.13 software.

At the end, additional point gauge measurements were collected to calibrate and validate the results obtained from the previous techniques.

## 3. Flow Mechanisms

As the geometric characteristics of the joint channel and total discharge remain constant during the experiments, the water surface profile was mainly determined by the position of the downstream tailgate. Previous results published in Alizadeh and Fernandes [21] revealed that the water surface profile was approximately the same for all flow cases.

Figure 3 presents the non-dimensional streamwise velocity magnitude $u/U$, where $u$ is the streamwise velocity and $U$ is the mean cross sectional streamwise velocity. These results are presented for the 6 cross sections identified in Figure 2d, ranging from CS1 in the top row to CS6 in the lower row, and for the 5 flow cases, from FC1 in the first column to FC5 in the last column. In the same plots, the vectors corresponding to the secondary flows ($v$ for spanwise and $w$ for vertical velocities) are also presented.

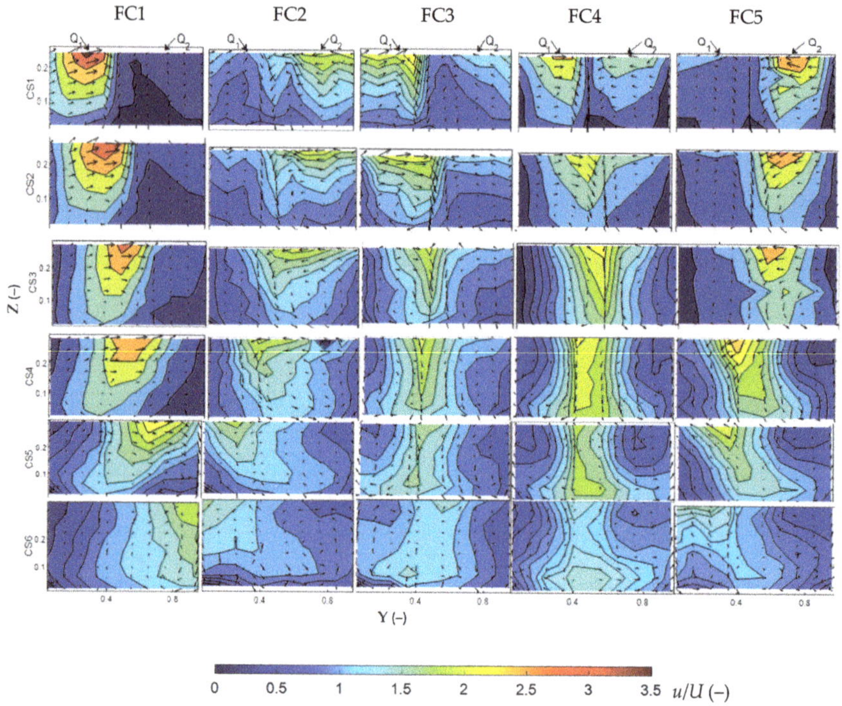


**Figure 3.** Non-dimensional streamwise velocity magnitude $u/U$ and secondary flow vectors (columns correspond to the 5 flow cases and rows stand for the 6 cross sections).

The secondary flows plotted in Figure 3 reveal that strong secondary currents are observed at the beginning of the joint channel (e.g., the first measurement cross section) for all flow cases. These currents are mainly driven by the momentum ratio. For FC1, only one secondary flow is observed, whereas FC2 to FC5 feature two secondary cells with opposite direction (one for each tributary flow). For these latter flow cases, the two secondary currents seem to be maintained until the last cross section. For FC1, it is observed that the flow velocity of tributary 1 enhances the magnitude of this secondary cell, which becomes dominant in the last cross sections.

Corroborating the conclusion from [4], velocity differences from the near bed to the surface are rather high, making depth average models (i.e., 2D modelling) not suitable for the complete description of the mixing processes. Even with the same junction angle, very different secondary patterns were obtained, which is in accordance to the findings of Bradrook et al. [7], who consider the velocity and momentum ratio as key factors influencing secondary currents.

Comparing FC1 and FC2 (same discharge distribution but different width), when lower discharge comes from a wider tributary (as in FC5), the jet flow is not so pronounced.

The distribution of the non-dimensional velocity magnitude $UV = \sqrt{(u/U)^2 + (v/U)^2}$ and the spanwise and streamwise velocity vectors u and v are presented in Figure 4a. To provide a clear presentation, only one of the 8 positions of the plan views for five flow cases are displayed at Z = 8.

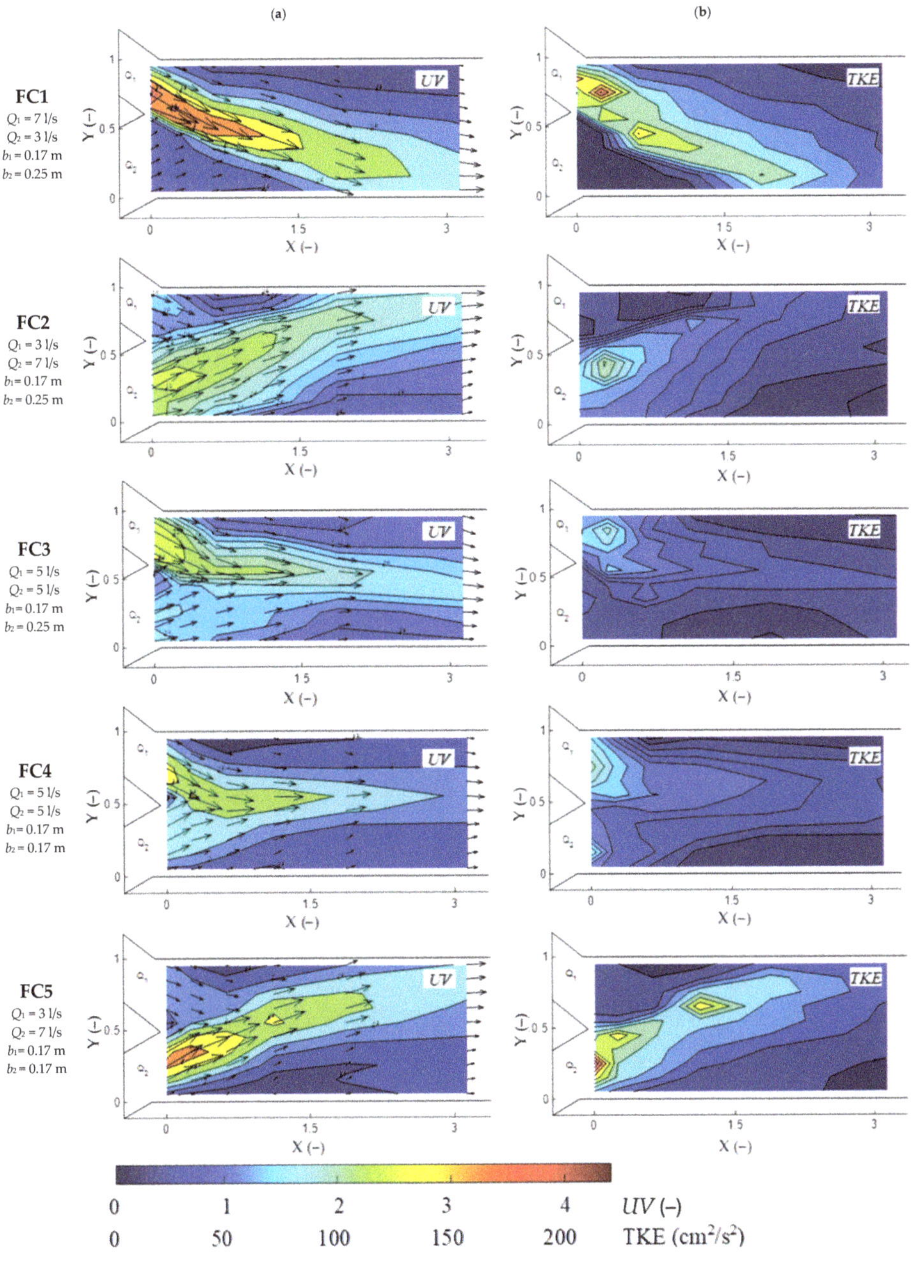


**Figure 4.** (**a**) Non-dimensional velocity magnitude *UV* and spanwise and streamwise velocity vectors *u* and *v* (in the left column) and (**b**) turbulent kinetic energy (in the left column).

The region of flow recovery is observed more downstream when the discharge ratio is far from 1. For flow cases 3 and 4, featuring $Q_2/Q_1 = 1$, despite some influence of spanwise flow, flow recovery is faster and transverse symmetry is achieved at the end of the flume.

Due to the abrupt change in the boundary geometry, tributary flows do not remain connected to the joint channel margins and flow separation is detected. This region is observed near the side walls for all flow cases along whole flow depth. Momentum ratio plays an important role in this flow element as it seems to influence its streamwise development. This is particularly evident for FC1 and FC5 that feature momentum ratios far from 1. Inverse flow in the flow separation region is not observed due to the relatively low angle between tributaries. As expected, boundary shear stress leads to the reduction of velocities near the bottom.

The interaction of the tributary flows at the confluence is one cause for the formation of the stagnation zone due to flow obstruction. This zone is obvious for FC4 near the surface, with both momentum and width ratio equal to one. Downstream to this stagnation region flow deflection is observed in joint channel. This region is visible over the whole junction depth but more evident and wider near the channel bed. Results seem to confirm Riley & Rhoads [6] observation that flow deflection is mainly controlled by the momentum ratio between the tributaries.

Regarding turbulence characteristics, for the same conditions as presented in Table 1, ref. [22] showed the cross-sectional distribution of turbulent kinetic energy ($TKE = 0.5\left(\overline{u'u'} + \overline{v'v'} + \overline{w'w'}\right)$, where $u'$, $v'$, and $w'$ stand for the fluctuating velocity components by Reynolds decomposition. Figure 4b presents the plan view of TKE for the vertical position $Z = 8$.

It was observed that TKE gradually decreased towards the channel wall and towards the bed. The maximum values of TKE occurred close to the joint of the tributaries (at approximately 10 cm). For flow cases with high momentum ratios (FC5 and FC1), there was a strong jet flow that hit the wall at cross-section 6, causing high gradients of momentum and velocity. The results are rather similar for FC3 and FC4 (FCs that feature momentum ratios close to 1). In these cases, a low magnitude of turbulent kinetic energy was observed.

In all FCs, as flow progresses in the joint channel, turbulence kinetic energy decreases, and at the end of the joint channel, a more stable flow is observed (flow recovery).

## 4. Morphodynamics

Sediments were steadily added at the upstream section (near the honeycombs in Figure 2d), flowing through each tributary toward the confluence. Sediment feeding lasted 1 h, simulating an 8 h real-world event based on Froude similarity. The simulation was derived from a 100-year return period and was applied equally to both tributaries. The sediments were eroded, transported, and deposited according to stream power and boundary shear stress. During the feeding period, the sedimentation pattern evolved and was monitored primarily through observation and photographs, as bathymetry measurements were not possible. It was clear that the equilibrium was not reached by the end of the feeding period, so experiments continued for approximately 6 more hours, until a constant sedimentation pattern indicated equilibrium. Control points were monitored to determine when this phase was achieved.

The final sedimentation configuration in the equilibrium phase was assessed using 3D reconstruction with photogrammetry and a kinetic sensor, with manual point gauge checks for control and monitoring. These river morphological patterns in confluences have significant implications for the management and design of river channels. Confluences are often sites for flooding, ice jams, and bed and bank instability, which are critical concerns, as they can threaten riverine infrastructure such as buildings and river docks.

Nazari-Giglou et al. [20] emphasize the importance of predicting scour depth for the design of bridges near confluences. Buildings and other structures located near riverbanks can also be affected by these morphodynamic processes.

Figure 5 illustrates the bed morphology for the five flow cases identified in Table 1. In the figure, the scale refers to the sediment layer in the channel.

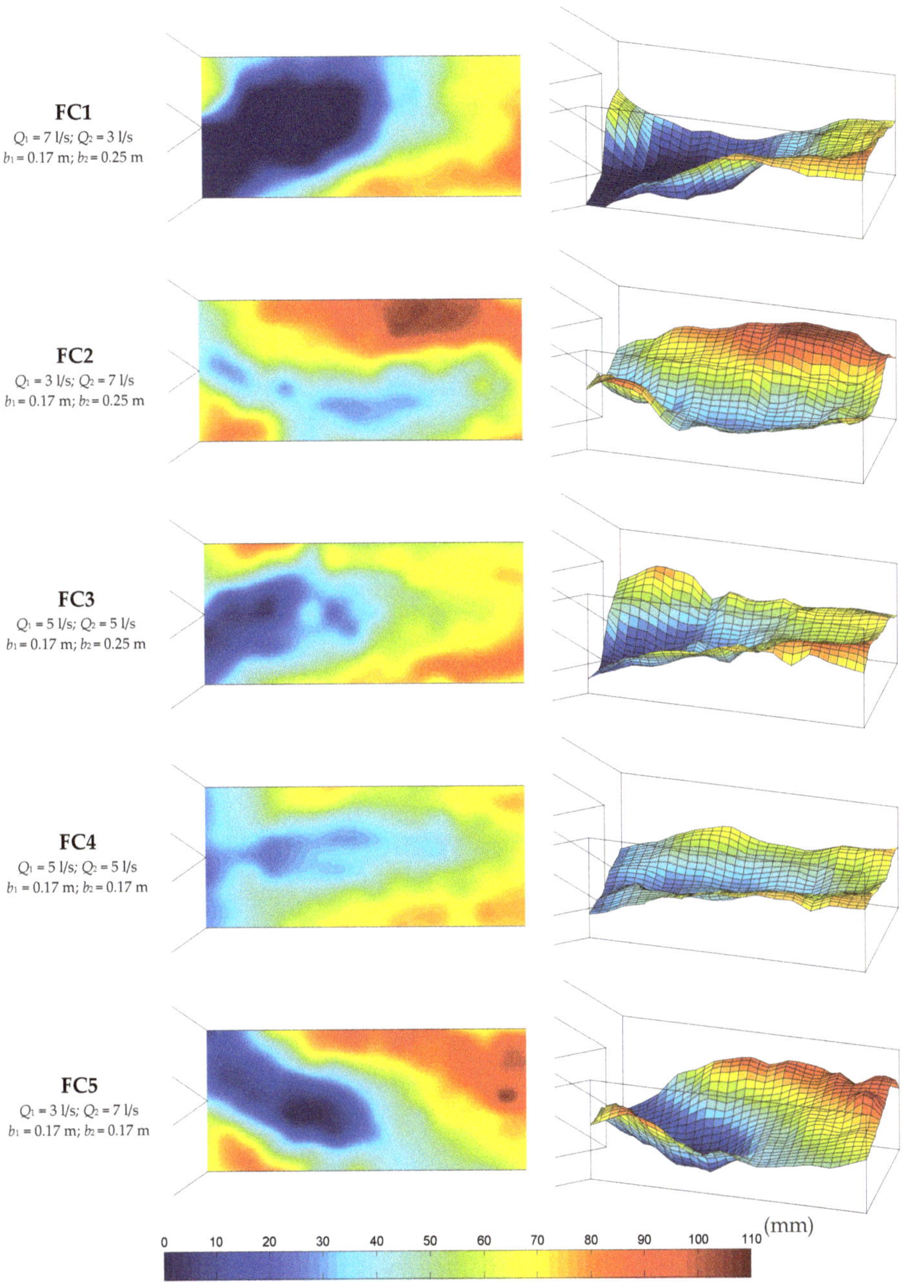


**Figure 5.** Bed morphology for five flow cases in equilibrium.

Understanding sediment transport and morphodynamic patterns in river confluences is essential for mitigating risks associated with flooding and structural instability. Effective management and design strategies rely on accurate predictions of these processes to protect infrastructure and ensure the safety and functionality of river channels.

Figure 6 shows the cumulative sediment volume along the longitudinal direction of the joint channel.

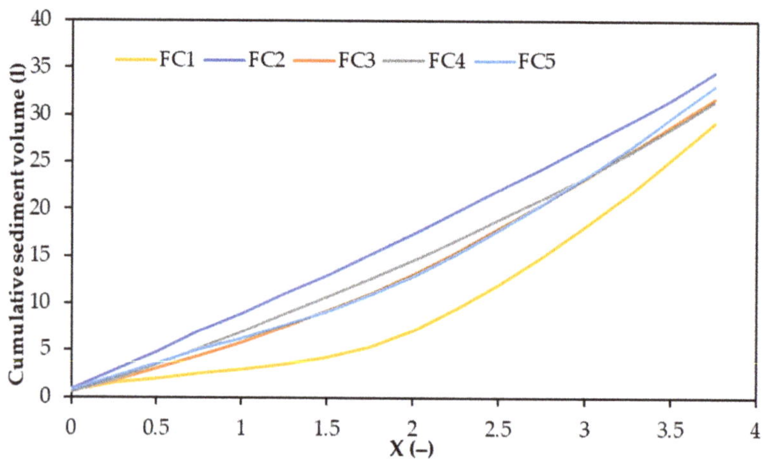


**Figure 6.** Cumulative sediment volume along the channel.

When in equilibrium, the typical morphological pattern in symmetric river confluences includes the following main characteristics:

(a) A bed transition between each tributary and the joint channel (this discordance can be so marked that it may be named avalanche slope);
(b) A deep local scour located right downstream of the confluence or aligned with the tributary direction;
(c) A dune or bar located in the middle of the joint channel;
(d) A dune or bar located in the lateral region of the joint channel downstream of the scour in the separation area;
(e) Sediment accumulation near the junction (in the flow stagnation area).

Despite the different patterns for each flow case, some of these elements can be globally identified in the morphological profile observed.

Figure 6 shows that, except for FC1, despite the different cross-sectional distributions along the channel, the results for the accumulated settled volume are generally comparable along the channel. These configurations, which involve different inlet configurations, promote similar sediment transport, and sediment is more evenly distributed along the channel length.

The typical features regarding the morphodynamics in sand- and gravel-bed confluences are a central scour hole, tributary mouth bars, and bank-attached bars in areas of flow recirculation and stagnation. The physical factors influencing confluence hydraulics and the resulting morphology include junction angle, bed discordance, discharge ratio, and upstream planform curvature.

The general pattern comprises a scour corridor aligned with the tributary with the highest momentum and velocity. This aspect is particularly evident for flow cases FC1 and FC5, which feature a high momentum ratio between the tributary flows. For the first flow case, it is observed that a very strong scour hole aligned with tributary 1 (highest discharge and lowest width). This configuration of the tributary leads to very high average velocity that seems to increase the stream power to lift, transport, and divert sediments towards downstream. This increase of the scour corridor from tributary 1 is complemented by a mouth bar in the alignment of tributary 2. The same pattern is also observed in FC5 and, to a lesser extent, in FC3 and FC2. However, in these two latter FCs, a pronounced bank-attached bar is observed near the wall of the lower average velocity.

The morphological configuration obtained in the equilibrium phase for the experiments with high discharge ratios seems to be characterized by strong scour leading to a excavated hole from one tributary and much less scour in the alignment of the second tributary. This process leads to an "avalanche" profile in this first tributary and it corresponds to the typical bed discordance confluences with uneven bed elevation and relatively steep slopes, identified in many cases with the beginning of systematic bathymetric surveys (e.g., [9]).

Together with the influence of the secondary currents, the shear layer formed in the interface of the tributary flows plays an important role in the increase of boundary shear stress and of the scouring process. From our experiments, it is important to point out that secondary currents were present with the channel even in the absence of sediments.

Characterized by symmetric geometry and equal discharge and momentum between the tributaries, FC4 is also characterized by a rather symmetric morphologic pattern. In this flow case, together with the central scour hole, banks are formed near the lateral walls. It is interesting to compare the formation of these side banks and their longitudinal evolution with the secondary current vectors (cf. Figure 3). Together with the increasing velocity magnitude, which leads to an increase in boundary shear stress, the downward flow in the center of the joint channel helps to lift sediments. Spanwise velocities transfer these sediments from the center towards the lateral banks which eventually deposit in the lateral bars. This mechanism can be observed in all flow cases, but in FC4, it is particularly evident, as there are no other spanwise differences or any other triggering factors influencing the morphological pattern. The influence of secondary currents seems to be crucial for this pattern.

Besides this feature, an additional interesting pattern is also clear in FC4. After the first central scour hole, a second one is formed downstream. Even though it is less evident, the same pattern also occurs for FC2 and FC3. This feature may be due to the deposit of the sediments diverted from the first main scour hole. With the decrease in velocity observed in Figure 4, these sediments form a small central bar. The eroded material that came from the scour hole was directly deposited downstream uniformly with respect to the alignment of the scour corridor.

The importance of the secondary currents for the formation of the lateral bars is also marked when one tries to link the direction and magnitude of these currents and the sedimentation near the banks. The strong secondary currents of FC2 and FC5 lead to the formation of high sedimentation in these cases, for instance. Following a numerical modelling approach, Bradbrook et al. [7] revealed the occurrence of secondary currents even without any topographic forcing. In the present work, it seems clear that these currents are a consequence of the mixture between the tributary flows and that they enhance the scour hole. With the increase in scouring, the secondary currents are also enhanced.

Discordant bed confluences are due to different conditions of sediment transport upstream. In this case, despite a concordant bed at the beginning of the experiments, very different bed elevations were obtained, depending on the flow and momentum ratios. Boyer et al. [26] studied the flow structure in a discordance confluence and investigated the effects on sediment transport by measuring near-bed flow turbulence, bed load transport rates, and changes in bed morphology for different flow conditions. High sediment transport rates are found at the edges of the shear layer region where the horizontal-vertical cross stresses are high. These features correspond to changes in bed morphology where erosion takes place along the shear layer. In our case, it seems that this pattern is followed, as high scour occurred typically in high shear regions (identified by high TKE in Figure 4). This feature highlights the importance of the shear layer for the sediment transport, as already suggested by Yuan et al. [15].

Flow hydrodynamics in a confluence plays a critical role in determining sediment scouring and deposition patterns. The distribution of flow across the cross sections, coupled with the influence of secondary flows, creates zones of high and low shear stresses that govern sediment behavior. For instance, areas with strong secondary currents, influenced by the tributary momentum ratio and flow deflection, enhance sediment scouring by increasing turbulence and boundary shear stress. Conversely, zones of flow stagnation or reduced velocity, such as those observed near the interface of tributaries or in downstream flow recovery regions, promote sediment deposition.

The shear layer contributes significantly to sediment redistribution. Its distortion, caused by the interaction between tributary flows, creates high-turbulence regions that drive sediment entrainment and transport. The interplay of these flow mechanisms supports the development of key morphological features, such as scour holes, sediment bars, and bed discordance, as highlighted in the experiments. In cases with higher tributary momentum, sediment scouring is more pronounced along the flow-aligned scour corridor, with eroded material being transported downstream before settling in lower-energy regions. In contrast, for more balanced discharge ratios, sediment distribution is more uniform, with deposition occurring in predictable patterns, such as mid-channel dunes and lateral bars. These findings align with previous studies on sediment transport in river confluences, reinforcing the importance of tributary flow conditions in determining confluence morphology. The formation of avalanche slopes and bed discordance in certain flow cases further supports the role of momentum asymmetry in shaping sedimentation patterns.

The findings align with previous studies, such as those by Yuan et al. [15] and Bombar and Cardoso [16], which emphasize the role of turbulence in sediment transport dynamics. Similar to these studies, the observed scouring and deposition patterns in the scale model reflect the influence of momentum and width ratios on hydromorphological processes. However, notable differences emerge, such as the specific length and depth of scour corridors, which may be attributed to variations in experimental setup, sediment characteristics, or confluence angles.

## 5. Discussion

### 5.1. Hydrodynamic Interactions and Flow Recovery

The results reveal an important influence of tributary momentum and discharge ratios on flow structures at the confluence. Velocity measurements demonstrated that stronger secondary currents developed when momentum ratios were high, with counter-rotating cells persisting downstream. In flow cases with balanced discharge ratios (such as FC3 and FC4), flow recovery occurred faster, leading to more uniform velocity distributions. In contrast, flow cases with significant momentum asymmetry (e.g., FC1 and FC5) exhibited prolonged turbulence, delaying flow recovery and intensifying sediment scouring.

The formation of flow separation zones and stagnation areas aligns with findings from past studies, confirming that momentum transfer between tributaries governs shear layer dynamics. Notably, for cases with extreme momentum asymmetry, jet flows persisted longer, deflecting towards the channel walls and driving turbulent kinetic energy peaks.

This extended turbulence intensified sediment transport, contributing to scour formation and sediment redistribution.

### 5.2. Sedimentation Patterns and Morphodynamic Features

The sedimentation experiments revealed distinctive morphological features that evolved based on flow conditions. The key features observed included:

- Central Scour Hole and Avalanche Slopes: High-momentum tributaries (as in FC1 and FC5) produced deep scour holes aligned with the dominant flow, with sediments diverted downstream. This pattern is consistent with real-world observations in mountain river confluences, where steep slopes and high stream power enhance sediment entrainment.
- Mouth and Bank-Attached Bars: Flow separation zones promoted sediment deposition, forming bars near the channel walls. These lateral features were more pronounced in flow cases with secondary currents of greater magnitude (e.g., FC2 and FC5), reinforcing the role of spanwise velocity components in sediment transport.
- Secondary Scour Holes: In symmetric flow conditions (FC4), a secondary scour hole formed downstream, likely due to sediment deposition from the initial scour zone. This finding highlights the cyclical nature of sediment transport, where flow deceleration downstream of high-energy zones promotes sediment settling and localized scour resurgence.

The results corroborate previous research, emphasizing the link between turbulence intensity and sediment dynamics. The shear layer distortion—exacerbated by momentum asymmetry—created high-energy zones that promoted sediment entrainment, while low-velocity regions acted as sediment sinks.

### 5.3. Practical Implications for River Management

These findings are critical for sustainable river management. Understanding the hydrodynamic forces driving sediment transport can inform strategies to mitigate erosion and infrastructure damage. For instance:

- Flood risk management: insights into scour hole formation can guide the placement of scour protection structures, reducing the risk of channel instability and bank collapse during extreme flood events.
- Sediment management strategies: predicting sediment deposition zones can help optimize dredging schedules, reducing maintenance costs and preventing channel blockages that could exacerbate flooding.
- Ecosystem restoration: Low-velocity regions downstream of the confluence may serve as natural habitats for aquatic species. Targeted habitat restoration efforts could enhance sediment retention and promote biodiversity, contributing to the overall resilience of river ecosystems.

### 5.4. Future Research

The high-resolution experimental data generated in this study provide a valuable benchmark for validating numerical models of sediment transport in river confluences. The detailed velocity fields, turbulence characteristics, and sedimentation patterns offer empirical evidence to refine model parameters, improving the predictive accuracy of hydro morphological simulations.

Future research could explore long-term morphological evolution under varying sediment loads and flow regimes, extending the experimental findings to more complex real-world scenarios. Additionally, the effect of the rigid boundaries and the influence of vegetation on sediment dynamics could offer new insights into nature-based solutions for flood mitigation and riverbank stabilization.

## 6. Conclusions

At river confluences, tributary flows interact to form a complex hydrodynamic interface, influenced by velocity, momentum, and direction differences. These interactions shape sedimentation patterns, impacting river structures and bank stability. This study used a 1:60 scale physical model of a 50° confluence on Madeira Island to analyze sedimentation patterns under 20- and 100-year flood scenarios. Five flow cases with varying flow and momentum ratios provided insights into hydrodynamics and sediment transport.

The model, designed using Froude similarity, replicates flow and sediment processes. The experiments revealed that momentum ratios strongly influence the size, location, and strength of secondary counter-rotating currents. These currents form due to flow deflection, reducing surface radial pressure via centrifugal forces. As a scour corridor develops, and secondary currents intensify, enhancing scour depth.

The experimental work highlighted the strong influence of tributary momentum ratios on the size, location, and strength of secondary counter-rotating currents. These currents initially form due to flow deflection, which lowers surface radial pressure through centrifugal forces, creating flow divergence—outward near the bottom and inward at the surface. As a scour corridor develops, secondary currents intensify, amplifying bed erosion and sediment transport. Flow recovery depends on discharge ratios, with near-equal flows stabilizing more quickly, while momentum asymmetry prolongs turbulence, leading to deeper scour holes, extended shear layers, and complex sediment deposition patterns. Downstream, velocities align longitudinally, and turbulence fades, gradually restoring flow stability.

These hydrodynamic processes shape sediment transport, influencing channel morphology and infrastructure stability. High-momentum tributaries create longer scour corridors, while secondary currents promote lateral bars and bank-attached deposits. Understanding these patterns enhances sediment management, informing strategies such as targeted dredging, structural reinforcements, and habitat restoration. Insights from this study can also guide flood defense design and port management, while experimental data offer a valuable benchmark for refining numerical models.

**Author Contributions:** J.N.F. proposed the theme and the setup and obtained the funding. L.A. conducted the experiments. J.N.F. drafted the first version of the paper. All authors contributed to the final writing and revision of this paper. All authors have read and agreed to the published version of the manuscript.

**Funding:** This research was funded by FEDER and Science and Technology Foundation under the scope of the Project MixFluv–Mixing Layers in fluvial systems (PTDC/ECI-EGC/31771/2017).

**Institutional Review Board Statement:** Not applicable.

**Informed Consent Statement:** Not applicable.

**Data Availability Statement:** The data that support the findings of this study are available from the corresponding author, J.N.F., upon reasonable request.

**Acknowledgments:** In addition to FEDER and FCT for funding this research, the authors would like to acknowledge Nuno Aido and César Costa for the valuable help during the experiments and Juana Fortes from the Harbours and Maritime Structures Division of LNEC for the support with the experimental facility.

**Conflicts of Interest:** The authors declare no conflicts of interest. The funders had no role in the design of this study; in the collection, analyses, or interpretation of data; in the writing of the manuscript, or in the decision to publish the results.

## References


1. Mazgareanu, I.; Biron, P.M.; Buffin-Bélanger, T. A fuzzy GIS model to determine confluence morphological sensitivity to tributary inputs at the watershed scale. *Geomorphology* **2020**, *357*, 107095. [CrossRef]
2. Roy, A.G. Introduction to part I: River channel confluences. In *River Confluences, Tributaries and the Fluvial Networks*; Rice, S.P., Roy, A.G., Rhoads, B.L., Eds.; Willy: Chichester, UK, 2008; pp. 13–15. [CrossRef]
3. Mosley, M.P. An Experimental Study of Channel Confluences. *J. Geol.* **1976**, *84*, 535–562. [CrossRef]
4. Rhoads, B.L.; Sukhodolov, A.N. Field investigation of three-dimensional flow structure at stream confluences: 1. Thermal mixing and time-averaged velocities. *Water Resour. Res.* **2001**, *37*, 2393–2410. [CrossRef]
5. Best, J.L. Flow dynamics at river confluences: Implications for sediment transport and bed morphology. In *Recent Developments in Fluvial Sedimentology*; Etheridge, F.G., Flores, R.M., Harvey, M.D., Eds.; GeoScienceWorld: McLean, VA, USA, 1987; Volume 39, pp. 27–35.
6. Riley, J.D.; Rhoads, B.L. Flow structure and channel morphology at a natural confluent meander bend. *Geomorphology* **2012**, *163–164*, 84–98. [CrossRef]
7. Bradbrook, K.F.; Lane, S.N.; Richards, K.S. Numerical simulation of three-dimensional, time-averaged flow structure at river channel confluences. *Water Resour. Res.* **2000**, *36*, 2731–2746. [CrossRef]
8. Nazari-Sharabian, M.; Karakouzian, M.; Hayes, D. Flow Topology in the Confluence of an Open Channel with Lateral Drainage Pipe. *Hydrology* **2020**, *7*, 57. [CrossRef]
9. Canelas, O.B.; Ferreira, R.M.L.; Guillén-Ludeña, S.; Alegria, F.C.; Cardoso, A.H. Three-dimensional flow structure at fixed 70° open-channel confluence with bed discordance. *J. Hydraul. Res.* **2019**, *58*, 434–446. [CrossRef]
10. Covelli, C.; Cimorelli, L.; Pagliuca, D.N.; Molino, B.; Pianese, D. Assessment of Erosion in River Basins: A Distributed Model to Estimate the Sediment Production over Watersheds by a 3-Dimensional LS Factor in RUSLE Model. *Hydrology* **2020**, *7*, 13. [CrossRef]
11. Imhoff, K.S.; Wilcox, A.C. Coarse bedload routing and dispersion through tributary confluences. *Earth Surf. Dyn.* **2016**, *4*, 591–605. [CrossRef]
12. Guillén-Ludeña, S.; Franca, M.J.; Cardoso, A.H.; Schleiss, A.J. Hydro-morphodynamic evolution in a 90° movable bed discordant confluence with low discharge ratio. *Earth Surf. Process. Landf.* **2015**, *40*, 1927–1938. [CrossRef]
13. Guillén-Ludeña, S.; Franca, M.; Cardoso, A.; Schleiss, A. Evolution of the hydromorphodynamics of mountain river confluences for varying discharge ratios and junction angles. *Geomorphology* **2016**, *255*, 1–15. [CrossRef]
14. Leite Ribeiro, M.; Blanckaert, K.; Roy, A.G.; Schleiss, A.J. Flow and sediment dynamics in channel confluences. *J. Geophys. Res. Earth Surf.* **2012**, *117*, F01035. [CrossRef]
15. Yuan, S.; Tang, H.; Xiao, Y.; Qiu, X.; Zhang, H.; Yu, D. Turbulent flow structure at a 90-degree open channel confluence: Accounting for the distortion of the shear layer. *J. Hydro-Environ. Res.* **2016**, *12*, 130–147. [CrossRef]
16. Bombar, G.; Cardoso, A. Effect of the sediment discharge on the equilibrium bed morphology of movable bed open-channel confluences. *Geomorphology* **2020**, *367*, 107329. [CrossRef]
17. Lewis, Q.W.; Rhoads, B.L. LSPIV Measurements of Two-Dimensional Flow Structure in Streams Using Small Unmanned Aerial Systems: 2. Hydrodynamic Mapping at River Confluences. *Water Resour. Res.* **2018**, *54*, 7981–7999. [CrossRef]
18. Fernandes, J.N.; Leal, J.B.; Cardoso, A.H. Apparent friction coefficient in straight compound channels. *J. Hydraul. Res.* **2011**, *49*, 836–838. [CrossRef]
19. Zhang, T.; Feng, M.; Chen, K. Hydrodynamic characteristics and channel morphodynamics at a large asymmetrical confluence with a high sediment-load main channel. *Geomorphology* **2020**, *356*, 107066. [CrossRef]
20. Nazari-Giglou, A.; Jabbari-Sahebari, A.; Shakibaeinia, A.; Borghei, S.M. An experimental study of sediment transport in channel confluences. *Int. J. Sediment Res.* **2016**, *31*, 87–96. [CrossRef]
21. Alizadeh, L.; Fernandes, J. Turbulent Flow Structure in a Confluence: Influence of Tributaries Width and Discharge Ratios. *Water* **2021**, *13*, 465. [CrossRef]
22. Heller, V. Scale effects in physical hydraulic engineering models. *J. Hydraul. Res.* **2011**, *49*, 293–306. [CrossRef]
23. Goring, D.G.; Nikora, V.I. Despiking Acoustic Doppler Velocimeter Data. *J. Hydraul. Eng.* **2002**, *128*, 117–126. [CrossRef]
24. Fernandes, J.; Jónatas, R. Experimental flow characterization in a spiral vortex dropshaft. *Water Sci. Technol.* **2019**, *80*, 274–281. [CrossRef] [PubMed]

25. Wu, C. Towards linear-time incremental structure from motion, in 3DV 2013. In Proceedings of the 2013 International Conferenc on 3D Vision, Seattle, WA, USA, 29 June–1 July 2013; pp. 127–134. [CrossRef]
26. Boyer, C.; Roy, A.G.; Best, J.L. Dynamics of a river channel confluence with discordant beds: Flow turbulence, bed load sedimer transport, and bed morphology. *J. Geophys. Res.* **2006**, *111*. [CrossRef]

*sustainability*

MDPI

*Article*

# Exploring the Interaction Between Landslides and Carbon Stocks in Italy

**Jibran Qadri and Francesca Ceccato ***

Department of Civil, Environmental and Architectural Engineering (DICEA), University of Padova, Via Ognissanti, 39, 35129 Padova, Italy; jibran.qadri@studenti.unipd.it
* Correspondence: francesca.ceccato.1@unipd.it; Tel.: +39-049-827-7991

**Abstract:** Landslides, as natural hazards, have far-reaching impacts beyond their immediate effects on human lives and infrastructure; landslides disrupt both carbon storage and ecosystem stability, and their role in the global carbon cycle cannot be underestimated. This study delves into the complex relationship between landslides and carbon stocks such as, in particular, soil organic carbon (SOC) and above-ground biomass (AGB), and outlines the spatial relationship between different types of landslides, soil organic carbon (SOC), and the carbon cycle, underscoring the importance of understanding these interconnections for environmental sustainability and climate change mitigation efforts. By employing machine learning algorithms on the Google Earth Engine platform, landslide susceptibility maps were created for different landslide types across Italy, and their spatial patterns with SOC accumulation were analyzed using the Python environment. The findings reveal a nuanced relationship between landslide hazard levels and SOC dynamics, with varying trends observed for different landslide types. In addition, this study investigates the potential impact of large-scale landslide events on carbon sequestration in the short term via a case study of the May 2023 landslide event in the Emilia Romagna region of Italy. The analysis reveals a substantial reduction in above-ground biomass by 35%, which approximately accounts for the loss of 0.133 MtC, and a decrease in SOC accumulation in 72% of the affected areas, indicating that landslides can transform carbon sinks into carbon sources, at least in the short term, and suggested that carbon released from extreme landslide events at a larger scale needs to be accounted for in regional or national carbon emissions. This research underscores the importance of considering landslides in carbon cycle assessments and emphasizes the need for sustainable land management strategies to protect and enhance carbon sinks, such as forests and healthy soils, in the face of increasing natural hazards and climate change impacts.

**Keywords:** landslide susceptibility map; machine learning; soil organic carbon; above-ground biomass; risk assessment; carbon stocks

## 1. Introduction

Landslides are natural hazards that can pose significant threats not only to human lives but also to ecosystems and can potentially become sources of carbon dioxide ($CO_2$) emissions. These mass movements of rock, debris, or earth down a slope can cause extensive damage to vegetation and disrupt soil cover, leading to various physical, chemical, and biological changes. Landslides can release soil organic carbon (SOC) stored in the soil and vegetation, contributing to increased atmospheric $CO_2$ levels [1,2]. SOC plays a crucial role in the global carbon cycle, acting as both a source and sink of atmospheric $CO_2$ through processes like photosynthesis, decomposition, and respiration [3–7]. However, both natural factors, like landslides and floods, and human activities like deforestation, agriculture, and urbanization often deplete this essential carbon pool [8]. Sustainable land management practices that promote SOC conservation and enhance carbon sequestration in soils, are critical for mitigating climate change and reducing anthropogenic $CO_2$ emissions [9].

Previous research suggests that landslides can act as both carbon sources and sinks, but their overall impact on the carbon cycle remains poorly understood. Some studies indicate that landslides can bury and sequester carbon by exposing carbon-rich subsoils and creating new spaces for vegetation growth [10–14]. However, other findings suggest that large-scale landslide events can release substantial amounts of stored carbon from soils and vegetation, potentially turning carbon sinks into sources [15]. Landslide deposits exhibit significant variability in soil properties, particularly in soil organic carbon (SOC), a critical component of global carbon cycling [16]. Soils in landslide scars are particularly vulnerable, with accelerated decomposition rates similar to those seen in surficial soil erosion [17,18]. This disruption extends to chemical and biochemical properties, emphasizing SOC's critical role in soil recovery and ecosystem restoration [19]. Forests are central to global carbon storage, holding 62–78% of terrestrial carbon, with 70% of this stored in soil [20]. Soil organic matter, a primary contributor to forest carbon stocks, ranks second only to above-ground biomass [21]. Yet, disturbances like deforestation and landslides lead to substantial carbon emissions and hinder reclamation, which is often slow due to long-term effects on SOC and nutrient availability [22,23]. The restoration of landslide-affected areas remains a challenge. Artificial restoration has been shown to yield immediate benefits for SOC recovery, but its long-term effects align with those of natural regeneration [24,25]. Effective forest management strategies, as discussed by Pang et al. (2024) [26], could enhance forest health and biodiversity, aiding carbon sequestration. Furthermore, Di et al. (2023) [27] emphasize the significance of landslide scale, particularly width, in determining SOC dynamics and the potential for post-disaster recovery.

Pandey Hari et al. (2022) [28] studied the recovery of vegetation after the landslides triggered by the 2015 Gorkha earthquake and found that by 2022, about 51% of the lost vegetation had been restored. Of this, only 9.5% was due to human-led restoration efforts, with the majority of regrowth occurring naturally. Their analysis suggests that it will take at least sixteen years for the vegetation to fully return to pre-landslide levels, highlighting the resilience of the ecosystem but also the long-term nature of recovery after such disasters.

In recent studies like the Fiordland region, multiple regional landslide events influenced carbon storage significantly. According to the study [29], such events should be included in national greenhouse gas inventories, and another study reveals that each large earthquake can erode millions of tonnes of carbon [30], highlighting the long-term impact of such disturbances on carbon export from mountain forests. Landslides significantly impact both carbon storage and ecosystem stability, making their role in the global carbon cycle critically important.

Most of the studies in the literature have focused on individual landslides, small regions [16], or earthquake-induced landslides over large areas [28,29]; this paper investigates the impact of rainfall-triggered landslides over a large region, which is a novelty in the literature, and contributes to advancing our understanding of the broader implications for the carbon cycle.

The purpose of this paper is to explore the interaction between landslides and carbon stocks, specifically soil organic carbon (SOC) and above-ground biomass (AGB). While it is well established that vegetation can reduce landslide susceptibility by improving soil stability and slope reinforcement through root systems [31,32], there is a lack of comprehensive studies analyzing the relationships between SOC and landslide hazard levels across different landslide types. This study aims to explore the spatial patterns between SOC accumulation and landslide susceptibility classes, providing insights into how SOC dynamics may influence or be influenced by different types of landslide events. To the authors' knowledge, this is a novelty in the literature. Understanding these relationships is important for sustainability as it highlights the interconnections between carbon stocks, natural hazards, and their potential impacts on climate change mitigation efforts.

Landslide susceptibility mapping (LSM) is a spatial tool employed to assess the probability of landslide occurrence and identify areas most vulnerable to such events within a given study region [33]. This mapping technique typically incorporates various causative factors that influence landslide occurrence, encompassing geological, hydrological, ecological, geomorphological, and anthropogenic variables. These factors may include, but are not limited to, precipitation patterns, land cover characteristics, and topographical slope.

Random Forest, a machine learning algorithm, has emerged as a particularly effective method for landslide susceptibility analysis. Its capacity to process large datasets and manage numerous variables simultaneously, coupled with its robustness against multicollinearity issues, renders it an advantageous choice for landslide studies [34]. This algorithm's ability to handle complex, multidimensional data makes it well suited for the intricate nature of landslide susceptibility assessment, which often involves the integration of diverse geospatial and environmental parameters.

Various types of landslides, such as rainfall-triggered (RT), slow (SL), complex (C), rotational/translational (RT), shallow (SH), fall (F), rapid (R), diffused fall (DF), and Deep-Seated Gravitational Slope Deformations (DSGSDs), were considered according to the distinction used in Table 1, which shows the Italian Landslide Inventory (IFFI). The definition of each landslide type is any given machine learning algorithm for landslide susceptibility mapping that uses 18 parameters derived from a digital elevation model (DEM) and satellite data. IFFI and ITALICA [35,36] catalogs (https://doi.org/10.5281/zenodo.8009366) are used for model training and testing. The landslide susceptibility map is used together with the SOC map available on GEE by OpenLandMap to identify patterns between landslide hazard levels and SOC. This is a novelty in the literature, and even if temporal changes in LS and SOC are not included, and despite the limitations deriving from the use of satellite data, it is meaningful to provide insights into the complex interplay between landslides, SOC dynamics, and their potential impacts on the carbon cycle and climate change mitigation strategies. We hypothesized that the influence on SOC and carbon stocks if landslides occur in the future, through patterns, could be instrumental in understanding the relative impact of different types of landslides on carbon stocks, and the pilot case of the May 2023 landslides in Emilia Romagna, Italy, is used to support this hypothesis.

**Table 1.** Definition of landslide types.

| Name | Definition | Figure |
|---|---|---|
| Rapid Flow (R) | This can be characterized by the fast movement of debris, often resembling a fluid-like motion. These landslides can occur in various forms, including debris flows, mudflows, and avalanches, depending on the composition and fluidity of the material involved [37,38]. |  (modified from [39]) |
| Fall (F) | Falls are landslides in which the fall of rock or debris occurs through free-fall motion. This type typically involves detached blocks of rock or soil that plunge or tumble downward due to gravitational forces. Falls often occur on steep slopes and are triggered by factors such as weathering, seismic activity, or erosion [40]. | 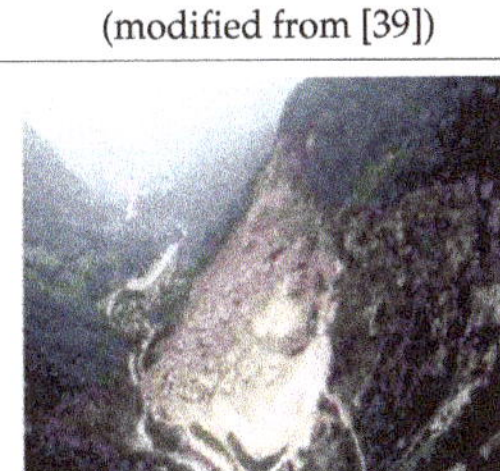 (modified from [41]) |

**Table 1.** *Cont.*

| Name | Definition | Figure |
| --- | --- | --- |
| Shallow (SH) | Shallow landslides occur within the upper layers of the slope, typically involving only a small depth of soil or rock material. These landslides often result from the failure of surface materials due to factors such as saturation, erosion, or changes in slope stability [38]. |  (modified from [42]) |
| Complex (C) | These landslides involve more than one failure mechanism occurring simultaneously or sequentially, making them multifaceted in nature. This type often includes a combination of sliding, flowing, and falling components [40]. |  (modified from [43]) |
| Rainfall-Triggered Landslide (RT) | Rainfall-triggered landslides are those induced or exacerbated by precipitation events. Heavy or prolonged rainfall can increase the pore water pressure within the slope material, reducing its shear strength and leading to slope failure. Rainfall-triggered landslides can induce both rapid flows (such as debris flows) and slow flows (such as creep or slumps) depending on the intensity and duration of the rainfall event and the characteristics of the slope material. |  (modified from [44]) |
| Diffused Fall (DF) | This category combines characteristics of falls and topples but is used when the precise initiation mechanism is unclear. It refers to landslides where the debris is evident, but the exact cause of the fall or topple cannot be distinctly identified, resulting in a generalized classification [45]. |  (modified from [46]) |
| Slow Flow (SL) | These landslides involve the gradual movement of material, usually containing clay, which moves slowly down a slope. This type includes non-liquefied flows of materials such as sand, silt, gravel, debris, and instances of lateral spreading [38]. |  (modified from [47]) |

**Table 1.** *Cont.*

| Name | Definition | Figure |
|---|---|---|
| Deep-Seated Landslides (DSGSDs) | These are characterized by their deep slip surfaces, often occurring within large volumes of soil or rock. They are typically slow-moving and involve significant depths below the surface, affecting large land areas [48]. |  (modified from [49]) |
| Translational/Rotational Landslides (TR) | These involve the downslope movement of material along a relatively planar surface. Translational landslides can include both translational sliding, where the material moves parallel to the slope, and rotational sliding, where the movement occurs along a curved surface, causing the sliding mass to rotate [38]. |  (modified from [43]) |

A case study of a big disaster event of landslides in Emilia Romagna (Italy) [50], which occurred in May 2023, is analyzed to evaluate the impact on vegetation and carbon storage. More than 60 thousand landslides of different typologies occurred and were mapped with polygons, over which the changes in NDVI and above-ground biomass is analyzed. This quantitative study proves that landslides reduce carbon sequestration and increase potential carbon release into the atmosphere, contributing to increased $CO_2$ levels, at least in the short term. This highlights the importance of considering landslides in carbon cycle assessments, especially as climate change induces more extreme weather events, potentially turning carbon-absorbing areas into carbon sources. Understanding these dynamics is essential for addressing climate change effectively.

This study presents a preliminary analysis of the relationship between SOC and landslide hazard levels across different landslide types in Italy. The overall goal is to provide insights into the complex interplay between landslides, carbon stocks, and the carbon cycle, which is crucial for sustainable land management, hazard mitigation, and climate change adaptation strategies.

## 2. Materials and Methods

In order to investigate the relation between hazard level and SOC, a landslide susceptibility map is prepared and then it is correlated with the SOC map (Figure 1). The methodology used to build the LSM is explained in Section 2.1. Section 2.2 illustrates the approach used to elaborate the SOC map and obtain the SOC for different hazard classes. Section 2.3 illustrates the methodology used to calculate the changes in AGB and SOC after a landslide event.

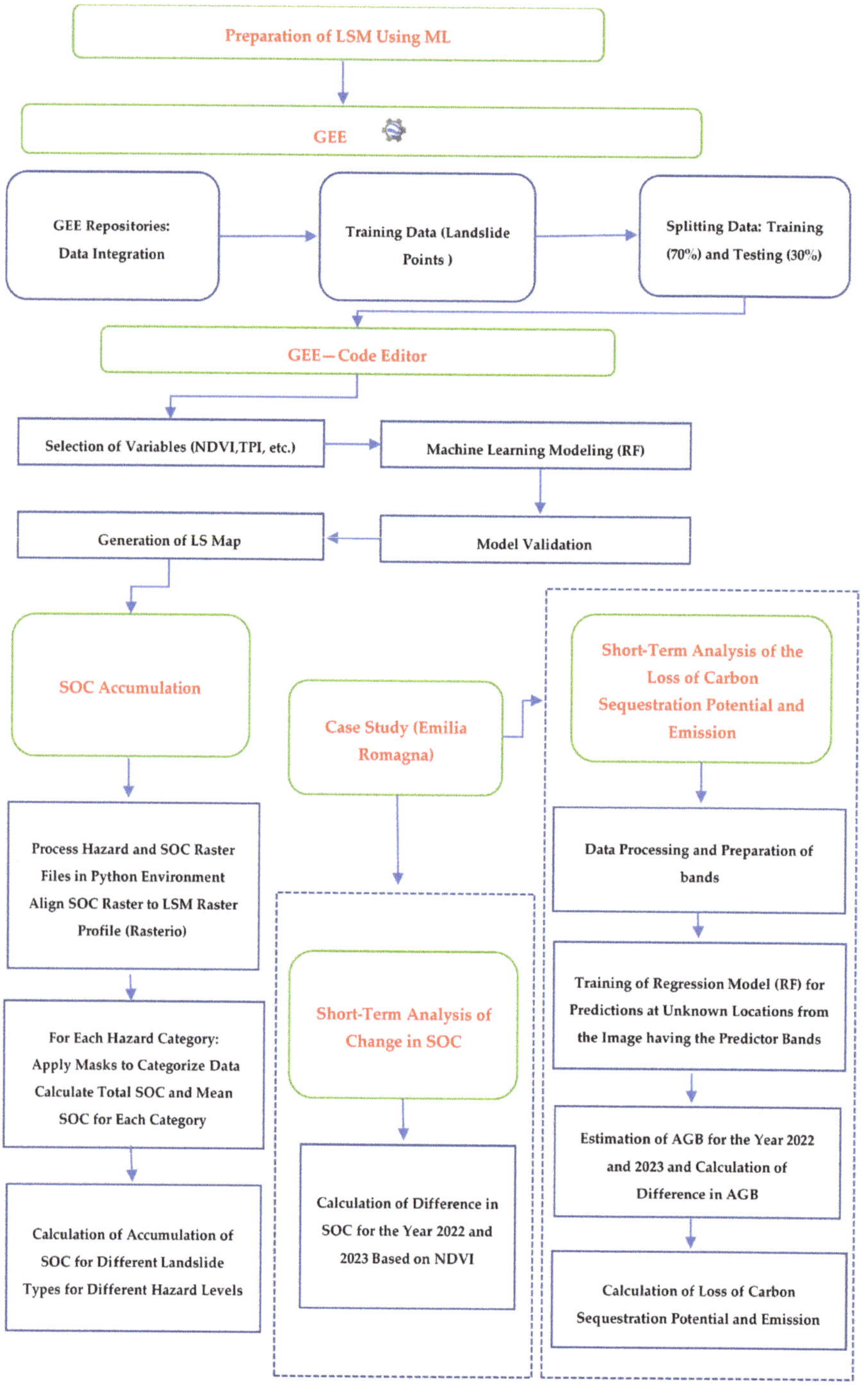


**Figure 1.** Schematic diagram of LSM framework and specific processing steps for each part (Sections 2.1–2.3).

## 2.1. Landslide Susceptibility Map

There are different methodologies that can be used to assess landslide susceptibility. In recent years, the use of traditional machine learning methods and deep learning methods has significantly increased and they have been shown to be robust and accurate [51]. The quality of an LS map produced by a model relies heavily on the accuracy, scale, and number of landslide-related factors considered as well as the completeness of the landslide inventory [52]. With the emergence of Google Earth Engine (GEE), there are abundant data resources and functions, which enable investigations at the regional and global scales [53,54].

This study employs the Random Forest (RF) machine learning method to assess landslide susceptibility on a large scale using the Google Earth Engine (GEE) platform at the pixel level [53]. The RF algorithm, renowned for its efficiency in managing high-dimensional datasets, is widely utilized in satellite-based applications due to its computational power and versatility [55]. This model is celebrated for its reliability, ease of interpretation, and robust predictive performance. By constructing multiple decision trees during training, the RF model determines classifications based on the majority vote across these trees. Its resistance to overfitting and multicollinearity makes it an excellent choice for landslide susceptibility assessment, a domain characterized by the interaction of numerous interrelated variables. Furthermore, the RF model provides valuable insights into feature importance, which enhances its interpretability and supports informed decision-making.

For this analysis, the RF model was trained and tested using data from two major landslide inventories: the Italian Landslide Inventory (IFFI) (https://www.progettoiffi.isprambiente.it/en/italian-landslide-inventory-iffi/) and ITALICA (https://doi.org/10.5281/zenodo.8009366) last accessed on 30 July 2024.

The IFFI database is a collaborative initiative led by national, regional, and provincial geological surveys to systematically document landslides across Italy [56]. This extensive inventory includes data on 622,439 landslides, covering a total area of approximately 22,500 km$^2$, which constitutes about 7.5% of Italy's total land area. The average density of landslides recorded is two per square kilometer. The data within IFFI are derived from diverse sources, including aerial photograph interpretation, analysis of historical records, field surveys, and pre-existing geological maps [56]. The inventory classifies landslides into several types, such as fall and/or topple (26,519 cases), slow (earth) flow (81,513 cases), rapid flow (81,213 cases), complex movement (59,472 cases), rotational/translational slides (222,653 cases), lateral spread (79 cases), sinkholes (409 cases), and undefined slope movements (36,379 cases).

ITALICA is a specialized inventory cataloging 6312 rainfall-triggered landslides that occurred in Italy between January 1996 and December 2021. Unlike the IFFI, ITALICA focuses exclusively on landslides caused by rainfall and provides precise temporal and spatial data. These landslides are evenly distributed across Italy's mountainous and hilly regions. ITALICA is distinctive in its emphasis on temporal accuracy and its exclusive focus on rainfall-induced landslides, setting it apart from IFFI, which covers landslides triggered by various causes but lacks detailed temporal data.

For this study, the two inventories were combined to create a comprehensive dataset for training and validating the RF model. By merging the strengths of both datasets—the IFFI's extensive spatial coverage and ITALICA's precise temporal focus—this approach ensures a robust and well-rounded dataset for landslide susceptibility analysis.

The combined dataset includes nine distinct landslide types (Table 1): slow (SL), complex (C), rotational/translational (RT), shallow (SH), fall (F), rapid (R), diffused fall (DF), Deep-Seated Gravitational Slope Deformations (DSGSDs) from the IFFI and rainfall-triggered (RT) landslides from ITALICA.

This integration offers a holistic view of landslide susceptibility by accounting for a wide variety of factors and types, providing a rich dataset for comprehensive analysis (Tables 2 and 3). In landslide studies, researchers emphasize the variability of factors influencing landslide occurrences across different areas. Identifying relevant causal factors is challenging, and there is no well-established rule regarding the necessary number of factors for landslide susceptibility analysis. The choice of predisposing factors relies on the unique characteristics of the study area, expert opinions, and the availability of data to generate suitable spatial and thematic patterns [33,57]. In this study, the selected variables for this analysis are listed in Table 2. Aspect, eastness, elevation, Gaussian curvature, HAND (Height Above Nearest Drainage), hillshade, horizontal curvature, maximal curvature, mean curvature, minimal curvature, northness, shape index, slope, TPI (Topographic Position Index), and vertical curvature are derived from a 30 m spatial resolution digital elevation model (DEM) using Google Earth Engine. The Normalized Difference Vegetation Index (NDVI) is derived at a 30 m spatial resolution from Landsat 8 imagery, accessed through the Google Earth Engine (GEE) data catalog. Soil density information, at a 250 m resolution, is sourced from OpenLandMap and accessed through GEE. Additionally, precipitation data from the Global Precipitation Measurement (GPM) Integrated Multi-satellitE Retrievals for GPM (IMERG) product, which provide global rainfall estimates at an 11,132 m resolution, are incorporated. This dataset, provided by NASA GES DISC, is also accessible via GEE. Table 3 summarizes the list of applied datasets and the source.

**Table 2.** Variables' definitions and rationale used in the model.

| Variable | Importance and Rationale | Variable | Importance and Rationale |
| --- | --- | --- | --- |
| Aspect | The compass direction that a slope faces. It impacts how much sun exposure the slope receives and the direction of water runoff, influencing the likelihood of landslides. Aspects facing the direction of frequent heavy rains might be more susceptible to landslides. | Mean Curvature | The average of the maximum and minimum curvature of the terrain. It provides an overall indication of the "bendiness" of the terrain, which can impact slope stability and susceptibility to landslides. |
| Eastness | A sine transformation of the aspect used to handle the circular nature of aspect data. It quantifies the degree to which the slope faces east. Eastern slopes might have different vegetation or water drainage patterns than western slopes, affecting the stability of the slopes. | Minimal Curvature | The minimum amount of curvature in any direction of the terrain. Areas with high minimal curvature might be more prone to landslides due to potentially weaker structural integrity. |
| Northness | A cosine transformation of the aspect used to handle the circular nature of aspect data. It quantifies the degree to which the slope faces north. Northern slopes might have different vegetation or water drainage patterns than southern slopes, impacting landslide susceptibility. | Vertical Curvature | Measurement of the curvature of the terrain in the vertical plane. This can influence how water flows down a slope, which could impact the saturation of the slope and, consequently, its stability. |

**Table 2.** *Cont.*

| Variable | Importance and Rationale | Variable | Importance and Rationale |
| --- | --- | --- | --- |
| Elevation | The height of an area above sea level. Higher areas tend to be more prone to landslides due to gravitational forces. Additionally, elevation might impact factors such as temperature and rainfall, which can affect slope stability. | HAND (Height Above Nearest Drainage) | The vertical distance of a location above the nearest water body. Locations closer to water bodies are more susceptible to landslides due to the possible saturation of soil. The greater the HAND, the less likely it is that the area will become saturated and prone to landslides. |
| Gaussian Curvature | A measure of the total curvature of the terrain. This is calculated as the product of the maximal and minimal curvature of the terrain. Extreme positive or negative Gaussian curvature can suggest that the terrain is convex or concave, respectively, which can influence water runoff and slope stability. | Hillshade | A measure that indicates the amount of sunlight a region receives. Regions receiving more sunlight might have lower soil moisture levels and different vegetation types compared to those receiving less sunlight. These factors can impact slope stability and susceptibility to landslides. |
| Horizontal Curvature | Measurement of the curvature of the terrain in the horizontal plane, potentially affecting the direction and speed of water flow. A high horizontal curvature might lead to faster water flow, possibly accelerating soil erosion and increasing landslide susceptibility. | Shape Index | Quantification of the shape of the terrain, giving an idea of landscape relief. This can help identify areas of potential water pooling or steep gradients, both of which can impact landslide occurrence. High shape index values typically correspond to ridges and peaks, while low values represent valleys and depressions. |
| Maximal Curvature | Reflection of the maximum amount of curvature in any direction of the terrain. Regions with high maximal curvature might be more prone to landslides due to potentially weaker structural integrity. | Rain (Mean Precipitation) | Mean monthly precipitation from IMERG GPM. The data were resampled to 30 m for landslide susceptibility model training. |
| TPI (Topographic Position Index) | A measure that indicates the position of an area in relation to its surroundings (e.g., valley, ridge, flat area). It helps understand the influence of landscape position on the direction and speed of water flow, which can affect soil saturation levels and landslide susceptibility. | Slope | The angle of the terrain, or steepness. Steeper slopes are generally more susceptible to landslides as the unstabilizing component of the gravitational force acting on the slope material is greater. |
| NDVI | NDVI is an indicator of Vegetation Health Assessment, Stability Indicator, and Rainfall and Erosion Control. High NDVI values indicate healthy vegetation, while low values suggest sparse or stressed vegetation. | Soil Density | Soil density may influence shear strength, water retention properties, and erosion susceptibility. |

**Table 3.** Data sources utilized in the study.

| Data | Resolution | Source | URL |
|---|---|---|---|
| NDVI (Normalized Difference Vegetation Index) | 30 m | Google Earth Engine | https://developers.google.com/earth-engine/datasets/catalog/landsat-8, last accessed on 30 July 2024 |
| NASA SRTM Digital Elevation 30 m | 30 m | Google Earth Engine | https://developers.google.com/earth-engine/datasets/catalog/USGS_SRTMGL1_003#description, last accessed on 30 July 2024 |
| Spatio-Temporal Catalog of Rainfall-Induced Landslides in Italy | | ITALICA | https://doi.org/10.5281/zenodo.8009366, last accessed on 30 July 2024 |
| Soil Density | 250 m | Google Earth Engine | https://developers.google.com/earth-engine/datasets/catalog/OpenLandMap_SOL_SOL_BULKDENS-FINEEARTH_USDA-4A1H_M_v02#description, last accessed on 30 July 2024 |
| Landslide Inventory in Italy | | IFFI catalog | https://www.progettoiffi.isprambiente.it/en/italian-landslide-inventory-iffi/, last accessed on 30 July 2024 |
| Rainfall (Precipitation) | 11,132 m | Google Earth Engine | https://disc.gsfc.nasa.gov/datasets/GPM_3IMERGM_07/summary, last accessed on 30 July 2024 |
| SOC | 250 m | Google Earth Engine | https://developers.google.com/earth-engine/datasets/catalog/OpenLandMap_SOL_SOL_ORGANIC-CARBON_USDA-6A1C_M_v02, last accessed on 15 October 2024 |

The landslide inventory is split into two sets: 70% for training and 30% for testing. The Random Forest model was trained on the 70% subset, optimized for minimum out-of-bag error, and the final model was selected based on this optimization. The final model, optimized through training on the 70% dataset, was employed to predict landslide susceptibility for the remaining 30% testing data subset. These predictions were then compared with actual values, and the accuracy of the model was assessed using the AUC-ROC (Area Under the Receiver Operating Characteristic) method. Subsequently, the chosen model was applied to the entire geographic extent of Italy. The outcome was a probability-based landslide susceptibility map, ranging from 0 to 100. This map was further classified into 6 classes indicating landslide susceptibility scores based on their likelihood of experiencing landslides: very low (0–15%), low (15–30%), low to moderate (30–45%), moderate to high (45–60%), high (60–80%), and very high susceptibility (80% and above). Different maps are prepared for each type of landslide (Figure 2a,b).

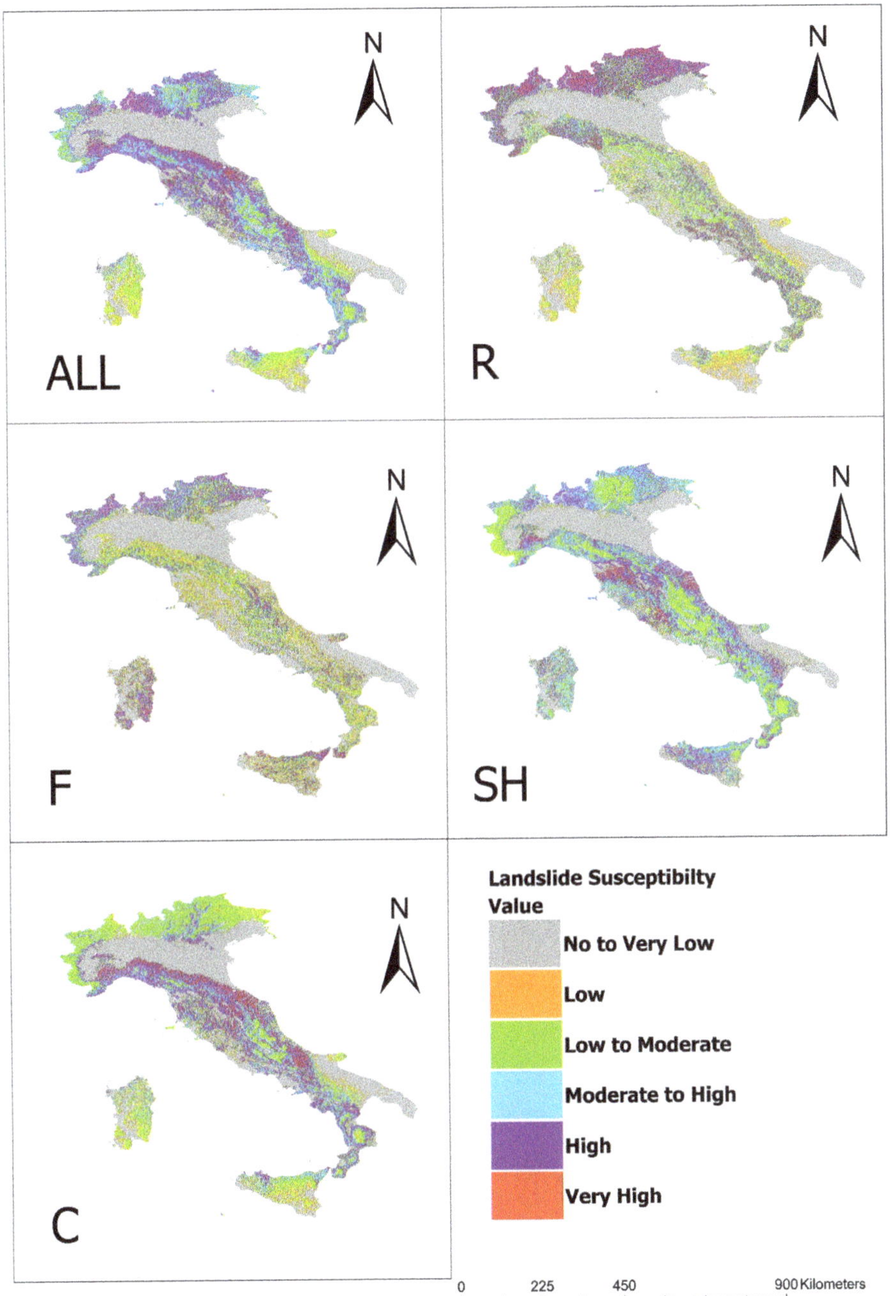


(a)

**Figure 2.** *Cont.*

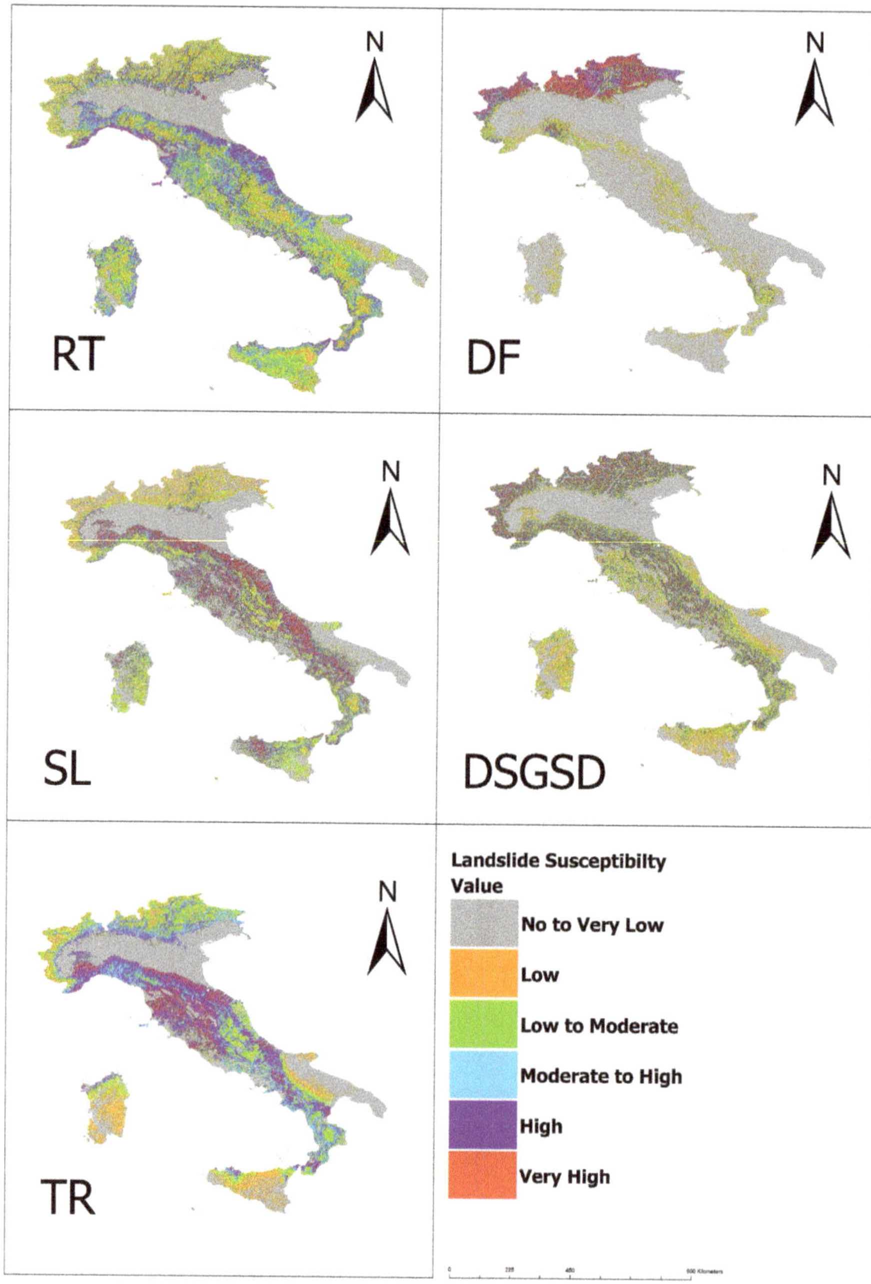


(b)

**Figure 2.** (**a**) Landslide susceptibility map for all, R, F, SH, and C landslide types. (**b**) Landslide susceptibility map for RT, DF, SL, DSGSD, and TR landslide type.

### 2.2. SOC and Hazard Categories

The map of SOC is taken from OpenLandMap available on GEE, and it represents soil organic carbon content in ×5 g/kg at 6 standard depths (0, 10, 30, 60, 100, and 200 cm) at a 250 m resolution. The algorithm processes different landslide hazard maps and SOC raster files in a Python environment to compute and visualize SOC values for different hazard categories. It aligns the SOC raster to the hazard raster profile using the rasterio library and calculates total and average SOC values for predefined hazard categories.

The primary function in the Python environment serves multiple purposes in analyzing the relationship between hazard levels and soil organic carbon (SOC). It begins by reading the input hazard raster and SOC raster files, which contain spatial data on hazard levels and soil organic carbon content, respectively. The function then applies masks to categorize the data based on predefined hazard levels. This categorization enables a detailed examination of how SOC is distributed across areas with varying hazard intensities.

Then, the function aggregates SOC values within each hazard category to calculate total soil organic carbon (total SOC) and mean soil organic carbon. Total SOC represents the cumulative sum of SOC across all soil areas within the specified hazard category while mean SOC is the average SOC value for the areas within each hazard category, providing a normalized measure of SOC distribution.

Finally, the function visualizes the results by plotting the total and mean SOC values, offering clear insights into the relationship between hazard levels and SOC distribution. This analysis highlights how soil organic carbon varies across different hazard intensities, aiding in understanding the interplay between environmental hazards and soil health.

Total SOC is typically expressed in mass units such as megagrams (Mg) or tons. Mean soil organic carbon is the average organic carbon content per unit area in the soil. It represents the concentration of soil organic carbon averaged across a specified area. It is typically expressed in mass per area units such as megagrams per hectare (Mg/ha) or tons per hectare (tons/ha) for each category and saves the plots for further analysis.

### 2.3. Above-Ground Biomass Density (AGBD) Calculation and Change in SOC

This section considers the case of the intense rainfall events that occurred in May 2023 in Emilia Romagna (Italy) causing more than 60 thousand landslides of different types (See Section 3.3). All landslide polygons are mapped including the source and deposition area (https://geoportale.regione.emilia-romagna.it/approfondimenti/emergenza-maggio-23/emergenza-rer-maggio-2023-servizi, accessed on 15 December 2024).

Two distinct methodological approaches are employed to quantify the impact of landslides on carbon stocks in the Emilia Romagna landslide-affected region:

1. The first method utilizes remote sensing data to find the change in SOC based on the Normalized Difference Vegetation Index (NDVI).
2. The second method leverages data from NASA's Global Ecosystem Dynamics Investigation (GEDI) mission to calculate AGB density. GEDI's lidar measurements, combined with Sentinel-2 satellite data and elevation models, are used to train a regression model for predicting AGB density.

Methodology (1) first defines pre- and post-landslide time periods and filters the Landsat collection to include only images with less than 20% cloud cover over the Emilia Romagna, then calculates NDVI for both periods, masks invalid NDVI values, and converts the NDVI to SOC and carbon stock estimates. The change in carbon sequestration is calculated by finding the difference between pre- and post-landslide carbon stocks.

Method (2) tries to quantify the loss in AGB. To evaluate and quantify the density and total stock of above-ground biomass, we selected only the landslide polygons out of the whole of Emilia Romagna with the aim of finding the difference between 2022 and 2023. We used GEDI mission data utilizing the MIT open-source data and adapted them according to our requirements. NASA's GEDI mission collects LIDAR measurements at a 30 m spatial resolution, capturing the vertical distribution of vegetation to estimate above-ground biomass density (AGBD) [58]. These point estimates serve as ground-truth data for regression models. The present study utilizes Sentinel-2 satellite data, including spectral bands and indices, which are cloud-masked using Google's Cloud Score dataset to create an annual cloud-free composite. Elevation and slope data from the Copernicus GLO-30 DEM are incorporated both for pre-processing GEDI data and as independent variables. Data preparation involves filtering, scaling, and creating mosaics for each dataset. GEDI data are filtered to remove unreliable measurements based on quality flags, relative error, and slope constraints, while Sentinel-2 images are processed to a median composite. Thereafter, all datasets are exported as assets to manage computational limitations.

In building the regression model, Sentinel-2 data, along with elevation and slope, are used as predictors, while GEDI AGBD values serve as the target variable. Resampling and stratified sampling techniques ensure alignment and adequate coverage of valid data points. The trained regression model predicts AGBD values, which are then used to estimate total AGB [59]. To refine this estimation, non-vegetated pixels are masked using the ESA World Cover dataset, and the total AGB is calculated by multiplying the AGBD values by each pixel's area and summing them across the region for both years (2022 and 2023), and the whole process is carried out on the GEE platform.

This model is then applied to the study area and the total AGB is quantified by integrating the AGBD values over the landslide-affected regions (polygons). By comparing the AGB estimates before and after the landslide event, the impact on above-ground carbon stocks can be assessed. The total AGB is calculated for the landslide area (polygons) by multiplying the AGBD values by each pixel's area under the landslide polygons and summing them across the landslide polygon regions for 2022 and 2023.

## 3. Results

This section illustrates and comments on the results first in terms of landslide susceptibility maps and then as SOC for each landslide hazard class.

### 3.1. Landslide Susceptibility Map

The landslide susceptibility maps obtained from the RF model introduced in the previous section are shown in Figure 2a,b, and the area falling in each susceptibility class for each type of landslide is summarized in Table 4 and graphically represented in Figure 3. The category "All" considers all types of landslides together without any specific distinction. A better-quality map can be downloaded.

**Table 4.** Areas (km$^2$) susceptible to different classifications of landslide types obtained from landslide susceptibility maps.

| Landslide Type | Very Low | Low | Low to Moderate | Moderate to High | High | Very High |
|---|---|---|---|---|---|---|
| All | 108,946.32 | 21,889.13 | 37,932.82 | 50,740.72 | 59,646.97 | 21,234.00 |
| C | 114,908.24 | 29,582.67 | 49,590.01 | 42,575.07 | 38,765.48 | 24,968.48 |
| R | 142,295.21 | 38,029.22 | 32,718.32 | 30,164.43 | 32,311.75 | 24,871.02 |
| SH | 97,312.387 | 23,854.49 | 49,941.89 | 62,858.51 | 46,691.48 | 19,731.20 |
| F | 139,696.10 | 45,507.46 | 34,386.96 | 28,975.88 | 28,991.85 | 22,831.71 |
| TR | 115,278.79 | 37,629.80 | 35,121.60 | 38,506.51 | 47,503.96 | 26,349.30 |
| RT | 81,825.44 | 53,948.82 | 56,061.36 | 49,942.25 | 39,319.29 | 19,292.79 |
| DSGSD | 141,084.88 | 51,165.93 | 34,987.85 | 24,515.30 | 24,594.20 | 24,041.80 |
| SL | 143,512.69 | 42,859.43 | 32,029.03 | 25,190.49 | 27,897.97 | 28,900.35 |
| DF | 212,932.59 | 25,752.76 | 13,077.50 | 9198.00 | 12,480.37 | 26,948.74 |

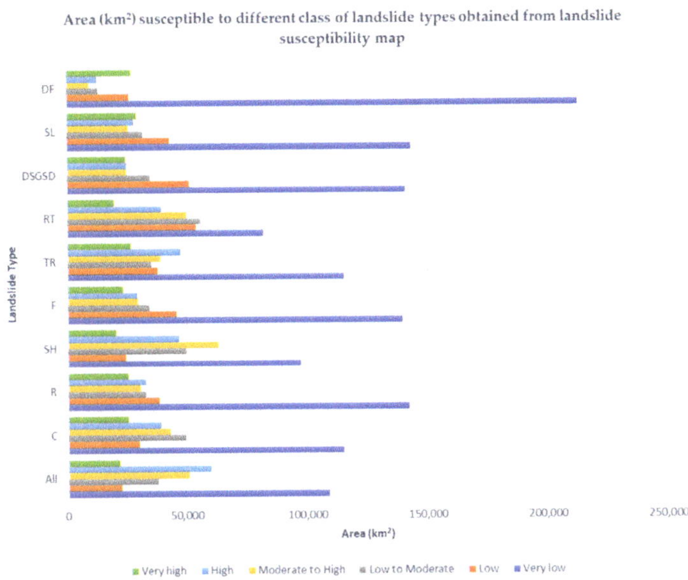


**Figure 3.** Areas (km$^2$) susceptible to different classifications of landslide types obtained from landslide susceptibility maps.

The model performances were evaluated using the Area Under the Receiver Operating Characteristic Curve (AUC-ROC), a metric that measures a model's ability to correctly distinguish between landslide and non-landslide areas (Figure 4). Specifically, the ROC curve plots the true positive rate (sensitivity) against the false positive rate (1 − specificity) at various classification thresholds. The area under this curve represents the model's ability to distinguish between positive (e.g., landslide) and negative (non-landslide) instances. An AUC-ROC score of 1 represents a perfect model, while a score of 0.5 implies performance no better than random chance. Generally, the higher the AUC-ROC value, the better the model's performance in separating the two classes. The models in this study achieved AUC scores ranging from 0.84 to 0.96, as shown in Figure 4, indicating strong overall accuracy and reliable predictions. Overall, higher AUC values reflect stronger model performance, with curves farther from the diagonal line indicating more accurate classification. Although not flawless, these scores suggest that the models are robust enough for assessing landslide susceptibility, effectively differentiating between areas at risk and those that are not.

Among the different landslide types, the model for diffused fall (DF) performed best with an AUC of 0.96, demonstrating excellent predictive power. The slow landslide classifier followed closely with an AUC of 0.91, reflecting strong accuracy, while the complex landslide model scored 0.87, indicating good but slightly lower performance. Other landslide types, such as rainfall-triggered (RT) and shallow landslides (SH), had AUC values of 0.85 and 0.84, respectively, indicating decent accuracy. The models for Deep-Seated Gravitational Slope Deformations (DSGSDs) and rapid landslides also performed well, both achieving AUC values between 0.91 and 0.92. The TR landslide model had an AUC of 0.86, indicating reasonable but improvable performance.

By this methodology, it is also possible to investigate variable importance (Figure 5); among the variables, rain and NDVI have the highest importance, followed by elevation in the case of all landslide types. HAND, slope, hillshade, soil density, and curvature metrics are moderately important. These are essential but not as critical as rain and NDVI. Variables like minimal curvature, shape index, and aspect have the least importance but still contribute to the model, while in the case of rainfall-triggered landslides (RT), elevation,

NDVI, and rain are the most important. It seems that when rain is explicitly modeled, elevation becomes significantly more critical, surpassing rain itself. This is because intense rainfall is the triggering factor of these mass movements, but the variable rainfall in the model represents a major causative factor and it is understood as a predisposing factor. Soil density, HAND, and slope follow, with hillshade and northness being slightly less important. Variables related to curvature and orientation (like curvature and aspect) have a lower influence.

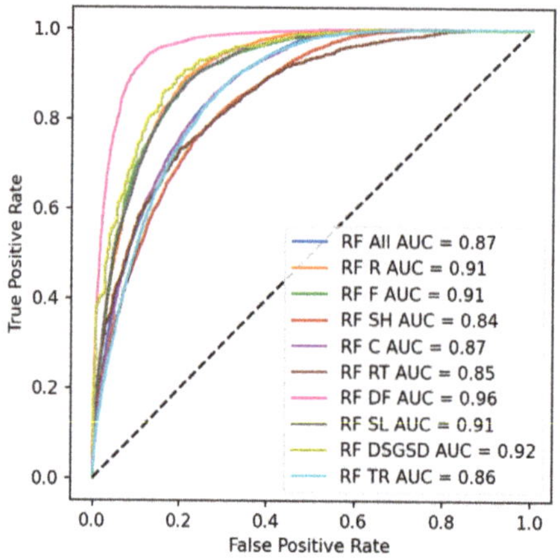


**Figure 4.** AUC-ROC for all, R, F, SH, C, RT, DF, SL, DSGSD, and TR landslide types.

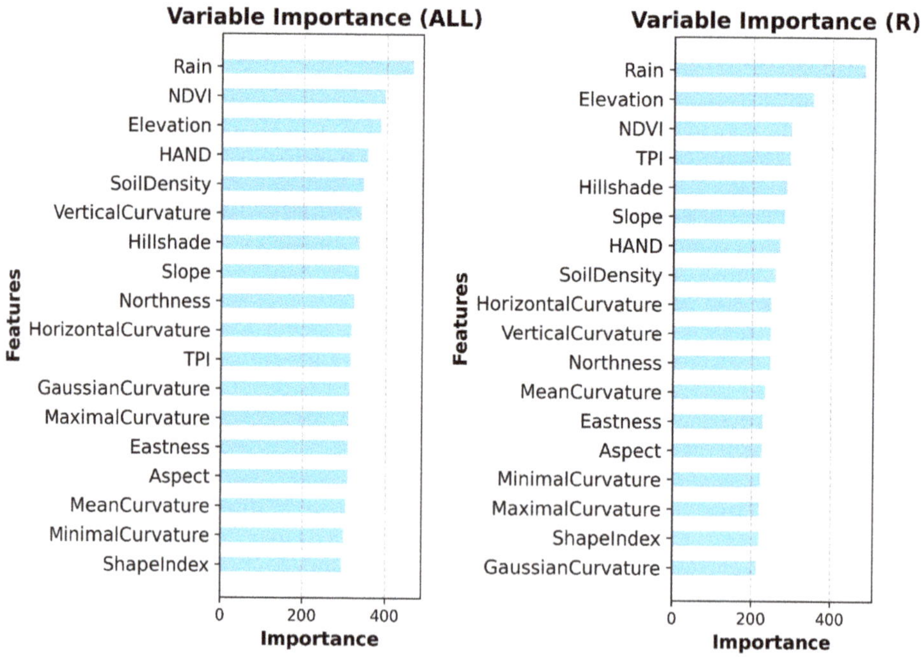


**Figure 5.** *Cont.*

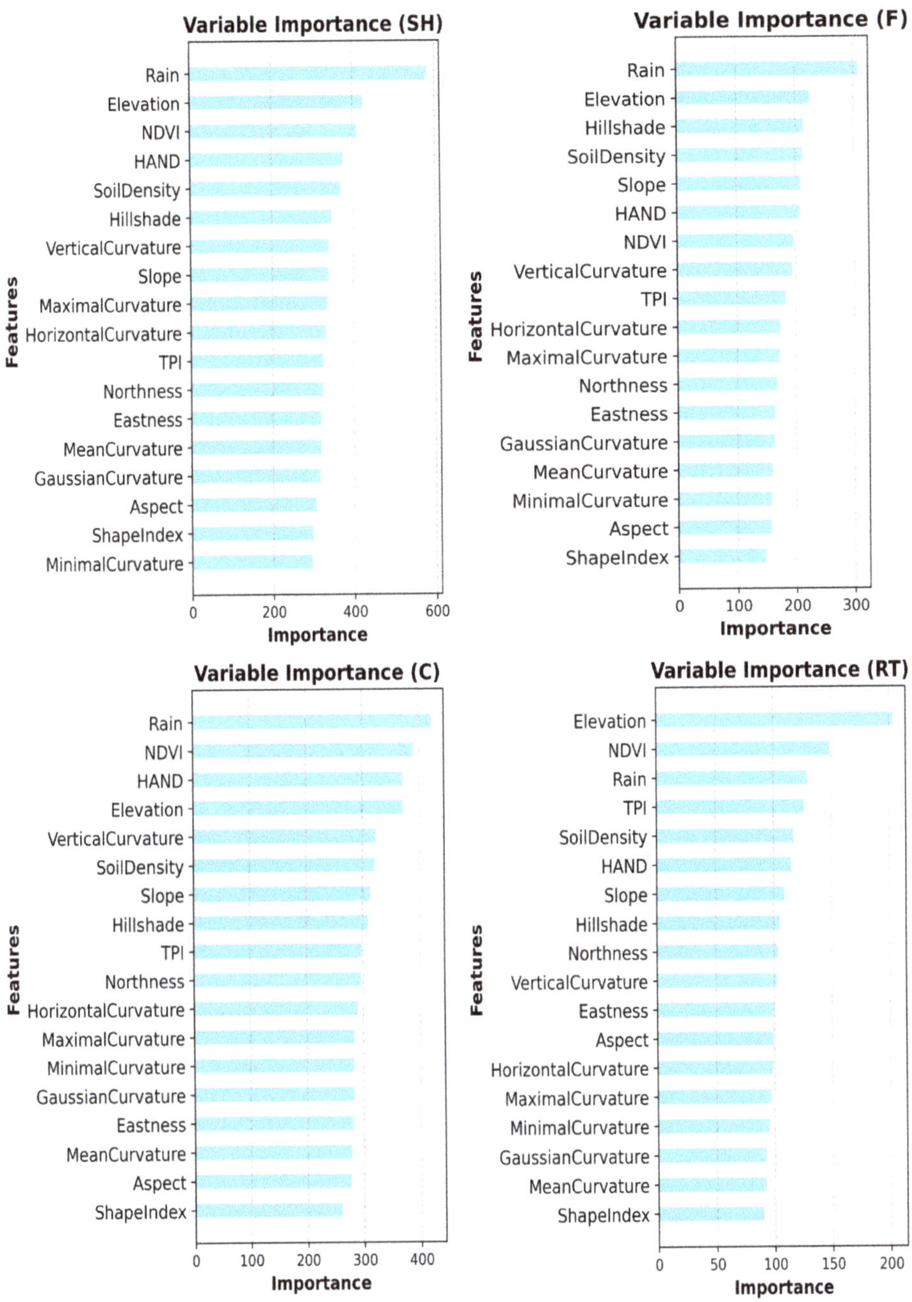


**Figure 5.** *Cont.*

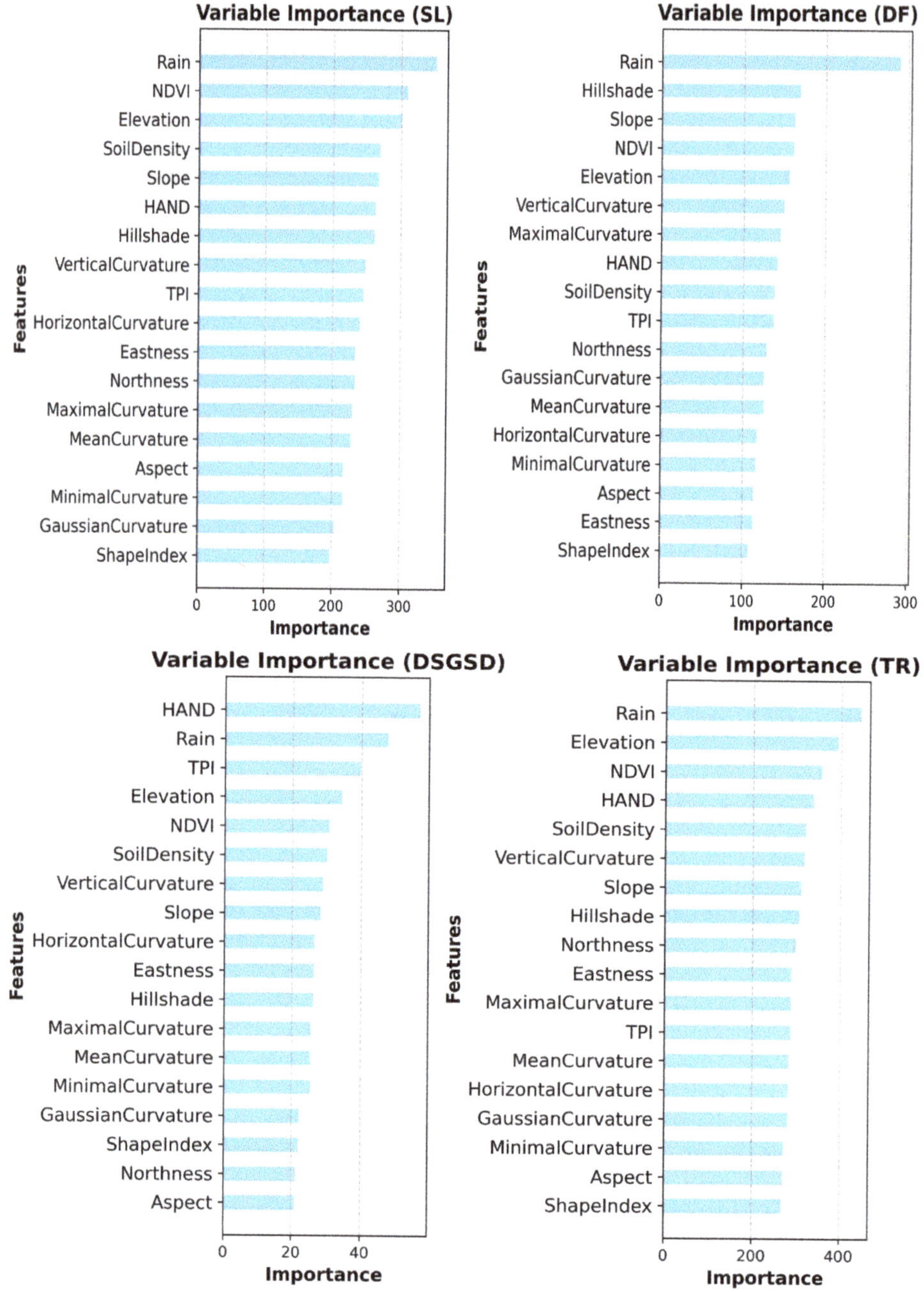


**Figure 5.** Feature importance for all, R, F, SH, C, RT, DF, SL, DSGSD, and TR landslide types.

In the case of the landslide type fall (F), rain dominates in importance, as comparable to other models. This suggests that rain is the most influential factor, which may be due to the weathering effect of rock slopes. Elevation and hillshade follow as the next most important. Soil density, HAND, and slope are important but significantly less than rain and elevation. Features like maximal curvature, northness, and eastness have relatively low importance. In the case of DSGSDs, HAND and rain are the most critical variables; TPI and elevation follow in importance but are significantly less impactful than HAND and rain. NDVI, soil density, and vertical curvature are moderately important but not as critical as the top features. Variables like northness, aspect, and shape index also have a significant impact.

We can draw an inference that each model gives a unique perspective on how environmental factors like terrain and climate variables influence outcomes. Rain is the most critical driver of landslide risk in most models, emphasizing that precipitation is not only a primary trigger for landslide events but also an important predisposing factor. High elevations and steep slopes consistently rank as important factors due to their impact on slope stability. Vegetation (NDVI) helps reduce landslide risk by stabilizing the soil, making this an essential factor in most models. Proximity to drainage networks is vital in landslide susceptibility (HAND), likely due to water's role in destabilizing slopes. Features like curvature, aspect, and TPI are less critical than rainfall and elevation but still influence landslide patterns by affecting runoff and erosion dynamics. Soil density and hillshade reflect soil properties and sunlight exposure, which indirectly impact landslide susceptibility. Each model gives the importance of variables based on seasonal or dynamic factors, reflecting how different environmental conditions influence landslide risk in various contexts.

### 3.2. SOC and Landslide Hazard Class

Figure 6 shows the total SOC and the mean SOC per unit of area in each landslide susceptibility class. Each subplot corresponds to a specific landslide type, and the x-axis represents the hazard level (ranging from 0 to 4), while the y-axis displays the total SOC (blue bars) and mean SOC (red line). Overall, this figure highlights the varying spatial relationships between SOC accumulation and landslide hazard levels across different landslide types. Some types, such as complex, shallow, rotational, and slow landslides, exhibit a decreasing trend in SOC with increasing hazard levels in classes 1 to 4; this may suggest that there is a beneficial effect of SOC in landslide hazard, but also that landslides that occurred in the past in highly susceptible areas had a negative effect of SOC. Other types of landslides, such as F, R DSGSD, and DF landslides, show an increasing trend. The elevated presence of SOC in highly susceptible areas may indicate that there are factors that increase both SOC and landslide hazards. For example, areas with high SOC often coincide with topographic features and climatic conditions that are inherently more susceptible to landslides, such as steep slopes, convergent topography where organic matter accumulates, or wet areas.

Considering all types of landslides together, low hazard levels (0 and 1) show limited mean and total SOC because these are typically urbanized and agricultural areas. Decreasing levels of mean and total SOC are observed for higher hazard levels (2 to 4), probably indicating that this factor may contribute positively to landslide hazard. At the same time, it is possible to observe that there is a significant amount of carbon stored in high or very high susceptibility classes which may become a carbon source if a landslide occurs.

These patterns suggest that the influence of landslides on SOC dynamics may vary depending on the specific landslide type and the environmental conditions associated with each type. Indeed, SOC is influenced by different factors, such as soil type, slope angle, rainfall patterns, and underlying geology, some of which are also relevant for landslide susceptibility, but in different measures depending on the landslide type. This can be useful for future studies related to soil health, carbon sequestration, and ecological balance. The interpretation provided and discussed in the present study should be regarded as a preliminary observation. While it offers initial insights, it requires further validation to ensure its accuracy and reliability. This can be achieved through additional field data collection, controlled experiments, or further supplementary studies. Such validation steps are essential to confirm the findings and strengthen the conclusions drawn from this study.

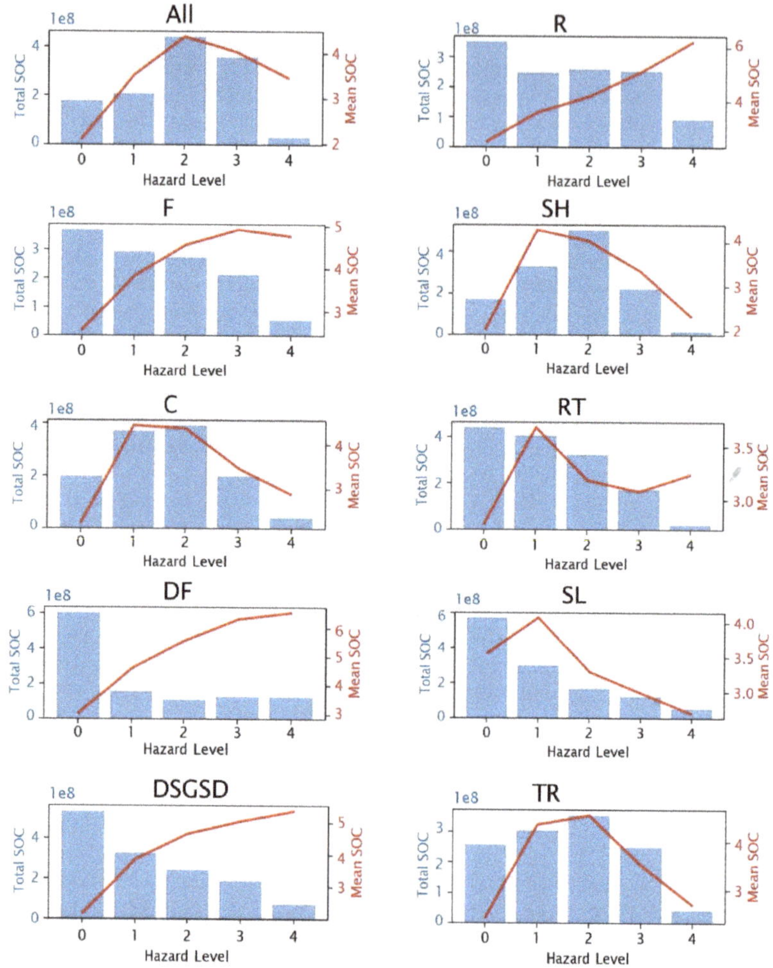


**Figure 6.** SOC accumulation trends and landslide type hazard level for all, R, F, SH, C, RT, DF, SL, DSGSD, and TR landslide types.

### 3.3. Short-Term Impact of Landslides on Carbon Stocks

The purpose of this section is to evaluate the impact of landslide events on carbon stocks, specifically soil organic carbon and above-ground biomass (AGB) in the study area (Figure 7). AGB, including leaves, stems, and branches, contributes to soil organic carbon through the addition of organic matter from litterfall, plant decomposition, and root exudates. Larger above-ground biomass often indicates a more extensive root system, which directly adds organic matter to the soil and supports microbial communities that decompose plant material into stable SOC forms. This biomass is crucial for carbon sequestration, as healthy vegetation captures atmospheric $CO_2$ and stores it in plant tissues, thus enhancing carbon storage in both the plant and soil. Changes in AGB can significantly impact SOC, influencing soil health, structure, fertility, and overall ecosystem productivity. Positive feedback loops can enhance biomass and SOC, improving soil conditions and promoting further growth, while negative feedback loops can degrade soil and reduce plant growth. Understanding the effects of landslides on these crucial carbon reservoirs is essential for assessing the overall influence on the carbon cycle and potential implications for climate change.

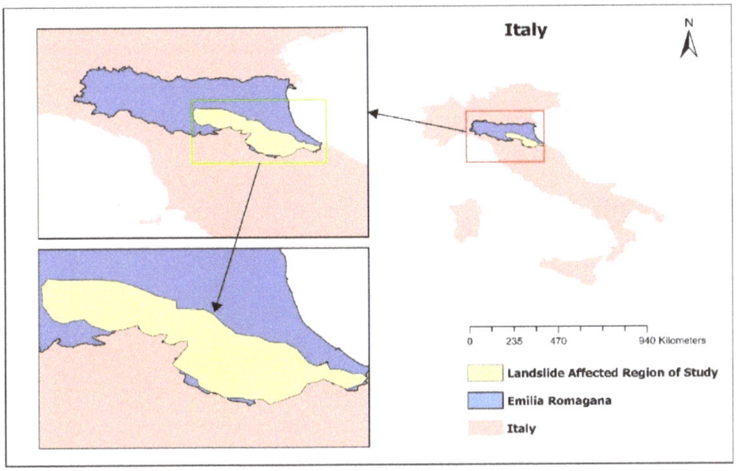


**Figure 7.** Landslide-affected region.

The analysis of SOC changes based on NDVI revealed that a larger area (approximately 21 million square meters) (Figure 8) experienced a decrease in SOC accumulation. However, a smaller area (around 8 million square meters) exhibited an increase in SOC, potentially due to environmental recovery or ecosystem shifts following the landslide event. In Figure 8, the red areas show negative carbon stock change, and the blue areas show positive carbon stock change. Negative sequestration change indicates a reduction in the area's ability to capture and store carbon. Positive sequestration change means that in this part of the landslide-affected area, there has been an increase in the ability to sequester carbon.

**Figure 8.** Impact of landslides on vegetation and carbon sequestration by comparing the NDVI (Normalized Difference Vegetation Index) and biomass levels before and after the landslides in the region.

To evaluate and quantify the density and total stock of above-ground biomass, we selected only the landslide polygons out of the whole of Emilia Romagna with the aim of finding the difference between 2022 and 2023. By comparing the AGB estimates before and after the landslide event, the impact on above-ground carbon stocks can be assessed. Figure 9 shows the AGBD difference before and after the event of the entire affected area.

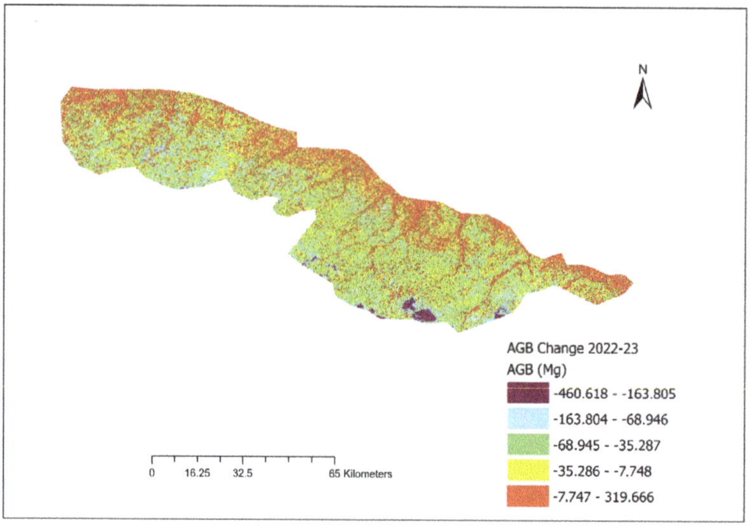


**Figure 9.** Above-ground biomass density difference between 2022 and 2023.

The comparison of AGB estimates before and after the landslide event showed a significant decrease in above-ground biomass, from 750,266.23 Mg in 2022 to 484,902.34 Mg in 2023. This reduction in AGB suggests that less carbon is being stored in the above-ground vegetation, potentially leading to the release of previously sequestered carbon into the atmosphere.

The loss of 265,364 Mg (megagrams) of AGB may lead to the release of 487,000 metric tons of $CO_2$ (see Appendix A), which could contribute to atmospheric greenhouse gases, exacerbating climate change. This is a significant loss of carbon sequestration potential from landslide-affected areas, contributing to the local and regional environmental imbalance.

## 4. Discussion

This section discusses the findings of this study in the context of the existing literature, highlights its limitations, and explores its implications for understanding the interplay between landslides, soil organic carbon (SOC), and the carbon cycle.

The analysis presented in Section 3.2 investigated, for the first time in the literature, patterns between landslide hazard levels and SOC accumulation. The results reveal that the trends depend on the landslide types. The overall trend indicates a decrease in SOC stocks with increasing landslide susceptibility, suggesting a correlation between environmental conditions and SOC accumulation, with higher susceptibility areas potentially experiencing lower SOC levels. This suggests that, on one hand, SOC may be beneficial for landslide hazard reduction, and on the other hand, there is a significant amount of SOC stored in highly susceptible areas, which may potentially become a source of $CO_2$ emission if a landslide occurs.

The results presented in Section 3.3 align with previous research by Cox et al. (2024) [29] and Frith et al. (2018) [30], which examined carbon losses associated with earthquake-induced landslides. Cox et al. [29] reported total carbon losses of 2.05 Mt and 0.217 Mt from major events in New Zealand, indicating significant short-term carbon disruptions relative to the region's annual carbon sequestration (6.3 Mt in 2020). Similarly,

Frith's simulations estimated a potential erosion of 14 ± 5 MtC from a single large earthquake. In contrast, the current study focused on rainfall-triggered landslides in Emilia Romagna and observed a loss of 265,364 Mg of above-ground biomass (AGB), equivalent to 0.133 MtC. While these losses are smaller in magnitude, it should be noted that our study is focused on the landslide polygons and not on the scale of the region as was examined in previous studies. This underscores the localized but significant impact of rainfall-triggered landslides on carbon dynamics.

Unlike earthquake-induced events, which are less frequent, rainfall-triggered landslides are expected to increase in frequency due to climate change. This cumulative effect represents a substantial risk to regional carbon storage, emphasizing the importance of rainfall-triggered events in long-term carbon assessments. The observed decrease in AGB, a proxy for carbon sequestration, further supports the conclusion that landslides disrupt above-ground carbon stocks and alter the balance between carbon storage and emissions.

The study also highlights the importance of considering recovery timelines, as vegetation regrowth and carbon stock restoration can span decades, influencing long-term carbon dynamics. Prior studies on post-landslide ecosystem recovery corroborate this finding, indicating that ecosystem resilience plays a critical role in mitigating the carbon impacts of landslides.

While this research provides valuable insights into the spatial relationship between landslides, soil organic carbon (SOC), and the carbon cycle, it is important to acknowledge the following limitations. The findings presented are based on remote sensing data and modeling techniques, which, while powerful, have inherent limitations. The results are preliminary and require further validation through field data collection and experimental studies to confirm the observed patterns and trends. This study focused on specific variables, such as SOC, above-ground biomass (AGB), and landslide susceptibility. However, additional environmental factors, including vegetation diversity, soil moisture, and land-use changes, were not incorporated, which might influence the observed relationships. The study has temporal constraints because the analysis centers on the May 2023 landslide event, providing a snapshot rather than a long-term perspective, which could reveal more about the resilience of ecosystems and carbon dynamics over time. Additionally, the spatial resolution may not fully capture small-scale heterogeneities in SOC dynamics. The estimation of carbon losses, especially $CO_2$ emissions, is based on established conversion factors and assumptions. These estimates could benefit from direct measurement and validation in affected areas to refine accuracy. The study's findings are based on landslides in Italy, particularly Emilia Romagna, which may limit the generalizability of the conclusions to other regions with different climatic, geological, and ecological conditions.

Future studies should address these limitations by incorporating field data, expanding variable selection, and examining long-term impacts to provide a more comprehensive understanding of landslide–carbon interactions.

## 5. Conclusions

This study contributes to advancing the current understanding of the relationship between different types of landslides, soil organic carbon (SOC), and the carbon cycle. This relationship has important implications for environmental sustainability and climate change mitigation efforts.

The analysis reveals a pattern between landslide hazard levels and SOC accumulation, with varying trends observed for different landslide types. The overall trend indicates a decrease in SOC stocks with increasing landslide susceptibility, suggesting a correlation between environmental conditions and SOC accumulation. This suggests that SOC may be beneficial for landslide hazard reduction, but there is also a significant amount of SOC in highly susceptible areas, which may potentially become a source of $CO_2$ emissions if landslides occur.

The case study of the May 2023 landslide event in the Emilia Romagna region of Italy highlights the potential impact of large-scale landslides on carbon sequestration. The observed decrease in SOC accumulation and AGB in the affected regions indicates that landslides can transform carbon sinks into carbon sources, at least in the short term. The result is a significant loss of carbon sequestration potential in Emilia Romagna and suggests that the effects of landslide events need to be accounted for in regional or national carbon emissions. This phenomenon is particularly concerning as climate change is expected to increase the frequency and intensity of extreme weather events, potentially triggering more landslides and exacerbating the release of stored carbon.

This study underscores the importance of considering landslides in carbon cycle assessments, especially as climate change is expected to induce more extreme weather events, potentially turning carbon-absorbing areas into carbon sources as these extreme events largely contribute to worsening the impacts of climate change. This creates a feedback loop where landslides amplify the climate changes that cause them and lead to environmental imbalance, severe disruptions to biodiversity, soil health, and local climate regulation, with wide-reaching consequences.

This study represents a preliminary step in quantifying the impact of landslides on carbon stocks, utilizing remote sensing data and modeling techniques. While the study provides valuable insights, the authors acknowledge that the interpretations presented are preliminary and require further validation through field data collection, experimental studies, and the incorporation of additional environmental parameters. Further research is necessary to fully understand the interactions between landslides, SOC dynamics, and the carbon cycle. This knowledge is crucial for improving hazard management, developing effective climate change mitigation strategies, and implementing soil conservation measures.

**Author Contributions:** Conceptualization, J.Q. and F.C.; methodology, J.Q. and F.C.; software, J.Q.; validation, J.Q.; writing—original draft preparation, J.Q. and F.C.; writing—review and editing, J.Q. and F.C.; visualization, J.Q. and F.C.; supervision, F.C.; funding acquisition, F.C. All authors have read and agreed to the published version of the manuscript.

**Funding:** This study was funded by the European Union—NextGenerationEU, Mission 4, Component 2, in the framework of the GRINS—Growing Resilient, INclusive and Sustainable project (GRINS PE00000018—(GRINS PE00000018—CUP C93C22005270001)). The views and opinions expressed are solely those of the authors and do not necessarily reflect those of the European Union, nor can the European Union be held responsible for them.

**Institutional Review Board Statement:** Not applicable.

**Informed Consent Statement:** Not applicable.

**Data Availability Statement:** The original contributions presented in the study are included in the article, further inquiries can be directed to the corresponding author.

**Conflicts of Interest:** The authors declare no conflicts of interest.

## Appendix A

This appendix illustrates how the amount of carbon dioxide ($CO_2$) released from the loss of AGB can be estimated. AGB, which refers to plant materials like trees and vegetation, contains around 50% carbon by dry weight [60]. In the considered area, the total AGB loss is 265,364 metric tons; to find the carbon content, we multiply the biomass amount by 0.5 (50%), which gives us 132,682 metric tons of carbon. Since burning biomass releases $CO_2$, we need to convert the carbon amount to $CO_2$. This is achieved by multiplying the carbon amount by the ratio of the molecular weights of $CO_2$ and carbon (44/12 or 3.67). The result is 486,941 metric tons of $CO_2$ released.

## References


1. Yi, C.; Ricciuto, D.; Li, R.; Wolbeck, J.; Xu, X.; Nilsson, M.; Aires, L.; Albertson, J.D.; Ammann, C.; Arain, M.A.; et al. Climate control of terrestrial carbon exchange across biomes and continents. *Environ. Res. Lett.* **2010**, *5*, 034007. [CrossRef]
2. Shiels, A.B.; Walker, L.R.; Thompson, D.B. Organic matter inputs create variable resource patches on Puerto Rican landslides. *Plant Ecol.* **2006**, *184*, 223–236. [CrossRef]
3. Shiels, A.B.; Walker, L.R. Landslides cause spatial and temporal gradients of multiple scales in the Laquillo Mountains of Puerto Rico. *Ecol. Bull.* **2013**, *54*, 211–221.
4. Lopes, L.F.; Oliveira, S.C.; Neto, C.; Zezere, J.L. Vegetation Evolution by Ecological Succession as a Potential Bioindicator of Landslides Relative Age in Southwestern Mediterranean Region. *Nat. Hazards* **2020**, *103*, 599–622. [CrossRef]
5. Gorham, E. Northern Peatlands: Role in the Carbon Cycle and Probable Responses to Climatic Warming. *Ecol. Appl.* **1991**, *1*, 182–195. [CrossRef]
6. Freeman, C.; Fenner, N.; Ostle, N.J.; Kang, H.; Dowrick, D.J.; Reynolds, B.; Lock, M.A.; Sleep, D.; Hughes, S.; Hudson, J. Export of dissolved organic carbon from peatlands under elevated carbon dioxide levels. *Nature* **2004**, *430*, 195–198. [CrossRef] [PubMed]
7. Lorenz, K.; Lal, R. Effects of Disturbance, Succession and Management on Carbon Sequestration. In *Carbon Sequestration in Forest Ecosystems*; Springer: Dordrecht, The Netherlands, 2010. [CrossRef]
8. Brevik, E.C. Soils and climate change: Gas fluxes and soil processes. *Soil Horiz.* **2012**, *53*, 12–23. [CrossRef]
9. Yazdanpanah, N.; Mahmoodabadi, M.; Cerdà, A. The impact of organic amendments on soil hydrology, structure and microbial respiration in semiarid lands. *Geoderma* **2016**, *266*, 58–65. [CrossRef]
10. Błońska, E.; Lasota, J.; Piaszczyk, W.; Wiecheć, M.; Klamerus-Iwan, A. The effect of landslide on soil organic carbon stock and biochemical properties of soil. *J. Soils Sediments* **2018**, *18*, 2727–2737. [CrossRef]
11. Booth, A.M.; Buma, B.; Nagorski, S. Effects of landslides onterrestrial carbon stocks with a coupledgeomorphic-biologic model: SoutheastAlaska, United States. *J. Ofgeophysical Res. Biogeosci.* **2023**, *128*, e2022JG007297. [CrossRef]
12. Clark, K.E.; West, A.J.; Hilton, R.G.; Asner, G.P.; Quesada, C.A.; Silman, M.R.; Saatchi, S.S.; Farfan-Rios, W.; Martin, R.E.; Horwath, A.B.; et al. Storm-triggered landslides in the Peru-vian Andes and implications for topography, carbon cycles, and biodiversity. *Earth Surf. Dyn.* **2016**, *4*, 47–70. [CrossRef]
13. Vascik, B.A.; Booth, A.M.; Buma, B.; Berti, M. Estimated amounts and rates of carbon mobilized by landsliding in old-growth temperate forests of SE Alaska. *J. Geophys. Res. Biogeosci.* **2021**, *126*, e2021JG006321. [CrossRef]
14. Wang, J.; Jin, Z.; Hilton, R.G.; Zhang, F.; Li, G.; Densmore, A.L.; Gröcke, D.R.; Xu, X.; West, A.J. Earthquake-triggered increase in biospheric carbon export froma mountain belt. *Geology* **2016**, *44*, 471–474. [CrossRef]
15. Ramos Scharrón, C.E.; Castellanos, E.J.; Restrepo, C. The transfer of modern organic carbon by landslide activity in tropical montane ecosystems. *J. Geophys. Res. Biogeosci.* **2012**, *117*. [CrossRef]
16. Błońska, E.; Lasota, J.; Zwydak, M.; Klamerus-Iwan, A.; Gołąb, J. Restoration of forest soil and vegetation 15 years after landslides in a lower zone of mountains in temperate climates. *Ecol. Eng.* **2016**, *97*, 503–515. [CrossRef]
17. Van Oost, K.; Six, J. Reconciling the paradox of soil organic carbon erosion by water. *Biogeosciences* **2023**, *20*, 635–646. [CrossRef]
18. Lal, R. Accelerated soil erosion as a source of atmospheric $CO_2$. *Soil Tillage Res.* **2019**, *188*, 35–40. [CrossRef]
19. Piaszczyk, W.; Lasota, J.; Gaura, G.; Błońska, E. Effect of Deadwood Decomposition on the Restoration of Soil Cover in Landslide Areas of the Karpaty Mountains, Poland. *Forests* **2021**, *12*, 237. [CrossRef]
20. Schimel, D.S. Terrestrial ecosystem and the carbon cycle. *Glob. Chang. Biol.* **1995**, *1*, 77–91. [CrossRef]
21. Lal, R. Forests soil and carbon sequestration. *For. Soils Carbon Sequestration* **2005**, *220*, 242–258. [CrossRef]
22. Page, S.E.; Siegert, F.; Rieley, J.O.; Boehm, H.D.V.; Jaya, A.; Limin, S. The amount of carbon released from peat and forest ires in Indonesia during 1997. *Nature* **2002**, *420*, 61–65. [CrossRef]
23. Goyal, D.; Joshi, V.; Gupta, N.; Cabral-Pinto, M.M.S. Soil Quality Assessment in a Landslide Chronosequence of Indian Himalayan Region. *Land* **2022**, *11*, 1819. [CrossRef]
24. Chen, S.; Hua, J.; Liu, W.; Yang, S.; Wang, X.; Ji, W. Effects of Artificial Restoration and Natural Recovery on Plant Communities and Soil Properties across Different Temporal Gradients after Landslides. *Forests* **2023**, *14*, 1974. [CrossRef]
25. Nicia, P.; Bejger, R.; Sterzyńska, M.; Zadrożny, P.; Parzych, P.; Bieda, A.; Kwartnik-Pruc, A. Recovery in soil cover and vegetation structure after ancient landslide in mountain fens under Caltho-Alnetum community and response of soil microarthropods (Hexapoda: Collembola) to natural restoration process. *J. Soils Sediments* **2019**, *20*, 714–722. [CrossRef]
26. Pang, D.; Xu, H. Carbon Sequestration and Stability and Soil Erosion in Forest Ecosystems. *Forests* **2024**, *15*, 1961. [CrossRef]
27. Kang, D.; Li, Y.; Ma, L.; Zou, S. Landslide scales affect soil organic carbon accumulation by influencing microbial decomposition of plant-derived carbon after earthquakes. *Ecol. Indic.* **2023**, *155*, 110949. [CrossRef]
28. Pandey, H.P.; Gnyawali, K.; Dahal, K.; Pokhrel, N.P.; Maraseni, T.N. Vegetation loss and recovery analysis from the 2015 Gorkha earthquake (7.8 Mw) triggered landslides. *Land Use Policy* **2022**, *119*, 106185. [CrossRef]
29. Cox, C. Carbon Loss from Earthquake Induced Landslides in Fiordland. *Pūhau Tailwinds* **2024**, *2*, 2. [CrossRef]
30. Frith, N.V.; Hilton, R.G.; Howarth, J.D.; Gröcke, D.R.; Fitzsimons, S.J.; Croissant, T.; Wang, J.; McClymont, E.L.; Dahl, J.; Densmore, A.L. Carbon export from mountain forests enhanced by earthquake-triggered landslides over millennia. *Nat. Geosci.* **2018**, *11*, 772–776. [CrossRef]
31. Peduzzi, P. Landslides and vegetation cover in the 2005 North Pakistan earthquake: A GIS and statistical quantitative approach. *Nat. Hazards Earth Syst. Sci.* **2010**, *10*, 623–640. [CrossRef]

32. Niraj, K.C.; Singh, A.; Shukla, D.P. Effect of the Normalized Difference Vegetation Index (NDVI) on GIS-Enabled Bivariate and Multivariate Statistical Models for Landslide Susceptibility Mapping. *J. Indian Soc. Remote Sens.* **2023**, *51*, 1739–1756. [CrossRef]
33. Shahabi, H.; Hashim, M. Landslide susceptibility mapping using GIS-based statistical models and Remote sensing data in tropical environment. *Sci. Rep.* **2015**, *5*, 9899. [CrossRef] [PubMed]
34. Sevgen, E.; Kocaman, S.; Nefeslioglu, H.A.; Gokceoglu, C. A novel performance assessment approach using photogrammetric techniques for landslide susceptibility mapping with logistic regression, ANN and random forest. *Sensors* **2019**, *19*, 3940. [CrossRef] [PubMed]
35. Peruccacci, S.; Gariano, S.L.; Melillo, M.; Solimano, M.; Guzzetti, F.; Brunetti, M.T. The ITAlian rainfall-induced LandslIdes CAtalogue, an extensive and accurate spatio-temporal catalogue of rainfall-induced landslides in Italy. *Earth Syst. Sci. Data* **2023** *15*, 2863–2877. [CrossRef]
36. IFFI. IFFI—Inventory of Landslides in Italy (isprambiente.it). Available online: https://idrogeo.isprambiente.it/app/pir?@=41.55172525858242,12.573501484000001,1 (accessed on 30 July 2024).
37. Hungr, O.; Leroueil, S.; Picarelli, L. The Varnes classification of landslide types, an update. *Landslides* **2014**, *11*, 167–194. [CrossRef]
38. Varnes, D.J. Slope Movement Types and Processes. *Spec. Rep.* **1978**, *176*, 11–33.
39. Triggering and Propagation of Rapid Flow-like Landslides—The Proceedings of the Second JTC-1 Workshop—The Landslide Blog—AGU Blogosphere. Available online: https://blogs.agu.org (accessed on 20 October 2024).
40. Larsen, M.C. Rainfall-triggered landslides, anthropogenic hazards, and mitigation strategies. *Adv. Geosci.* **2008**, *14*, 147–153 [CrossRef]
41. Colombo, A.; Lanteri, L.; Ramasco, M.; Troisi, C. Systematic GIS-based landslide inventory as the first step for effective landslide-hazard management. *Landslides* **2005**, *2*, 291–301. [CrossRef]
42. Apip; Takara, K.; Yamashiki, Y.; Sassa, K.; Ibrahim, A.B.; Fukuoka, H. A distributed hydrological–geotechnical model using satellite-derived rainfall estimates for shallow landslide prediction system at a catchment scale. *Landslides* **2010**, *7*, 237–258. [CrossRef]
43. Conforti, M.; Muto, F.; Rago, V.; Critelli, S. Landslide inventory map of north-eastern Calabria (South Italy). *J. Maps* **2013**, *10*, 90–102. [CrossRef]
44. Evaluating the Pre and Syn-Collapse Conditions of Rainfall-Triggered Landslides in Indonesia: Steps Toward Improved Forecasting—Envision—Developing Next Generation Leaders in Environmental Science. Available online: https://www.envision-dtp.org/2023/evaluating-the-pre-and-syn-collapse-conditions-of-rainfall-triggered-landslides-in-indonesia-steps-toward-improved-forecasting/ (accessed on 20 October 2024).
45. Loche, M.; Alvioli, M.; Marchesini, I.; Bakka, H.; Lombardo, L. Landslide susceptibility maps of Italy: Lesson learnt from dealing with multiple landslide types and the uneven spatial distribution of the national inventory. *Earth-Sci. Rev.* **2022**, *232*, 104125. [CrossRef]
46. Application Landslides Archives—Sisgeo. Available online: https://sisgeo.com/category/applications/landslides/ (accessed on 20 October 2024).
47. Slow-Moving Landslides Are a Growing Threat to Mountain Communities, Say Researchers | Technology Networks. Available online: https://www.technologynetworks.com/applied-sciences/news/slow-moving-landslides-are-a-growing-threat-to-mountain-communities-say-researchers-391105 (accessed on 20 October 2024).
48. Guerricchio, A.; Simeone, V. Preliminary remarks on gravitational tectonic deformations and DSGSD in mount Poro headland (Calabria—South Italy). *Rend. Online Soc. Geol. Ital.* **2012**, *21*, 370–371.
49. Deep-Seated Landslide on Mountain in Sehara, Kihō, Beside of Kumano River between Wakayama and Mie Japan Caused by Torrential Rain of Tropical Storm Talas. 2011. Available online: https://www.unsustainablemagazine.com/landslide-prevention/ (accessed on 20 October 2024).
50. Flood in Emilia-Romagna in May 2023, Cartographic Services to Support Emergency Management and Reconstruction Activities—Geoportale. Available online: https://geoportale.regione.emilia-romagna.it/approfondimenti/emergenza-maggio-23/emergenza-rer-maggio-2023-servizi (accessed on 15 October 2024).
51. Huang, F.; Zhang, J.; Zhou, C.; Wang, Y.; Huang, J.; Zhu, L. A deep learning algorithm using a fully connected sparse autoencoder neural network for landslide susceptibility prediction. *Landslides* **2020**, *17*, 217–229. [CrossRef]
52. Mahalingam, R.; Olsen, M.J. Evaluation of the influence of source and spatial resolution of DEMs on derivative products used in landslide mapping. *Geomat. Nat. Hazards Risk* **2016**, *7*, 1835–1855. [CrossRef]
53. Gorelick, N.; Hancher, M.; Dixon, M.; Ilyushchenko, S.; Thau, D.; Moore, R. Google Earth Engine: Planetary-scale geospatial analysis for everyone. *Remote Sens. Environ.* **2017**, *202*, 18–27. [CrossRef]
54. Jin, Z.; Azzari, G.; You, C.; Di Tommaso, S.; Aston, S.; Burke, M.; Lobell, D.B. Smallholder maize area and yield mapping at national scales with Google Earth Engine. *Remote Sens. Environ.* **2019**, *228*, 115–128. [CrossRef]
55. Prakash, N.; Manconi, A.; Loew, S. Mapping landslides on EO data: Performance of deep learning models vs. Traditional machine learning models. *Remote Sens.* **2020**, *12*, 346. [CrossRef]
56. Trigila, A.; Iadanza, C.; Spizzichino, D. Quality assessment of the Italian Landslide Inventory using GIS processing. *Landslides* **2010**, *7*, 455–470. [CrossRef]
57. Kavzoglu, T.; Sahin, E.K.; Colkesen, I. An assessment of multivariate and bivariate approaches in landslide susceptibility mapping: A case study of Duzkoy district. *Nat. Hazards* **2015**, *76*, 471–496. [CrossRef]

8. Kellner, J.R.; Armston, J.; Duncanson, L. Algorithm theoretical basis document for GEDI footprint aboveground biomass density. *Earth Space Sci.* **2023**, *10*, e2022EA002516. [CrossRef]
9. Shendryk, Y. Fusing GEDI with earth observation data for large area aboveground biomass mapping. *Int. J. Appl. Earth Obs. Geoinf.* **2022**, *115*, 103108. [CrossRef]
0. 2006 IPCC Guidelines for National Greenhouse Gas Inventories, Volume 4, Agriculture, Forestryand Other Land Use. Available online: https://www.ipcc-nggip.iges.or.jp/public/2006gl/vol4.html (accessed on 20 October 2024).

sustainability

*Article*

# The Effect of Soil Tillage Systems on the Soil Microbial and Enzymatic Properties Under Soybean (*Glycine max* L. Merrill) Cultivation—Implications for Sustainable Soil Management

**Jacek Długosz, Bożena Dębska and Anna Piotrowska-Długosz ***

Department of Biogeochemistry and Soil Science, Faculty of Agriculture and Biotechnology, Bydgoszcz University of Science and Technology, 6 Bernardyńska St, 85-029 Bydgoszcz, Poland; jacekd@pbs.edu.pl (J.D.); debska@pbs.edu.pl (B.D.)

* Correspondence: apiotr@pbs.edu.pl; Tel.: +48-52-374-95-55

**Abstract:** Reducing soil tillage with the application of catch-crop green mass as a mulch is a conservation practice that is used in agriculture to improve soil ecosystem functioning. Such a cultivation method enhances soil organic matter quantity and quality through the improvement of soil biological activity and nutrient availability, while reducing soil disturbance. Therefore, a three-year field experiment was conducted in the years 2017–2019 to evaluate the effect of three tillage methods (TMs) (conventional, CT; reduced, RT; and strip tillage, ST) on soil microbial and specific enzyme properties (microbial C and N content, the activity of dehydrogenases—DHA, the rate of fluorescein sodium salt hydrolysis—FDAH, CMC-cellulase—Cel and β-glucosidase—Glu) and certain basic soil properties. The study was performed in a field; it was a one-factor experiment that was carried out in a randomized block design. The soil samples were collected from the upper soil layer five times a year: in April (before the sowing of soybean), May, June, August and September (before soybean harvesting). The tillage methods or sampling dates used had no significant effect on the organic carbon and total nitrogen levels. Most of the C-related properties (the content of microbial biomass and the C-cycling enzymatic activity such as Cel and Glu) and microbial activity bioindicators (DHA activity, FDAH rate) revealed significant seasonal changes, whereby each variable was affected in a different order (e.g., the Cel activity was significantly higher in April and September than in other months—22%, while the DHA activity was significantly higher in June and August compared to other months—18%). RT significantly increased the enzymatic activity as compared to CT and ST, and the difference was between 8 and 33% (with a mean of 18%). The exception was the β-glucosidase activity as determined in 2019, which was significantly higher in the case of CT (1.02 mg *p*NP $kg^{-1}$ $h^{-1}$) and ST than in RT (0.705 mg *p*NP $kg^{-1}$ $h^{-1}$). However, the explanation for such phenomenon could not possibly be based on the available data. Our results suggested that the response of the enzyme activities toward the same factor may be due to the inherent variability in enzyme response associated with the spatial variability in soil properties as well as the properties of the enzyme itself and changes in the periodic occurrence of its substrates in the soil. Generally, the reduced tillage combined with plant residues return could be recommended for enhancing soil health and quality by improving its microbial and enzymatic features. The findings above suggest that a reduced tillage system is an important component of soil management in sustainable agriculture.

**Keywords:** cover crops; microbial biomass; soil enzymes; reduced tillage; strip-till; soybean; sustainable soil management

## 1. Introduction

Soil is considered to be a key, unrenewable resource for agricultural production. Moreover, soil also performs other essential ecosystem services like water and nutrient storage and regulation, C sequestration, the degradation and immobilization of pollutants and the reduction in greenhouse gases ($CO_2$ and $N_2O$) emissions [1]. However, the resource of healthy and high-quality soils is decreasing globally due to widespread land degradation [2]. Although about 85% of land degradation is caused by water and wind erosion [3], soil degradation is also caused by unsustainable soil management practices, which include intensive farming practices such as conventional tillage, improper application of inorganic fertilizers and plant protection products, soil compaction caused by agricultural machinery and soil salinization due to the improper irrigation; all of these also contribute to a decrease in soil quality [4,5]. Since these unsustainable farming practices have a potent impact on soil biological, physical and chemical properties, there is an increasing tendency for the application of a tillage system which results in sustainable, healthy and quality soil, which is of great significance in modern, sustainable agriculture [6]. In recent years, certain conservation tillage systems, like the no-till or reduced tillage systems, have often been practiced as the objectives of a sustainable approach to agriculture [7,8]. These tillage systems are often preferred over conventional tillage due to their limited physical disturbance of the soil and the presence of plant residues (treated as mulch) on the soil surface, which improve its hydrothermal conditions (decrease soil temperature and limit water losses). This, in turn, increases soil organic matter accumulation [9–11]. One of the reduced tillage techniques used for soil cultivation, which limit both soil reversing and disturbance, is strip tillage. This technique integrates some benefits of no-till (soil and water conservation) and conventional tillage (e.g., improvement of seedbed conditions, enhancement of seed germination) and indicates a positive impact on numerous soil properties, including those essential to increasing the content of organic matter by improving air–water conditions and enhancing soil biological activity [12,13]. Over the last few decades, strip tillage has been increasingly adopted as a technique of conservation tillage worldwide [14,15]. Nevertheless, the reaction of plant and soil features to soybean cultivation is little known as of yet and needs additional research. That is why further research should be conducted to explain the impact of various tillage systems, especially in the view of changeable climatic conditions and the growing interest in replacing conventional tillage with alternative systems such as ploughless or simplified tillage regimes [16]. It would be particularly valuable to conduct studies concerning local conditions in Northern and Central Poland. To date, only a restricted number of Polish studies have been conducted with reference to the influence of reduced tillage on plant and soil variables under soybean cultivation; at present, there are no national literature sources related to the cultivation of soybean in strip tillage [17]. Globally, inconsistent results have been presented with reference to the impact of various tillage methods concerning the soil environment status under soybean cultivation [18]. Soybean has traditionally been cultivated using conventional tillage (CT), which requires many operations that increase the cost of cultivation enormously, and also cause soil properties to worsen due to soil compaction in addition to the depletion of the soil organic matter content. Conflicting data have been obtained with regards to the effect of various tillage methods on the soil environment under soybean cultivation, although many of the results indicate the advantages of the no-till or reduced tillage methods/systems on certain soil properties as compared to conventional tillage, especially in the long term [16]. As was noted by Weidhuner et al. [19] and Alhameid et al. [20], long-term conventional tillage (CT) system alters both the physical and chemical properties of the soil, which results in an aggregate breakdown, a decrease in soil organic matter (SOM) and a degradation of soil quality [21]. As a consequence, there is a growing interest in using alternative conservation technologies that can improve soil quality by decreasing the mineralization of soil organic matter and increasing the content of soil organic carbon [22]. Conservation technologies such as reduced tillage (RT) also improves soil physical environment of the soil via reducing soil disturbance, improving aggregate stability and nutrient availability and when applied with catch crops such as mulch, the technique also protects soil against wind and water erosion [23,24]. Moreover, some research has reported that RT beneficially affects the quality of the soil by improving its water-related properties such as the soil water-retention capacity and water infiltration, increasing its microbial and biochemical activity, these factors play a key role in SOM transformation and nutrient cycling [23].

Among the indicators of soil biological status, soil enzymes play a significant role in soil mineralization processes, which determine nutrient availability to plants and microorganisms [25]. Intracellular enzymes such as dehydrogenases are oxidoreductases which are commonly used as an important indicator of the total oxidative activity of soil microbial populations and are general indicators of soil metabolic activity [26]. Non-specific soil hydrolases like proteases, lipases and esterases, which have been shown to hydrolyse fluorescein sodium salt (FDA), take part in the decomposition of a range of substrates. The rate of fluorescein sodium salt hydrolysis is used as a measure of total microbial activity, and it is also an indicator of soil quality [27]. The enzymes involved in the transformation of C compounds in the soil such as cellulases and β-glucosidase, which take part in the degradation of cellulose, the main plant polysaccharide, are crucial for the maintenance of the overall microbial activity, since the product of their hydrolysis (glucose) is the main source of energy for soil microbial populations [28,29]. Since enzymes have various origins and are located in different soil components as well as playing different roles in soil processes, their different responses to various environmental factors and conditions are to be expected for the most part [30]. In fact, soil enzymes are extremely sensitive to various agricultural practices (including TMs) and rapidly react to different natural agents like soil humidity and temperature [31]. Therefore, soil enzymatic activity is a suitable indicator of soil status and is readily applied in order to assess the effect of management practices on soil properties with particular concern being paid to soil microbiology and biochemistry [32]. Various studies have been undertaken to assess the effect of different tillage practices on soil enzymatic activity [18,31–34]. Piazza et al. [18] found that a reduction in tillage intensity from conventional tillage to minimum tillage significantly increased soil C-, P- and S-cycle enzyme activities at the 0–15 cm depth (by between 57 and 92%) and at the 15–30 cm depth (by between 36 and 58%). In the case study of He et al. [35] and a meta-analysis approach [36], the enzymatic activity was markedly higher using the no-tillage method as opposed to the conventional and minimum tillage methods, while in other studies, enzymes levels were higher in soil that had been conventionally tilled than in soil with plant residues incorporated onto its surface [34].

This study was undertaken to determine the response of soil C- and N-related properties and microbial activity indicators under soybean cultivation to three tillage systems: conventional tillage (CT), reduced tillage with the application of a catch crop (RT) and strip tillage (ST). It was initiated as a response to the increasing interest in soybean cultivation in Poland. At present, there are limited and opposing data regarding the effect of various tillage systems (especially strip tillage) applied for soybean cultivation on plant and soil variables under the climatic conditions of Poland. It was assumed that conservation agricultural practices such as strip tillage and/or reduced tillage with the introduction of catch crops in soybean cultivation could provide more resilience to changing climatic conditions and, at the same time, improve soil properties as compared to CT. It was also assumed that greater influence would be expected for RT than ST concerning catch crops' implementation. The input of catch-crop green mass incorporated into the soil may improve soil variables, mainly the microbial and enzymatic ones. Moreover, it was anticipated that the activity of different enzymes in response to the same tillage technique may be various and related to the types and properties of the particular enzymes as well as their substrates. Accordingly, the aim of the study was to evaluate the influence of the different tillage types on selected soil properties under variable humidity conditions during the growing season in the years 2017–2019.

The originality of the research approach described above, as compared to the conventional approach presented in the literature, is based on the proposal to conduct a wide range of research, including the examination of plant properties, as well as the determination of a range of soil properties under different tillage regimes and with climatic variations occurring over the years of the study. This seems to be a very significant approach, not only from an agronomic point of view but also with regard to soil resource protection. When one considers the negative consequences for the soil environment that may result from the application of very intensive tillage regimes, the results of this study have the potential to benefit farmers who are willing to adopt a reduced tillage system, which is an important component of soil management in sustainable agriculture.

## 2. Materials and Methods

### 2.1. Study Site and Experimental Design

The subject of the research was soybean (*Glycine max.* L. Merrill.), which can be cultivated throughout Poland and is characterized by a very favourable ability to regenerate after damage (frost, spraying, animals feeding on it), and it is noted for its tolerance to agronomic mistakes. The effect of the three tillage systems in concert with changing weather conditions as determined by certain selected soil properties were studied for three years (2017–2019) in a field. The study involved a one-factor experiment that was carried out in a randomized block design with four replications at the Experimental Station of the Bydgoszcz University of Science and Technology in Mochełek (53°12′37″ N and 17°52′12″ E) near Bydgoszcz (Midwestern Poland). The Experimental Station is located in a region with a temperate, changeable climate where the marine air from the North Atlantic and the continental air from the east converge, thereby causing frequent day-to-day and year-to-year variabilities in the weather patterns. The soil selected for the study was Haplic Luvisols (Cutanic) with a sandy loam texture which was formed from a fluvioglacial deposit [37].

Data concerning the weather conditions during the sampling periods were obtained from a local weather station that is located at the Experimental Station at Mochełek (53°12′37″ N and 17°52′12″ E) near Bydgoszcz (Midwestern Poland). The meteorological conditions in the study area are presented in Figure 1. The most important line (the highest one in the overall data hierarchy) in Walter and Lieth's climatogram is the temperature line because it provides the basis for distinguishing the decades/months with dry conditions (the precipitation lines are below the temperature line) and decades/months with humid conditions (the precipitation lines are above the temperature line). An analysis of the climatograms showed that the two years of the study (2018–2019) were very dry and that the vegetation period in these years was characterized by the occurrence of a large precipitation deficit. Only May and July 2018 and also the first ten days of September 2019 did not have water shortages. On the other hand, 2017 was a normal year in which only the first ten days of May and in a period of June, a minor precipitation deficit occurred (Figure 1). The relatively high total rainfall observed in the growing season of soybean (474.8 mm) in 2017 had an average air temperature which facilitated the very favourable growth of plants, as well as their development and yield. The low total rainfall from the beginning of June until the end of August in the years 2018 and 2019 substantially limited flowering, pod setting and filling. While the rainfall in September no longer produced an increase in soybean yield, in the years 2018 and 2019, it significantly postponed the harvest date due to the high air humidity; the pods and seeds dried slowly, which extended the growing season.

The following tillage systems were tested:

- Conventional tillage (CT)—ploughing after harvesting the spring wheat; this was performed to a depth of 25–30 cm and was carried out in late autumn in the years 2016, 2017 and 2018; each year in spring before sowing, mineral fertilizers were sown (30 kg N, 80 kg $P_2O_5$ and 70 kg $K_2O$ per ha).
- Reduced tillage (RT)—location as above; a mixture composed of lacy phacelia (10 kg·ha$^{-1}$) and white mustard (8 kg·ha$^{-1}$) was sown in the second half of August in the years 2016, 2017 and 2018; the field was left to mulch the catch crops for winter, and in the spring of 2017, 2018 and 2019, mineral fertilizers were sown as above; the soil was prepared for sowing using a disc harrow.
- Strip tillage (ST)—in the years 2016, 2017 and 2018 after the spring wheat was harvested, the field was left until spring as an approx. 15 cm high stubble; mineral

fertilizers were applied together as above with sowing, using a seeder for strip sowing, Czajkowski STZ 300.

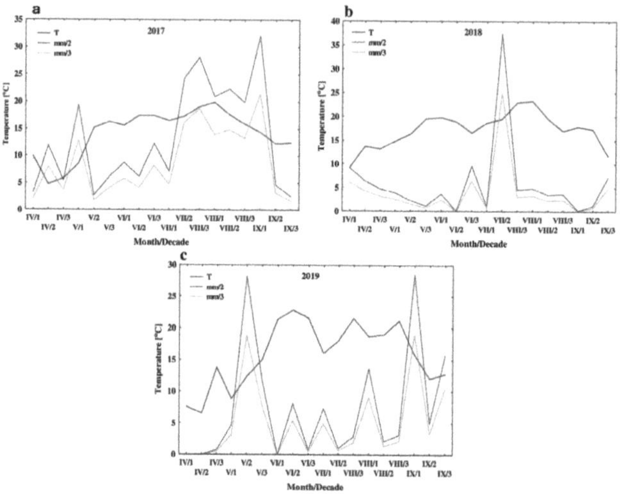


**Figure 1.** Climatograms presenting meteorological conditions in the study area: (**a**) 2017, (**b**) 2018 and (c) 2019. IV—April, V—May, VI—June, VII—July, VIII—August, IX—September, 1—the 1st decade, 2—the 2nd decade, 3—the 3rd decade.

The planned density was 90 plants per 1 m$^2$. Between 182 and 202 kg of seeds was sown per ha to a depth of 3–5 cm on three experimental fields with an area of 1440 m$^2$ each; the width of the interrows was 25 cm, and the working width of the seeders used was 3 m. Fungicides were applied to each field: after sowing, Sencor Liquid 600 S.C. (550 mL·ha$^{-1}$) and after emergence, Fusilade Forte 150 EC (0.8 L·ha$^{-1}$). In 2019, the insecticide Proteus 110 OD was also used due to the occurrence of *Vanessa cardui* in BBCH 51–55.

The soil samples were taken five times a year as follows: before the soybean was sown (3–15 of April) and during the four development stages of soybean (BBCH) in BBCH 10–25 (18–21 of May), in BBCH 55–61 (25–26 of June), in BBCH 67 (1–3 of August) and before the soybean was harvested (BBCH 95–97) (19–24 of September) in 2017, 2018 and 2019. Such a sampling schedule was proposed in order to better track the changes which occurred in the studied properties under the applied tillage systems. At times, the assessed factor affects soil microbial properties intensively but only for a short time, and in a study carried out over the course of one or two years, it is not possible to detect every change which occurs. Moreover, the response in terms of soil enzyme activity toward different factors is often contradictory and may be associated with the types and properties of the studied enzymes and their substrates, which only occur in the soil. Ten soil samples were collected from the upper soil layer (0–30 cm) from each field (with different tillage systems being applied). Each of these samples consisted of ten sub-samples that were bulked in order to obtain one representative sample. Changes in the soil enzymatic activity and the content of microbial biomass C and N were determined in the samples that were collected during each of these terms. In turn, the content of TOC, Nt and pH was analysed in the soil samples that were taken before the soybean was sown and before the harvest.

### *2.2. Laboratory Analysis*

Changes in the soil enzymatic activity and the content of microbial biomass C and N (MBC, MBN) were determined in the fresh soil samples. In turn, the content of TOC, DOC, Nt, DNt as well as pH and texture were analysed in the soil samples dried at room temperature (Figure 2).

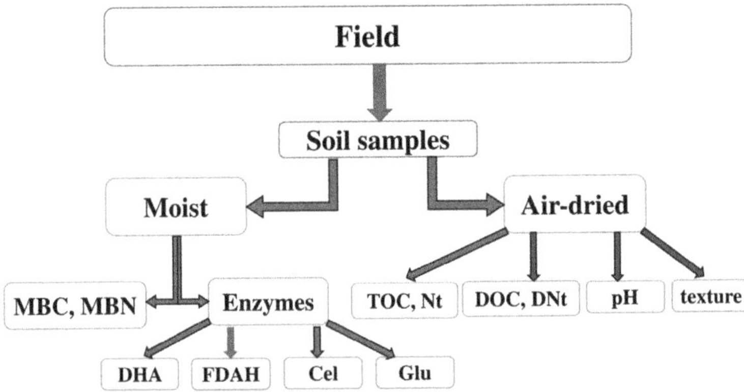


**Figure 2.** Block diagram of the soil properties determined in the laboratory. TOC—total organic carbon, DOC—dissolved organic carbon, Nt—total nitrogen, DNt—dissolved nitrogen, DHA—dehydrogenase, FDAH—rate of fluorescein sodium salt hydrolysis, Cel—CMC-cellulase, Glu—β-glucosidase.

2.2.1. Determination of Soil C and N Content and Other Soil Abiotic Properties

The abiotic properties were analysed using standard soil analytical methods [38]. Soil moisture was evaluated gravimetrically by drying the soil samples at 105 °C for 24 h. The distribution of soil particle-size fraction was assessed according to the Casagrande protocol which was modified by Prószyński. The pH of the solution of 1 M KCl was measured using the potentiometric method in 1:2.5 soil–solution suspensions using a pH meter (S210 Seven Compact™, Mettler-Toledo, Switzerland). The content of total organic carbon (TOC) and total nitrogen (Nt) was determined using a dry combustion CN analyser (Vario Max CN) provided by the Elementar company (Langenselbold, Germany). In order to determine the concentration of dissolved organic carbon (DOC) and dissolved nitrogen (DNt), soil samples were extracted with 0.004 mol $dm^{-3}$ $CaCl_2$, with a ratio of soil sample–extractant of 1:10. The extraction was carried out for an hour, and then, the solution resulting from the extraction was centrifuged, and the concentration of DOC and DNt was evaluated using a Muli N/C 3100 Analytik Jena analyser. The percentage share in the pool of TOC and Nt was also calculated. The chloroform fumigation–extraction method was used to evaluate the content of microbial biomass C and N [39]. Fresh soil samples (25 g oven-dry) were adjusted to 50% of their water-holding capacity (WHC) (to ensure a uniform chloroform distribution) and fumigated in a desiccator with ethanol-free chloroform at 25 °C for 24 h. After that time, the beaker with chloroform was removed and the chloroform vapour was separated from the soil with 5–7 repeated vacuum extractions). Then, all soil samples were extracted with 0.5 M of $K_2SO_4$ for 30 min at 200 rpm, and the concentration of soluble C and total N was evaluated according to a protocol proposed by Vance et al. [39] and Bremner and Mulvaney [40], respectively. Concerning the calculation of any incomplete recovery of the MBC and MBN content, the difference between the fumigated and the control samples was corrected using factors of 0.38 (kEC) and 0.54 (kEN), respectively [39,41].

2.2.2. Selected Enzyme Activities

Analyses of the soil enzyme activities were completed using fresh, moist soil samples sieved through a mesh sieve <2 mm. The enzymatic activity values were calculated on the basis of the oven-dried (105 °C) weight of the soil. The soil dehydrogenase activity (DHA) was determined according to a protocol proposed by Thalmann [42]. Briefly, one gram of soil was added to a Tris-HCl buffer (100 Mm, pH 7.6) and 1% TTC solution (triphenyltetrazolium chloride), and the mixture was incubated at 37 °C for 24 h. Afterwards, acetone (analytical grade) was added to each flask, and the mixture from all flasks was incubated at room temperature for 2 h in the dark. After incubation, the soil suspension was filtered, and the optical density of the product (triphenylophormazan, TPF) was measured at 546 nm. In order to perform the analysis of the microbial indicator based on measuring the activity of fluorescein diacetate sodium salt hydrolysis, the method elaborated by Adam and Duncan was used [43]. One gram of soil was mixed with fluorescein diacetate (substrate) and 60 mM phosphate buffer (pH 7.6). The mixture was incubated for 60 min at 37 °C. Then, the ongoing enzymatic reaction was completed by the addition of a mixture of $C_2H_5OH$ and $CHCl_3$ (1:2). Finally, the supernatant from each sample was filtered, and the concentration of fluorescein (a coloured end product) was measured at 490 nm.

A method proposed by Eivazi and Tabatabai [44] was used to assess the β-glucosidase (Glu) activity. Testing was performed on 1 g wet soil samples, which were incubated with a modified universal buffer (0.05 M, pH 6.0) and with a substrate of 50 mM p-nitrophenyl-α-D-glucopyranoside at 37 °C for a period of 60 min. After stopping the reaction through the addition of a Tris/NaOH buffer (100 mm, pH 12.0) and $CaCl_2$, the soil suspension was filtered and the concentration of product (*p*-nitrophenol, pNP) in the filtrate was determined spectrophotometrically at 400 nm. The method of Schinner and von Mersi [45] was adopted in order to determine the endo-cellulase activity (CEL). The method involved the determination of glucose produced during the incubation of soil samples with an acetate buffer (2 M, pH 5.5) and a carboxymethyl cellulose sodium salt (CMC, 0.7% *w*/*v*) as a substrate (24 h at 50 °C). After incubation the mixture was filtered immediately, and the resulting filtrate was analysed for its glucose concentration using the Prussian blue method, which is based on the formation of Prussian blue (it results from the reduction of ferricyanide ions in an alkaline environment), which was evaluated spectrophotometrically (690 nm).

#### 2.2.3. Statistical Analysis

The data of all of the studied properties were checked for normal distribution using the Shapiro–Wilk test. Because the data were normally distributed, a statistical analysis was performed using the untransformed data. A two-way analysis of variance was used to determine the effect of the tillage systems (conventional tillage—CT, reduced tillage—RT and strip tillage—ST) and the sampling dates (April, May, June, August, September) on the studied soil properties. In the event that significant treatment effects were found, Tukey's test with a significance level $p \leq 0.05$ was used to compare the means. A multiple regression analysis was carried out to present the association between the studied soil properties. The differences between the soil properties were also determined using a principal component analysis (PCA) based on the mean data values of all of the studied properties. The first two principal components (PC1 and PC2) were chosen for the ordination of the cases.

## 3. Results

### 3.1. Soil Physicochemical Properties

Neither the tillage methods applied, nor the sampling months used, significantly affected the basic soil properties (Table 1). The TOC content was significantly higher in 2017 as compared to 2018 and 2019. In turn, the Nt content was significantly higher in 2017 and 2018 than in 2019. The soil reaction was slightly acid and did not change significantly during the entire experimental period (2017–2019). Similarly, the content of clay and silt was not significantly differentiated according to the tillage methods used.

The content of dissolved carbon (DOC) for the entire experimental period was found to be within a range of 83.7–135.3 mg $kg^{-1}$, while the content of dissolved nitrogen (DNt) ranged from 7.5 to 17.2 mg $kg^{-1}$. The content of soil DOC and DNt depended to a significant extent on the sampling date in 2017 and to a degree in 2019 (Table 2). In soil samples collected in 2017, the content of DOC and DNt was significantly higher in April than in September. In 2019, in turn, such a relationship was only found with regards to DOC when the ST method was applied. The DNt content was found to be significantly higher in September than in April with regard to the CT tillage method, while no significant seasonal differences were found in the RT and ST methods. With regard to the tillage methods, the content of DOC and DNt was found to be significantly higher in the RT tillage system, as compared to the CT and ST systems in April 2017 and in both study months in 2019. Finally, no significant effect was found with regard to the DOC and DNt content in 2018 in relation to the tillage practices applied.

**Table 1.** Basic soil properties during the experiment, mean values (±SE).

| Year | TOC | Nt | C/N | $pH_{KCl}$ | Clay | Silt |
|---|---|---|---|---|---|---|
| | [g·kg$^{-1}$] | | | | [%] | |
| 2017 | 9.17 a (±0.77) | 1.05 a (±0.11) | 8.7 | 5.82 a (±0.24) | 6.3 a (±0.4) | 20.2 a (±1.4) |
| 2018 | 7.80 b (±0.87) | 0.96 a (±0.07) | 8.1 | 5.56 a (±0.47) | 6.7 a (±0.6) | 19.3 a (±1.1) |
| 2019 | 7.79 b (±0.63) | 0.72 b (±0.05) | 10.8 | 5.76 a (±0.66) | 6.1 a (±0.4) | 20.6 a (±1.7) |
| Mean | 7.92 (±0.53) | 0.91 (±0.08) | 9.2 | 5.71 (±0.53) | 6.4 (±0.4) | 20.3 (±1.4) |

TOC—total organic carbon, Nt—total nitrogen. Different letters indicate significant differences ($p < 0.05$) between the study years.

**Table 2.** The content of dissolved organic carbon (DOC) and dissolved nitrogen (DNt) depending on the tillage system and sampling dates; mean values (±SE).

| Year | Tillage | DOC [mg kg$^{-1}$] | | | DNt [mg kg$^{-1}$] | | |
|---|---|---|---|---|---|---|---|
| | | April | September | Mean | April | September | Mean |
| 2017 | CT | 124.8 ± 3.0 ABa | 97.9 ± 3.8 Ab | 111.3 ± 3.5 A | 15.1± 0.6 Ba | 10.1 ± 0.5 Ab | 12.6 ± 0.5 AB |
| | RT | 135.3 ± 6.4 Aa | 100.4 ± 3.5 Ab | 117.8 ± 5.0 A | 17.2 ± 0.8 Aa | 11.0 ± 0.7 Ab | 14.1 ± 0.8 A |
| | ST | 117.6 ± 3.3 Ba | 94.5 ± 2.1 Ab | 106.1 ± 2.9 A | 13.0 ± 0.3 Ca | 9.7 ± 0.2 Ab | 11.3 ± 0.3 B |
| 2018 | CT | 89.7 ± 2.7 Aa | 84.9 ± 5.6 Aa | 87.3 ± 4,3 A | 9.2 ± 0.4 Aa | 7.5 ± 0.6 Aa | 8.3 ± 0.5 A |
| | RT | 93.7 ± 3.7 Aa | 83.7 ± 4.2 Aa | 88.7 ± 3.9 A | 9.2 ± 0.5 Aa | 8.2 ± 0.4 Aa | 8.7 ± 0.5 A |
| | ST | 88.9 ± 2.3 Aa | 87.5 ± 1.0 Aa | 88.2 ± 1.7 A | 8.1 ± 0.4 Aa | 8.6 ± 0.2 Aa | 8.4 ± 0.3 A |
| 2019 | CT | 81.3 ± 0.9 Ba | 81.0 ± 2.4 ABa | 81.1 ± 1.5 A | 7.7 ± 0.3 Bb | 8.8 ± 0.2 ABa | 8.3 ± 0.3 B |
| | RT | 93.4 ± 3.1 Aa | 84.2 ± 2.8 Aa | 88.8 ± 3.1 A | 9.2 ± 0.6 Aa | 10.1 ± 0.8 Aa | 9.7 ± 0.7 A |
| | ST | 89.8 ± 3.5 Aa | 75.3 ± 1.6 Bb | 82.5 ± 2.5 A | 8.3 ± 0.2 ABa | 8.4 ± 0.3 Ba | 8.3 ± 0.3 B |
| 2017–2019 | CT | 98.6 ± 2.4 Aa | 87.9 ± 4.0 Aa | 93.3 ± 3.1 A | 10.7 ± 0.4 Aa | 8.8 ± 0.4 Ab | 9.7 ± 0.4 AB |
| | RT | 107.6 ± 4.2 Aa | 89.4 ± 3.4 Ab | 98.4 ± 4.1 A | 11.9 ± 0.6 Aa | 9.8 ± 0.6 Ab | 10.8 ± 0.7 A |
| | ST | 98.8 ± 3.2 Aa | 85.8 ± 1.6 Ab | 92.3 ± 2.5 A | 9.8 ± 0.3 Aa | 8.9 ± 0.3 Aa | 9.3 ± 0.3 B |

CT—conventional tillage, RT—reduced tillage, ST—strip tillage. Different capital letters indicate significant differences ($p < 0.05$) between tillage systems within the same year. Different small letters indicate significant differences ($p < 0.05$) between the sampling months within the same tillage system.

The contribution of DOC in TOC was found to be within a range of 0.97–1.49% (Table 3). A significantly higher DOC/TOC ratio was observed in April than in September, with the exception of 2018. The same relationship was also found for the DNt/Nt ratio calculated in 2017 and for the CT method in 2018. The opposite result was obtained when the CT system was applied in 2019.

**Table 3.** The contribution of dissolved organic carbon (DOC) and dissolved nitrogen (DNt) with respect to total organic carbon (DOC) and total nitrogen (Nt) depending on the tillage system and sampling dates; mean values (±SE).

| Year | Tillage | DOC [%] | | | DNt [%] | | |
|---|---|---|---|---|---|---|---|
| | | April | September | Mean | April | September | Mean |
| 2017 | CT | 1.32 ± 0.03 Aa | 1.05 ± 0.02 Ab | 1.19 ± 0.02 A | 1.35 ± 0.09 Ba | 0.91 ± 0.06 Ab | 1.13 ± 0.06 AB |
| | RT | 1.49 ± 0.05 Aa | 1.09 ± 0.05 Ab | 1.29 ± 0.05 A | 1.68 ± 0.08 Aa | 1.06 ± 0.06 Ab | 1.37 ± 0.07 A |
| | ST | 1.36 ±0.05 Aa | 1.08 ± 0.03 Ab | 1.22 ± 0.05 A | 1.25 ± 0.05 Ba | 0.99 ± 0.03 Ab | 1.12 ± 0.05 B |
| 2018 | CT | 1.19 ± 0.03 Aa | 1.06 ± 0.09 Aa | 1.13 ± 0.06 A | 0.99 ± 0.03 Aa | 0.76 ± 0.07 Bb | 0.87 ± 0.05 A |
| | RT | 1.17 ± 0.08 Aa | 1.02 ± 0.07 Aa | 1.10 ± 0.08 A | 0.95 ± 0.08 Aa | 0.83 ± 0.05 ABa | 0.89 ± 0.08 A |
| | ST | 1.21 ± 0.03 Aa | 1.18 ± 0.03 Aa | 1.20 ± 0.03 A | 0.89 ± 0.04 Aa | 0.94 ± 0.03 Aa | 0.92 ± 0.04 A |

**Table 3.** *Cont.*

| Year | Tillage | DOC [%] | | | DNt [%] | | |
|---|---|---|---|---|---|---|---|
| | | April | September | Mean | April | September | Mean |
| 2019 | CT | 1.05 ± 0.01 Ba | 1.02 ± 0.03 Aa | 1.03 ± 0.03 B | 1.05 ± 0.02 Bb | 1.19 ± 0.05 Ba | 1.12 ± 0.04 B |
| | RT | 1.27 ± 0.06 Aa | 1.11 ± 0.04 Ab | 1.19 ± 0.05 A | 1.31 ± 0.08 Aa | 1.43 ± 0.09 Aa | 1.37 ± 0.09 A |
| | ST | 1.09 ± 0.04 Ba | 0.97 ± 0.04 Ab | 1.03 ± 0.04 B | 1.12 ± 0.02 Aa | 1.17 ± 0.04 Ba | 1.15 ± 0.04 B |
| 2017–2019 | CT | 1.19 ± 0.03 Aa | 1.04 ± 0.08 Ab | 1.11 ± 0.06 A | 1.13 ± 0.05 ABa | 0.95 ± 0.06 Ab | 1.04 ± 0.06 B |
| | RT | 1.31 ± 0.06 Aa | 1.07 ± 0.05 Ab | 1.19 ± 0.05 A | 1.31 ± 0.08 Aa | 1.11 ± 0.04 Ab | 1.21 ± 0.05 A |
| | ST | 1.22 ± 0.04 Aa | 1.08 ± 0.04 Ab | 1.15 ± 0.04 A | 7.29 ± 0.04 Ba | 1.03 ± 0.03 Aa | 1.06 ± 0.03 B |

CT—conventional tillage, RT—reduced tillage, ST—strip tillage. Different capital letters indicate significant differences ($p < 0.05$) between tillage systems within the same year. Different small letters indicate significant differences ($p < 0.05$) between the sampling months within the same tillage system.

### 3.2. Soil Microbial Biomass Carbon (MBC) and Nitrogen (MBN)

The soil tillage methods and sampling months significantly affected the content of MBC and MBN; however, no clear trends were found for the changes which occurred in the subsequent study years when detailed results associated with TMs and seasonal changes were considered (Figure 3). The MBC content for the entire experimental period occurred within a range of 100.1–211.6 mg kg$^{-1}$, while the MBN content ranged from 19.5 to 31.4 mg kg$^{-1}$. The content of both properties was the highest in the RT plots, followed by the CT and ST methods.

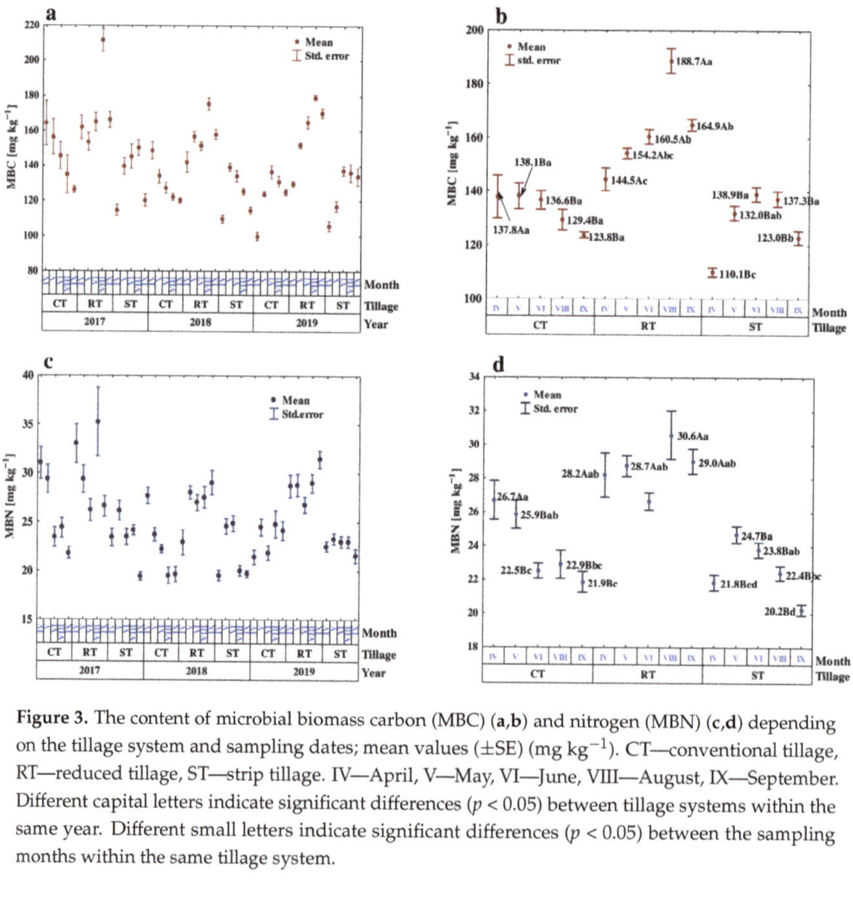


**Figure 3.** The content of microbial biomass carbon (MBC) (**a**,**b**) and nitrogen (MBN) (**c**,**d**) depending on the tillage system and sampling dates; mean values (±SE) (mg kg$^{-1}$). CT—conventional tillage, RT—reduced tillage, ST—strip tillage. IV—April, V—May, VI—June, VIII—August, IX—September. Different capital letters indicate significant differences ($p < 0.05$) between tillage systems within the same year. Different small letters indicate significant differences ($p < 0.05$) between the sampling months within the same tillage system.

Although significant seasonal changes in the C and N biomass content were noted, no clear trends were found with regard to the following months and tillage methods. The average MBN content, independently of the TMs used, revealed seasonal changes in the following order: May > April, August, June > September. The MBC content was in turn the highest in August, followed by June, May, September and April.

*3.3. Soil Enzymatic Variables*

With regard to the mean values, the soil tillage methods and sampling months significantly affected the activity of the soil enzymes, while no clear trends were found for the changes which occurred in the subsequent study years when detailed results associated with TMs and seasonal changes were considered. The DHA activity, ranged from 1.90 to 4.64 mg TPF $kg^{-1}$ $h^{-1}$, while the FDAH ratio was within the range of 14.1–39.9 mg F $kg^{-1}$ $h^{-1}$. Both the DHA activity and the FDAH ratio behaved similarly with regard to the effect of TMs (Tables 4 and 5), and they were significantly higher in the RT system compared to the other tillage systems. Generally, the DHA activity was not statistically significant between CT and ST, except for the last year of the study (2019). In that study year, a significantly higher level of DHA activity was found in the ST than in the CT method. When all the years of the study were considered together, the DHA activity revealed a seasonally based variation in the following order: August > June > April, September > May. In turn, the FDAH level was seasonally affected in the following order: May > August > April > June, September.

**Table 4.** Soil dehydrogenase activity depending on the tillage system and sampling dates; mean values (±SE) (mg TPF $kg^{-1}$ $h^{-1}$).

| Year | Tillage | April | May | June | August | September | Mean |
|---|---|---|---|---|---|---|---|
| 2017 | CT | 2.45 ± 0.14 Bb | 1.90 ± 0.15 Bc | 2.56 ± 0.07 Bb | 3.41 ± 0.11 Ca | 2.35 ± 0.04 Cb | 2.54 ± 0.12 B |
| | RT | 3.20 ± 0.18 Ab | 2.35 ± 0.10 Ac | 3.20 ± 0.14 Ab | 4.64 ± 0.16 Aa | 2.93 ± 0.09 Ab | 3.27 ± 0.11 A |
| | ST | 2.38 ± 0.13 Bb | 1.99 ± 0.07 Bc | 2.65 ± 0.10 Bb | 4.00 ± 0.16 Ba | 2.58 ± 0.11 Bb | 2.72 ± 0.16 B |
| 2018 | CT | 2.89 ± 0.16 Cb | 2.18 ± 0.08 Bc | 2.86 ± 0.08 Bb | 3.45 ± 0.09 Ba | 3.13 ± 0.05 Ba | 2.90 ± 0.16 B |
| | RT | 4.34 ± 0.14 Aa | 3.28 ± 0.12 Ac | 4.18 ± 0.16 Aa | 4.11 ± 0.08 Aa | 3.81 ± 0.10 Ab | 3.94 ± 0.13 A |
| | ST | 3.18 ± 0.09 Ba | 2.21 ± 0.05 Bc | 2.66 ± 0.10 Bb | 3.01 ± 0.10 Ca | 2.67 ± 0.06 Cb | 2.74 ± 0.09 B |
| 2019 | CT | 1.89 ± 0.04 Bc | 2.88 ± 0.06 Ba | 3.11 ± 0.08 Ba | 2.25 ± 0.10 Cb | 3.15 ± 0.10 ABb | 2.46 ± 0.13 C |
| | RT | 2.42 ± 0.13 Ac | 3.67 ± 0.13 Ab | 4.22 ± 0.11 Aa | 3.44 ± 0.11 Ab | 3.31 ± 0.11 Ab | 3.41 ± 0.16 A |
| | ST | 2.54 ± 0.04 Abc | 2.95 ± 0.20 Bb | 3.97 ± 0.10 Aa | 2.77 ± 0.08 Bbc | 2.45 ± 0.10 Bc | 2.93 ± 0.18 B |
| 2017–2019 | CT | 2.41 ± 0.11 Bc | 2.32 ± 0.30 Bc | 2.85 ± 0.11 Bb | 3.04 ± 0.15 Ba | 2.55 ± 0.13 Bbc | 2.63 ± 0.19 B |
| | RT | 3.32 ± 0.20 Ac | 3.10 ± 0.12 Ac | 3.87 ± 0.09 Ab | 4.06 ± 0.21 Aa | 3.35 ± 0.20 Ac | 3.54 ± 0.10 A |
| | ST | 2.70 ± 0.14 Bb | 2.38 ± 0.12 Bc | 3.09 ± 0.12 Bab | 3.26 ± 0.09 Ba | 2.57 ± 0.09 Bb | 2.80 ± 0.12 B |

CT—conventional tillage, RT—reduced tillage, ST—strip tillage. Different capital letters indicate significant differences ($p < 0.05$) between tillage systems within the same year. Different small letters indicate significant differences ($p < 0.05$) between the sampling months within the same tillage system.

The activities of cellulases (Cels) ranged between 5.05 and 10.4 mg Gl $kg^{-1}$ $h^{-1}$ during the study period (Table 6). Generally, the highest Cel activity was noted in the soil samples taken in the RT plots as compared to the CT and ST. There were no clear trends for the changes in the Cel activity in the subsequent study years when the sampling months were considered. However, when all the years of the study were considered together, the Cel activity was affected in the following order: April = August = September > June > May.

The soil β-glucosidase activity (Glu) for the entire experimental period was within the range of 0.25–1.22 mg *p*NP $kg^{-1}$ $h^{-1}$ with a mean value of 0.58 mg *p*NP $kg^{-1}$ $h^{-1}$ (Table 7). When the mean values of the sampling months were considered, the effect of TMs on the Glu activity was similar to that found for the Cel activity in 2017 and 2018. In 2019 in turn, the RT system reduced the Glu activity significantly (by about 30%) as compared to the CT and ST systems. The Glu activity behaved differently than the Cel activity with regard to the effect of the sampling times and revealed the occurrence of seasonal changes in the following order: June > May > August > September > April. The greatest difference in Glu activity (64–72%) was noted between April and May 2019.

**Table 5.** Soil FDAH level depending on the tillage system and sampling dates; mean values (±SE) (mg F kg$^{-1}$ h$^{-1}$).

| Year | Tillage | April | May | June | August | September | Mean |
|---|---|---|---|---|---|---|---|
| 2017 | CT | 14.5 ± 1.4 Bb | 15.4 ± 1.1 Ab | 14.1 ± 1.0 Bb | 17.8 ± 1.1 ABb | 15.4 ± 2.0 Bb | 15.5 ± 1.1 B |
| | RT | 16.3 ± 1.8 Bb | 16.0 ± 0.9 Ab | 16.3 ± 2.1 Ab | 18.0 ± 1.2 Aa | 17.8 ± 1.2 Aa | 16.9 ± 2.1 A |
| | ST | 13.5 ± 1.0 Bb | 15.2 ± 1.0 Ab | 14.4 ± 1.1 Bc | 17.0 ± 1.2 Ba | 15.9 ± 1.1 Bb | 15.2 ± 0.16 B |
| 2018 | CT | 30.1 ± 1.6 Ba | 28.3 ± 1.1 Bb | 26.2 ±3.1 Bb | 30.3 ± 3.1 Ba | 26.2 ± 3.2 Bb | 28.2 ± 3.2 B |
| | RT | 33.8 ± 2.4 Ab | 34.6 ± 3.1 Ab | 34.0 ± 3.2 Aa | 39.9 ± 3.2 Aa | 32.6 ± 4.1 Ab | 35.0 ± 3.1 A |
| | ST | 26.8 ± 1.9 Bd | 31.5 ± 2.0 ABa | 28.6 ± 2.1 Bc | 29.6 ± 2.1 Bb | 26.4 ± 3.1 Bd | 28.6 ± 2.1 B |
| 2019 | CT | 20.4 ± 1.4 Bb | 24.9 ± 1.1 Ca | 20.7 ± 2.1 Bb | 20.6 ± 1.1 Bb | 20.1 ± 1.1 Bb | 21.4 ± 2.1 B |
| | RT | 23.5 ± 1.3 Ab | 27.8 ± 2.1 Aa | 23.2 ± 2.1 Ab | 23.1 ± 1.1 Ab | 21.5 ± 1.1 Ab | 23.8 ± 2.2 A |
| | ST | 22.4 ± 1.4 Ab | 27.0 ± 2.0 Ba | 20.7 ± 1.1 Bc | 20.4 ± 1.1 Bc | 19.3 ± 1.1 Bc | 21.0 ± 1.7 B |
| 2017–2019 | CT | 21.7 ± 0.1 Aab | 22.9 ± 0.3 Ba | 20.3 ± 1.1 Bb | 22.9 ± 1.1 Ba | 20.8 ± 1.1 Ab | 21.7 ± 1.7 B |
| | RT | 24.5 ± 2.1 Ab | 26.1 ± 0.1 Aa | 24.5 ± 1.1 Ab | 27.0 ± 2.2 Aa | 24.0 ± 2.3 Ab | 25.2 ± 2.1 A |
| | ST | 20.9 ± 1.1 Ac | 24.6 ± 0.1 ABa | 21.2 ± 2.1 Bc | 22.3 ± 1.2 Bb | 20.5 ± 2.1 Ac | 21.9 ± 2.1 B |

CT—conventional tillage, RT—reduced tillage, ST—strip tillage. Different capital letters indicate significant differences ($p < 0.05$) between tillage systems within the same year. Different small letters indicate significant differences ($p < 0.05$) between the sampling months within the same tillage system.

**Table 6.** Soil cellulase activity depending on the tillage system and sampling dates; mean values (±SE) (mg Gl kg$^{-1}$ h$^{-1}$).

| Year | Tillage | April | May | June | August | September | Mean |
|---|---|---|---|---|---|---|---|
| 2017 | CT | 10.4 ± 1.1 Ba | 5.05 ± 0.3 Bd | 6.37 ± 0.4 Bc | 8.74 ± 1.0 Bb | 9.00 ± 0.5 Bb | 7.91 ± 0.6 B |
| | RT | 12.4 ± 0.9 Aa | 6.07 ± 0.5 Ad | 7.70 ± 0.9 Ac | 9.53 ±1.1 Ab | 9.82 ± 1.2 Ab | 9.10 ± 0.7 A |
| | ST | 9.44 ± 0.7 Ba | 5.17 ± 0.4 Bd | 6.68 ± 0.5 Bc | 8.26 ± 0.8 Bb | 8.45 ± 1.3 Bb | 7.60 ± 0.5 B |
| 2018 | CT | 8.13 ± 0.7 Ba | 6.22 ± 0.5 ABb | 7.05 ± 0.6 Ab | 8.26 ± 0.7 Ba | 8.23 ± 1.1 Ba | 7.58 ± 1.1 B |
| | RT | 9.20 ± 0.5 Aa | 6.26 ± 0.3 Ac | 7.29 ± 1.1 Ab | 8.90 ± 0.6 Aa | 9.30 ±1.0 Aa | 8.19 ± 0.9 A |
| | ST | 7.07 ± 0.6 Ca | 5.63 ± 0.7 Bc | 6.64 ± 0.4 Bb | 7.50 ± 1.0 Ca | 7.60 ±1.3 Ca | 6.89 ± 0.8 B |
| 2019 | CT | 5.64 ± 0.3 Bb | 6.56 ± 0.3 Ba | 5.32 ± 0.3 Bb | 6.19 ± 0.6 Ba | 6.50 ± 0.5 Ba | 6.04 ± 0.7 B |
| | RT | 6.12 ± 0.7 Ac | 7.12 ± 0.5 Ab | 6.32 ± 0.5 Ac | 7.06 ± 0.5 Ab | 8.06 ± 1.0 Aa | 6.94 ± 0.6 A |
| | ST | 5.67 ± 0.5 Bbc | 5.60 ± 0.8 Cbc | 5.37 ± 0.6 Bc | 6.09 ± 1.0 Bab | 6.36 ± 0.5 Ba | 5.82 ± 0.5 B |
| 2017–2019 | CT | 8.06 ± 1.1 Ba | 5.94 ± 0.8 Bb | 6.25 ± 1.1 Bb | 7.73 ± 1.1 ABa | 7.91 ± 0.6 Ba | 7.18 ± 1.1 B |
| | RT | 9.11 ± 1.2 Aa | 6.48 ± 1.0 Ab | 7.10 ± 1.0 Ab | 8.50 ± 1.2 Aa | 9.06 ± 1.1 Aa | 8.05 ± 0.6 A |
| | ST | 7.39 ± 0.9 Ca | 5.47 ± 0.6 Cc | 6.23 ± 0.5 Bb | 7.29 ± 1.2 Ba | 7.47 ± 0.8 Ba | 6.77 ± 0.5 B |

CT—conventional tillage, RT—reduced tillage, ST—strip tillage. Different capital letters indicate significant differences ($p < 0.05$) between tillage systems within the same year. Different small letters indicate significant differences ($p < 0.05$) between the sampling months within the same tillage system.

**Table 7.** Soil β-glucosidase activity depending on the tillage system and sampling dates; mean values (±SE) (mg *p*NP kg$^{-1}$ h$^{-1}$).

| Year | Tillage | April | May | June | August | September | Mean |
|---|---|---|---|---|---|---|---|
| 2017 | CT | 0.49 ± 0.02 Ab | 0.43 ± 0.02 Ab | 0.61 ± 0.03 Aa | 1.97 ± 0.05 Bc | 1.07 ± 0.02 Ce | 0.91 ± 0.12 B |
| | RT | 0.50 ± 0.02 Ab | 0.45 ± 0.02 Ab | 0.64 ± 0.09 Aa | 2.78 ± 0.15 Aab | 1.82 ± 0.04 Ac | 1.24 ± 0.10 A |
| | ST | 0.50 ± 0.04 Aa | 0.42 ± 0.02 Ab | 0.54 ± 0.03 Aa | 1.73 ± 0.03 Ba | 0.97 ± 0.04 Bb | 0.83 ± 0.05 C |
| 2018 | CT | 0.52 ± 0.03 Ab | 0.61 ± 0.02 Aa | 0.6 8± 0.03 Ba | 0.76 ± 0.01 ABc | 0.61 ± 0.02 Bd | 0.60 ± 0.07 B |
| | RT | 0.50 ± 0.04 Ac | 0.65 ± 0.02 Ab | 0.75 ± 0.09 Aa | 0.85 ± 0.02 Ad | 0.81 ± 0.02 Ad | 0.71 ± 0.10 A |
| | ST | 0.43 ± 0.03 Bc | 0.53 ± 0.02 Bb | 0.66 ± 0.02 Cb | 0.64 ± 0.02 Bc | 0.61 ± 0.02 Bc | 0.57 ± 0.03 B |
| 2019 | CT | 0.44 ± 0.01 Ac | 1.22 ± 0.04 Aa | 0.95 ± 0.03 Ab | 1.36 ± 0.05 Ab | 1.15 ± 0.06 Ac | 1.02 ± 0.07 A |
| | RT | 0.25 ± 0.02 Ce | 0.91 ± 0.01 Ba | 0.71 ± 0.02 Cb | 0.72 ± 0.02 Cb | 0.61 ± 0.01 Cc | 0.64 ± 0.04 C |
| | ST | 0.31 ± 0.01 Be | 0.98 ± 0.01 Ba | 0.80 ± 0.04 Bb | 0.92 ± 0.03 Ba | 0.82 ± 0.01 Bb | 0.77 ± 0.03 B |

Table 7. *Cont.*

| Year | Tillage | April | May | June | August | September | Mean |
|---|---|---|---|---|---|---|---|
| 2017–2019 | CT | 0.48 ± 0.07 Ab | 0.76 ± 0.05 Ab | 0.75 ± 0.07 Aa | 1.36 ± 0.12 Ab | 0.94 ± 0.06 ABc | 0.86 ± 0.07 A |
| | RT | 0.42 ± 0.28 Bc | 0.67 ± 0.04 Ab | 0.70 ± 0.10 Aa | 1.45 ± 0.23 Ab | 1.08 ± 0.13 Ac | 0.86 ± 0.10 A |
| | ST | 0.41 ± 0.09 Bc | 0.64 ± 0.03 Bc | 0.67 ± 0.08 Aa | 1.10 ± 0.08 Bab | 0.80 ± 0.03 Bc | 0.72 ± 0.15 B |

CT—conventional tillage, RT—reduced tillage, ST—strip tillage. Different capital letters indicate significant differences ($p < 0.05$) between tillage systems within the same year. Different small letters indicate significant differences ($p < 0.05$) between the sampling months within the same tillage system.

### 3.4. Relationship Between the Studied Variables—Analysis of Regression and PCA

According to the multiple linear regression analysis, the studied soil enzymes were significantly correlated with certain chemical properties (Table 8). The enzyme activities (except for the Glu activity) were significantly and positively correlated with MBC content, while only the DHA activity was significantly but negatively related to the MBN content. All of the studied enzymes correlated significantly with DOC and DNt content; however, both positive and negative correlation coefficients were calculated. Only the activity of DHA and Glu was positively correlated with $pH_{KCl}$.

Table 8. Regression summary for dependent variables (y).

| Variables | Regression | *p* | $r^2$ |
|---|---|---|---|
| Cel | y = 0.048 MBC + 0.232 DOC − 1.295 DNt + 0.375 clay + 14.748 | 0.00063 | 0.990 |
| Glu | y = 0.099 $pH_{KCl}$ − 0.033 DOC + 0.139 DNt ± 1.683 | 0.00152 | 0.964 |
| FDAH | y = 0.329 MBC + 0.919 DOC − 9.436 DNt ± 16.428 | 0.00267 | 0.886 |
| DHA | y = 0.102 MBC + 0.095 DOC + 0.361 DNt + 0.128 $pH_{KCl}$ − 0.169 clay − 0.447 MBN ± 6.631 | 0.02067 | 0.972 |

DHA—dehydrogenase, FDAH—fluorescein sodium salt hydrolysis, Cel—cellulase, Glu—β-glucosidase.

A PCA analysis recognized four main components which described 94% of the total variance. However, most of this variance (69%) was explained by the first two component (PC1 and PC2) (Figure 4a,b). The first principal component had the largest share in explaining the variance of the studied variables, but did it not exceed 48%. The highest negative correlation (>0.700) was found between PC1 and the following properties: DOC, DNt, TOC, Cel, Nt and $pH_{KCl}$. In turn, PC1 was only significantly but positively correlated with the Glu activity (0.653) (Figure 4a). In general terms, PCA 1 explained from 60 to 95% of the variance in the variables above. The largest share in explaining the variance of PCA 1 was found for DOC (95%) and DNt (92%). A PCA analysis showed that PC2 only had a small share in explaining the variance of most of the studied variables and generally did not exceed 50%. This was confirmed by the values of the loading scores presented in Figure 4a. Only in the case of the DHA activity did PCA 2 explain 85% of the variance of this variable.

A PCA analysis of the cases showed significant differences in the properties determined in the soil samples collected in particular years of the study. The results of the properties assessed in samples taken in 2017 were positively correlated with PC1, while the variables determined in the samples collected in 2018 and 2019 revealed a negative relationship (or the lack of a relationship) with that main component (Figure 4b). On the other hand, the PCA 2 analysis showed the occurrence of significant differences between RT and the others tillage systems (Figure 4b). The data indicated a highly positive correlation of RT with PCA 2, while a negative correlation or its absence existed in the case of the other examined tillage systems (CT and ST). The exception was the samples collected in 2017 from the RT tillage system. The results obtained indicated the significant influence of climatic conditions, especially precipitation, on the studied properties; this factor could have caused the lack of a clear influence of the studied cultivation systems on the assessed variables.

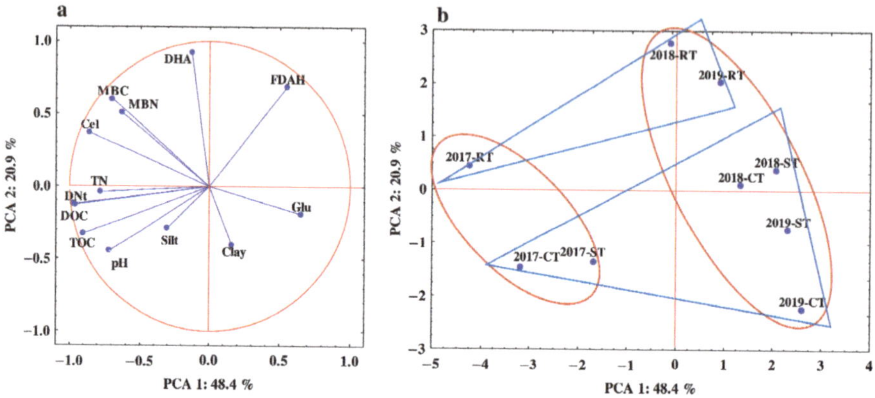


**Figure 4.** Principal component analysis derived from the studied soil properties; (**a**) plot of the first two principal components (PC) for the assessed soil variables; TOC—total organic carbon, Nt—total nitrogen, DHA—dehydrogenase, FDAH—fluorescein sodium salt hydrolysis, Cel—cellulase, Glu—β-glucosidase, DOC—dissolved organic carbon, DNt—dissolved nitrogen, MBC—microbial biomass carbon, MBN—microbial biomass nitrogen, (**b**) principal component analysis of the properties determined in the individual study years (2017, 2018, 2019); CT—conventional tillage, RT—reduced tillage, ST—strip tillage.

## 4. Discussion

### *4.1. Dynamics of Soil Carbon and Nitrogen Contents*

During the three-year-long experiment, the tillage methods had no significant impact on the organic carbon (TOC) and total nitrogen (Nt) content. The lack of accumulation and significant changes in terms of TOC and Nt in our study may have been caused by the comparatively short-term treatments that were applied on different parts of the field each year. The influence of TMs on the physical and chemical properties of the soil is generally observed after a long application time in the same field [46]. Additionally, the significant rainfall deficit in 2018 and 2019 could have slowed down the degradation of plant residues, e.g., by the negative impact on soil microbial and enzymatic activity. In fact, soil moisture is one of the main factors affecting the microbial decomposition of plant residues [47]. Too much or too little moisture in the soil significantly affects the activity of soil microorganisms, which are the main source of enzymes in soil. It was noted that lower moisture diminished the soil enzymatic activity due to the decreased diffusion of soluble substrates and enzyme mobility, resulting in a poor contact of the substrate with enzymatic protein [48]. The most important enzymes involved in the decomposition of plant residues are hydrolases, which require water directly in order to carry out the transformation of substrates [30,32].

Many studies have indicated that a reduction in the soil tillage intensity leads to an increase in soil organic carbon as well as nitrogen content, e.g., [33,49]. Reduced tillage, which diminishes soil disturbance, inhibits soil organic matter (SOM) decomposition and carbon dioxide emission from the soil, thereby resulting in an increase in SOM content [50]. In fact, the use of continuous reduced tillage for several years increases the TOC content as compared to a conventional ploughing system [51]. Similarly, Dębska et al. [12] recently noted that ploughless tillage as well as the strip-till technique could enhance the content of TOC and can also lead to C sequestration. In addition, Fernandez et al. [52] found that after five years of strip tillage, the content of organic C increased by almost 9%. Other authors, however, found no substantial changes in C accumulation in bulk soil in wheat–soybean rotation under different tillage systems [18]. Cover crops left on the soil surface and their residues incorporated into the soil are known to enhance mainly the labile fractions of soil organic matter, such as particulate organic matter (POM). These labile fractions are considered as the main sources of microbial activity and decomposition as well as heterotrophic soil respiration [53]. In turn, other authors have noted that the SOM decomposition caused by this effect also mineralizes from a persistent C pool and that a more stable organic matter compounds is formed when decomposing more labile fractions [54].

Due to its slow transformation rate, changes in the soil TOC content are only a poor reflection of the status of organic matter, as it is affected by various management practices [55]. Hence, in order to assess the quality of SOM, the content of the dissolved organic carbon (DOC) and nitrogen (DNt) as well as the C and N in the soil microbial biomass (MBC, MBN) was assessed. These C and N forms are particularly important because of their crucial role in the SOM transformation and their high turnover rate. Given the dynamic nature of these fractions, they are easily affected by land management practices such as soil tillage and the addition of various amendments [56]. Previously, some authors [12,57] have reported that water dissolved organic matter (DOM) serves key roles in many C- and N-related biogeochemical processes occurring in soils. DOM is an important source of nutrients (C, N, P, K, S) for soil microbial and plant growth and is a medium responsible for the transport of these nutrients in the soil [58]. In addition, even minor changes in DOM concentration may reflect the changes in soil status to a significant extent, particularly under the influence of anthropogenic factors [59]. Soil tillage has been observed to alter the content of soil dissolved organic matter (DOM) and it should be noted that different viewpoints with regard to these changes have been found in the literature. However, the amount of DOM is usually found to be lower in conventionally tilled surface soils as compared to no-till soils [12,57,60]. Some authors [61,62] have found that conventional tillage (as defined by intensive ploughing) enhances the activity of microbial populations, which in turn increases the microbial degradation of plant residues and as a consequence, increases the DOM content. Also, Dębska et al. [12] reported the highest DOC content in soil under conventional tillage, while the highest contribution of DOC in TOC was noted in soil under strip tillage. In turn, Liu et al. [63] and Wright et al. [64] observed twofold and one-third increases in DOC content in the surface soil horizon with no tillage as compared with a conventional tillage system.

The content and quality of the post-harvest residue or other organic amendments introduced into the soil is considered to be an important agent determining the DOM in arable soils [12,57]. It should be stressed that DOM from post-harvest residues, as well as being present in organic and natural fertilizers, is also susceptible to decomposition and rapidly utilized by microorganisms, and that is why only a short-term increase in the content of soil dissolved organic matter was observed [65]. Comparably, Rochette and Gregorich [66], in their field experiments, noted a very rapid increase in soluble forms of organic carbon compounds shortly after the application of farmyard manure and then a gradual decrease throughout the vegetation period. Significant seasonal changes in DOM content reported in the literature are consistent with the results of our experiment. The content of DOC and DNt also changed with time and was significantly higher in April and lower in September. Additionally, the greatest differences in these properties occurred between April and September, as was noted in 2017, a year with the highest precipitation, suggesting the significant influence of humidity conditions on the DOM status.

### 4.2. Soil Enzyme Activities

We assumed that the year-to-year and seasonal changes in enzymatic activity revealed by our study were affected in two ways, (1) by weather conditions, mainly the level of precipitation distributions, which affected soil humidity and (2) by substrate availability (activation) or the accumulation of reaction end products (feedback inhibition).

It is commonly known that when the soil moisture level is low, enzymatic activity, especially that of the hydrolases, which indirectly require water to conduct their reactions, is low, and long droughts usually inhibit the production of enzymes and consequently, their activity [67–69]. An increase in soil moisture activates organic matter transformation and its availability because of cell lysis and the disruption of soil aggregates. This may, in turn, activate the transformation of the microbial biomass, thereby contributing to a temporary increase in the soil enzymatic activity [70]. With regard to the year-to-year and seasonal changes in enzymatic activity, no clear relationship between the studied enzymes and soil humidity was found in our study. Only the Glu activity was markedly higher in 2017 (when the highest level of precipitation occurred) compared to its activity found in 2018 and 2019. In general terms, months with the highest level of soil humidity did not coincide with months that had the highest level of enzymatic activity; however, there were some exceptions, e.g., Glu, DHA and FDAH in August 2017 and the FDAH ratio in May 2019. The relatively high level of enzymatic activity, which occurred despite unfavourable humidity conditions, may be related to soil microorganisms, which, through their adaptation to dry conditions, are able to permanently produce enzymatic proteins, even in small quantities, which may help to maintain the total enzymatic pool size even during a dry period [71]. In addition, the root morphology, which determines the quantity, quality and spatial distribution of root exudates, is also responsible for the specified level of microbial biomass content and enzymatic activity in the rhizosphere soil [72–74].

The opposite seasonal changes in the activity of Cel and Glu (months with the highest Cel activity did not coincide with the months having the highest Glu activity) may be at least partially explained by the fact that during the enzymatic hydrolysis of lignocellulose, some products of the reaction can cause the feedback inhibition of hydrolytic enzymes [75]. Thus, as was stated earlier [76], the accumulation of glucose and cellobiose (end products of hydrolysis) probably inhibited the Cel activity in this study.

Many studies have shown the significant influence of different tillage methods on soil microbial and enzymatic properties (e.g., [34,77]). Conventional tillage practices contribute to weed control and prepare the soil for plant growth and development, but in the long term, these practices change the physical, chemical and biological properties of the soil due to soil erosion, by disturbing soil structure, producing unfavourable changes in soil moisture levels and temperature and also producing further decreases in organic matter content, as well as making soil microbial and enzymatic properties worse [78]. That is why there is a need for alternative, conservative tillage methods that minimize soil deterioration and improve the conditions of the soil environment. Finding an improvement in soil enzymatic properties was expected in soil that had been subjected to conservation tillage (reduced tillage, RT); this was mainly due to the application of a mixture of catch crops (lacy phacelia and white mustard), which were sown in the autumn and left to mulch over the winter. In fact, the activity of the studied enzymes was significantly higher for the RT than for the other tillage methods (with the exception of the Glu activity that was assessed in 2019), which suggests that this tillage method enriched the soil with organic matter, which in turn resulted in increasing microbial abundance and diversity, as well as substrate availability [79]. Our results are in agreement with some previous theories that the application of crop residues increases the activity of C-hydrolysing enzymes [80,81].

It was established by the present study that the content of soil organic carbon (TOC) and total nitrogen (TN) was generally unaffected by tillage practices; differences in enzymatic activity between the tillage systems were rather determined by the content of dissolved organic matter and soil microbial biomass, which were in turn affected by the fresh organic materials (plant residues) entering the soil and by the improved conditions of the soil environment. This statement was confirmed by the regression analysis (Table 8). Previous research has shown that the increases in enzyme activities may be attributed to the increased microbial biomass content which occurs as a result of catch crop application [81,82]. Soil enzymes are mainly of microbial origin, and microbial populations are a potential source of enzyme production and release into the soil [25,83]. In the present study, a significant and positive correlation was obtained between the content of microbial biomass C and enzyme activities (Table 8). Additionally, it was noted that the green mass of catch crops used not only improved the soil microbial activity but also may have contained enzymatic proteins themselves since plants are also a substantial source of soil enzymes [84].

In agreement with the finding of Bielińska and Mocek-Płóciniak [85], the activity of dehydrogenase was 50–70% higher in the no-tillage than in the classical (ploughing) tillage system. Also, other researchers [31,86] reached similar conclusions and reported the statistically significant higher activity of dehydrogenases and β-glucosidase with the protective tillage treatment (no tillage with mulch) as compared to the conventional tillage system. Similarly, Alhameid et al. [20] found that the β-glucosidase activity was higher under the 2- and 3-year-long rotation with no tillage than the same rotations under conventional tillage (CT).

In general terms it was found that reduced tillage with less soil disturbance and the application of a mixture of catch crops improved soil properties compared to conventional tillage and strip tillage. However, the soil β-glucosidase activity was significantly higher in the soil subjected to conventional tillage (CT) than in those subjected to the other tillage methods. It was explained previously that ploughing increased the enzymatic activity in agricultural soils due to the breakdown of soil aggregates, this causes new enzymatic active centres to be exposed [33]. Likewise, Seifert et al. [87] showed that conventional tillage, because of the high degree of aeration in the ploughed soil, accelerated the oxidation of organic matter and thus increased the soil microbial activity. In turn, in a study by Acosta-Martinez et al. [88], it was established that the application of CT had no significant influence on enzymatic activity in some field studies that were carried out with a loam soil in different locations in the USA as compared to no tillage. The findings above suggest that the response of soil enzyme activity toward different tillage systems is often contradictory and may be associated with the types and properties of the studied enzymes and their substrates [89].

### 4.3. Overall Significance of the Study

Considering the general research finding, it should be noted that differences in microbial and enzymatic properties were observed between different tillage practices, but other driving factors such as soil water content also modulated their response to various tillage practices. The reduced tillage with the return of plant residues effectively enhanced the activity of soil enzymes and microbial biomass C content. Although the content of dissolved C and N was not significantly affected by the tillage systems, it was significantly correlated with microbial and enzymatic properties, forming a close relationship among these properties. The results indicate that long-term field trials should be conducted to explain tillage-induced changes in soil properties, especially the physico-chemical ones.

Short-term conservation tillage enhances enzymatic activity by short-term soil organic C accumulation, which in turn stimulates soil microbial activity and the secretion of soil enzymes [90].

Taking into account the environmental benefits of conservation tillage (e.g., reduction in soil erosion, increase in water availability and enhancement of soil biological activity) in soybean cultivation, this tillage technique has been more and more adopted across the globe, especially in warm regions with high rainfall and risk of soil erosion [91]. The higher adoption of conservation tillage methods in regions with greater rainfall has been noted earlier [92]. Previously, soil enzymatic activity was used to explain the effect of tillage methods on soil environment, showing the main role of enzymes in the transformation of soil organic matter and nutrient acquisition [93,94]. Therefore, the level of their activity could successfully reflect soil biological activity, soil health and quality, as well as sensitivity to various management practices [35,95]. Since improving agricultural soil quality is universally desired, our results can provide deeper insight into the interaction between tillage methods and soil properties and can be applied for soybean cultivation under different climatic conditions and in various regions.

## 5. Conclusions

Generally, reduced tillage (RT) combined with a return of plant residues could be recommended for improving the soil by modifying its biological properties, and the tested enzymes turned out to be useful indicators by differentiating between the examined tillage systems. This tillage system, which produced less soil disturbance, and the application of a mixture of catch crops significantly increased the activity of three of the four measured enzymes (dehydrogenases, FDAH rate, cellulases) (average 18%) as well as the microbial biomass content (almost 20%) as compared to conventional tillage and strip tillage. Only the β-glucosidase activity behaved differently in 2019 (significantly higher β-glucosidase activity was found in CT method, than in RT and ST systems). However, it was not possible to explain such a phenomenon in terms of enzyme response based on the available data. The findings above suggest that the response of the given soil enzyme activity toward different tillage systems is often contradictory and may be due to the inherent variability in enzyme response associated with the spatial variability in soil properties as well as the types and properties of the studied enzymes and their substrates occurring periodically in the soil. Neither the tillage methods nor the sampling months used significantly affected basic soil properties (e.g., the content of TOC and Nt); however, this may be due to the relatively short-term treatments that were applied on different sections of the field each year.

Since improving soil quality is the main goal in modern sustainable agriculture, our data could provide deeper insight into the interaction between tillage methods and soil properties and can be applied for soybean cultivation in different regions of the globe and under different climatic conditions. Soil microbial and enzymatic variables are commonly accepted as universal indicators of changes in the soil environment as they are affected by a range of different biotic and abiotic factors, such as management practices, geographical location, climate and initial soil properties.

**Author Contributions:** Conceptualization, A.P.-D., J.D. and B.D.; methodology, A.P.-D., J.D. and B.D.; software, A.P.-D. and J.D.; investigation, A.P.-D., J.D. and B.D.; resources, A.P.-D. and J.D.; writing—original draft preparation, A.P.-D., B.D. and J.D.; writing—review and editing, A.P.-D. and B.D.; visualization, A.P.-D. and J.D.; project administration, A.P.-D.; funding acquisition, A.P.-D. and J.D. All authors have read and agreed to the published version of the manuscript.

**Funding:** This research was funded by Ministry of Agriculture and Rural Development, Poland, grant number 559 HOR 3.6.2/2017 560.

**Institutional Review Board Statement:** Not applicable.

**Informed Consent Statement:** Not applicable.

**Data Availability Statement:** The data used to support the findings of this study are available from the authors upon request.

**Conflicts of Interest:** The authors declare no conflicts of interest.

## References

1. Vogel, H.J.; Eberhardt, E.; Franko, U.; Lang, B.; Ließ, M.; Weller, U.; Wiesmeier, M.; Wollschläger, U. Quantitative evaluation of soil functions: Potential and state. *Front. Environ. Sci.* **2019**, *7*, 164. [CrossRef]
2. Lal, R. Restoring soil quality to mitigate soil degradation. *Sustainability* **2015**, *7*, 5875–5895. [CrossRef]
3. Oldeman, L.R. *Global Extent of Soil Degradation*; ISRIC Bi-Annual Report 1991–1992; International Soil Reference and Information Centre: Wageningen, The Netherlands, 1992; pp. 19–36.
4. Saljnikov, E.; Mirschel, W.; Prasuhn, V.; Keller, T.; Blum, W.E.H.; Chumbaev, A.S.; Zhang, J.; Abuduwaili, J.; Lavrishchev, A.; Schindler, U.; et al. Types of physical soil degradation and implications for their prevention and monitoring. In *Advances in Understanding Soil Degradation. Innovations in Landscape Research*; Saljnikov, E., Mueller, L., Lavrishchev, A., Eulenstein, F., Eds.; Springer: Cham, Switzerland, 2022; pp. 43–73. [CrossRef]
5. Jocien, S.K.; Nfor, F. Unsustainable agricultural practices as a driver of soil degradation in Santa Sub-Division, North West Region of Cameroon. *Asian Soil Res. J.* **2023**, *7*, 19–33. [CrossRef]
6. Dalal, R.C.; Patra, A.K.; Lal, R. Sustainable soil management: Challenges, prospects and benefits. In *Sustainable Soil Management-Beyond Food Production*; Jayaraman, S., Dalal, R.C., Lal, R., Eds.; Cambridge Scholars Publishing: Newcastle upon Tyne, UK, 2023; pp. 1–22.

7. Friedrich, T.; Derpsch, R.; Kassam, A.H. Overview of the global spread of conservation agriculture. *Field Actions Sci. Rep.* **2012**, *6*, 1–7.
8. Lozano-Garcia, B.; Parras-Alcantara, L. Land use and management effects on carbon and nitrogen in Mediterranean Cambisols. *Agric. Ecosyst. Environ.* **2013**, *179*, 208–214. [CrossRef]
9. Pittelkow, C.M.; Linquist, B.A.; Lundy, M.E.; Liang, X.; van Groenigen, K.J.; Lee, J.; van Gestel, N.; Six, J.; Venterea, R.T.; van Kessel, C. When does no-till yield more? A global meta-analysis. *Field Crops Res.* **2014**, *183*, 156–168. [CrossRef]
10. Wezel, A.; Casagrande, M.; Celette, F.; Vian, J.F.; Ferrer, A.; Peigne, J. Agroecological practices for sustainable agriculture. A review. *Agron. Sustain. Dev.* **2014**, *34*, 1–20. [CrossRef]
11. Williams, A.; Davis, A.S.; Ewing, P.M.; Grandy, A.S.; Kane, D.A.; Koide, R.T.; Mortensen, D.A.; Smith, R.G.; Snapp, S.S.; Spokas, K.A.; et al. A comparison of soil hydrothermal properties in zonal and uniform tillage systems across the US Corn Belt. *Geoderma* **2016**, *273*, 12–19. [CrossRef]
12. Dębska, B.; Jaskulska, I.; Jaskulski, D. Method of tillage with the factor determining the quality of organic matter. *Agronomy* **2020**, *10*, 1250. [CrossRef]
13. Morrison, J.E. Strip tillage for "no-till" row crop production. *Appl. Eng. Agric.* **2002**, *18*, 277–284. [CrossRef]
14. Reicosky, D.C. Conservation tillage is not conservation agriculture. *J. Soil Water Conserv.* **2015**, *70*, 103A–108A. [CrossRef]
15. Jaskulska, I.; Jaskulski, D. Strip-till one-pass technology in Central and Eastern Europe: A MZURI Pro-Til Hybrid Machine Case Study. *Agronomy* **2020**, *10*, 925. [CrossRef]
16. Kiryluk, A. Changes in technologies soil and plant cultivation in the province Podlaskie and their impact on environment. *Econ. Environ.* **2016**, *2*, 287–301.
17. Różewicz, M. Review of current knowledge on strip-till cultivation and possibilities of its popularization in Poland. *Pol. J. Agron.* **2022**, *49*, 20–30. [CrossRef]
18. Piazza, G.; Pellegrino, E.; Moscatelii, M.C.; Ercoli, L. Long-term conservation tillage and nitrogen fertilization effects on soil aggregate distribution, nutrient stocks and enzymatic activities in bulk soil and occluded microaggregates. *Soil Tillage Res.* **2020**, *196*, 104482. [CrossRef]
19. Weidhuner, A.; Hanauer, A.; Krausz, R.; Crittenden, S.J.; Gage, K.; Sadeghpour, A. Tillage impacts on soil aggregation and aggregate-associated carbon and nitrogen after 49 years. *Soil Tillage Res.* **2021**, *208*, 104878. [CrossRef]
20. Alhameid, A.; Singh, J.; Sekaran, U.; Ozlu, E.; Kumar, S.; Singh, S. Crop rotational diversity impacts soil physical and hydrological properties under long-term no- and conventional-till soils. *Soil Res.* **2020**, *58*, 84. [CrossRef]
21. Álvaro-Fuentes, J.; Morell, F.J.; Madejón, E.; Lampurlanés, J.; Arrúe, J.L.; Cantero-Martínez, C. Soil biochemical properties in a semiarid Mediterranean agroecosystem as affected by long-term tillage and N fertilization. *Soil Tillage Res.* **2013**, *129*, 69–74. [CrossRef]
22. Swędzyńska, D.; Małecka, I.; Blecharczyk, A.; Swędrzyńskki, A.; Starzyk, J. Effects of various long-term tillage systems on some chemical and biological properties of soil. *Pol. J. Environ. Stud.* **2013**, *22*, 1835–1844.
23. Blanco-Canqui, H.; Ruis, S.J. No-tillage and soil physical environment. *Geoderma* **2018**, *326*, 164–200. [CrossRef]
24. Sałata, A.; Pandino, G.; Buczkowska, H.; Lombardo, S. Influence of catch crops on yield and chemical composition of winter garlic grown for bunch harvesting. *Agriculture* **2020**, *10*, 134. [CrossRef]
25. Furtak, K.; Gałązka, A. Enzymatic activity as a popular parameter used to determine the quality of the soil environment. *Pol. J. Agron.* **2019**, *37*, 22–30. [CrossRef]
26. Medo, J.; Maková, J.; Medová, J.; Lipková, N.; Cinkocki, R.; Omelka, R.; Javoreková, S. Changes in soil microbial community and activity caused by application of dimethachlor and linuron. *Sci. Rep.* **2021**, *11*, 12786. [CrossRef]
27. Patle, P.N.; Navnage, N.P.; Barange, P.K. Fluorescein diacetate (FDA): Measure of total microbial activity and as indicator of soil quality. *Int. J. Curr. Microbiol. Appl. Sci.* **2018**, *7*, 2103–2107. [CrossRef]
28. López-Mondéjar, R.; Zühlke, D.; Becher, D.; Riedel, K.; Baldrian, P. Cellulose and hemicellulose decomposition by forest soil bacteria proceeds by the action of structurally variable enzymatic systems. *Sci. Rep.* **2016**, *6*, 25279. [CrossRef]
29. Minelli-Oliveira, C.; Farias Miyamoto, M.S.; Filho, S.A.; de Oliveira, L.A.; Pereira, J.O. Characterization of α-amylases complex produced by a *Bradyrhizobium* sp. isolated from Amazonian soil. *Int. J. Dev. Res.* **2019**, *9*, 32868–32871.
30. Daunoras, J.; Kácergius, A.; Gudiukaite, R. Role of soil microbiota enzymes in soil health and activity changes depending on climate change and the type of soil ecosystem. *Biology* **2024**, *13*, 85. [CrossRef] [PubMed]
31. Mikanová, O.; Javůrek, M.; Šimon, T.; Friedlová, M.; Vach, M. The effect of tillage systems on some microbial characteristics. *Soil Tillage Res.* **2009**, *105*, 72–76. [CrossRef]
32. Piotrowska-Długosz, A. Significance of enzymes and their application in agriculture. In *Biocatalysis, Enzymatic Basics and Applications*, 1st ed.; Husain, Q., Ullath, M.F., Eds.; Springer Nature: Cham, Switzerland, 2019; pp. 277–308.
33. Ulrich, S.; Tischer, S.; Hofmann, B.; Christen, O. Biological soil properties in a long-term tillage trial in Germany. *J. Plant Nutr. Soil Sci.* **2010**, *173*, 483–489. [CrossRef]
34. Huang, B.; Wang, M.; Jin, X.; Zhang, Y.; Hu, G. Effects of different tillage measures on soil microbes and enzymatic activity. *Nat. Environ. Pollut. Technol.* **2020**, *19*, 1443–1452. [CrossRef]
35. He, L.; Lu, S.; Wang, C.; Mu, J.; Zhang, Y.; Wang, X. Changes in soil organic carbon fractions and enzyme activities in response to tillage practices in the Loess Plateau of China. *Soil Till. Res.* **2021**, *209*, 104940. [CrossRef]

36. Zuber, S.M.; Villamil, M.B. Meta-analysis approach to assess effect of tillage on microbial biomass and enzyme activities. *Soil Biol. Biochem.* **2016**, *97*, 176–187. [CrossRef]
37. IUSS Working Group WRB. World reference base for soil resources 2014, update 2015. In *International Soil Classification System for Naming Soils and Creating Legends for Soil Maps*; World Soil Resources Reports No. 106; FAO: Rome, Italy, 2015.
38. *Soil Survey Staff, Keys to Soil Taxonomy*, 11th ed.; USDA Natural Resources Conservation Service: Washington, DC, USA, 2010.
39. Vance, E.D.; Brookes, P.C.; Jenkinsen, D.S. An extraction method for measuring soil microbial biomass C. *Soil Biol. Biochem.* **1987** *19*, 703–707. [CrossRef]
40. Bremner, J.M.; Mulvaney, C.S. Nitrogen—Total. In *Methods of Soil Analysis Part 2*; Page, A.L., Miller, R.H., Keeny, D.R., Eds. American Society of Agronomy: Madison, WI, USA, 1982; pp. 594–624.
41. Joergensen, R.G.; Mueller, T. The fumigation-extraction method to estimate soil microbial biomass: Calibration of the kEN value *Soil Biol. Biochem.* **1996**, *28*, 33–37. [CrossRef]
42. Thalmann, A. Zur Methodik der Bestimmung der Dehydrodgenaseaktivität im Boden mittels Triphenyltetrazoliumchlorid (TTC) *Landwirtsch. Forsch.* **1968**, *21*, 249–258.
43. Adam, G.; Duncan, H. Development of a sensitive and rapid method for the measurement of total microbial activity using fluorescein diacetate (FDA) in a range of soils. *Soil Biol. Biochem.* **2001**, *33*, 943–951. [CrossRef]
44. Eivazi, F.; Tabatabai, M.A. Glucosidases and galactosidases in soils. *Soil Biol. Biochem.* **1988**, *20*, 601–606. [CrossRef]
45. Schinner, F.; von Mersi, W. Xylanase, CM-cellulase and invertase activity in soil: An improved method. *Soil Biol. Biochem.* **1990**, *22*, 511–515. [CrossRef]
46. Yusuf, A.A.; Abaidoo, R.C.; Iwuafor, E.N.O.; Olufajo, O.O.; Sanginga, N. Rotation effects of grain legumes and fallow on maize yield, microbial biomass and chemical properties of an *Alfisol* in the Nigerian savanna. *Agric. Ecosyst. Environ.* **2009**, *129*, 325–331. [CrossRef]
47. Grzyb, A.; Wolna-Maruwka, A.; Niewiadomska, A. Environmental factors affecting the mineralization of crop residues. *Agronomy* **2020**, *10*, 1951. [CrossRef]
48. Zuccarini, P.; Sardas, J.; Asensio, L.; Peñuelas, J. Altered activities of extracellular soil enzymes by the interacting global environmental changes. *Glob. Chang. Biol.* **2023**, *29*, 2067–2091. [CrossRef] [PubMed]
49. Adamczyk, B. How do boreal forest soils store carbon? *BioEssays* **2021**, *43*, 2100010. [CrossRef] [PubMed]
50. Lal, R. Soil carbon sequestration to mitigate climate change. *Geoderma* **2004**, *123*, 1–22. [CrossRef]
51. Scharlemann, J.P.W.; Tanner, E.V.J.; Hiederer, R.; Kapos, V. Global soil carbon: Understanding and managing the largest terrestrial carbon pool. *Carbon Manag.* **2014**, *5*, 81–91. [CrossRef]
52. Fernandez, F.G.; Sorensen, B.A.; Villamil, M.B. A comparison of soil properties after five years of no-till and strip-till. *Agron. J.* **2015**, *107*, 1339–1346. [CrossRef]
53. Jakab, G.; Madarász, B.; Masoudi, M.; Karlik, M.; Király, C.; Zacháry, D.; Filep, T.; Dekemati, I.; Centeri, C.; Al-Graiti, T. Soil organic matter gain by reduced tillage intensity: Storage, pools, and chemical composition. *Soil Tillage Res.* **2023**, *226*, 105584. [CrossRef]
54. Lehmann, J.; Kleberg, M. The contentious nature of soil organic matter. *Nature* **2015**, *528*, 60–68. [CrossRef]
55. Amin, M.; Salamba, H.N.; Juita, N. Role of labile fraction of carbon for soil quality assessment (A Review). *IOP Conf. Ser. Earth Environ. Sci.* **2021**, *807*, 032095. [CrossRef]
56. Murindangabo, Y.T.; Kopecky, M.; Hoang, T.N.; Bernas, J.; Parajuli, T.; Dhakal, S.; Konvalina, P.; Ufitikirezi, J.M.; Kaneza, G.; Khanal, B.R.; et al. Comparative analysis of soil organic matter fractions, lability, stability ratios, and carbon management index in various land use types within bharatpur catchment, Chitwan District, Nepal. *Carbon Balance Manag.* **2023**, *18*, 21. [CrossRef]
57. Rosa, E.; Debska, B. Seasonal changes in the content of dissolved organic matter in arable soils. *J. Soils Sediments* **2018**, *18*, 2703–2714. [CrossRef]
58. Qualls, R.G.; Haines, B.L.; Swank, W.T. Fluxes of dissolved organic nutrients and humic substances in a deciduous forest. *Ecology* **1991**, *72*, 254–266. [CrossRef]
59. Bolan, N.S.; Adriano, D.C.; Kunhikrishnan, A.; James, T.; McDowell, R.; Senesi, N. Dissolved organic matter: Biogeochemistry, dynamics, and environmental significance in soils. *Adv. Agron.* **2011**, *110*, 1–75. [CrossRef]
60. Schmidt, M.P.; Martinez, C.E. The influence of tillage on dissolved organic matter dynamics in a Mid-Atlantic agroecosystem. *Geoderma* **2009**, *344*, 63–73. [CrossRef]
61. Leinweber, P.; Schulten, H.R.; Kalbitz, K.; Meissner, R.; Jancke, H. Fulvic acid composition in degraded fenlands. *J. Plant Nutr. Soil Sci.* **2001**, *164*, 371–379. [CrossRef]
62. Andruschkiewitsch, R.; Geisseler, D.; Koch, H.J.; Ludwig, B. Effects of tillage on contents of organic carbon, nitrogen, water-stable aggregates and light fraction four different long-term trials. *Geoderma* **2013**, *192*, 368–377. [CrossRef]
63. Liu, E.; Teclemariam, S.G.; Yan, C.; Yu, J.; Gu, R.; Liu, S.; He, W.; Liu, Q. Long-term effects no-tillage management practice on soil organic carbon and its fractions in the northern China. *Geoderma* **2014**, *213*, 379–384. [CrossRef]
64. Wright, A.; Dou, F.; Hons, M.F. Soil organic C and N distribution for wheat cropping systems after 20 years of conservation tillage in central Texas. *Agric. Ecosyst. Environ.* **2007**, *121*, 376–382. [CrossRef]
65. Singh, S.; Dutta, S.; Inamdar, S. Land application of poultry manure and its influence on spectrofluorometric characteristics of dissolved organic matter. *Agric. Ecosyst. Environ.* **2014**, *193*, 25–36. [CrossRef]

66. Rochette, P.; Gregorich, E.G. Dynamics of soil microbial biomass C, soluble organic C and $CO_2$ evolution after three years of manure application. *Can. J. Soil Sci.* **1998**, *78*, 283–290. [CrossRef]
67. Qu, Q.; Wang, Z.; Gan, Q.; Liu, R.; Xu, H. Impact of drought on soil microbial biomass and extracellular enzyme activity. *Front. Plant Sci.* **2023**, *14*, 1221288. [CrossRef] [PubMed]
68. Geisseler, D.; Horwath, W.R.; Scow, K. Soil moisture and plant residue addition interact in their effect on extracellular enzyme activity. *Pedobiologia* **2013**, *54*, 71–78. [CrossRef]
69. Bogati, K.; Walczak, M. The impact of drought stress on soil microbial community, enzyme activities and plants. *Agronomy* **2022**, *12*, 189. [CrossRef]
70. Spohn, M.; Kuzyakov, Y. Distribution of microbial- and root-derived phosphatase activities in the rhizosphere depending on P availability and C allocation—Coupling soil zymography with 14C imaging. *Soil Biol. Biochem.* **2013**, *67*, 106–113. [CrossRef]
71. Burns, R.G.; DeForest, J.L.; Marxsen, J.; Sinsabaugh, R.L.; Stromberger, M.E.; Wallenstein, M.D.; Weintraub, M.N.; Zoppini, A. Soil enzymes in a changing environment: Current knowledge and future directions. *Soil Biol. Biochem.* **2013**, *58*, 216–234. [CrossRef]
72. Datta, S.; Kim, C.M.; Pernas, M.; Pires, N.D.; Proust, H.; Tam, T.; Vijayakumar, P.; Dolan, L. Root hairs: Development, growth and evolution at the plant-soil interface. *Plant Soil* **2011**, *346*, 1–14. [CrossRef]
73. Poirier, V.; Roumet, C.; Munson, A.D. The root of the matter: Linking root traits and soil organic matter stabilization processes. *Soil Biol. Biochem.* **2018**, *120*, 246–259. [CrossRef]
74. Gómez, E.J.; Delgado, J.A.; Gonzalez, J. Persistence of microbial extracellular enzymes in soils under different temperatures and water availabilities. *Ecol. Evol.* **2020**, *10*, 10168. [CrossRef] [PubMed]
75. Puurand, M.; Väljamäe, P.; Pettersson, G.; Johansson, G. Inhibition of the *Trichoderma reesei* cellulases by cellobiose is strongly dependent on the nature of the substrate. *Biotechnol. Bioeng.* **2004**, *86*, 503–511. [CrossRef]
76. Hsieh, C.W.; Cannella, D.; Jørgensen, H.; Felby, C.; Thygesen, L.G. Cellulase inhibition by high concentrations of monosacchari-des. *J. Agric. Food Chem.* **2014**, *30*, 3800–3805. [CrossRef]
77. Szostek, M.; Szpunar-Krok, E.; Pawlak, R.; Stanek-Tarkowska, J.; Ilek, A. Effect of different tillage systems on soil organic carbon and enzymatic activity. *Agronomy* **2022**, *12*, 208. [CrossRef]
78. Lupwayi, N.Z.; Lafond, G.P.; Ziadi, N.; Grant, C.A. Soil microbial response to nitrogen fertilizer and tillage in barley and corn. *Soil Tillage Res.* **2012**, *118*, 139–146. [CrossRef]
79. Wang, X.L.; Ma, K.; Wang, Z.Q.; Li, Y.; Wei, C.H. Effects of no-tillage, mulching and organic fertilization on soil microbial composition in winter wheat field. *Chin. J. Eco-Agric.* **2019**, *27*, 267–276. [CrossRef]
80. Piotrowska-Długosz, A.; Wilczewski, E. Influences of catch crop and its incorporation time on soil carbon and carbon related enzymes. *Pedosphere* **2015**, *25*, 569–579. [CrossRef]
81. Piotrowska-Dlugosz, A.; Wilczewski, E. Effects of catch crops cultivated for green manure on soil C and N content and associated enzyme activities. *Agronomy* **2024**, *14*, 898. [CrossRef]
82. Strickland, M.S.; Thomason, W.E.; Avera, B.; Franklin, J.; Minick, K.; Yamada, S.; Badgley, B.D. Short-term effects of cover cropson soil microbial characteristics and biogeochemical processes across actively managed farms. *Agrosyst. Geosci. Environ.* **2019**, *2*, 180064. [CrossRef]
83. Vranova, V.; Rejsek, K.; Formanek, P. Proteolytic activity in soil: A review. *Appl. Soil Ecol.* **2013**, *70*, 23–32. [CrossRef]
84. Brennan, E.B.; Acosta-Martinez, V. Cover crops and compost influence soil enzymes during six years of tillage-intensive, organic vegetable production. *Soil Sci. Soc. Am. J.* **2019**, *83*, 624–637. [CrossRef]
85. Bielińska, J.; Mocek-Płóciniak, A. Impact of the tillage system on the soil enzymatic activity. *Arch. Environ. Prot.* **2012**, *38*, 75–82. [CrossRef]
86. Pandey, D.; Agrawal, M.; Bohra, J. Effects of conventional tillage and no tillage permutations on extracellular soil enzyme activities and microbial biomass under rice cultivation. *Soil Tillage Res.* **2014**, *136*, 51–60. [CrossRef]
87. Seifert, S.; Shaw, D.R.; Zablatowicz, R.M.; Wesley, R.A.; Kingery, W.L. Effect of tillage on microbial characteristics and herbicide degradation in a Sharkey clay soil. *Weed Sci.* **2001**, *49*, 685–693. [CrossRef]
88. Acosta-Martinez, V.; Mikha, M.M.; Sistani, K.R.; Stahlman, P.W.; Benjamin, J.G.; Vigil, M.F.; Erickson, R. Multi-location study of soil enzyme activities as affected by types and rates of manure application and tillage practices. *Agriculture* **2011**, *1*, 4–21. [CrossRef]
89. Wen, L.; Peng, Y.; Zhou, Y.; Cai, G.; Lin, Y.; Li, B. Effects of conservation tillage on soil enzyme activities of global cultivated land: A meta-analysis. *J. Environ. Manag.* **2023**, *345*, 118904. [CrossRef] [PubMed]
90. Guo, L.; Zhang, L.; Liu, L.; Sheng, F.; Cao, C.; Li, C. Effects of long-term no tillage and straw return on greenhouse gas emissions and crop yields from a rice-wheat system in central China. *Agric. Ecosyst. Environ.* **2021**, *322*, 107650. [CrossRef]
91. Ogierriokhi, M.O.; Woodward, R.T. Understanding why farmers adopt soil conservation tillage: A systematic review. *Soil Secur.* **2022**, *9*, 100077. [CrossRef]
92. Pautsch, G.; Kurkalova, L.; Babcock, B.; Kling, C. The Efficiency of Sequestering Carbon in Agricultural Soils. *Contemp. Econ. Policy* **2021**, *19*, 123–134. [CrossRef]
93. Acosta-Martínez, V.; Cruz, L.; Sotomayor-Ramírez, D.; Pérez-Alegría, L. Enzyme activities as affected by soil properties and land use in a tropical watershed. *Appl. Soil Ecol.* **2007**, *35*, 35–45. [CrossRef]

94. Gispert, M.; Emran, M.; Pardini, G.; Doni, S.; Ceccanti, B. The impact of land management and abandonment on soil enzymati activity, glomalin content and aggregate stability. *Geoderma* **2013**, *202–203*, 51–61. [CrossRef]
95. Zheng, F.; Wu, X.; Zhang, M.; Liu, X.; Song, X.; Lu, J.; Wang, B.; Jan van Groenigen, K.; Li, S. Linking soil microbial communit traits and organic carbon accumulation rate under long-term conservation tillage practices. *Soil Tillage Res.* **2022**, *220*, 10536 [CrossRef]

sustainability

MDPI

*Article*

# Investigation and Simulation Study on the Impact of Vegetation Cover Evolution on Watershed Soil Erosion

Dandan Shen [1], Yuangang Guo [2], Bo Qu [3,4,*], Sisi Cao [1], Yaer Wu [1], Yu Bai [1], Yiting Shao [1] and Jinglin Qian [1]

1 School of Hydraulic Engineering, Zhejiang University of Water Resources and Electric Power, Hangzhou 310018, China
2 Zhejiang Design Institute of Water Conservancy & Hydro-Electric Power Co., Ltd., Hangzhou 310016, China
3 Yellow River Institute of Hydraulic Research, YRCC, Zhengzhou 450003, China
4 Key Laboratory of Lower Yellow River Channel and Estuary Regulation, MWR, Zhengzhou 450003, China
* Correspondence: qubo_edu_hohai@163.com

**Abstract:** Soil erosion has always been a critical issue confronting watershed environments, impacting the progress of sustainable development. As an increasing number of countries turn their attention to this problem, numerous policies have been enacted to halt the progression of soil erosion. However, policy-driven interventions often lead to significant changes in watershed vegetation coverage, under which circumstances, the original sediment erosion models may fall short in terms of simulation accuracy. Taking the Kuye River watershed as the research subject, this study investigates soil erosion data spanning from 1981 to 2015 and utilizes the Revised Universal Soil Loss Equation (RUSLE) model to simulate soil erosion. It is found that the extensive planting of vegetation after 2000 has led to a rapid reduction in soil erosion within the Kuye River watershed. The original vegetation cover and management factor ($C$) proves inadequate in predicting the abrupt changes in vegetation coverage. Consequently, this study adopts two improved plant cover and management factor equations. We propose two new methods for calculating the vegetation cover and management factor, one using machine learning techniques and the other employing a segmented calculation approach. The machine learning approach utilizes the Eureqa software (version11.0, Cornell University, New York, American) to search for the relationship between Normalized Difference Vegetation Index ($NDVI$) and $C$, ultimately establishing an equation that describes this relationship. On the other hand, the piecewise method determines critical values based on data trends and provides separate formulas for $C$ above and below these critical values. Both methods have achieved superior calculation accuracy. Specifically, the overall data calculation using the machine learning method achieved an determined coefficient ($R^2$) of 0.5959, while the segmented calculation method achieved an $R^2$ of 0.6649. Compared to the $R^2$ calculated by the traditional RULSE method, these two new methods can more accurately predict soil erosion. The findings of this study can provide valuable theoretical reference for water and soil prediction in watersheds.

**Keywords:** RULSE; soil erosion; vegetation cover factor; machine learning method

## 1. Introduction

Soil erosion is a critical factor influencing the environment of watersheds, restricting societal economic development and food security [1,2]. Soil erosion and water loss within watersheds is a complex process related to both natural and anthropogenic factors, primarily encompassing vegetation cover, land use, geological conditions, watershed characteristics, water and soil temperature, and land management practices [3,4]. The area of soil erosion in China has far exceeded 4 million square kilometers, accounting for about half of the total land area. In China, the issue of water and sediment in the Yellow River Basin has long been a focal point of scientific research, with soil erosion in this basin posing a major obstacle to the achievement of sustainable agricultural development [5,6]. The area and intensity of water and soil loss in the Yellow River basin have been somewhat controlled over the past few decades, yet they still face severe challenges. According to publicly available information, the area and intensity of water and soil loss in the Yellow River basin have achieved a "double decline" through continuous governance. However, the Loess Plateau, which suffers from the most severe water and soil loss, still accounts for a large proportion of the total area affected in the Yellow River basin.

In recent years, an increasing number of scholars have embarked on research efforts to study soil and water loss within watersheds and analyze the main factors that influence it [7]. Numerous studies have indicated that rainfall is one of the primary factors contributing to soil erosion. Raindrops impact the soil, disrupting the particle structure and causing splash erosion of soil particles [8,9]. Vegetation has been proven to be an important part of protecting the ecological environment [10]. Vegetation, on the other hand, reduces the direct impact of raindrops on the soil due to its foliage and branches covering the soil surface, thereby mitigating soil erosion [11,12]. Therefore, improving the vegetation structure and increasing the vegetation coverage in watersheds can effectively slow down the current situation of soil and water loss. Zaimes et al. [13] studied the amount of soil erosion under four vegetation cover forms and proposed protective management measures to reduce soil erosion. To describe the influence of vegetation on soil erosion, many scholars have employed the Normalized Difference Vegetation Index (*NDVI*) as a generalized indicator of vegetation's effect on soil erosion [14]. Zaimes et al. [15] evaluated the soil erosion risk of the semi-arid Greek island of Rhodes using a combination of Analytic Hierarchy Process and Weighted Linear Combination Method based on multi criteria decision analysis.

With advancements in science and technology, the research on simulating and predicting soil erosion has become increasingly sophisticated [16,17]. The development of soil erosion models can facilitate the study of soil erosion patterns in river basins. There are various methods for simulating soil erosion, among which the commonly used ones include empirical equation methods (such as the Universal Soil Loss Equation (USLE)), physical models, and Multi-Criteria Decision Making (MCDM) techniques to identify erosion-prone areas [18]. The USLE and its derivative models, such as the Revised Universal Soil Loss Equation (RUSLE), are practical models for estimating soil loss at the watershed scale [19]. The RUSLE model is based on five primary factors of soil erosion: rainfall erosivity, soil erodibility, slope length and steepness factor, cover-management factor, and conservation practice factor. By integrating these factors, the RUSLE model can estimate the average annual soil loss. Golkarian et al. [18] combine the Frequency Ratio (FR) model with the RULSE model, aiming to determine the relationship between historical SWE (Snow Water Equivalent) data and controlling geographical and environmental factors at 116 locations within the Nowroud watershed in northern Iran. Zhou et al. [20] research was conducted in the dry-hot valley region of the Jinsha River, which is one of the major sediment-producing areas in the Yangtze River Basin. It took into account topographic factors and land cover, and applied the RUSLE combined with Geographic Information System (GIS) to quantify the contributions and variation patterns of different factors to soil erosion in the dry-hot valley. The development of these models contributes to the study of soil erosion in regional watersheds and provides theoretical references for soil and water conservation measures in these areas.

Although numerous studies have been conducted on traditional sediment erosion models, the vegetation cover coefficients established in these studies are mostly validated in areas where the vegetation cover area does not change significantly. This paper innovatively proposes two new methods for calculating the vegetation cover coefficient. Compared to the traditional, relatively modified methods for determining the vegetation cover coefficient, these new methods take into account the impact on the coefficient following drastic changes in vegetation cover area. The improved vegetation cover coefficient has a significantly broader range of applicability. With the deepening of sustainable development, China has placed increasing emphasis on the issue of soil erosion and water loss in the Yellow River Basin. Large-scale vegetation planting in this basin began after 2000, leading to significant changes in vegetation cover. Given this rapid change in vegetation cover, it remains to be seen whether the findings of traditional models are still applicable. Taking the Kuye River Basin, a sub-basin of the Yellow River, as the research object, this paper investigates the variation patterns of rainfall, runoff, vegetation cover, and soil erosion in the Kuye River from 1981 to 2015. The study aims to explore the mechanisms that need improvement in the RUSLE under such conditions, providing a theoretical reference for predicting soil loss in similar watersheds.

## 2. Materials and Methods

### 2.1. Study Area

Kuye River, a first-order tributary of the middle reaches of the Yellow River, is situated between 38°22′ N and 39°30′ N (as illustrated in Figure 1). With a main stream length of 242 km and a drainage area of 8706 square kilometers, the Kuye River boasts a straightforward hydrological system characterized by short tributaries on both banks. There are 21 relatively significant tributaries, among which 9 are located on the northeastern bank. The Kuye River Basin suffers from severe soil erosion, leading to significant variations in both flood and dry flow rates, and thus, frequent floods. The vegetation in the Kuye River basin is sparse, with shrubs as the main vegetation in the upstream and herbaceous plants in the downstream. The watershed is severely eroded by water and wind erosion, and the proportion of soil erosion area to the total area reaches 95%.

### 2.2. Data Source and Processing Method

The runoff and soil erosion data used in this study were monitored by the Wenjiachuan station from 1981 to 2015, which is the hydrological control station for the Kuye river, with a catchment area of 8515 square kilometers upstream and a riverbed slope of 3.44‰. The data are primarily excerpted from hydrological yearbooks and observation records from hydrological stations provided by the Yellow River Institute of Hydraulic Research and the Bureau of Hydrology of the Yellow River Conservancy Commission. Additionally, data are extracted from publications such as the "China River Sediment Bulletin", the "Yellow River Sediment Bulletin", and the "Yellow River Water Resources Bulletin".

The *NDVI* data utilized in this research originate from a dataset provided by the Institute of Geographic Sciences and Natural Resources Research, Chinese Academy of Sciences. The data span from 1981 to 2015, with an *NDVI* data resolution of 8 km for the period 1981–1997 and 1 km for the period 1998–2015. With the implementation of policies such as the "Decision of the State Council on Soil and Water Conservation in the Middle Reaches of the Yellow River" and the "Guiding Opinions on Promoting High-Quality Development of Soil and Water Conservation in the Yellow River Basin", the vegetation coverage in the Yellow River Basin has seen a significant increase. From Table 1, it can be seen that grassland is the dominant land use type in the Kuye river basin, occupying more than half of the area in each period.

**Table 1.** Area proportion of different land use types in the study catchment.

| | 1986 | 1996 | 2006 | 2011 |
|---|---|---|---|---|
| Grass | 58.6% | 65.2% | 52.7% | 65.3% |
| Tree | 1.5% | 1.5% | 7.8% | 6.7% |
| Farmland | 21% | 14.2% | 18.5% | 11.8% |
| Other | 18.9% | 19.1% | 21% | 16.2% |

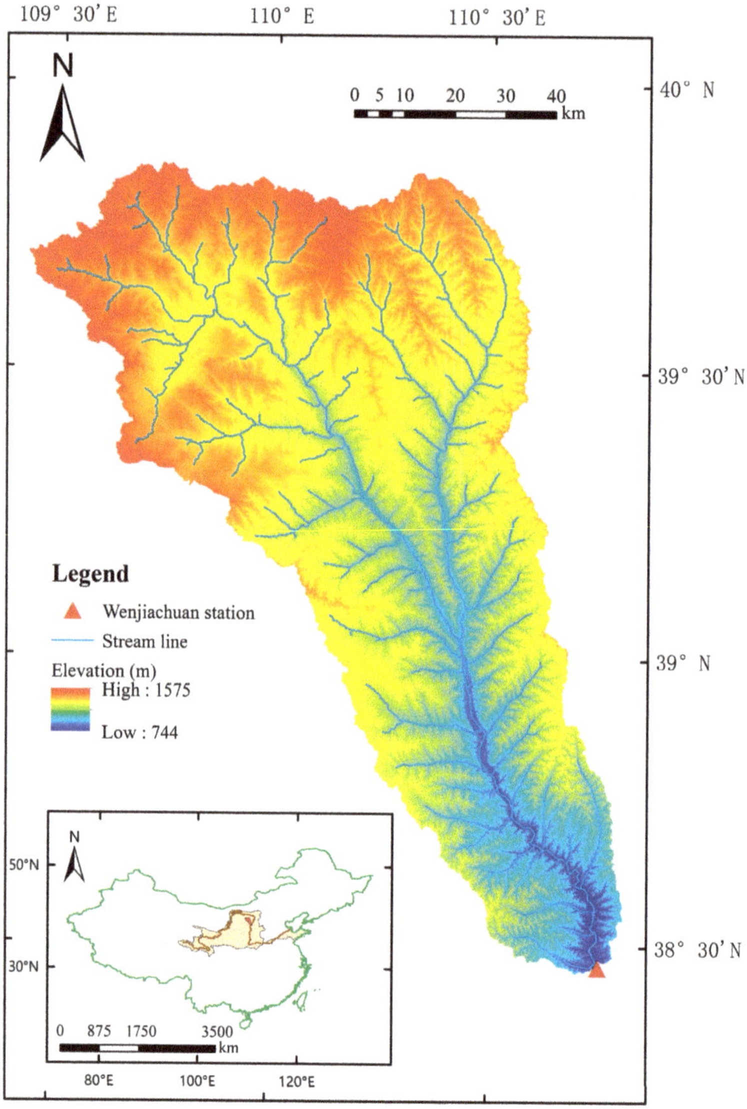


**Figure 1.** The study area.

### 2.3. RULSE Model

The soil erosion prediction model used in this study is the widely applied RULSE model [21,22]. The RULSE model is built upon the foundation of the Universal Soil Loss Equation (USLE) originally developed in the United States [23]. It is one of the widely used soil erosion prediction models both domestically and internationally. The specific formula of the RULSE model is as follows [21]:

$$A = R \times K \times L \times S \times C \times P \tag{1}$$

where $A$ denotes the amount of the average soil loss (t ha$^{-1}$ year$^{-1}$); $R$ represents the rainfall erosivity factor (mm ha$^{-1}$ h$^{-1}$ year$^{-1}$); $K$ is the soil erodibility factor (t ha h ha$^{-1}$ MJ$^{-1}$ mm$^{-1}$);

$L$ and $S$ respectively represent the measurements of slope steepness and slope length; $C$ is the Vegetation cover and management factor; $P$ indicates the conservation practice factor.

#### 2.3.1. Slope Length ($L$) and Steepness Factor ($S$)

The slope and slope length factors in the RUSLE model were calculated in accordance with the methodologies outlined by Beskow et al. [24] and Ganasri and Ramesh [25], as follows:

$$S = \begin{cases} 10.8 sin\theta + 0.036 & \theta < 5^\circ \\ 16.8 sin\theta - 0.5 & 5^\circ \leq \theta < 10^\circ \\ 21.9 sin\theta - 0.96 & \theta \geq 10^\circ \end{cases} \quad (2)$$

$$L = (\lambda/22.13)^m,\ m = \begin{cases} 0.2 & \theta < 1^\circ \\ 0.3 & 1^\circ \leq \theta < 3^\circ \\ 0.4 & 3^\circ \leq \theta < 5^\circ \\ 0.5 & \theta \geq 5^\circ \end{cases} \quad (3)$$

where $\theta$ denotes the slope; $\lambda$ represents the slope length; $m$ is a dimensionless constant depending on the percent slope ($\theta$).

#### 2.3.2. Rainfall Erosivity Factor ($R$)

The average annual erosivity ($R$) equation was formulated based on the research conducted by Golkarian et al. [18]:

$$R = 0.0483(p)^{1.61} \quad (4)$$

where $p$ denote the annual precipitations (mm).

#### 2.3.3. Soil Erodibility Factor ($K$)

Soil erodibility encapsulates the extent of soil erosion attributable solely to rainfall erosivity, highlighting variations in erosion rates while other erosion factors remain constant. For the present investigation, the soil erosion factor was determined utilizing the Erosion/Productivity Impact Calculator model, which relies on soil texture as outlined by Williams [26].

$$K = (0.2 + 0.3exp(-0.0256SAN(1 - SIL)/100))\left(\frac{SIL}{CLA + SIL}\right)^{0.3} \times \left(1.0 - \frac{0.25CA}{CA + \exp(3.72 - 2.95CA)}\right)\left(1.0 - \frac{0.7SN1}{SN1 + \exp(-5.51 + 22.9SN1)}\right) \times 0.1317 \quad (5)$$

where $SAN$, $SIL$ and $CLA$ respectively denote the sand fraction (%), silt fraction (%), and clay fraction (%); $CA$ represents the soil organic carbon content (%).

#### 2.3.4. Vegetation Cover and Management Factor ($C$)

In traditional studies, $NDVI$ serves as a prevalent and effective tool for monitoring surface vegetation cover. The C factor in soil erosion models also ranges from 0 to +1. Consequently, many researchers have adhered to this principle to estimate the C factor from $NDVI$. Typically, this estimation is carried out using the following formula [27]:

$$C = \exp[-\alpha \frac{NDVI}{(\beta - NDVI)}] \quad (6)$$

where, $\alpha = 1$ and $\beta = 2$ [27].

2.3.5. Conservation Practice Factor ($P$)

The erosion control practice factor ($P$) represents the proportion of soil loss with implemented conservation practices compared to soil loss without any conservation measures (i.e., upslope and downslope cultivation). Its value ranges from 0 to 1, where 0 signifies an area completely free from soil erosion, and 1 represents an area with no water conservation measures in place. Land use information can effectively indicate the level of water conservation measures employed. In this study, we assume $P = 1$ for bare land, following Fang et al. [28], and $P = 0.2 + 0.03 \times$ slope degrees for other land types, as suggested by Wen et al. [29].

### 2.4. Data Analysis

For numerical models, main requirement is accuracy. In order to evaluate the accuracy of numerical model simulation, certain data analysis methods and evaluation indexes are usually used. The coefficient of determination ($R^2$) and mean absolute error ($MAE$) were adopted to evaluate the simulation performance [30]:

$$R^2 = 1 - \frac{SSE}{SST'} \tag{7}$$

$$SST = \sum_{i=1}^{N}(Y_i - meanY)^2 \tag{8}$$

$$SSE = \sum_{i=1}^{N}(Y_i - X_i)^2 \tag{9}$$

$$meanY = \frac{1}{N}\sum_{i=1}^{N} Y_i \tag{10}$$

$$MAE = \frac{\sum_1^N |Y_i - X_i|}{N} \tag{11}$$

where $N$ denotes the number of lateral measuring points; $X$ and $Y$ represent the calculated and measured values, respectively, $SSE$ is Sum of Squares for Error, $SST$ is Sum of Squares for total.

## 3. Results

The changes in $NDVI$ data are depicted in Figure 2, which reveals that $NDVI$ was relatively low before 2000, generally below 0.35, and gradually increased after 2000, with a variation range of 0.292 to 0.478. The interannual variations of precipitation and runoff are shown in Figure 3. It can be observed that the variation range of precipitation is relatively small, ranging from 147 to 466 mm, whereas the variation range of runoff is significantly larger, varying between 14 and 90 mm. The interannual changes in sediment erosion are presented in Figure 4. Prior to 2000, the soil erosion rate was relatively high, reaching a maximum of 177.63 t/ha/year. However, after 2000, the sediment erosion rate dropped sharply, with the maximum decreasing to only 14.91 t/ha/year. From Table 2, it can be observed that $NDVI$ exhibits a significantly positive correlation with rainfall, while displaying a significantly negative correlation with sediment erosion and runoff. Rainfall, on the other hand, shows a significant positive correlation with runoff but does not have a significant relationship with sediment erosion. Furthermore, runoff demonstrates a significant positive correlation with sediment erosion.

**Table 2.** The correlation between different factors.

| | *NDVI* | Rainfall | Soil erosion | Runoff |
|---|---|---|---|---|
| *NDVI* | 1 | 0.449481 ** | −0.54034 ** | −0.48906 ** |
| Rainfall | | 1 | 0.247428 | 0.390712 * |
| Soil erosion | | | 1 | 0.913491 ** |
| Runoff | | | | 1 |

Note: * presents significance $p$-value < 0.05, ** presents significance $p$-value < 0.01.

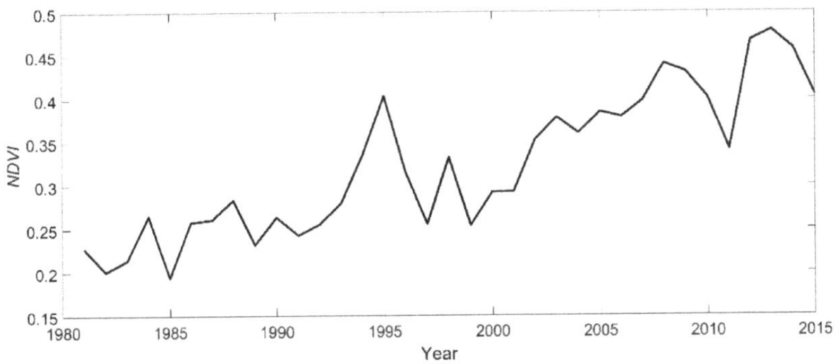


**Figure 2.** The annual variation of *NDVI*.

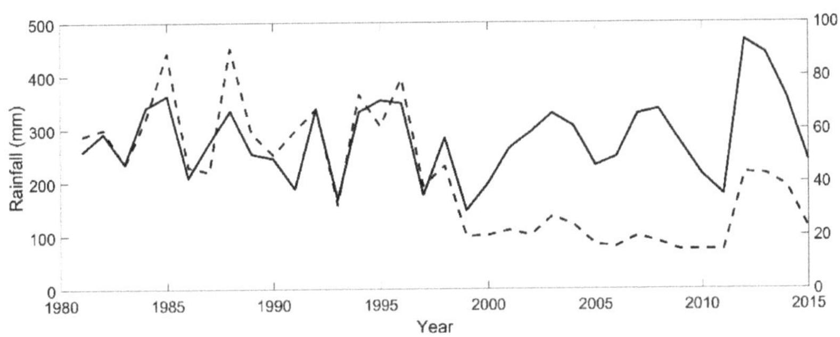


**Figure 3.** The annual variation of rainfall and runoff (The solid line represents rainfall data, while the dashed line represents runoff data).

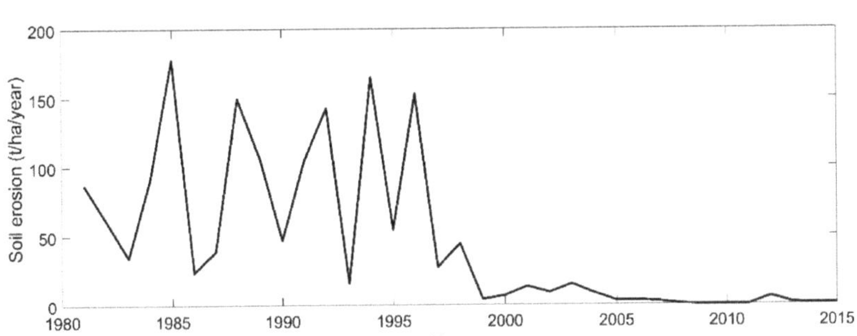


**Figure 4.** The annual variation of soil erosion.

Using the RULSE model, predictions of sediment erosion were made, and the results are shown in Figure 5. It can be observed that the simulation accuracy is relatively low.

Most of the predicted values are higher than the measured values, which may be attributed to the sharp increase in *NDVI* values after 2000, resulting in reduced simulation accuracy of vegetation cover and management factor in Equation (6).

In this study, we propose two new methods for determining the vegetation cover and management factor. Firstly, it was hypothesized that *NVDI* and *p* are closely associated with the vegetation cover and management factor, respectively. To determine the appropriate formula form, an optimization algorithm (Eureqa) was employed in conjunction with measured data. Eureqa is an innovative computer program designed to search for numbers in raw data and discover their correlations, thereby proposing a series of linear equations to describe these connections [30]. We establish the relationship between *NDVI*, *p*, and *C* through Eureqa and obtain the formula form for *C*.

$$C = f(NVDI,\ p) \tag{12}$$

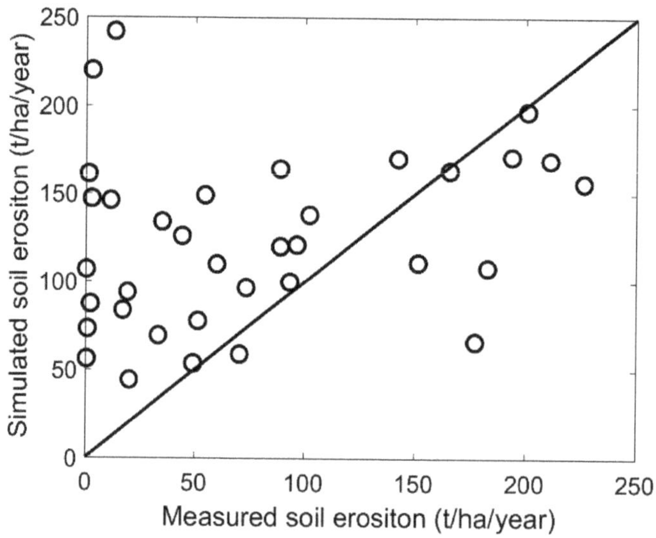


**Figure 5.** The simulation result of soil erosion by RULSE model.

Utilizing the machine learning method, Eureqa, to solve for the equation form, with 2/3 of the data used for training and 1/3 of the data used for validation, the simulation results are presented in Table 3. It can be observed that as the complexity increases, the formula form gradually becomes more intricate. Correspondingly, the accuracy of the model simulation also improves gradually (Figure 6), the formula with a complexity of 14 was selected. Then use this formula to verify the simulation effect of all data. The simulation accuracy for the training data reaches 0.7189, the accuracy for validation data is 0.3617, while the accuracy for all data is 0.5959 (Figure 7). Compared to Equation (6), the predicted values from the improved formula are more evenly distributed above and below the 1:1 line with the simulated values.

**Table 3.** The correlation between solution and complexity.

| Complexity | Solution | MAE |
|---|---|---|
| 1 | $C = 0.2207$ | 1 |
| 3 | $C = 2.47NVDI$ | 0.648 |
| 5 | $C = 1.19 - 2.93NVDI$ | 0.646 |
| 6 | $C = 0.25/NVDI - 0.75$ | 0.607 |
| 7 | $C = 0.5681 + 0.0016p - 2.47NVDI$ | 0.582 |
| 9 | $C = 0.5928 + 0.0016p - 2.5688NVDI$ | 0.579 |
| 14 | $C = 0.5434 + 0.0018p + 0.1087\cos(p) - 2.47NVDI$ | 0.46 |
| 19 | $C = 0.5187 + 0.0016p - 0.0395/(0.078 + \cos(0.19p)) - 2.47NVDI$ | 0.384 |
| 21 | $C = 0.5928 + 0.0019p - 0.0395/(0.075 + \cos(0.19p)) - 2.8652NVDI$ | 0.382 |
| 24 | $C = 0.5928 + 0.0018p - 0.0395/(0.076 + \cos(0.19p)) - 2.8899\sin(NVDI)$ | 0.38 |

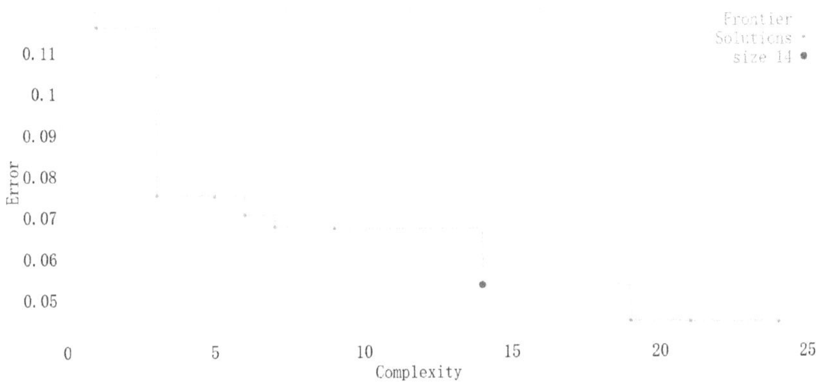


**Figure 6.** The correlation between complexity and error.

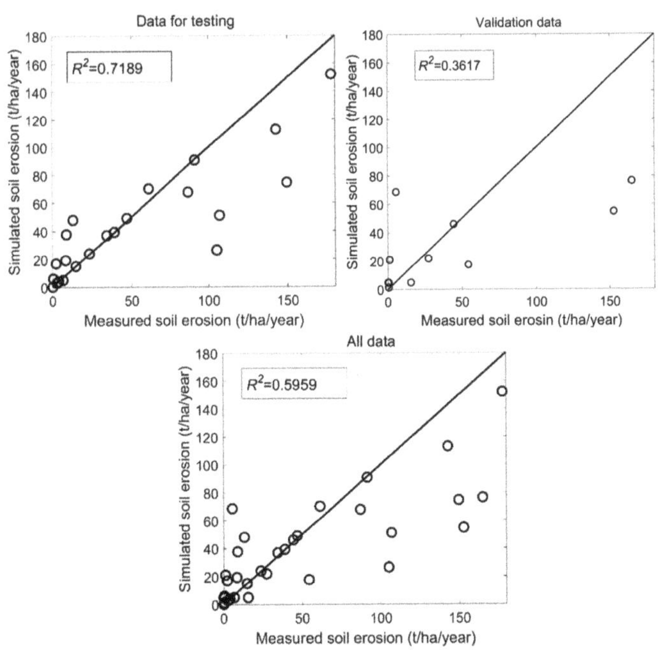


**Figure 7.** The simulation result of soil erosion by improved RULSE model by machine learning.

The second approach involves stratifying the vegetation cover and management coefficient. The basic pattern of the model is the same as Equation (6), but the original equation is divided into two parts. Calculate $C$ values greater than and less than the critical value using equations. As can be seen from the Figure 4, the year 2000 represents a significant change point in sediment erosion. Therefore, we adopt the *NDVI* data (0.3) from 2000 as the critical value, and the proposed formula form is divided into two parts as follows:

$$\begin{cases} C = \exp\left[-\alpha_1 \frac{NDVI}{(\beta_1 - NDVI)}\right] & NDVI < 0.3 \\ C = \exp\left[-\alpha_2 \frac{NDVI}{(\beta_2 - NDVI)}\right] & NDVI \geq 0.3 \end{cases} \quad (13)$$

where $\alpha_1$, $\beta_1$, $\alpha_2$ and $\beta_2$ are constant coefficients.

Based on 35 years of data, the constant parameters in Equation (13) were solved, resulting in the following formula. The accuracy of the simulated data is shown in Figure 8, where it can be observed that the simulated values and measured values are well distributed on both sides of the 1:1 line, with an $R^2$ of 0.6649, indicating a good fit between the model and the observed data.

$$\begin{cases} C = \exp\left[-3 \frac{NDVI}{(2 - NDVI)}\right] & NDVI < 0.3 \\ C = \exp\left[-12 \frac{NDVI}{(2 - NDVI)}\right] & NDVI \geq 0.3 \end{cases} \quad (14)$$

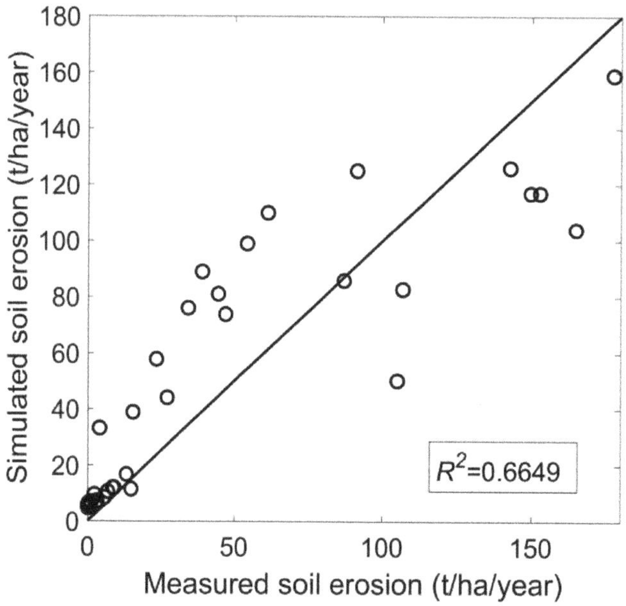


**Figure 8.** The simulation result of soil erosion by improved RULSE model by Equation (14).

## 4. Discussion

As shown in Figure 3, the runoff in Kuye river has significantly decreased since 2000, which might be related to the sudden increase of vegetation coverage. The enhanced vegetation can effectively increase the drag force on fluids during the runoff formation process, prolonging the infiltration time of runoff [31]. Furthermore, vegetation roots contribute to increasing soil porosity, thereby enhancing the infiltration rate of rainfall [32]. As shown in Table 1, there is a significant negative correlation between *NDVI*, soil erosion and runoff, the strong correlation between *NDVI*, soil erosion and runoff formation aligns with previous studies [33,34]. As shown in Table 3, the positive correlation between *NDVI* and rainfall can be attributed to several factors. One possible reason is that rainfall aids in plant growth. Research by Ratzmann et al. [35] indicates that rainfall can enhance vegetation productivity in arid and semi-arid regions, enabling them to better withstand periods of below-average rainfall. Another explanation could be the vegetation-precipitation feedback mechanism resulting from the evapotranspiration caused by vegetation, which contributes to precipitation recycling [36]. Therefore, there exists a significant positive correlation between rainfall and *NDVI*.

The data on sediment erosion, as depicted in Figure 4, reveal a significant decrease in soil erosion after 2000, which is correlated with the sharp increase in *NDVI*. Numerous studies have indicated that vegetation can effectively mitigate the splash erosion of soil by raindrops, with higher vegetation cover efficiency leading to more effective reduction in soil loss [37]. While the RULSE model traditionally demonstrates excellent predictive performance [38], in this study, its accuracy decreases when predicting scenarios with substantial variations in *NDVI*. This discrepancy might stem from China's policies aimed at the Yellow River governance, which have resulted in a rapid and substantial increase in vegetation cover within the basin over a short period. This abrupt change in *NDVI* contrasts with previous applications of the model [39], where the original vegetation cover and management factor formulations yield relatively lower simulation accuracy under such circumstances (Figure 5). The relationship between C and *NDVI* based on Formula (6) is shown in Figure 9. The vegetation cover and management factor obtained from Equation (6) decreases gradually with the increase of *NDVI*, but in actual processes, the amount of soil erosion decreases slowly as *NDVI* increases. However, when *NDVI* reaches a critical value, the amount of soil erosion significantly decreases. According to the research of Chen et al. [40], it was found that when the vegetation coverage exceeds 30%, the amount of sediment erosion decreases significantly, making traditional formulas that are close to linear relationships unusable.

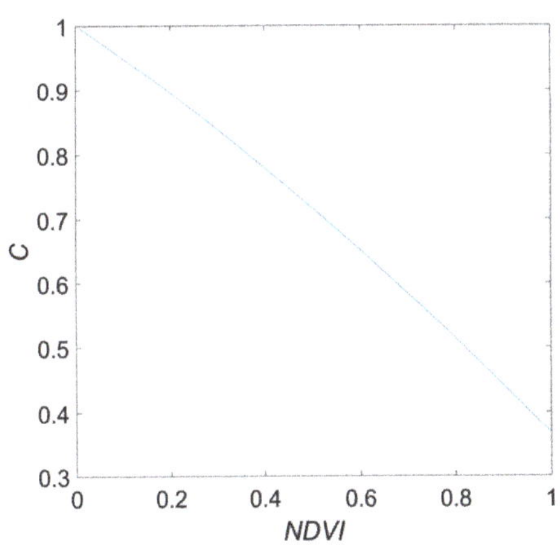


**Figure 9.** The relationship between *NDVI* and C by Equation (6).

At this point, it may be necessary to adopt a new formula form. Many scholars have suggested that the vegetation cover and management coefficient (C) is correlated with the *NDVI* and have derived equations for c based on *NDVI* [20]. As research progresses, Bai and Cui [41] found that in scenarios where *NDVI* undergoes significant temporal variations, precipitation also plays a pivotal role in influencing the vegetation coefficient. In the context of substantial variations in *NDVI*, this study incorporates the impact of precipitation on c within the formula. By employing machine learning techniques to simulate the vegetation cover and management coefficient (C), we have achieved promising results that account for the influence of rainfall. Alternatively, we can separately solve for the vegetation cover and management coefficient above and below the critical *NDVI* value. By comparing the two proposed new formula forms, the overall simulation effect of Equation (8) is found to be higher. This indicates that Equation (8) provides a more accurate representation of the relationship between soil erosion and *NDVI*, especially when considering the distinct behavior above and below the critical *NDVI* threshold.

As can be seen from Figure 4, a significant decline in sediment loss began around 1999. Meanwhile, Figure 2 indicates a substantial increase in *NDVI* around the same period. Therefore, we hypothesize that the average *NDVI* value (0.3) from 1998 to 2000 can be taken as the critical value. Upon testing, we found that the simulation results were satisfactory, confirming that using 0.3 as the critical value is reasonable. The first prediction method for soil and water loss is more easily generalized to other watersheds, providing a basis for soil and water loss prediction in those areas. The second method has good application potential in watershed management, and watershed managers should strive to maintain the *NDVI* of the watershed close to the critical value to ensure minimal sediment loss.

The advancement of meteorological forecasting technology has made it possible for us to obtain future meteorological data [42], and the development of optimization algorithms has enabled people to acquire optimized management systems more swiftly [43]. The integration of precise meteorological forecasting methods, optimization algorithms, and soil erosion prediction models can better provide a basis for watershed vegetation management. As shown in Figure 10, this is our flowchart for optimizing future vegetation management systems in watersheds. Using the genetic algorithm as the optimization algorithm, vegetation management data is randomly generated into the model as optimization items, while sediment loss is incorporated into the model as the optimization function. Simultaneously, meteorological forecasting data and basic watershed parameters are input into the improved RULSE model. Then, the genetic algorithm is run, and the optimization results are output when the genetic generations are completed.

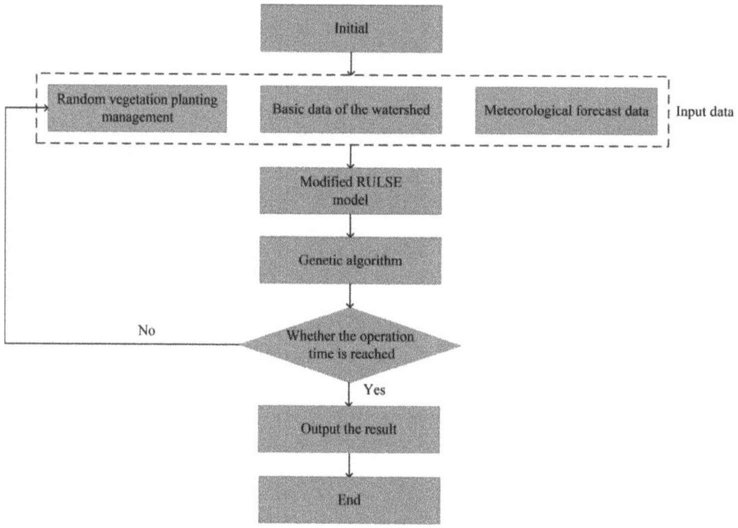


**Figure 10.** Schematic diagram of vegetation management measures optimization.

The current translation also has the following limitations. Firstly, the scope of validation is limited to only one watershed, and further verification across different watersheds is necessary to ascertain the model's accuracy. Secondly, despite the various improvements made to the RULSE model based on the USLE model, there are still certain limitations. For example, some parameters within the model still rely on empirical data for estimation, and the scope of application of the model is still constrained by the availability of data and the adjustment of model parameters. Therefore, in practical applications, it is necessary to validate and adjust the model in light of specific circumstances. Additionally, the model has the potential to incorporate a meteorological forecasting module, which could enable predictions of future soil erosion and water loss [44].

## 5. Conclusions

This study investigated the soil erosion patterns in the Kuye River using the RULSE model, yielding the following key findings:

(1) The planting of vegetation has led to a drastic reduction in both runoff and soil erosion within the Kuye River basin, demonstrating the significant role that policy guidance plays in promoting soil and water conservation efforts.

(2) The traditional formula for the vegetation cover and management factor (C) used in the RULSE model proved to have relatively low accuracy in simulating sediment erosion in the Kuye River basin. To address this, this study employed machine learning techniques to derive a new formula for the vegetation cover factor, which subsequently resulted in improved simulation outcomes and a more precise representation of soil erosion dynamics.

(3) Through data analysis, it was found that an *NDVI* value of 0.3 serves as a critical threshold. Applying different vegetation cover and management formulas above and below this critical value yielded better simulation results. The two proposed methods provide a theoretical reference for predicting sediment erosion in watersheds with significant variations in *NDVI*.

(4) In engineering practice, meteorological prediction methods can be combined to use the simulation method proposed in this article to predict the amount of sediment erosion in the watershed, providing reliable warnings for soil erosion. Simultaneously incorporating optimization algorithms to simulate the future sediment erosion in the watershed under different vegetation cultivation measures, and optimizing vegetation cultivation strategies.

**Author Contributions:** Conceptualization, D.S. and Y.G.; methodology, D.S. and Y.B.; software, Y.B.; validation, D.S. and Y.G.; formal analysis, S.C. and Y.W.; investigation, B.Q.; resources, Y.G. and B.Q.; data curation, B.Q.; writing—original draft preparation, D.S.; writing—review and editing, D.S. and B.Q.; visualization, Y.S.; project administration, J.Q.; funding acquisition, B.Q. and J.Q. All authors have read and agreed to the published version of the manuscript.

**Funding:** National Natural Science Foundation of China (U2243601), Students' innovation and entrepreneurship training program of Zhejiang University of Water Resources and Electric Power (S202311481001), Zhejiang Provincial Joint Fund Key Projects, Study on the dynamic forecasting and early warning of flash floods caused by heavy rainfalls (LZJWZ24E090003).

**Institutional Review Board Statement:** Not applicable.

**Informed Consent Statement:** Not applicable.

**Data Availability Statement:** The data presented in this study are available on request from the corresponding author.

**Conflicts of Interest:** Author Yuangang Guo was employed by Zhejiang Design Institute of Water Conservancy & Hydro-Electric Power Co., Ltd. The remaining authors declare that the research was conducted in the absence of any commercial or financial relationships that could be construed as a potential conflict of interest.

## References

1. Deng, J.; Jing, Y.; Yin, D. Soil erosion studies should consider the effect of water erosion on tillage erosion. *J. Soil Water Conserv.* **2017**, *72*, 38A–41A. [CrossRef]
2. Kuhwald, M.; Busche, F.; Saggau, P.; Duttmann, R. Is soil loss due to crop harvesting the most disregarded soil erosion process? A review of harvest erosion. *Soil Tillage Res.* **2022**, *215*, 105213. [CrossRef]
3. Yao, X.; Yu, J.; Jiang, H.; Sun, W.; Li, Z. Roles of soil erodibility, rainfall erosivity and land use in affecting soil erosion at the basin scale. *Agric. Water Manag.* **2016**, *174*, 82–92. [CrossRef]
4. Wuepper, D.; Borrelli, P.; Finger, R. Countries and the global rate of soil erosion. *Nat. Sustain.* **2020**, *3*, 51–55. [CrossRef]
5. Zhang, X.S.; Hao, F.H.; Cheng, H.G.; Li, D.F. Application of SWAT model in the upstream watershed of the Luohe River. *Chin. Geogr. Sci.* **2003**, *13*, 334–339. [CrossRef]

6. Qin, H.; Chen, F.; Liu, Y. Study on wave-influenced resistance to erosion of silty soil in Huanghe (Yellow) River Delta. *Acta Oceanol. Sin.* **2010**, *29*, 53–57. [CrossRef]
7. Zaimes, G.N.; Tamparopoulos, A.E.; Tufekcioglu, M.; Schultz, R.C. Understanding stream bank erosion and deposition in Iowa USA: A seven year study along streams in different regions with different riparian land-uses. *J. Environ. Manag.* **2021**, *287*, 112352 [CrossRef]
8. Ohnuki, Y.; Shimizu, A. Experimental studies on rain splash erosion of forest soils after clearing in Okinawa using an artificial rainfall apparatus. *J. For. Res.* **2004**, *9*, 101–109. [CrossRef]
9. Koch, T.; Chifflard, P.; Aartsma, P.; Panten, K. A review of the characteristics of rainfall simulators in soil erosion research studies. *MethodsX* **2023**, *12*, 102506. [CrossRef]
10. Zaimes, G.N.; Gounaridis, D.; Fotakis, D. Assessing riparian land-uses/vegetation cover along the Nestos river in Greece. *Fresenius Environ. Bull.* **2011**, *20*, 3217–3225.
11. Wang, Z.J.; Jiao, J.Y.; Rayburg, S.; Wang, Q.L.; Su, Y. Soil erosion resistance of "Grain for Green" vegetation types under extreme rainfall conditions on the Loess Plateau, China. *Catena* **2016**, *141*, 109–116. [CrossRef]
12. Boukhari, Y.; Ginovart, M.; Verdú, A.M.; Mas, M.T.; Mederbal, K.; Josa, R. Hydrological soil behavior in areas with semi-arid vegetation (Beni Chougrane Mountains, Algeria). *Biologia* **2016**, *71*, 1131–1136. [CrossRef]
13. Zaimes, G.N.; Emmanouloudis, D.; Iakovoglou, V. Estimating soil erosion in Natura 2000 areas located on three semi-arid Mediterranean islands. *J. Environ. Biol.* **2012**, *33*, 277.
14. Aslam, B.; Maqsoom, A.; Alaloul, W.S.; Musarat, M.A.; Jabbar, T.; Zafar, A. Soil erosion susceptibility mapping using a GIS-based multi-criteria decision approach: Case of district Chitral, Pakistan. *Ain Shams Eng. J.* **2021**, *12*, 1637–1649. [CrossRef]
15. Zaimes, G.N.; Gounaridis, D.; Iakovoglou, V.; Emmanouloudis, D. Assessing soil erosion risk for Rhodes Island, Greece with a GIS-based multi-criteria decision analysis. In Proceedings of the IASTED International Conference on Water Resource Management, Gaborone, Botswana, 3–5 September 2012; pp. 317–324.
16. Zhang, K.; Li, S.; Peng, W.; Yu, B. Erodibility of agricultural soils on the Loess Plateau of China. *Soil Tillage Res.* **2004**, *76*, 157–165. [CrossRef]
17. Tarek, Z.; Elshewey, A.M.; Shohieb, S.M.; Elhady, A.M.; El-Attar, N.E.; Elseuofi, S.; Shams, M.Y. Soil erosion status prediction using a novel random forest model optimized by random search method. *Sustainability* **2023**, *15*, 7114. [CrossRef]
18. Golkarian, A.; Khosravi, K.; Panahi, M.; Clague, J.J. Spatial variability of soil water erosion: Comparing empirical and intelligent techniques. *Geosci. Front.* **2023**, *14*, 101456. [CrossRef]
19. Das, B.; Paul, A.; Bordoloi, R.; Tripathi, O.P.; Pandey, P.K. Soil erosion risk assessment of hilly terrain through integrated approach of RUSLE and geospatial technology: A case study of Tirap District, Arunachal Pradesh. *Model. Earth Syst. Environ.* **2018**, *4*, 373–381. [CrossRef]
20. Zhou, Y.; Yi, Y.; Liu, H.; Tang, C.; Zhang, S. Spatiotemporal dynamic of soil erosion and the key factors impact processes over semi-arid catchments in Southwest China. *Ecol. Eng.* **2024**, *201*, 107217. [CrossRef]
21. Mouris, K.; Schwindt, S.; Haun, S.; Morales Oreamuno, M.F.; Wieprecht, S. Introducing seasonal snow memory into the RUSLE. *J. Soils Sediments* **2022**, *22*, 1609–1628. [CrossRef]
22. Räsänen, T.A.; Tähtikarhu, M.; Uusi-Kämppä, J.; Piirainen, S.; Turtola, E. Evaluation of RUSLE and spatial assessment of agricultural soil erosion in Finland. *Geoderma Reg.* **2023**, *32*, e00610. [CrossRef]
23. Islam, M.R.; Jaafar, W.Z.W.; Hin, L.S.; Osman, N.; Din, M.A.M.; Zuki, F.M.; Adham, M.I. Soil erosion assessment on hillslope of GCE using RUSLE model. *J. Earth Syst. Sci.* **2018**, *127*, 50. [CrossRef]
24. Beskow, S.; Mello, C.R.; Norton, L.D.; Curi, N.; Viola, M.R.; Avanzi, J.C. Soil erosion prediction in the Grande River Basin, Brazil using distributed modeling. *Catena* **2009**, *79*, 49–59. [CrossRef]
25. Ganasri, B.P.; Ramesh, H. Assessment of soil erosion by RUSLE model using remote sensing and GIS-A case study of Nethravathi Basin. *Geosci. Front.* **2016**, *7*, 953–961. [CrossRef]
26. Williams, J.R. Chapter 25: The EPIC model. In *Computer Models of Watershed Hydrology*; Singh, V.P., Ed.; Water Resource Publication: Highlands Ranch, CO, USA, 1995; pp. 909–1000.
27. Alexakis, D.D.; Hadjimitsis, D.G.; Agapiou, A. Integrated use of remote sensing, GIS and precipitation data for the assessment of soil erosion rate in the catchment area of "Yialias" in Cyprus. *Atmos Res.* **2013**, *131*, 108–124. [CrossRef]
28. Fang, G.; Yuan, T.; Zhang, Y.; Wen, X.; Lin, R. Integrated study on soil erosion using RUSLE and GIS in Yangtze River Basin of Jiangsu Province (China). *Arab. J. Geosci.* **2019**, *12*, 173. [CrossRef]
29. Wen, X.; Deng, X.; Zhang, F. Scale effects of vegetation restoration on soil and water conservation in a semi-arid region in China: Resources conservation and sustainable management. *Resour. Conserv. Recycl.* **2019**, *151*, 104474. [CrossRef]
30. Wang, H.; Bai, Y.; Man, X.; Tang, Z.; Zhang, S. Improved RUSLE model to simulate the effect of slope forest area on soil and water conservation. *Water Supply* **2023**, *23*, 2799–2813. [CrossRef]
31. Zhang, J.; Zhang, S.; Li, G.; Liu, M.; Chen, S. Effects of vegetation lodging on overland runoff flow regime and resistance. *Water Supply* **2020**, *20*, 1463–1473. [CrossRef]
32. Dunne, T.; Zhang, W.; Aubry, B.F. Effects of rainfall, vegetation, and microtopography on infiltration and runoff. *Water Resour. Res.* **1991**, *27*, 2271–2285. [CrossRef]
33. Kim, H.W.; Kim, J.H.; Li, W.; Yang, P.; Cao, Y. Exploring the impact of green space health on runoff reduction using NDVI. *Urban For. Urban Green.* **2017**, *28*, 81–87. [CrossRef]

34. Yue, S.; Huang, J.; Zhang, Y.; Chen, W.; Guo, Y.; Cheng, M.; Ji, G. Quantitative evaluation of the impact of vegetation restoration and climate variation on runoff attenuation in the Luan River Basin based on the extended Budyko Model. *Land* **2023**, *12*, 1626. [CrossRef]
35. Ratzmann, G.; Gangkofner, U.; Tietjen, B.; Fensholt, R. Dryland vegetation functional response to altered rainfall amounts and variability derived from satellite time series data. *Remote Sens.* **2016**, *8*, 1026. [CrossRef]
36. Zhou, J.; Jia, L.; Menenti, M.; van Hoek, M.; Lu, J.; Zheng, C.; Yuan, X. Characterizing vegetation response to rainfall at multiple temporal scales in the Sahel-Sudano-Guinean region using transfer function analysis. *Remote Sens. Environ.* **2021**, *252*, 112108. [CrossRef]
37. Jomaa, S.; Barry, D.A.; Brovelli, A.; Heng, B.C.P.; Sander, G.C.; Parlange, J.Y.; Rose, C.W. Rain splash soil erosion estimation in the presence of rock fragments. *Catena* **2012**, *92*, 38–48. [CrossRef]
38. Cheng, Q.; Cai, Q.; Ma, W. Comparative study on rain splash erosion of representative soils in China. *Chin. Geogr. Sci.* **2008**, *18*, 155–161. [CrossRef]
39. Kumar, N.; Singh, S.K. Soil erosion assessment using earth observation data in a trans-boundary river basin. *Nat. Hazards* **2021**, *107*, 1–34. [CrossRef]
40. Chen, J.; Xiao, H.; Li, Z.; Liu, C.; Wang, D.; Wang, L.; Tang, C. Threshold effects of vegetation coverage on soil erosion control in small watersheds of the red soil hilly region in China. *Ecol. Eng.* **2019**, *132*, 109–114. [CrossRef]
41. Bai, Y.; Cui, H. An improved vegetation cover and management factor for RUSLE model in prediction of soil erosion. *Environ. Sci. Pollut. Res.* **2021**, *28*, 21132–21144. [CrossRef]
42. Pérez-Vega, A.; Travieso-González, C.M.; Hernández-Travieso, J.G. An approach for multiparameter meteorological forecasts. *Appl. Sci.* **2018**, *8*, 2292. [CrossRef]
43. Burks, A.R.; Punch, W.F. An analysis of the genetic marker diversity algorithm for genetic programming. *Genet. Program. Evolvable Mach.* **2017**, *18*, 213–245. [CrossRef]
44. Rao, W.; Shen, Z.; Duan, X. Spatiotemporal patterns and drivers of soil erosion in Yunnan, Southwest China: RULSE assessments for recent 30 years and future predictions based on CMIP6. *Catena* **2023**, *220*, 106703. [CrossRef]

*sustainability*

MDPI

*Article*

# Assessment of Total Mercury Levels Emitted from ASGM into Soil and Groundwater in Chami Town, Mauritania

**Mohamed Mamoune Maha [1,*], Akito Matsuyama [2], Takahiko Arima [3] and Atsushi Sainoki [4,5]**

1 Department of Advanced Industrial Science, Graduate School of Science and Technology, Kumamoto University, Kumamoto 860-8555, Japan
2 National Institute for Minamata Disease, 4058-18 Hama, Minamata-City 867-0008, Kumamoto, Japan; matsuyam@nimd.go.jp
3 Division of Sustainable Resources Engineering, Graduate School of Engineering, Hokkaido University, Sapporo 060-8628, Japan; takahiko_arima@eng.hokudai.ac.jp
4 Faculty of Advanced Science and Technology, Kumamoto University, Kumamoto 860-8555, Japan; atsushi_sainoki@kumamoto-u.ac.jp
5 International Research Organization for Advanced Science and Technology, Kumamoto University, Kumamoto 860-8555, Japan
* Correspondence: 248d9201@st.kumamoto-u.ac.jp or mln95@outlook.fr

**Abstract:** Artisanal and small-scale gold mining (ASGM) is a serious growing concern in Sub-Saharan Africa. In Mauritania, recent gold discoveries in the north and northwest have led to an increase in ASGM centers, reflecting trends across the region and posing considerable risks of mercury (Hg) contamination. Notwithstanding this fact, the extent of mercury contamination in the region remains unclear due to insufficient knowledge on the mechanisms of Hg dispersion in hyper-arid regions. In light of this, the present study aimed to acquire fundamental knowledge to elucidate the dispersion mechanism of mercury through conducting soil and groundwater sampling in and around Chami town, Mauritania, where ASGM activities have intensified. We analyzed 180 soil samples and 5 groundwater samples for total mercury (total Hg) using cold vapor atomic absorption spectrometry (CVAAS) and atomic fluorescence spectrometry (AFS) methods. The total Hg levels in soil samples ranged from 0.002 to 9.3 ppm, with the highest concentrations found at ASGM sites. Groundwater samples exhibited low total Hg levels (0.25–1.25 ng/L). The total Hg content in soil and groundwater samples was below Japanese standards, yet soil samples from hotspot points exceeded other international standards. Our study emphasizes the Hg dispersion patterns around Chami town, suggesting a gradual decrease in total Hg with increasing distance from ASGM sites and a potential influence of wind dynamics. The knowledge accumulated in this study provides essential insights into the Hg dispersion mechanisms in Chami town, laying the foundation for establishing a predictive model of Hg contamination in hyper-arid regions.

**Keywords:** mercury pollution; artisanal and small-scale gold mining; Mauritania; Banc d'Arguin National Park; elemental mercury

## 1. Introduction

Mercury (Hg), a fundamental element in artisanal mining, has been employed since antiquity. Its application in the amalgamation process has precipitated considerable environmental degradation and poses severe health threats to humans. The artisanal and small-scale gold mining (ASGM) sector, which extends across more than 80 nations, employs an estimated 15 million miners and supports the livelihoods of nearly 100 million individuals [1]. Despite its socio-economic significance, it is also the leading global contributor to Hg emissions. ASGM operations require water resources for the processing of gold ores, which can lead to the potential hazard of Hg contamination for both the local population and the environment. This is exemplified by the context of Egypt, where research has demonstrated the dispersion of Hg originating from ASGM centers situated in proximity to the Nile Valley, with a high risk of methylmercury (MeHg) contamination [2]. Additionally, in Mongolia, total Hg levels in the sediments of the upper Boroo have surpassed 2 ppm dry weight as a result of artisanal gold mining activities [3]; meanwhile, the levels of mercury in soils have exceeded the standard limit in Peru, with industrial areas exhibiting an average concentration of 72.6 ppm compared to the standard level of 24 ppm, while residential areas revealed an average of 9.5 ppm against the standard limit in residential areas of 6.6 ppm [4]. Furthermore, cyanidation processing has been initiated to extract the residual gold using cyanide from the remediation of waste tailings that contain elemental mercury (Hg(0)). During this process, cyanide–mercury complexes are formed, which enable the transportation of Hg over extended distances. This increases the potential impact area of Hg contamination [5].

The Minamata Convention on Mercury—an international treaty—aimed to curtail and, if possible, eradicate the use of Hg in ASGM. However, the exact quantity of Hg utilized worldwide in ASGM remains unclear, and the adoption of Hg-free techniques is still inadequate [6–13]. Article 7 of the internationally acknowledged Minamata convention explicitly proscribes the application of cyanide to Hg-rich tailings, marking a significant stride towards environmental conservation. Moreover, the convention endorses the development of an exhaustive national action plan addressing Hg utilization. This plan is designed to regulate and oversee the employment of Hg, particularly within the ASGM sector [14]. Recent investigations have indicated escalating emissions from ASGM into the environment over the past decade [15,16]. This increase can be attributed to the pervasive application of Hg in processing centers, particularly in the whole amalgamation method [17–19]. Amalgamation is a prevalent technique for gold extraction that is linked with significant environmental harm. The process begins with the hauling of gold ore from extraction sites, leading to habitat degradation and soil erosion [20]. The subsequent crushing phase, which is essential for reducing the particle size, is energy-intensive and intensifies greenhouse gas emissions [21,22].

In 2005, archival data suggested that the levels of Hg emissions in West Africa were rather low. However, these emissions have progressively escalated over time, transforming West Africa into the most significant contributor to global Hg emissions within the ASGM sector. Hg emissions from ASGM have witnessed a consistent rise, with an estimated discharge of 727 tonnes in the year 2013. These emissions from ASGM have now become the most substantial source worldwide, contributing to 37% of the total emissions [23]. Moreover, the most recent data indicate an increase in emissions, with ASGM releasing 838 tonnes of Hg, of which 80% originated from Sub-Saharan Africa [24]. This surpasses the previously recorded amount, signifying a growing environmental concern in Mauritania and contributing to an overall increase in Hg emission in Africa [25]. A selective extraction study conducted in Senegal revealed that soils in the vicinity of ASGM activities were contaminated with Hg(0). Additionally, it was observed that, along a stream, the concentrations of total Hg in the sediment decreased with increasing distance from the ASGM sites. The study estimated that MeHg is commonly found at locations at a distance from the ASGM sites, suggesting that the transported Hg(0) transforms into MeHg as it moves away from the source. Interestingly, the study unveiled a trend of decreasing concentrations of MeHg as one approaches the ASGM site [26]. In addition, recent investigations focused on the assessment of trace metals in soil, wastewater, and borehole groundwater within the precincts of ASGM processing centers in Chami town, Mauritania, and indicated that the concentrations of essential trace metal elements (ETMs) and Hg were elevated in samples of process water, while being significantly lower in samples obtained from drilled water wells [27]. The discernible escalation in Hg usage in West Africa implies an increasing growth in the environmental hazards linked to its application.

It is worth noting, for areas with similar characteristics to our study area, that the extreme hyper-arid climate and infrequent rainfall in the northern countries of Sub-Saharan Africa could pose distinct environmental challenges, particularly in the context of natural resource extraction and processing. These challenges can exacerbate the environmental impacts of these activities, leading to the dispersion of large-scale Hg contamination. This can have detrimental effects on both the environment and the health of local communities [28]. Sudan has experienced an increase in mercury emissions, where recent studies revealed 34.8 ppm in soil samples near amalgam burning shops [29]. Additionally, observations from the United States Geological Survey's research initiative at the Amargosa Desert Research Site (ADRS) in southwestern Nevada revealed a potential conduit that facilitates the extensive lateral migration of Hg(0) through the unsaturated zone. This suggests that Hg(0) can travel significant distances horizontally in such arid environments, highlighting the need for a nuanced understanding of its mobility and distribution [30].

Notwithstanding this fact, previous studies on Hg dispersion in Sub-Saharan Africa predominantly focused on the investigation of Hg contamination in the vicinity of ASGM sites or on underground water and/or streams as migration paths [31]. According to a series of studies, the ASGM center in Chami town exhibits a significant accumulation of metals in the soil, which can persist in the environment [32]. Importantly, even though Hg can disperse, diffuse, and migrate to distant locations—particularly in its gaseous form—its mechanisms of dispersion remain to be elucidated [33–35].

Consequently, to address this problem, our research scope was expanded to the surrounding region of Chami town, including sensitive areas such as the Banc d'Arguin National Park (PNBA), the most extensive protected natural zone in Mauritania, which is located in proximity to Chami town, where researchers discovered that 53% of the 233 bird species associated with the PNBA exhibit a high degree of dependency on the park and its coastal and marine ecosystems during their life cycle [36].

This research endeavor sought to elucidate the extent of Hg dispersion in and around Chami town, including a comprehensive assessment of total Hg concentrations in soil and groundwater samples collected from former and active ASGM centers in Chami town, its surrounding areas, and parts of the PNBA. Hence, the present study represents a significant step forward in deepening our understanding of the dispersion of Hg in extreme hyper-arid climates.

## 2. Materials and Methods

### 2.1. Zone of Study

Miners are engaged in the extraction of gold ores from mining sites located within the Tasiast–Tijirit Terrane around 300 km north of the capital city Nouakchott, a region situated in the western segment of the Rguibatt Shield [37]. Notably, these extraction activities are predominantly concentrated in the Precambrian greenstone belts of Ahmeyim, Tasiast, and Tijirit, and extend to the northeast parts of the country. These areas are recognized for their rich deposits of gold ores, making them prime locations for mining operations [38].

The ASGM activity began in 2016 and, due to the scarcity of water resources within the mining areas, miners abstain from processing gold ores in the vicinity of the mining sites. Instead, they transport the gold ores from the mining site to Chami town, a newly constructed city. This city is strategically located on the principal national road that connects the capital cities of Nouakchott and Nouadhibou. This connectivity has catalyzed an unprecedented transformation in the city, leading to a surge in population and urban development, thereby earning it the reputation of being the 'Capital Gold City' in Mauritania, and the development of Chami town has been closely tied to the growth of ASGM activities in the region [39,40].

Chami town is located 250 km north of the capital, Nouakchott, and has coordinates between latitude $-15°55'$ and $-16°6'$ W and longitude $20°4'$ and $20°11'$ N, with 13,000 inhabitants in 2021 [41]. The geographical region under study is characterized by a relatively flat topography, with altitudes ranging from sea level to a maximum of 50 m. This landscape is further distinguished by the presence of partially encircling dunes, adding a unique element to the area's geography. This region is geologically significant, as it forms part of the Meso-Cenozoic Atlantic Coastal Basin in Mauritania [42]. This basin is a noteworthy feature, due to its extensive coverage and the variety of geological periods that it represents. The research area is situated in a desert region where human habitation has evolved along a gold-laden dyke, now lined with white tents and gray drilled holes indicating ongoing extraction, potentially composed of granodiorite rocks. This location is exploited for ASGM, a process that involves subsistence miners employing their resources—typically manual labor and rudimentary tools—to extract gold from the rock formations [43].

### *2.2. Data Collection, Preparation, and Analysis*

Soil and water samples were collected and prepared for the measurement of total Hg, leaching tests, and mineralogical analysis using the X-ray fluorescence (XRF) method, as well as organic material testing using the loss on ignition test (LOI).

#### 2.2.1. Measurement of Total Hg Content and Moisture Ratio in Soil

The sampling area encompassed a broad selection within the PNBA and the surrounding regions of Chami town for the evaluation of Hg levels. Soil samples were collected from natural soils, residential areas, and ASGM sites. A Garmin 66i GPS was used to determine and save the coordinates of each sampling site precisely, and the open space allowed for perfect satellite connection; however, the alignment of the grid sampling points was affected by the existence of expanded dunes and private land. The samples were taken from a depth of 10 to 20 cm. At each of the 36 sampling points illustrated in Figure 1, five samples were collected: from the center and the four cardinal directions (north, south, east, and west). Hotspots 1 and 2 represent active and former ASGM sites in Chami town, while Hotspot 3 is located in a semi-industrial gold mining area, comprising recently installed cyanide-based processing plants alimented by ASGM residues.

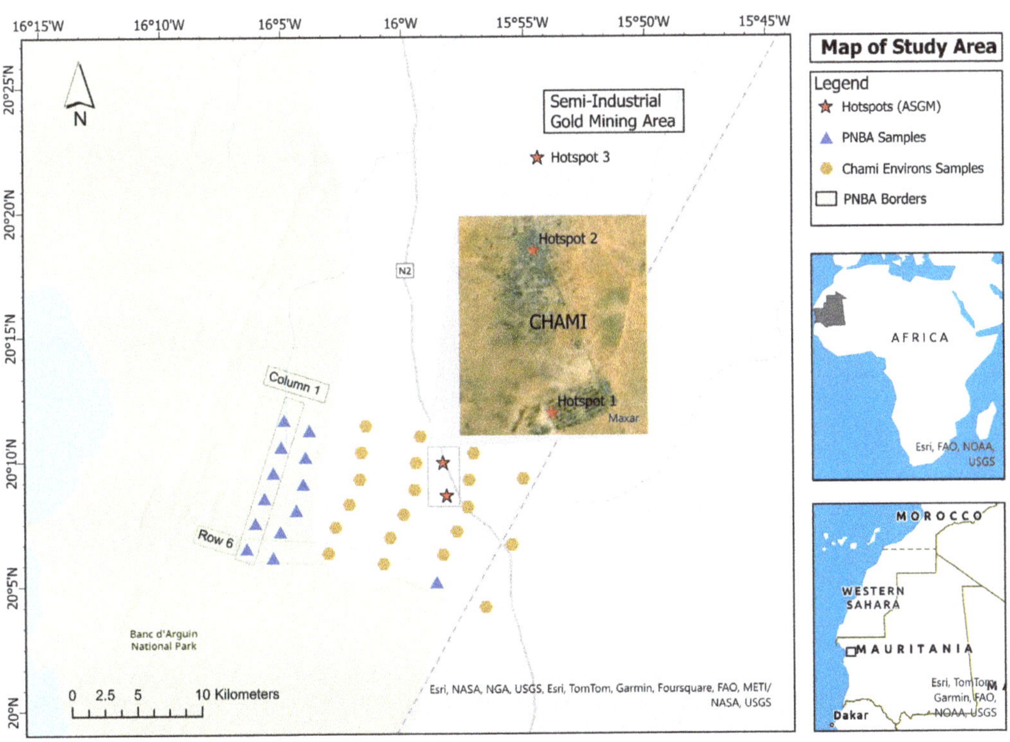


**Figure 1.** Location of the sampling sites within the zone of study.

The collected samples were mixed to enhance the representativeness of the soil samples at each point. The sampling area comprised the control point (CP), the PNBA (with samples labeled PA, A1, A2, A3, A4, PB, B1, B2, B3, B4, B5, B17, and F1), the environs of Chami town (with samples labeled F2, B6, B7, B8, B9, B10, B11, B12, B13, B14, B15, B16, B18, B19, B20, B21, B22, PC, C1, and PD), and the active site named 'Grillage', as well as the former and the semi-industrial ASGM sites (with samples labeled Hotspot 1, Hotspot 2, and Hotspot 3, respectively). In addition, a control sample was collected from a pristine area. Before being placed into a cooler and frozen in preparation for Hg analysis, all samples were precisely weighed using a precise electronic scale, labeled, and preserved in Ziploc and Lami Zip bags.

To measure the total Hg content, the soil samples were prepared for digestion at the International Research Cooperation laboratory at the National Institute for Minamata Disease (NIMD) in Minamata, Japan. The preparation process involved the removal of pebbles, wood particles, and shells from the soil samples using the quarter method. Specific weights of the prepared samples were then introduced into a 50 mL volume metric flask for digestion. In this experiment, sample weights of 100 and 200 mg were weighed and digested for each soil sample using pure strong acids: 1 mL of distilled water was added, followed by 2 mL of a (1:1) $HNO_3 - HClO_4$ solution and 5 mL $H_2SO_4$, and the solution was heated to 200–230 °C for 30 min continuously. After cooling, the digestion flask was filled using pure water. The samples were digested using pure strong acids under identical conditions, to prevent any inconsistencies and contamination during the measurement of total Hg.

This experiment was repeated several times to ensure adequate measurement and analysis. This rigorous and systematic approach to sample preparation and analysis ensured the reliability and accuracy of the Hg measurements obtained from the soil samples [44,45]. The total Hg levels in the soil were determined using the cold vapor atomic absorption spectrometry (CVAAS) method, according to the method developed in [46] and modified in [47]. A volume of 5 mL of the sample test solution was introduced into the Hg analyzer. The total Hg levels were calculated using the peak heights obtained from the blank, standard, and sample test solutions, and recorded using an Hg-201 semi-automated mercury analyzer (Sanso seisakusho Co., Ltd., Tokyo, Japan). The added solution was reacted with a $SnCl_2$ solution in a reaction vessel connected to the Hg analyzer, resulting in the formation of Hg(0). These atoms were then transported into an open-ended quartz tube in the Hg analyzer's optical path using an air stream.

In the field of total Hg and MeHg analysis, adhering to strict quality control (QC) measures is crucial, as mandated by EU legislation. To ensure QC in chemical analysis, the use of certified reference materials (CRMs) has been recognized as an effective approach. In this study, the CRM 580 standards were used for the measurement of total Hg concentrations in soil samples. The moisture contents were determined as proposed by the Ministry of Environment in Japan [45], measuring the weight and dry weight (WW/DW) using a porcelain crucible with a known weight, to which 1 g of soil sample was added. This setup was then placed in a drying oven and kept overnight at a temperature of 105 °C. Then, the crucible porcelains were allowed to cool in a desiccator for 30 min to prevent moisture absorption from the surrounding air. After cooling down, the dried soil samples were reweighed and the WW/WD ratios were calculated accordingly.

### 2.2.2. Soil Mineralogical Characteristics

Samples were collected from five locations within the study area, ensuring their representativeness of the region of interest. The collection process focused on maintaining the consistency and integrity of the samples, in order to preserve their natural state as much as possible. Following collection, the samples were stored in Ziploc and Lamizip bags in a freezer and prepared for analysis by ensuring the appropriate size and purity, in order to avoid any inconsistency that could interfere with the XRF analysis. The analysis was conducted using a Rigaku Co., Ltd., Tokyo, Japan, ZSX Primus II device, which is optimized for light element-based substances, such as polymers, and has a measurable element range from Beryllium (4Be) to Uranium (92U). The device features a minimum analysis diameter of $\varphi = 0.5$ mm, allowing for the analysis of small sample areas, and employs a semi-quantitative analysis method based on fundamental parameters.

Additionally, the device was equipped with a top-illuminated wavelength dispersion type and a CCD camera for enhanced precision and accuracy in elemental detection. To ensure that the samples fit the analysis chamber of the ZSX Primus II device, they were prepared to be flat and homogeneous, and were ground and pressed into pellets to achieve a suitable surface [48].

#### 2.2.3. Soil Loss on Ignition Test

LOI is described as a measure that quantifies the weight reduction experienced by a sample when subjected to high temperatures; namely, a temperature of 550 °C was used in this experiment. The observed weight loss typically results from the expulsion of volatile substances, moisture, or organic matter present within the sample. In this study, a LOI test was implemented to evaluate the accuracy of the total volatile measurement. The 2 mm sieved samples were dried in an oven at 100 °C for 24 h minimum and then heated inside a furnace at 550 °C for 60 min.

#### 2.2.4. Soil Leaching Experiment

The leaching test was performed following the Japanese Standard Notification JLT. No. 46. [49], with the goal of determining the leachability of Hg in the soil samples. Approximately 10 grams of each sample from each location were combined to create a single representative sample for the five sampling locations. During the experiment, the soil samples were prepared in duplicate. Each soil sample, weighing precisely 5 g, was transferred to a 50 mL glass bottle and filled with distilled water with the pH adjusted from 5.8 to 6.3 using hydrochloric acid. Three standards (precisely, 0, 0.1, and 0.2 ng/L Hg) were prepared to establish a baseline for Hg content analysis. The samples were shaken using a Recipro Shaker (TAIYO company Co., Ltd., Tokyo, Japan, SR-1) for 6 hrs and then centrifuged at 2500 rpm for 25 min. Next, 0.45 µm filters were used for filtration of the samples, to ensure their purity and consistent texture and composition.

The filtered samples were transferred to a 200 mL funnel, and around 90 to 95 mL of each sample was added to the separatable funnel. Finally, the funnel was topped up with pre-measured pH distilled water until it reached a volume of 200 mL. Several reagents were then prepared and added one by one to the sample funnel. First, 1 mL of 20N $H_2SO_4$ was added, followed by 0.5 mL of 0.5% $KMnO_4$, then, 2 mL of 10N NaOH and, finally, 0.5 mL of 10% $NH_2OH - HCl$ solution. After mixing, another addition was made after 20 min; namely, 0.5 mL of 10% EDTA solution and 5 mL of 0.01% dithizone–toluene were combined and mixed thoroughly with the samples. The samples were left undisturbed overnight while covered with a black plastic sheet. The following day, the aqueous phase was discarded from the funnel, and 4.5 mL of the remaining organic phase was transferred to a 25 mL volumetric flask that was for digestion. The sample was then evaporated to dryness using a rotary evaporator floating on a 60 °C hot water plate. Then, 1 mL of distilled water was added, followed by pure strong acids: 2 mL of a (1:1) $HNO_3 - HClO_4$ solution and 5 mL $H_2SO_4$. The mixture was heated at 200–230 °C for 30 min for digestion. After heating, the samples were allowed to cool at room temperature and topped up to 25 mL with pure water.

The solubility of Hg in the leached samples was measured using the AFS method employing an RA-3F Mercury Analyzer (Nippon Instruments Corporation, Osaka, Japan). A fixed volume of 10 mL of each sample test solution and 0.2 mL of 10% $SnCl_2$ solution were introduced into the Hg analyzer for the measurement of total Hg.

2.2.5. Quality Assurance

At the same time as the preparation of total Hg measurements for water samples and the leaching test, standard samples were prepared under the same conditions to ensure the reliability of the analysis, as mentioned in the section detailing the preparation and analysis of water samples. Calibration curves were carefully created for the low and high concentration ranges in water samples and the leachate, showing outstanding linearity, with coefficients of determination ($R^2$) of 0.9846, 0.9999, and 0.9905, respectively. To check the accuracy of the analytical method, recovery tests were conducted through injecting known quantities of Hg into the samples, followed by conducting a duplicate test. Furthermore, each sample was analyzed in triplicate, with all samples displaying coefficients of variation below 10%. For soil samples, analyses were performed in triplicate, resulting in a coefficient of variation of less than 10%. CRMs were utilized, specifically the CRM 580 from NIMD in Japan, which was designed for Hg in marine sediment. The obtained values ranged from 129 to 134 ppm, corresponding to an average recovery rate of 99.5%, and further validations of the accuracy and precision of the analytical method were implemented in these analyses.

2.2.6. Measurement of Total Hg in Water

The process began with the thorough collection of groundwater samples from wells located near Chami town. The samples were collected, and the pH, electric conductivity (EC), and oxidation–reduction potential (ORP) were measured on-site. To further ensure the accuracy of these measurements, the samples were filtered using PVFD 0.45 μm Glass-fiber filters, which were essential to ensure the purity of the samples through removing impurities, particles, and micro-organisms from the water samples. Following these initial preparations, the samples were carefully preserved in 1000 mL polyethylene bottles. These bottles were then placed in a cooler and frozen at −80 °C to prepare them for subsequent Hg measurements. The water sample preparation involved the following steps. In the beginning, 500 mL of the sample was transferred into a separating funnel. Then, 2.5 mL of 20N $H_2SO_4$ followed by 1.5 mL of 0.5% $KMnO_4$ solution were added. The mixture was shaken and left to stand for 5 min. After that, the mixture was neutralized by adding 5 mL of 10N NaOH and 1.5 mL of 10% $NH_2OH - HCl$ solution. Again, the mixture was shaken. The mixture was then allowed to stand for another 20 min. Then, 1.5 mL of a 10% EDTA solution was added and shaken. Next, 10 mL of purified 0.01% dithizone–toluene was added to the mixture to extract the Hg in the solution, and vigorously shaken. Finally, we left the mixture undisturbed overnight covered by a black sheet, to ensure complete extraction. After separating the organic and water phases, the water phase was discarded, and the toluene phase was added to a 10 mL conical centrifuge tube fitted with a glass stopper. The tube was centrifuged at 1200 rpm for 3 min. Once centrifuged, a fixed volume of 8 mL of the toluene phase was transferred into a 50 mL volume metric flask for digestion. A rotating evaporator and a hot water bath set at 60 °C were used to completely evaporate only toluene in the dithizone–toluene phase. After evaporation, the samples were digested, and 1 mL of deionized water was added to the residue, followed by 2 mL of $HNO_3 - HClO_4$ (1:1) and 5 mL of $H_2SO_4$. The mixture was then heated on a plate at temperatures between 200 and 230 °C for around 30 min. Once it had cooled down, water was added to reach a volume of 50 mL. The resulting solution was thoroughly mixed and used as a sample test solution. In parallel, a sample with a lower Hg content was selected, due to the scarcity of Hg in the groundwater samples. To each of these selected 500 mL water samples, 0, 0.2, or 0.4 mL of MeHg-cysteine solution (100 ng Hg/mL corresponding to 0, 20, or 40 ng Hg) were added. The above preparation procedures for sample test solutions were followed to obtain blank and standard test solutions for the measurement of total Hg. The total Hg levels were measured using the atomic fluorescence spectrometry (AFS) method, which is mainly used for the detection of low total Hg levels in water samples. This measurement involved transferring 10 mL of pure water, blank, and standard test solutions, followed by 10 mL of the digested sample test solution into the RA-3FG+ mercury analyzer. In parallel, the CVAAS method was also employed. In both methods, 1 mL of $SnCl_2$ solution was added and automatically and manually employed as a reductant for total Hg measurement [50–53].

### 2.3. Geo-Accumulation Index

The quantitative assessment of Hg contamination in soils entailed the calculation of geo-accumulation indices (Igeo) using Equation (1), as proposed by Muller [54]. This calculation employed background values established for the Earth's continental crust (Hg = 0.04 ppm) [55], in conjunction with the total Hg concentration observed in the PNBA, Chami town environs, and ASGM site soil samples. The results were categorized into the classes presented in Table 1.

$$I_{geo} = \log_2 \left( C_{Hg} / 1.5 \times B_{Hg} \right) \quad (1)$$

where

- Igeo: geo-accumulation index for Hg;
- $C_{Hg}$: Hg concentration in the soil samples;
- $B_{Hg}$: Hg background level (0.04 ppm);
- 1.5: the factor employed to correct for lithogenic effects.

**Table 1.** Geo-accumulation indices (Igeo) and the corresponding contamination levels, as defined by Martínez and Poleto (2014) [56].

| Class | Value | Soil Quality |
|---|---|---|
| 0 | $I_{geo} \leq 0$ | Unpolluted |
| 1 | $0 < I_{geo} \leq 1$ | Unpolluted to moderately polluted |
| 2 | $1 < I_{geo} \leq 2$ | Moderately polluted |
| 3 | $2 < I_{geo} \leq 3$ | Moderately to heavily polluted |
| 4 | $3 < I_{geo} \leq 4$ | Heavily polluted |
| 5 | $4 < I_{geo} \leq 5$ | Heavily to extremely polluted |
| 6 | $I_{geo} > 5$ | Extremely polluted |

Note: Source in the references.

## 3. Results

This study revealed a strong link between ASGM activities and increased levels of Hg in soil samples. ASGM mainly uses Hg to extract gold from milled ore, leading to severe Hg leakage into the environment. This causes Hg levels to rise, especially in the soil. The environment in the area, marked by high temperatures and a hyper-arid climate, increases the evaporation of Hg, thus reducing its presence in water. Hg tends to spread through the air over large distances, with this phenomenon occurring due to its capacity to evaporate. The combination of high temperatures and dryness speeds up the evaporation of Hg, resulting in lower concentrations in water bodies. Moreover, the physical and chemical properties of soil and water can greatly affect the behavior of Hg. Certain chemicals in the soil can cause Hg to be trapped and then leak out, possibly leading to long-term contamination of groundwater. The amount of organic matter, pH levels, and various soil characteristics can affect how easily Hg moves and how available it is for uptake by organisms.

### 3.1. Total Hg in the Soil

The concentrations of total Hg measured at various sampling points within the PNBA area are shown in Figure 2.

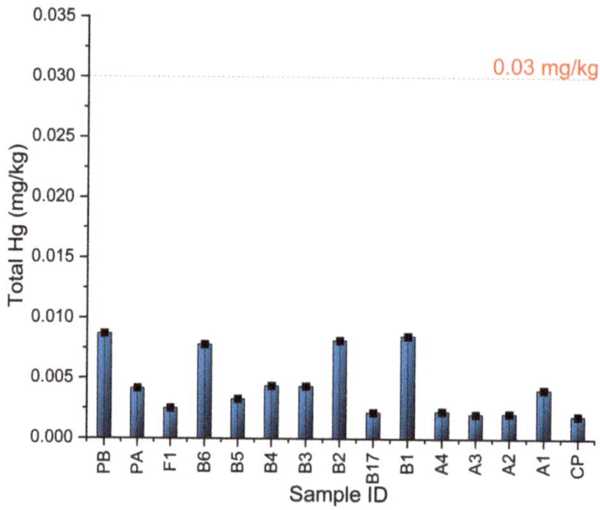


**Figure 2.** Chart of the total Hg content in soil samples within the PNBA area.

A number of sampling points demonstrated higher concentrations of total Hg in comparison to the control point, which had a concentration of 0.002 ppm. The soil surrounding Chami town displayed similar characteristics to those observed in the PNBA area, but with slightly elevated concentrations of total Hg, ranging from 0.003 to 0.02 ppm, as illustrated in Figure 3. This increase was due to being adjacent to the hotspot points.

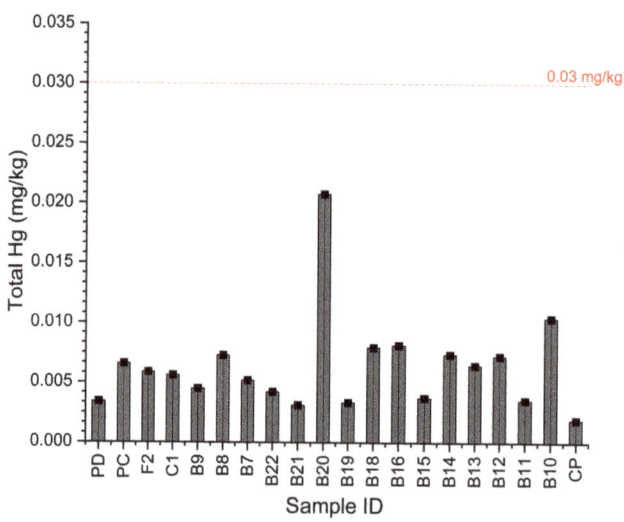


**Figure 3.** Chart of the total Hg in soil samples within the Chami town environs.

Despite this increase, it is crucial to note that these levels remained below the average natural Hg concentrations typically found in sedimentary rocks, approximately 0.03 ppm [57] and the Hg concentration in the Earth's crust, 0.04 ppm [55]. The varying heights of the bars in the chart indicate that the total Hg concentrations were not uniform across the PNBA area, which could be attributed to various environmental factors such as the soil composition, rainfall, and human activity.

Figure 4 presents a distribution map for the total Hg concentration in soil samples from the PNBA area, Hotspots from ASGM sites, and Chami town environs (excluding the control point, which was located around 100 km south of the sampling grid), displaying the fluctuating total Hg levels across the different sampling points. The largest points on the map represent the ASGM sites, particularly the antecedent ASGM site in the central market of Chami town and the active ASGM centers in 'Grillage'. A remarkable observation was the elevated concentrations of total Hg at a specific sampling point, designated as B20, which was located east but slightly to the south from the ASGM sites.

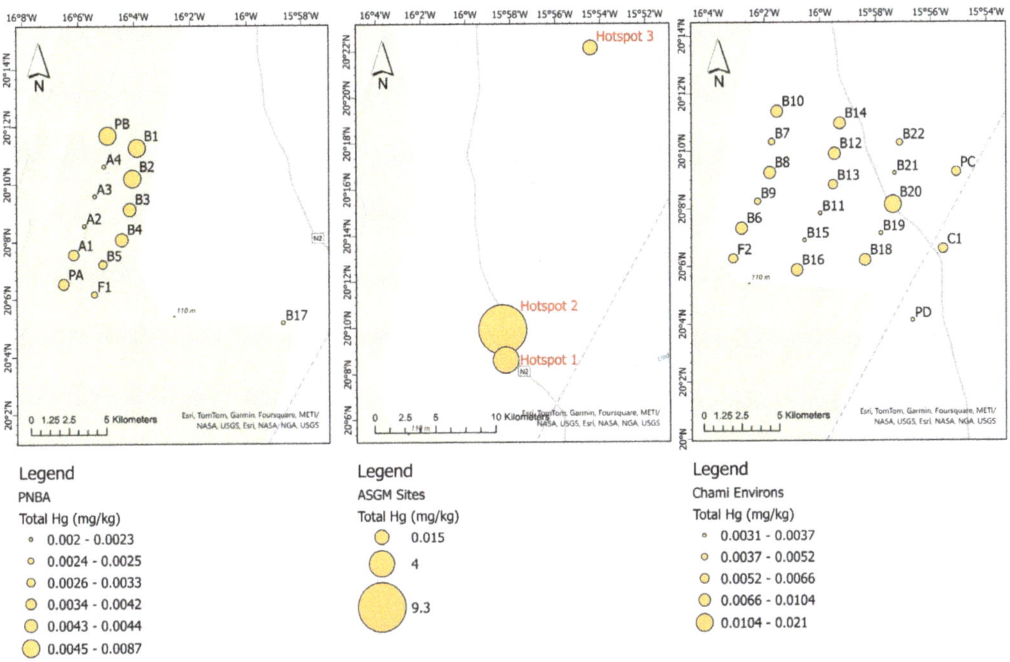


**Figure 4.** Distribution map of the total Hg content in soil samples.

The data further revealed significant variability in total Hg concentrations between locations, emphasizing the complex spatial distribution of Hg in the region. These figures provide a comprehensive overview of the spatial distribution of total Hg concentrations, highlighting the need for further research to understand the factors contributing to this variability and its potential environmental implications. This heightened concentration of total Hg could potentially be attributed to the interplay of several factors. The dynamics of wind patterns, as a natural phenomenon, could significantly contribute to this. Wind can transport and deposit dust particles, including those containing Hg, over large distances [58]. In the study area designated as Hotspot 3—a semi-industrial gold mining zone situated roughly 30 km away from Chami town—the total Hg concentration was measured at 0.015 ppm. This contrasted starkly with the findings from Grillage, depicted in Figure 5, which is an active artisanal gold mining site, where the total Hg levels soared to a substantial 4 ppm. However, the most concerning data emerged from the Chami town market, previously an artisanal gold mining center, where total Hg concentrations reached an alarming 9.3 ppm, far exceeding the critical Hg threshold of 3 to 5 ppm established for environmental and health safety protocols [59].

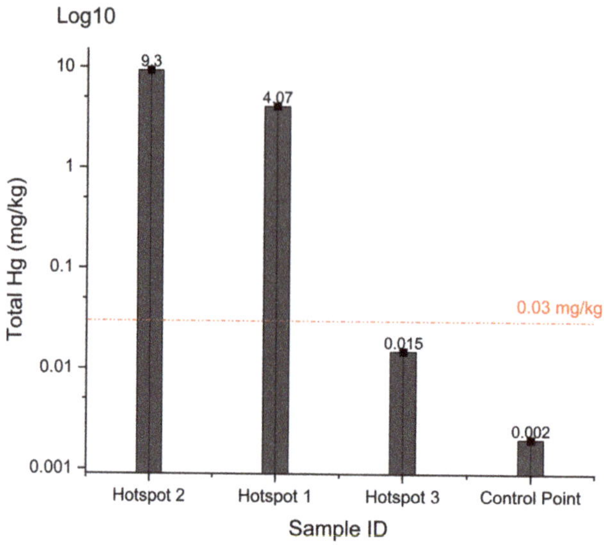


**Figure 5.** Chart of the total Hg content in the ASGM sites (Hotspots).

### 3.2. Chemical Composition and Loss on Ignition

The XRF results revealed that, among the five samples, significant differences were not observed. Regarding the content ratio, $SiO_2$ was the dominant component, whereas $Al_2O_3$, CaO, $Fe_2O_3$, $Na_2O$, and MgO were present in small quantities in all five samples. The findings of the XRF analysis are shown in Table 2, which indicate the variation in the elemental composition between samples from different sampling locations, reflecting the complex geochemical processes shaping the region's landscape. The high levels of light elements such as carbon and oxygen across all samples are indicative of the prevalence of organic matter and silicate minerals. The differences in trace element concentrations can provide insights into the past environmental conditions, geological history, and mineral resource potential of the area under study.

**Table 2.** Chemical composition of soil samples from XRF analysis and average LOI in different sampling locations.

| Location | $SiO_2$ (wt%) | $TiO_2$ (wt%) | $Al_2O_3$ (wt%) | $Fe_2O_3$ (wt%) | MnO (wt%) | MgO (wt%) | CaO (wt%) | $Na_2O$ (wt%) | $K_2O$ (wt%) | $P_2O_5$ (wt%) | S (wt%) | N (wt%) | Cl (wt%) | C (wt%) | LOI (%) |
|---|---|---|---|---|---|---|---|---|---|---|---|---|---|---|---|
| PNBA | 63.1 | 0.501 | 11.4 | 1.84 | 0.03 | 1.29 | 2.38 | 1.48 | 2.17 | 0.070 | 0.02 | - | 0.09 | 3.17 | 0.84 |
| CE | 83.9 | 0.167 | 4.48 | 0.79 | - | 0.38 | 0.84 | 1.62 | 1.93 | 0.007 | - | - | 0.05 | 1.01 | 1.49 |
| HS1 | 69.3 | 0.167 | 5.16 | 0.99 | - | 1.36 | 5.46 | 1.33 | 1.45 | 0.050 | 0.03 | - | 0.05 | 3.60 | 0.83 |
| HS2 | 58.9 | 0.167 | 7.31 | 1.46 | 0.03 | 1.16 | 8.40 | 1.56 | 2.41 | 0.070 | 0.05 | - | 0.04 | 3.99 | 1.92 |
| HS3 | 68.7 | 0.334 | 12.3 | 2.04 | 0.03 | 1.21 | 0.98 | 1.33 | 3.13 | 0.050 | 0.01 | 0.48 | - | 1.75 | 1.40 |

Note: CE: Chami town environs. HS: Hotspot. Source: Authors' data.

### 3.3. Total Hg in Soil Leachate

The analysis of total Hg levels in soil leachates revealed a correlation with the total Hg levels in the corresponding soil samples, suggesting that the Hg present in the soil is being leached out, potentially affecting the surrounding environment within Chami town. Notably, the highest concentrations of total Hg in the leachate were observed in leached samples from hotspots at both former and active ASGM sites. These leachates (HS1 and HS2) revealed total Hg concentrations higher than the WHO standard limit in drinking water of 500 ng/L [60]. These findings, as detailed in Table 3, highlight the significant impact of ASGM activities on the Hg concentrations in these areas.

Table 3. Total Hg concentrations in leachates.

| Locations | Samples | Total Hg (ng/L) |
|---|---|---|
| PNBA | 1 | 1.04 |
| Chami town environs | 1 | 6.30 |
| Hotspot 1 | 1 | 559.30 |
| Hotspot 2 | 1 | 1338.03 |
| Hotspot 3 | 1 | 61.28 |

Source: Authors' data.

The results of the leaching experiment revealed a resemblance between the total Hg concentrations in the soil and those in the corresponding soil leachates. This observation underscores the efficacy of the extraction process employed for the quantification of total Hg. Furthermore, similarly to the total Hg levels in soil, the levels in soil leachates demonstrated a marked elevation in Hg concentrations. These levels were significantly higher in comparison to the soil leachates of the PNBA and Chami town environs, which are areas devoid of any ASGM activities.

### 3.4. Total Hg in Water/Groundwater Samples

The analysis of tap water from the capital Nouakchott (NKC) revealed the highest total Hg concentrations, with values ranging from 0.33 to 1.32 ng/L. Interestingly, these levels are significantly lower than the typical Hg level found in groundwater, which is less than 500 ng/L according to the World Health Organization [60]. In addition, the groundwater samples (labeled WT1, WT2, WT3, WT4, and WT5) also revealed low total Hg levels. In the first analysis performed using the CVAAS method, the concentrations ranged from 0.29 to 0.54 ng/L. In the subsequent analyses, the range was slightly larger, from 0.1 to 1.25 ng/L. These results are represented in Table 4.

Table 4. Total Hg concentrations in water samples according to CVAAS and AFS measurement methods.

| Sample | pH | Ec (ms/cm) | ORP (mV) | CVAAS (ng/L) | AFS No. 1 (ng/L) | AFS No. 2 (ng/L) |
|---|---|---|---|---|---|---|
| WT1 | 7.60 | 1.07 | 204 | 0.12 | 0.25 | 0.54 |
| WT2 | 7.99 | 1.49 | 203 | 0.20 | 0.30 | 0.53 |
| WT3 | 6.86 | 1.46 | 214 | 0.10 | 1.25 | 0.32 |
| WT4 | 5.55 | 19.1 | 356 | 0.11 | 0.23 | 0.38 |
| WT5 | 7.21 | 3.90 | 334 | 0.11 | 0.46 | 0.29 |
| NKC | 7.73 | 0.13 | 312 | 0.33 | 0.21 | 1.32 |

Source: authors' data.

## 4. Discussion

### 4.1. Fluctuating Total Hg Levels in Soil

The analysis of total Hg levels in soil indicated considerably elevated total mercury concentrations in the hotspots located at the ASGM sites. However, total Hg levels found in the vicinity of Chami town and PNBA were below the Hg world median value of 0.04 ppm in stream sediments [55]. The results of this study suggest that the fluctuating total Hg levels at the ASGM sites were comparable to those observed in the Tinkoto, Kharakhena, and Sambarabougou villages in Senegal, ranging from 0.95 to 2.21 ppm [61]. The elevated total mercury levels observed at the hotspots have had an impact on the surrounding soil. Specifically, the sampling points B20, B6, and B10 exhibited higher total mercury levels when compared to other samples, as illustrated in Figure 4.

#### 4.1.1. Average Total Hg in Sample Grid's Columns and Rows

Figure 6 presents a comparison of the distribution of average total Hg concentrations across different columns (Figure 1) of the sampling grid, excluding the total Hg concentrations in the hotspots; here, Column 1 comprises the samples PB, A4, A3, A2, A1, and P1; Column 2 comprises samples B1, B2, B3, B4, B5, and F1; Column 3 comprises samples B10, B9, B8, B7, B6, and F2; Column 4 comprises samples B11, B12, B13, B14, B15, and B16; Column 5 comprises samples B22, B21, B20, B19, B18, and B17; and Column 6 comprises samples PC, C1, and PD. The results suggest that the average total Hg levels were notably higher in the columns located to the west of the ASGM site, compared to those on the east.

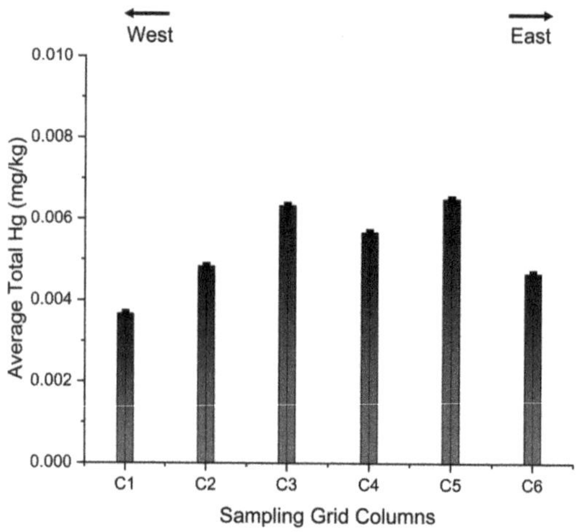


**Figure 6.** Total Hg average concentrations across the columns of the sampling grid, with Column 6 representing the east direction and Column 1 positioned to the west.

The sampling grid rows, as illustrated in Figure 1, encompass six rows: Row 1 (PA, F1, F2, B16, B17, and PD), Row 2 (A1, B5, B6, B15, and B18), Row 3 (A2, B4, B7, B14, BA, and B19), Row 4 (A3, B3, B9, B13, B20, and C1), Row 5 (A4, B2, B8, B12, and B21), and Row 6 (PB, B1, B10, B11, B22, and PC). The rows to the south of the ASGM sites exhibited a higher distribution of average total Hg concentrations, which might be due to the influence of the frequent north-eastern wind in the area. The distribution of the average total Hg concentrations within the sampling grid rows is shown in Figure 7.

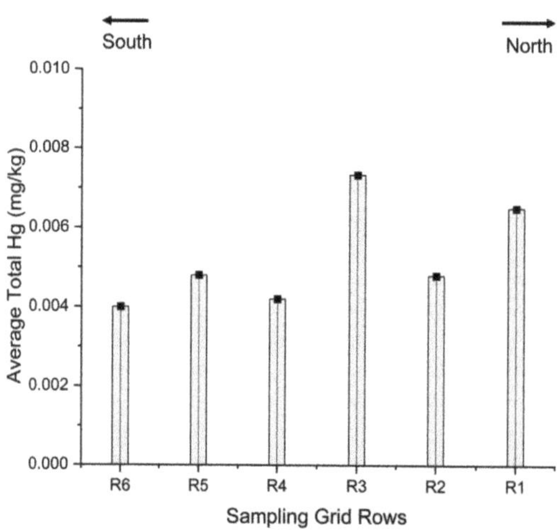


**Figure 7.** Total Hg average concentrations across the rows of the sampling grid, where Row 1 represents the north direction and Row 6 is positioned to the south.

4.1.2. Total Hg and Distance from ASGM Site

The measurement of the distance of each sampling point from Hotspot 2 is illustrated in Figure 8, which provides insight regarding the effect of ASGM activity, with higher total Hg levels in sampling points near the ASGM sites, especially sample B20 (located close to Hotspots 1 and 2). Additionally, the inherent characteristics associated with dune mobility could also play a crucial role. Dunes are not static; they shift and move due to wind and other environmental factors. This mobility could lead to a redistribution of sandy material and, consequently, Hg within these dunes [62].

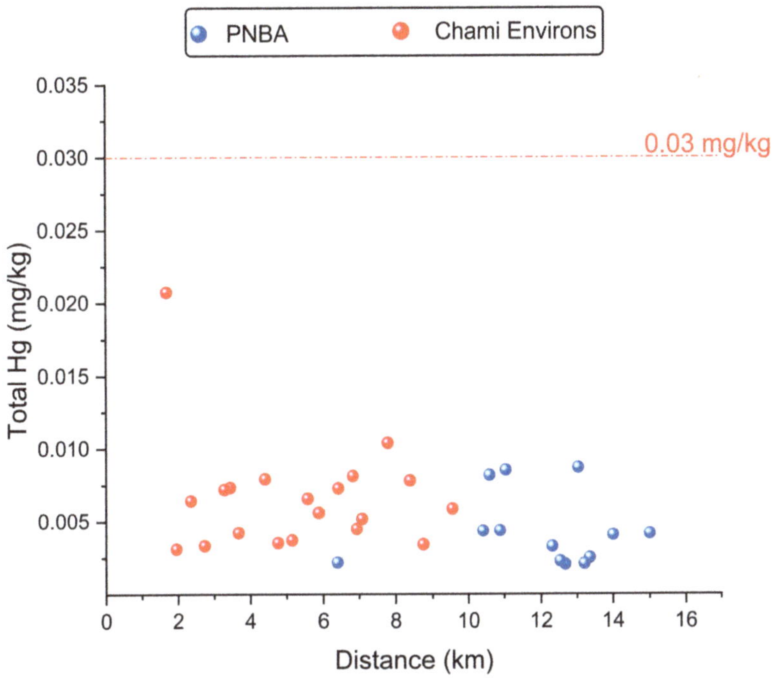


**Figure 8.** Correlation between total Hg content and distance from the main ASGM site 'Grillage' of PNBA and Chami town environs soil samples.

*4.2. Moisture Ratio*

The results of the moisture content analysis revealed a notably low wet-weight-to-dry-weight (WW/DW) ratio. This outcome is attributed to the hyper-arid climate prevalent in the study area, as represented in Figure 9. The low WW/DW ratio is indicative of minimal moisture content, a characteristic feature of hyper-arid climates. This finding aligns with the climatic conditions of the study zone, thereby providing a plausible explanation for the observed total Hg content.

The correlation between the moisture content and the climatic conditions underscores the influence of environmental factors on the physical properties of the samples. We calculated the correlation coefficient using the two variables of total Hg in ppm (mg/kg) and moisture ratio in (%) via Equation (2):

$$r = \frac{n(\Sigma xy) - (\Sigma x)(\Sigma y)}{\sqrt{[n\Sigma x^2 - (n\Sigma y^2 - (\Sigma y)^2)]}}, \quad (2)$$

where

- $n$ is the number of data points;
- $\Sigma xy$ is the sum of the product of $x$ and $y$;

- $\Sigma x$ and $\Sigma y$ are the sums of $x$ and $y$, respectively;
- $\Sigma x^2$ and $\Sigma y^2$ are the sums of the squares of $x$ and $y$, respectively.

The obtained coefficient factor was 0.4, which implies a weak positive linearity between the total Hg levels and the moisture content. This means an increase or decrease in moisture content within the region does not necessarily correspond to an increase or decrease in Hg levels. For example, sample B20 had a low moisture content but a relatively high Hg level. This suggests that, even though this area has a lower moisture content, the Hg concentration can still be high compared to other sampling points; for instance, sample B10 had a relatively high moisture and total Hg content. The results indicating the correlation between moisture and total Hg content are illustrated in Figure 10.

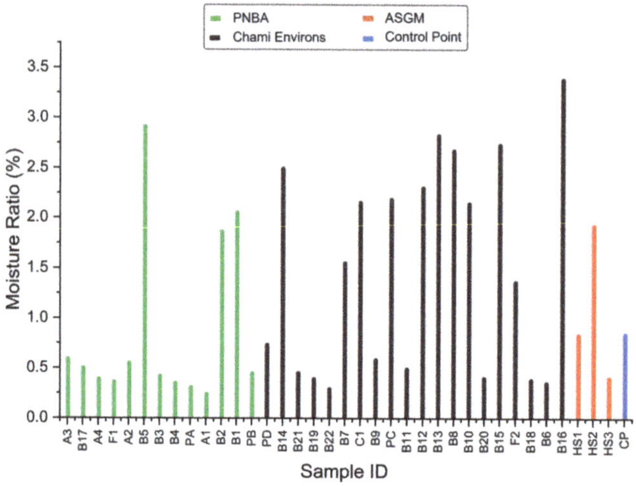


**Figure 9.** Chart for moisture ratio in soil samples.

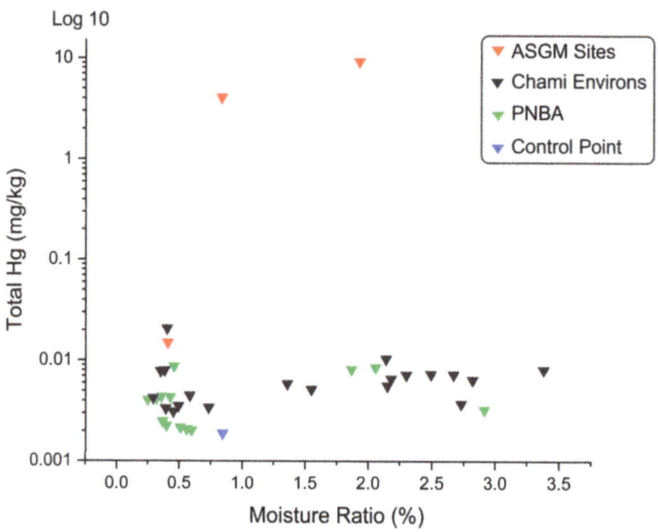


**Figure 10.** Correlation chart between the total Hg concentrations and moisture content in soil samples.

These results contradict the interpretation reported in [63] regarding Indiana soil in the United States, where the Hg concentrations were found to be relatively high in areas with high moisture content.

## 4.3. Hg Methylation

The results obtained from the XRF analysis, as presented in Table 2, revealed the existence of specific elements that are conducive to microbial activity. This microbial activity plays a crucial role in the methylation of Hg. Notably, the elevated levels of iron detected may significantly stimulate the activity of iron-reducing bacteria and enhance the oxidation of iron, which are recognized for their ability to methylate Hg [64].

## 4.4. Hg Extraction

The total Hg levels in soil leachate were consistent with those in the soil samples, indicating an effective extraction process of total Hg in the soil samples. This result suggests that specific chemicals in the leachate facilitate Hg binding and extraction, potentially leading to Hg contamination in groundwater over time, as shown in Figure 11.

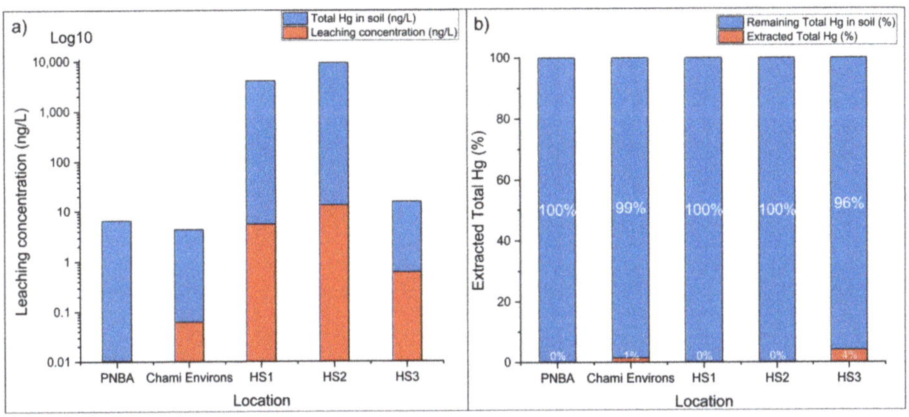


**Figure 11.** Comparison chart: (**a**) The percentage of total Hg in leachate versus the remaining total Hg in soil samples; (**b**) the total Hg content in soil samples compared to soil sample leachates.

## 4.5. Geo-Accumulation Index (Igeo)

The Hg concentration in the continental crust (0.04 ppm) was employed as a benchmark to evaluate contamination levels in the study area. The geo-accumulation index (Igeo) categories in Table 1 were used to classify individual samples according to their contamination levels. The cumulative distribution chart of the geo-accumulation index (Igeo) in Figure 12 displays the distribution of contamination levels across various sample types.

The geo-accumulation index analysis revealed that most samples—particularly those from PNBA and the Chami town environs—were classified as class 0, indicating unpolluted conditions. However, samples HS2 and HS1 from the ASGM sites showed a stark contrast, falling into class 6, signifying extreme pollution. This disparity underscores the extensive impact of the sampling grid. Notably, the HS2 sample, classified as class 6 and indicating severe pollution, was situated in the central market of Chami town, thereby exposing the local population to significant Hg pollution risks. In this study, it is important to note that samples were not collected from tailings, despite the Igeo index results indicating a level of pollution comparable to that of the tailing samples in Darmali, Sudan [65].

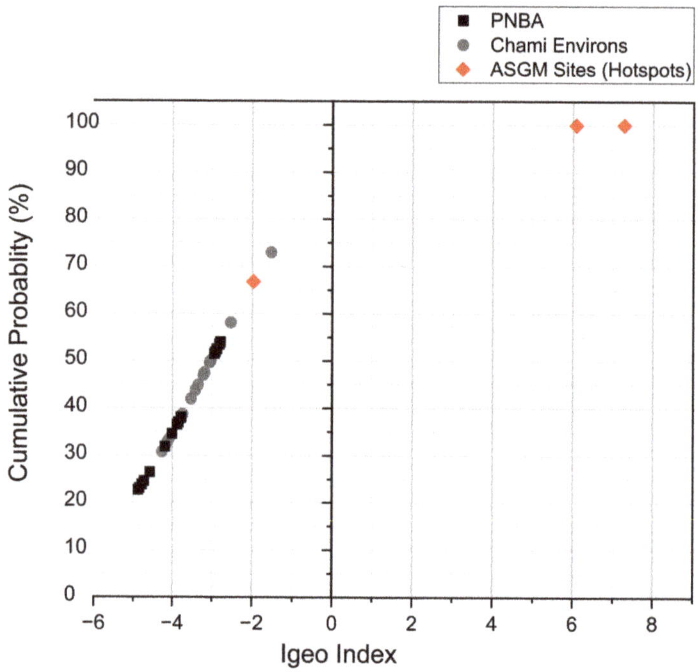


**Figure 12.** Distribution chart of the geo-accumulation index (Igeo) for the various sampling locations.

*4.6. Implication of the Study Findings*

The geo-accumulation index revealed an extreme risk of pollution. Although there was no apparent correlation between soil moisture content or LOI and the total Hg levels in soil, the presence of iron in the soil raised concerns about the possibility of microbial activation causing transformation of Hg into MeHg. Furthermore, based on the results of leaching experiments, the dissolved total Hg levels in the soil samples from Chami town and PNBA presented safe levels, although high concentrations were observed at the hotspots, raising concerns regarding the contamination of groundwater. However, an insignificant impact on groundwater around Chami town was observed in the analyzed samples. This is because the study area is a hyper-arid region with scarce rainfall. The above results confirmed that the impacts of Hg on soil at the hotspots were generated by ASGM activities and the diffusion trend of Hg around the hotspots.

The finding of the total Hg content in water suggests that, despite the presence of gold mining activities in the region, the Hg levels in the groundwater are within safe limits as per WHO guidelines. However, it is important to note that these results are based on the specific samples analyzed and may not represent the overall water quality in the region. This interpretation assumes that the necessary precautions were taken during the data collection and analysis stages to ensure the accuracy and reliability of the total Hg measurements. The low Hg levels in these samples were considered to be due to Mauritania's climate; in particular, the region's high temperatures and hyper-arid conditions could enhance the evaporation of Hg(0), thereby reducing its concentration in the groundwater [66]. This, along with the coherence observed between the total Hg analysis and the leaching test, contributes to the validity of the total Hg measurements. However, the inconsistency in LOI and iron content in PNBA and Chami town environs samples, along with fluctuating levels in total Hg, despite the distance from the ASGM site, necessitates further investigation to comprehend these phenomena and their underlying reasons. Future research could take the effects of dune movement and wind dynamics on the dispersion of Hg vapors in hyper-arid regions into consideration, where additional parameters could influence Hg emission and dispersion patterns. Furthermore, similarly to the outcomes of research conducted in Tanzania, Zimbabwe, and South Africa, our findings revealed that the total Hg concentrations were higher within the artisanal mining site [67]. In future studies, we plan to develop a mercury dispersion model for Hg emitted from ASGM sites in hyper-arid climates, as well as to model the influence of factors that potentially affect the diffusion and deposition of mercury, such as wind dynamics and dune movement.

## 5. Conclusions

This study investigated the environmental impacts of ASGM activities, with a specific focus on developing a predictive model for mercury contamination in hyper-arid regions. The research was conducted in Chami town, Mauritania—a region characterized by a hyper-arid climate. The analysis revealed elevated total Hg contents in local soils, particularly within identified ASGM hotspots. Meanwhile, groundwater samples exhibited relatively low mercury concentrations, suggesting the limited and slow infiltration of Hg into the groundwater due to the distance from ASGM sites. However, the study of soil mineralogy and organic matter composition indicated conditions that could facilitate the methylation of mercury, posing additional environmental and health risks. The extreme pollution levels detected within ASGM sites, as evidenced by the calculated geo-accumulation index values, underscored the urgent need for further investigations. Furthermore, the findings for Hotspot 3 suggested that the transport of waste tailings from ASGM operations to semi-industrial mining locations is a significant contributor to the widespread mercury contamination, which could be potentially affected and enhanced by the strong wind dynamics within the area. The significant use of Hg(0) in the amalgamation process, coupled with uncertainties regarding its environmental fate, highlight the need for a detailed mercury dispersion model tailored to hyper-arid regions. This study provides important insights regarding the mechanisms driving mercury contamination in hyper-arid environments, which can be useful in the development of predictive models and guide targeted interventions to mitigate the environmental and health impacts of ASGM activities in Chami town, similar regions, and worldwide.

**Author Contributions:** Writing of the original manuscript, M.M.M.; Critical review and editing of the manuscript, M.M.M., A.M., T.A. and A.S.; Data collection, M.M.M.; Data curation, M.M.M., A.S. and A.M.; Measurement and analysis, M.M.M. and A.M.; Approval, A.M.; Data visualization, M.M.M. and A.S.; Supervision, A.S., A.M. and T.A. All authors have read and agreed to the published version of the manuscript.

**Funding:** This research received no external funding.

**Institutional Review Board Statement:** Not applicable.

**Informed Consent Statement:** Not applicable.

**Data Availability Statement:** The authors confirm that the data supporting the findings of this study are available within the article.

**Acknowledgments:** The authors would like to express their sincere gratitude to the National Institute for Minamata Disease for their invaluable support in preserving the samples and providing the essential materials and equipment necessary for conducting total Hg measurements. We also thank our colleagues from Nouakchott University in Mauritania for their contributions. We would also like to express our gratitude to the Mauritanian Minister of Petroleum, Mines, and Energy, as well as the Mauritanian customs for facilitating the sampling process and for providing the required support in the export process.

**Conflicts of Interest:** The authors declare no conflicts of interest.

## References


1. Basu, N.; Clarke, E.; Green, A.; Calys-Tagoe, B.; Chan, L.; Dzodzomenyo, M.; Fobil, J.; Long, R.N.; Neitzel, R.L.; Obiri, S.; et al Integrated assessment of artisanal and small-scale gold mining in Ghana—Part 1: Human health review. *Int. J. Environ. Res. Public Health* **2015**, *12*, 5143–5176. [CrossRef] [PubMed]
2. Abdelaal, A.; Sultan, M.; Abotalib, A.Z.; Bedair, M.; Krishnamurthy, R.; Elhebiry, M. Emerging mercury and methylmercury contamination from new artisanal and small-scale gold mining along the Nile Valley, Egypt. *Environ. Sci. Pollut. Res.* **2023**, *30*, 52514–52534. [CrossRef] [PubMed]
3. Udodenko, Y.G.; Robinson, C.T.; Choijil, J.; Badrakh, R.; Munkhbat, J.; Ivanova, E.S.; Komov, V.T. Mercury levels in sediment, fish and macroinvertebrates of the Boroo River, northern Mongolia, under the legacy of gold mining. *Ecotoxicology* **2022**, *31*, 312–323. [CrossRef]
4. Aquino, R.Q.; Malone, A.; Smith, N.M.; Zúñiga, F.F.G. Perceptions and realities of mercury contamination in a Peruvian artisanal and small-scale gold mining (ASGM) community. *Environ. Res.* **2022**, *214*, 114092. [CrossRef]
5. Drace, K.; Kiefer, A.M.; Veiga, M.M. Cyanidation of mercury-contaminated tailings: Potential health effects and environmental justice. *Curr. Environ. Health Rep.* **2016**, *3*, 443–449. [CrossRef]
6. United Nations (UN). UN Environment Agency Tackles Mercury-Tarnished Gold Mining Industry. Available online: https://news.un.org/en/story/2023/02/1133587 (accessed on 10 January 2024).
7. United States Environmental Protection Agency (USEPA). Artisanal and Small-Scale Gold Mining without Mercury. Available online: https://www.epa.gov/international-cooperation/artisanal-and-small-scale-gold-mining-without-mercury (accessed on 10 January 2024).
8. Global Environment Facility (GEF). Making Mercury History in the Artisanal & Small-Scale Gold Mining Sector. Available online: https://www.thegef.org/news/making-mercury-history-artisanal-small-scale-gold-mining-sector (accessed on 10 January 2024).
9. Nriagu, J.O. Mercury pollution from the past mining of gold and silver in the Americas. *Sci. Total Environ.* **1994**, *149*, 167–181. [CrossRef]
10. UN Environment Programme (UNEP). Annual Report 2016 Empowering People to Protect the Planet. Available online: https://www.unep.org/resources/unep-annual-report-2016 (accessed on 20 April 2024).
11. Bhan, A.; Sarkar, N. Mercury in the environment: Effect on health and reproduction. *Rev. Environ. Health* **2005**, *20*, 39–56. [CrossRef]
12. UN Environment Programme. Global Mercury Assessment 2018, UN Environment Programme, Chemicals and Health Branch Geneva, Switzerland. Available online: https://www.unep.org/resources/publication/global-mercury-assessment-2018 (accessed on 5 January 2024).
13. Obrist, D.; Kirk, J.L.; Zhang, L.; Sunderland, E.M.; Jiskra, M.; Selin, N.E. A review of global environmental mercury processes in response to human and natural perturbations: Changes of emissions, climate, and land use. *Ambio* **2018**, *47*, 116–140. [CrossRef] [PubMed]
14. Cerceau, J.; Maman-Illatou, O.E.F.; Abrami, G.; Bouillet, P.; Vinches, M. LIPTAKOR, a participatory approach for increasing risk awareness and facilitating operational changes at artisanal gold mines. *Extr. Ind. Soc.* **2024**, *17*, 101385. [CrossRef]
15. Pacyna, J.M. Recent advances in mercury research. *Sci. Total Environ.* **2020**, *738*, 139955. [CrossRef]
16. Keane, S.; Bernaudat, L.; Davis, K.J.; Stylo, M.; Mutemeri, N.; Singo, P.; Twala, P.; Mutemeri, I.; Nakafeero, A.; Etui, I.D. Mercury and artisanal and small-scale gold mining: Review of global use estimates and considerations for promoting mercury-free alternatives. *Ambio* **2023**, *52*, 833–852. [CrossRef] [PubMed]
17. Tsang, V.W.; Lockhart, K.; Spiegel, S.J.; Yassi, A. Occupational health programs for artisanal and small-scale gold mining: A systematic review for the WHO global plan of action for workers' health. *Ann. Glob. Health* **2019**, *85*, 128. [CrossRef]
18. Veiga, M.M.; Angeloci, G.; Hitch, M.; Velasquez-Lopez, P.C. Processing centres in artisanal gold mining. *J. Clean. Prod.* **2014**, *64*, 535–544. [CrossRef]
19. World Health Organization (WHO). Environmental and Occupational Health Hazards Associated with Artisanal and Small-Scale Gold Mining. Available online: https://www.who.int/publications/i/item/9789241510271 (accessed on 13 May 2024).
20. Odumo, B.O.; Nanos, N.; Carbonell, G.; Torrijos, M.; Patel, J.P.; Rodríguez Martín, J.A. Artisanal gold-mining in a rural environment: Land degradation in Kenya. *Land Degrad. Dev.* **2018**, *29*, 3285–3293. [CrossRef]
21. Yoshimura, A.; Suemasu, K.; Veiga, M.M. Estimation of mercury losses and gold production by artisanal and small-scale gold mining (ASGM). *J. Sustain. Metall.* **2021**, *7*, 1045–1059. [CrossRef]
22. Pure Earth. Global Mercury Program: Strategy and Approach. Available online: https://www.pureearth.org/global-mercury-program/strategy-and-approach/ (accessed on 15 January 2024).
23. UN Environment Programme (UNEP). Global Mercury Assessment 2013: Sources, Emissions, Releases, and Environmental Transport. Available online: https://www.unep.org/resources/report/global-mercury-assessment-2013-sources-emissions-releases-and-environmental (accessed on 25 April 2024).
24. Saito, M. Minamata convention on mercury-our challenges and its future. *J. Environ. Saf.* **2019**, *10*, 95–97.
25. Arab World Press (AWP). Miners in Mauritania Search for Gold Using Traditional, Dangerous Methods. Available online: https://awp.net/en/stories/miners-in-mauritania-search-for-gold-using-traditional-dangerous-methods/ (accessed on 15 January 2024).

26. Niane, B.; Guédron, S.; Feder, F.; Legros, S.; Ngom, P.M.; Moritz, R. Impact of recent artisanal small-scale gold mining in Senegal: Mercury and methylmercury contamination of terrestrial and aquatic ecosystems. *Sci. Total Environ.* **2019**, *669*, 185–193. [CrossRef]
27. El Hadi, S.M.; Hamoud, A.; Chakiri, S.; El Hadi, H.; Sabbar, M.S.; Abdeina, E.H.; Baghdad, B.; Zerdeb, M.A.; Vall, M. Environmental Assessment by Metallic Trace Elements of Gold Soils–Case of Chami (Dakhlet Nouadhibou, Mauritania). *Ecol. Eng. Environ. Technol.* **2023**, *24*, 239–249. [CrossRef]
28. African Development Bank. National Climate Change Profile in Mauritania. Available online: https://www.afdb.org/en/documents/mauritania-national-climate-change-profile (accessed on 15 March 2024).
29. Elwaleed, A.; Jeong, H.; Abdelbagi, A.H.; Quynh, N.T.; Agusa, T.; Ishibashi, Y.; Arizono, K. Human Health Risk Assessment from Mercury-Contaminated Soil and Water in Abu Hamad Mining Market, Sudan. *Toxics* **2024**, *12*, 112. [CrossRef]
30. Walvoord, M.A.; Andraski, B.J.; Krabbenhoft, D.P.; Striegl, R.G. Transport of elemental mercury in the unsaturated zone from a waste disposal site in an arid region. *Appl. Geochem.* **2008**, *23*, 572–583. [CrossRef]
31. Mulenga, M.; Ouma, K.O.; Monde, C.; Syampungani, S. Aquatic mercury pollution from artisanal and small-scale gold mining in sub-Saharan Africa: Status, Impacts, and Interventions. *Water* **2024**, *16*, 756. [CrossRef]
32. El Hadi, S.M.; Hamoud, A.; Chakiri, S.; El Hadi, H.; Abdeina, E.H.; Sabbar, M.S.; Baghdad, B.; Vall, M.; Zerdeb, M.A.; Dahi, S.M. Assessment of Groundwater Contamination by Trace Metals in the Chami Gold Area (Mauritania). *Ecol. Eng. Environ. Technol.* **2023**, *24*, 139–147.
33. Lindberg, S.; Stratton, W. Atmospheric mercury speciation: Concentrations and behavior of reactive gaseous mercury in ambient air. *Environ. Sci. Technol.* **1998**, *32*, 49–57. [CrossRef]
34. Esbrí, J.M.; Higueras, P.L.; Martínez-Coronado, A.; Naharro, R. 4D dispersion of total gaseous mercury derived from a mining source: Identification of criteria to assess risks related to high concentrations of atmospheric mercury. *Atmos. Chem. Phys.* **2020**, *20*, 12995–13010. [CrossRef]
35. Fantozzi, L.; Guerrieri, N.; Manca, G.; Orrù, A.; Marziali, L. An Integrated investigation of atmospheric gaseous elemental mercury transport and dispersion around a chlor-alkali plant in the Ossola Valley (Italian Central Alps). *Toxics* **2021**, *9*, 172. [CrossRef]
36. Cornet, C.C.; Trégarot, E.; Failler, P. Birds of the National Park of Banc d'Arguin, Mauritania: a panel on ecosystem services. *J. Sustain. Res.* **2023**, *5*, e230009.
37. Heron, K.; Jessell, M.; Benn, K.; Harris, E.; Crowley, Q.G. The Tasiast deposit, Mauritania. *Ore Geol. Rev.* **2016**, *78*, 564–572. [CrossRef]
38. Tait, J.; Straathof, G.; Söderlund, U.; Ernst, R.E.; Key, R.; Jowitt, S.M.; Lo, K.; Dahmada, M.; N'Diaye, O. The Ahmeyim great dyke of Mauritania: A newly dated Archaean intrusion. *Lithos* **2013**, *174*, 323–332. [CrossRef]
39. Gagnol, L.; Magrin, G.; Chevrillon-Guibert, R. Chami, ville Nouvelle et Ville de l'or. Une Trajectoire Urbaine Insolite en Mauritanie, 2020. Available online: http://journals.openedition.org/espacepolitique/6562 (accessed on 25 February 2023).
40. Gagnol, L.; Afane, A. De sable, d'or et de mercure: Note sur la production urbaine contrastée de la ruée vers l'or au Sahara. *Afr. Contemp.* **2019**, 225–248. . [CrossRef]
41. Radio France Internationale (RFI). Mauritanie: Les Orpailleurs de Chami Confrontés au VIH (2/4). Available online: https://www.rfi.fr/fr/podcasts/reportage-afrique/20211206-mauritanie-les-orpailleurs-de-chami-confront%C3%A9s-au-vih-2-4 (accessed on 23 August 2024).
42. Ennih, N.; Liégeois, J.P. The boundaries of the West African craton, with special reference to the basement of the Moroccan metacratonic Anti-Atlas belt. *Geol. Soc. Lond. Spec. Publ.* **2008**, *297*, 1–17. [CrossRef]
43. Verbrugge, B.; Geenen, S. The gold commodity frontier: A fresh perspective on change and diversity in the global gold mining economy. *Extr. Ind. Soc.* **2019**, *6*, 413–423. [CrossRef]
44. Hilson, G.; Zolnikov, T.R.; Ortiz, D.R.; Kumah, C. Formalizing artisanal gold mining under the Minamata convention: Previewing the challenge in Sub-Saharan Africa. *Environ. Sci. Policy* **2018**, *85*, 123–131. [CrossRef]
45. Akagi, H.; Suzuki, T.; Arimura, K.; Ando, T.; Sakamoto, M.; Satoh, H.; Matsuyama, A. *Mercury Analysis Manual*, 105th ed.; Ministry of the Environment: Tokyo, Japan, 2004. Available online: http://nimd.env.go.jp/kenkyu/docs/march_mercury_analysis_manual(e).pdf (accessed on 28 May 2023).
46. Akagi, H.; Nishimura, H. Speciation of mercury in the environment. In *Advances in Mercury Toxicology*; Springer: Boston, MA, USA, 1991; pp. 53–76. ._3. [CrossRef]
47. Akagi, H.; Malm, O.; Branches, F.J.P.; Kinjo, Y.; Kashima, Y.; Guimaraes, J.R.D.; Oliveira, R.B.; Haraguchi, K.; Pfeiffer, W.C.; Takizawa, Y.; et al. Human exposure to mercury due to goldmining in the Tapajos River basin, Amazon, Brazil: Speciation of mercury in human hair, blood and urine. *Water Air Soil Pollut.* **1995**, *80*, 85–94. [CrossRef]
48. Yatabe, Z.; Nishiyama, K.; Tsuda, T.; Nakamura, Y. Structural characterization of mist chemical vapor deposited amorphous aluminum oxide films using water-free solvent. *J. Ceram. Soc. Jpn.* **2019**, *127*, 590–593. [CrossRef]
49. Japanese Ministry of the Environment. Environmental Agency Notifcations No. 46 (Environmental Standards Relating to Contamination of Soil). Available online: http://www.env.go.jp/kijun/dojou.html (accessed on 3 October 2023).
50. Agemian, H.; Chau, A. An improved digestion method for the extraction of mercury from environmental samples. *Analyst* **1976**, *101*, 91–95. [CrossRef] [PubMed]

51. Horvat, M. Determination of mercury and its compounds in water, sediment, soil and biological samples. In *Dynamics of Mercury Pollution on Regional and Global Scales: Atmospheric Processes and Human Exposures around the World*; Springer: Boston, MA, USA, 2005; pp. 153–190. [CrossRef]
52. Matsuyama, A.; Yano, S.; Matsunoshita, K.; Kindaichi, M.; Tada, A.; Akagi, H. The spatial distribution of total mercury in sediments in the Yatsushiro Sea, Japan. *Mar. Pollut. Bull.* **2019**, *149*, 110539. [CrossRef]
53. Logar, M.; Horvat, M.; Akagi, H.; Ando, T.; Tomiyasu, T.; Fajon, V. Determination of total mercury and monomethylmercury compounds in water samples from Minamata Bay, Japan: An interlaboratory comparative study of different analytical techniques. *Appl. Organomet. Chem.* **2001**, *15*, 515–526. [CrossRef]
54. Muller, G. Index of geoaccumulation in sediments of the Rhine river. *Geo J.* **1969**, *2*, 108–118.
55. Reimann, C.; De Caritat, P. *Chemical Elements in the Environment: Factsheets for the Geochemist and Environmental Scientist*; Springer Science & Business Media: Berlin/Heidelberg, Germany, 2012.
56. Martínez, L.L.G.; Poleto, C. Assessment of diffuse pollution associated with metals in urban sediments using the geoaccumulation index (I geo). *J. Soils Sediments* **2014**, *14*, 1251–1257. [CrossRef]
57. Kabata-Pendias, A. Trace Elements in Plants. In *Trace Elements in Soils and Plants, Fourth Edition*; Kabata-Pendias, A., Ed.; CRC Press (Taylor and Francis Group): Boca Raton, FL, USA, 2010; pp. 93–122, ISBN 9781420093704.
58. Turekian, K.K.; Wedepohl, K.H. Distribution of the elements in some major units of the earth's crust. *Geol. Soc. Am. Bull.* **1961**, *72*, 175–192. [CrossRef]
59. Pirrone, N.; Mason, R. *Mercury Fate and Transport in the Global Atmosphere*; Springer: Dordrecht, The Netherlands, 2009.
60. Edition, F. Guidelines for drinking-water quality. *WHO Chron.* **2011**, *38*, 104–108.
61. Thiombane, M.; De Vivo, B.; Niane, B.; Watts, M.J.; Marriott, A.L.; Di Bonito, M. A new hazard assessment workflow to assess soil contamination from large and artisanal scale gold mining. *Environ. Geochem. Health* **2023**, *45*, 5067–5091. [CrossRef] [PubMed]
62. Neary, D.G.; Garcia-Chevesich, P. Hydrology and erosion impacts of mining derived coastal sand dunes, Chanaral Bay, Chile. *Hydrol. Water Resour. Ariz. Southwest* **2008**, *38*, 47–52.
63. Briggs, C.; Gustin, M.S. Building upon the conceptual model for soil mercury flux: Evidence of a link between moisture evaporation and Hg evasion. *Water Air Soil Pollut.* **2013**, *224*, 1744. [CrossRef]
64. Kodamatani, H.; Maeda, C.; Balogh, S.J.; Nollet, Y.H.; Kanzaki, R.; Tomiyasu, T. The influence of sample drying and storage conditions on methylmercury determination in soils and sediments. *Chemosphere* **2017**, *173*, 380–386. [CrossRef]
65. Elwaleed, A.; Jeong, H.; Abdelbagi, A.H.; Thi Quynh, N.; Nugraha, W.C.; Agusa, T.; Ishibashi, Y.; Arizono, K. Assessment of Mercury Contamination in Water and Soil from Informal Artisanal Gold Mining: Implications for Environmental and Human Health in Darmali Area, Sudan. *Sustainability* **2024**, *16*, 3931. [CrossRef]
66. Biester, H.; Müller, G.; Schöler, H. Binding and mobility of mercury in soils contaminated by emissions from chlor-alkali plants. *Sci. Total Environ.* **2002**, *284*, 191–203. [CrossRef] [PubMed]
67. Van Straaten, P. Mercury contamination associated with small-scale gold mining in Tanzania and Zimbabwe. *Sci. Total Environ.* **2000**, *259*, 105–113. [CrossRef]

sustainability MDPI

Article

# A Scenario-Based Multi-Criteria Decision-Making Approach for Allocation of Pistachio Processing Facilities: A Case Study of Zarand, Iran

Mohammad Ebrahimi Sirizi [1], Esmaeil Taghavi Zirvani [2], Abdulsalam Esmailzadeh [3], Jafar Khosravian [4], Reyhaneh Ahmadi [5], Naeim Mijani [4,*], Reyhaneh Soltannia [6,7] and Jamal Jokar Arsanjani [8]

1 Department of Urban Planning, Faculty of Architecture and Urban Planning, Islamic Azad University Kerman Branch, Kerman 1477893855, Iran; mohammadebrahimi505@yahoo.com
2 Department of Urban Planning and Rural, Faculty of Social Science, University of Mohaghegh Ardabili, Ardabil 5619911367, Iran; esmaeiltaghavi34@gmail.com
3 Department of Social Planning, Faculty of Social Science, Allameh Tabataba'i University, Tehran 1485643449, Iran; a.esmailzadeh2000@gmail.com
4 Department of Remote Sensing and GIS, Faculty of Geography, University of Tehran, Tehran 141556311, Iran; jafarkhosravian@ut.ac.ir
5 Department of Regional and City Planning, Christopher C. Gibbs's College of Architecture, University of Oklahoma, Norman, OK 73019, USA; reyhaneh.ahmadi-1@ou.edu
6 Department of Human Geography, Faculty of Geography, University of Tehran, Tehran 1417613131, Iran; reyhan.soltannia@yahoo.com
7 Department of Urban Planning, Islamic Azad University Science and Research Branch, Tehran 1477893855, Iran
8 Geoinformatics Research Group, Department of Planning and Development, Aalborg University Copenhagen, A.C. Meyers Vænge 15, DK-2450 Copenhagen, Denmark; jja@plan.aau.dk
* Correspondence: naeim.mijani@ut.ac.ir

**Abstract:** Site selection and allocation of manufacturing and processing facilities are essential to sustainable economic productivity of a given product while preserving soil, the environment, and biodiversity. An essential criterion when evaluating various approaches to model land suitability for pistachio processing facilities is their adaptability to accommodate diverse perspectives and circumstances of managers and decision makers. Incorporating the concept of risk into the decision-making process stands as a significant research gap in modeling land suitability for pistachio processing facilities. This study presents a scenario-based multi-criteria decision-making system for modeling the land suitability of pistachio processing facilities. The model was implemented based on a stakeholder analysis as well as inclusion of a set of influential criteria and restrictions for an Iranian case study, which is among the top three producers. The weight of each criterion was determined based on the best-worst method (BWM) after the stakeholder analysis. Then, the ordered weighted averaging (OWA) model was used to prepare maps of spatial potential for building a pistachio processing factory in different decision-making scenarios, including very pessimistic, pessimistic, intermediate, optimistic, and very optimistic attitudes. Finally, the sensitivity analysis of very-high- and high-potential regions to changes in the weight of the effective criteria was evaluated and proved that the most important criteria were proximity to pistachio orchards, proximity to residential areas, proximity to the road network, and proximity to industrial areas. Overall, 327 $km^2$ of the study area was classified as restricted, meaning that they are not suitable locations for pistachio processing. The average estimated potential values based on the proposed model for very pessimistic, pessimistic, intermediate, optimistic, and very optimistic scenarios were 0.19, 0.47, 0.63, 0.78, and 0.97, respectively. The very-high-potential class covered 0, 0.41, 8.25, 39.64, and 99.78 percent of the study area based on these scenarios, respectively. The area of suitable regions for investment decreased by increasing risk aversion in decision making. The model was more sensitive to changes in the weights of proximity to residential areas, proximity to pistachio orchards, and proximity to transportation hubs. The proposed approach and the achieved findings could be of broader use to respective stakeholders and investors. Given the suitability of arid regions for planting pistachio and its relatively high profitability, the local authorities and decision makers can promote further expansion of the orchards, which can lead to better welfare of farmers and reducing rural-urban migration in the region.

**Keywords:** facilities' allocation assessment; GIS-MCDA; risk based; pistachio; sensitivity analysis

## 1. Introduction

The sustainable use and development of land are vital factors in economic growth [1–3]. One of the goals of land suitability and ultimately sustainable development programs is the selection and evaluation of different locations for the development of different uses [4–7]. This evaluation is an effective step for devising a sustainable development plan because by identifying and evaluating land suitability, development programs can be adapted to suit regional characteristics [8]. Optimal land use can help to achieve economic, social, ecological, and environmental efficiency for optimal use of land resources [9]. Therefore, it is necessary to determine suitability maps for different applications before development.

Industry can be considered the leading sector in regional economic development [10]. The location of manufacturing and processing facilities is of great importance for optimal land use, sustainable economic development, and increased productivity [11]. In other words, optimizing land use and determining optimal locations for the development of industrial areas can enhance sustainable development and reduce the waste of resources at the macro level. The feasibility of industrial development leads to cost reduction and the success of industrial units [12]. Conducting proper and appropriate feasibility studies, in addition to the economic impact on the performance of the industrial unit, will have social, environmental, cultural, and economic effects on the area of its construction.

Location selection for facilities is a multi-criteria decision-making problem. Determining the optimal location for the construction of facilities depends on a set of different factors, including accessibility and environmental, economic, social, etc., criteria [13,14]. Spatial decision making prevents decision makers from focusing on one criterion and losing sight of the others [15]. The integration of GIS and MCDA helps the decision maker to perform decision analysis functions such as location ranking [16]. GIS is used as a powerful and integrated tool for storing, manipulating, and analyzing industrial development criteria [17]. Considering that many criteria can affect location selection, the use of multi-criteria decision-making methods can facilitate the decision-making process by considering the key criteria [18–21].

The combination of GIS-MCDA has been used for various purposes such as the assessment of land suitability [11,22–25], renewable energy [26–29], vulnerability to natural hazards [30,31], socio-environmental resilience [32], agricultural management [33], habitat suitability [34], spatial modeling of migration [35], and urban management [1,36,37]. In past studies, various GIS-MCDA methods such as AHP, Fuzzy AHP, Fuzzy TOPSIS, ANP, VIKOR, ELECTRE, and PROMETHEE have been used for these applications. In some studies, GIS-MCDA capabilities were used in sustainable development of agriculture [38–40]. The focus of these studies has been on identifying and determining agricultural suitable land. In a number of other studies, attention has been paid to determining the optimal location for the development of agricultural product transformation facilities, which is one of the basic prerequisites for achieving sustainable development of agriculture. In addition, some studies used MCDA-based GIS to assess land suitability for the construction of facilities. Ramya and Devadas [10] used a hierarchical model combined with GIS to evaluate suitable locations for industrial development in the Tehri Garhwal district, India. Benaissa and Khalfallah [41] developed a model based on Delphi-AHP and GIS to assess land suitability for industrial activities to prevent land degradation in Bordj Bouarreridj, Algeria. Baghel [42] presented an integrated approach using GIS-based expert methods to prepare a mechanism for allocating land to industrial parks. The sites chosen for future industrial development were mostly near roads and railroads.

As a strategic product, pistachio is one of the most precious agricultural and medicinal products in the world [43]. Due to its high resistance to drought and salinity, it has been able to play a great role in improving the social and economic status of rural and urban areas in the central regions of Iran, especially the province of Kerman, which is facing many limitations in agricultural production [44]. Pistachio is called green gold thanks to its economic significance and holds a special, strategic place among horticultural products [45]. Therefore, it is highly regarded in Iran given its contribution to securing foreign currency, job creation, added value creation, and other economic aspects. Hard access to pistachio processing facilities in the province of Kerman has led orchardists to sell their products raw with low economic returns. Therefore, the development of pistachio processing facilities in suitable locations can boost the economic situation of these regions.

An essential criterion when evaluating various approaches to model land suitability for pistachio processing facilities is their adaptability to accommodate diverse perspectives and circumstances of managers and decision makers. Incorporating the concept of risk into the decision-making process stands as a significant research gap in modeling land suitability for pistachio processing facilities. Evaluating land suitability while considering the element of risk in decision making and the manager's specific approach and planning can be highly beneficial in allocating financial resources and time effectively to achieve sustainable development. The mental attitudes influencing decision making encompass a spectrum ranging from very optimistic and optimistic to neutral, pessimistic, and very pessimistic. Mapping the suitability of land for pistachio processing facilities holds great significance for each of these mental attitudes, aligning with sustainable development based on distinct timeframes and financial conditions. In optimistic and pessimistic mental attitudes, the level of risk in decision making is comparatively low and high, respectively. For instance, when financial and time constraints are absent, managers can incorporate the results derived from an optimistic outlook into their planning. Conversely, when faced with financial and time limitations, the results from a pessimistic perspective can inform their planning decisions.

This study aims to present a scenario-based multi-criteria decision-making system for measuring the spatial potential of pistachio processing facilities in the city of Zarand, the province of Kerman. A risk-based expert system based on GIS was used to achieve this goal. The advantage of this system is flexibility in creating different scenarios with an emphasis on decision-making risk.

The structure of the rest of the paper is as follows: In Section 2, the characteristics of the study area are described. In Section 3, the details of the data and methods used to model land suitability for pistachio processing facilities in different scenarios are stated. In Section 4, the results obtained from different implementations are presented. In Section 5, various dimensions of the method used and the results obtained in this study are discussed. Finally, the general conclusion of the study is presented in Section 6.

## 2. Study Area

The area studied in this research is the city of Zarand, located in the north of the province of Kerman, covering latitude 30°49′ N and longitude 56°34′ E. According to the latest census of the Iran Statistics Center in 2016, its population is 60,370. Zarand has a semi-desert climate with relatively hot summers and cold winters. Its average rainfall is 140 mm. The highest and lowest elevation of the studied area is 1608 and 3125 m above sea level, respectively. Recently, the entry of industry into the city of Zarand, as one of the less privileged areas of the province of Kerman, has been an effective step towards the all-round development of this city because economic enterprises have improved the cultural and economic infrastructure indicators of the region with a supportive view and in the form of fulfilling social responsibilities. The city covers over 43,000 hectares of pistachio orchards, of which 35,000 hectares have been harvested. Thanks to grafting, there are many pistachio varieties, the most famous of which are called Mumtaz, Ohadi, Kale Ghochi, Akbari, and Fandgi. Zarandi Mumtaz pistachio has the first rank in the world in terms of quality. In this city, a significant portion of the population consists of individuals engaged in farming, particularly those involved in pistachio-related occupations. Consequently, the establishment and enhancement of pistachio processing facilities can significantly contribute to elevating the quality of life for these residents. Considering that the pistachio processing facilities should not be far from the city due to easy access to the labor force and sales market, a buffer of 10 km from the border of the city of Zarand was assumed in this study. Figure 1 shows the geographic location of the study area.

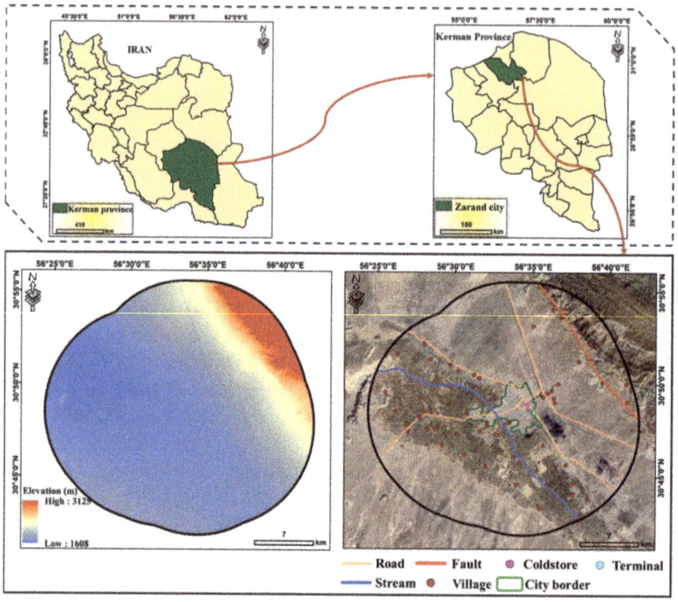


**Figure 1.** Geographical location of the study area.

## 3. Data and Methods

### 3.1. Data

Data collection is an integral yet time-consuming step in spatial potential assessment. The research data were obtained from various sources such as organizations and departments, satellite images, and statistical data. The primary data types included DEM data to obtain elevation and slope (https://earthexplorer.usgs.gov/ accessed on 24 August 2023), maps of urban and rural areas, gas and electricity lines, industrial areas, and coldstores (https://ncc.gov.ir/ accessed on 24 August 2023), a map of faults and landslide points (https://gsi.ir/ accessed on 24 August 2023), a map of pistachio orchards (https://www.maj.ir/ accessed on 24 August 2023), maps of the road network, a railway station, a terminal (https://www.mrud.ir/ accessed on 24 August 2023), and the stream network map (https://frw.ir/ accessed on 24 August 2023). Finally, after collecting the primary data layers, a spatial analysis was performed in ArcGIS to produce and analyze the information layers and prepare the criteria map according to their capabilities in each stage.

### 3.2. Methods

The flowchart of the main stages of the proposed method for spatial potential assessment of pistachio processing facilities is shown in Figure 2. The proposed method consists of five steps: (1) effective criteria were selected using the opinion of experts and specialists, and then, criteria and restriction maps were prepared using the spatial analysis; (2) Min and Max functions were used to standardize the values of the criteria map and Boolean logic was used to classify the study area into suitable and unsuitable classes; (3) using the

BWM method, the weight and importance of the criteria were calculated; (4) potential maps under different scenarios were prepared using the OWA model, and the OR operator was used for obtaining the maps of restrictions; and (5) finally, the sensitivity analysis of the area of classes with high and very high potential was performed by changing the weights of the criteria.

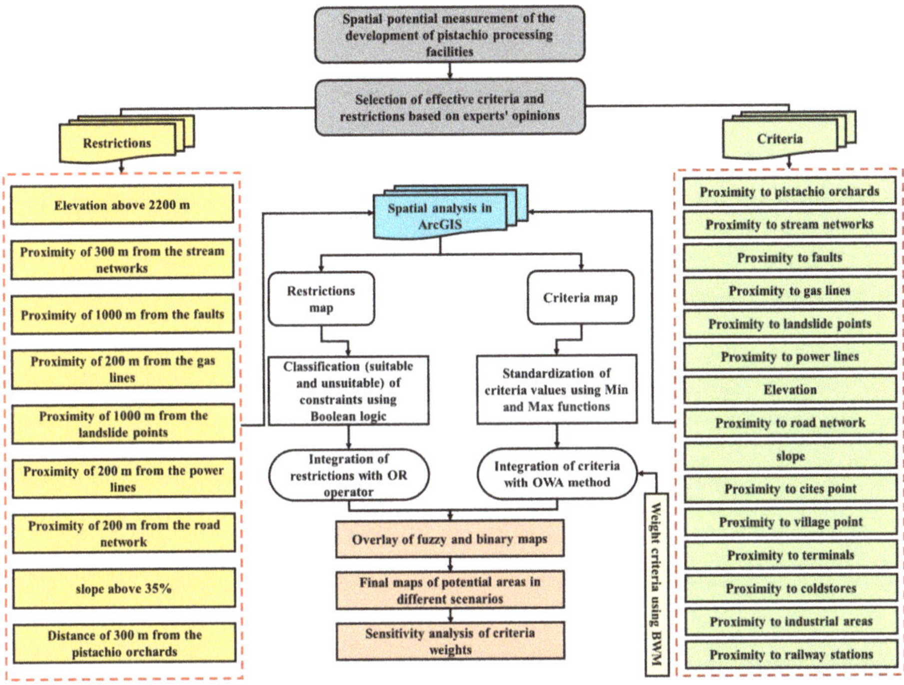


**Figure 2.** Flowchart of the main stages of the proposed method for spatial potential assessment of pistachio processing facilities.

### 3.2.1. Criteria and Restrictions

Choosing the suitable areas for measuring the spatial potential of the development of pistachio processing facilities is a multi-criteria decision-making process that requires careful consideration. The criteria were selected by interviewing experts and specialists and considering the social, geographical, and physical conditions of the study area. In this study, the opinions of 45 experts and specialists in the field of sustainable development, GIS, site selection, agriculture, and industrial facilities and places have been used. The selection criteria of these experts included practical and research experience, especially in relation to the location of transformation facilities.

Criteria were divided into two groups, namely restrictions and evaluation factors. Restrictions define areas that are not suitable for the development of pistachio processing facilities based on national and international standards. To determine such restricted areas, the study area was divided into suitable and unsuitable classes using Boolean logic. The evaluation factors were checked according to the suitable ranges for all possible locations. Finally, 15 spatial criteria and 9 restricted criteria were used here. Table 1 shows the list and description of criteria identified for the development of pistachio processing facilities in the study area.

**Table 1.** The list and description of criteria identified for the development of pistachio processing facilities in the study area.

| Criteria | Description |
| --- | --- |
| Proximity to city and village | Pistachio processing facilities should not be far from urban and rural areas, as they host the workforce and central markets [46]. Therefore, location suitability increases as the proximity decreases. However, in order to reduce the negative effects that industrial areas may have on the urban environment, a proximity of less than 300 m from the city and 200 m from the village was considered a restriction. |
| Proximity to fault and landslide points | As the proximity to fault and landslide points increases, the location suitability increases since such events can cause serious damage [47]. Therefore, locations with a proximity of less than 1000 m were considered restricted areas. |
| Elevation and slope | The higher the slope and elevation, the higher the cost of energy transfer, construction, and equipment transport [28]. Therefore, a lower slope and elevation means higher suitability. Here, slopes over 35% and elevations over 2200 m were considered restricted areas. |
| Proximity to power transmission lines | Considering both easy access and cost saving [48], the lesser the proximity to the power grid, the higher the suitability. However, in order to reduce electrical hazards, areas with a proximity of less than 200 m were considered restricted areas. |
| Accessibility | Easy accessibility means higher investment opportunities and economic growth. In addition, the longer the proximity, the higher the transportation costs and production time [49]. |
| Proximity to stream network | Being too close to water bodies can cause flooding hazards and economic loss [50]. According to the regulations of Iran's Environment and Forestry Organization, a proximity of less than 300 m from rivers was considered a restriction. |
| Proximity to coldstores | Proximity to coldstores is very important given their necessity for maintaining product properties and quality and increasing added value [41]. |
| Proximity to pistachio orchards | A fundamental criterion in determining a suitable location for pistachio processing facilities is proximity to pistachio orchards [10,42]. Still, a proximity of less than 300 m was considered a restriction to prevent possible damage to pistachio orchards. |
| Proximity to industrial areas | Given the existence of suitable infrastructure, access to raw materials and equipment, and access to technology in industrial areas, the shorter the proximity, the higher the suitability [51]. |

3.2.2. Standardization of Criteria Values

To evaluate the location suitability, it is necessary to convert different indicators into one unit for the sake of comparison. The criteria may have a positive or negative relationship with the purpose of the research. Positive criteria indicate that higher values are preferred, such as proximity to fault and proximity to landslide points, whereas in negative criteria, lower values are preferred, such as proximity to gas lines and proximity to road networks. Therefore, the minimum (Equation (1)) and maximum (Equation (2)) methods, respectively, were used to standardize these (positive and negative) criteria.

$$Negative = \frac{C_j^{max} - C_{ij}}{C_j^{max} - C_j^{min}} \tag{1}$$

$$Positive = \frac{C_{ij} - C_j^{min}}{C_j^{max} - C_j^{min}} \tag{2}$$

where $C_j^{min}$ and $C_j^{max}$ are the minimum and maximum values of the $j$th criterion, respectively, and $C_{ij}$ is the value of the ith position relative to the $j$th criterion [52,53].

3.2.3. Criteria Weight Calculation

The pairwise comparisons used by the expert are intended to reveal the relative preferences of m options or actions in situations where it is not possible to assign values to such options or actions based on specific criteria [54]. For example, under the method based on a relative scale, in the AHP, which was first introduced by Saaty [55], weights are derived through a pairwise comparison of criteria, and scores are derived through a pairwise comparison of options based on criteria. The main challenge in the pairwise comparison method is the lack of compatibility of the matrices, which often occurs in practice. Rezaei [56] states that inconsistencies occur for various reasons, including changes/corrections in expert opinions, and the order in which comparisons are made. Moreover, inconsistencies may lead to the loss of validity in previously used methods. For this purpose, Rezaei [56] introduced BWM. The BWM method is based on the assumption that decisions are generally evaluated based on the best and worst options, and other options are scored or weighted accordingly. As a result, instead of a pairwise comparison of all indicators, it is enough to compare all indicators with the best (most important) and worst (least important) ones [57]. When we have $n$ indices, the total number of pairwise comparisons is $n^2$. Here, the $n$th comparisons are also avoided because the comparison of each index is with itself and $a_{ij} = 1$, and so the remaining comparisons become equal to $n(n-1)$ [28]. There are two types of comparisons in pairwise comparisons: reference comparison and secondary comparison. Reference comparison $a_{ij}$ is a comparison where option or criterion $i$ is the best or $j$ is the worst indicator. Meanwhile, comparison aij is regarded as a secondary comparison if neither $i$ nor $j$ are the best or the worst elements and $aij \geq 1$. As a result, in the BWM method, the total number of comparisons is equal to $2n-3$ [56]. The following are the steps of the BWM method [58]:

Step 1: Specify the set of decision criteria.
Step 2: Determine the best (most preferred or most important) and worst (least important or weakest) criteria.
Step 3: The preference of the best criteria over other criteria and also the preference of other criteria over the worst criteria are determined by assigning numbers 1 to 9. This best-to-others vector can be written as Equation (3):

$$A_B = (a_{B1}, a_{B2}, \ldots, a_{Bn}) \tag{3}$$

where $a_{Bn}$ represents the preference degree of the criterion $j$ compared to the worst criterion $(B)$, and $a_{BB}$ is equal to one. This others-to-worst vector can be written as Equation (4):

$$a_w = (a_{1w}, a_{2w}, \ldots, a_{nw})^T \tag{4}$$

where $a_{jW}$ represents the preference degree of the criterion $j$ compared to the worst criterion $(W)$, and $a_{WW}$ is equal to one.
Step 4: Calculating optimal criteria weights.

To determine the optimal values, the weight assigned to each $\frac{w_B}{w_j}$ and $\frac{w_j}{w_w}$ pair must yield $\frac{w_B}{w_j} = a_{Bj}$ and $\frac{w_j}{w_w} = a_{jw}$. To satisfy these conditions for all js, the solution must have the lowest maximum absolute value for $\frac{w_j}{w_w} = a_{jw}$ and $\left|\frac{w_j}{w_w} = a_{jw}\right|$ differences for all criteria. Considering the non-negativity restrictions and weight normality, the problem can be modeled as Equation (5):

$$\begin{gathered} min \overset{max}{j} \left\{ \left| \frac{\omega_B}{\omega_j} - a_{Bj} \right| . \left| \frac{\omega_j}{\omega_W} - a_{jW} \right| \right\} \\ s \cdot t \\ \sum_{j=1}^{n} w_j = 1 \\ w_j \geq 0.\ for\ all\ j \end{gathered} \tag{5}$$

The value of the compatibility rate in the best-worst method can be calculated using the value of $\xi$ and the compatibility index, through Equation (6):

$$Consistency\ Ratio = \frac{\xi}{Consistency\ index} \quad (6)$$

where $\xi$ represents the compatibility ratio. Consistency values close to 1 indicate less consistent and stable comparisons, whereas those close to 0 indicate more consistency and stability in the comparisons. Table 2 shows the compatibility index values:

**Table 2.** Compatibility index values concerning the priority of the best criterion over the worst criterion.

| $a_{BW}$ | 1 | 2 | 3 | 4 | 5 | 6 | 7 | 8 | 9 |
|---|---|---|---|---|---|---|---|---|---|
| $CI\ (max\xi)$ | 0.00 | 0.44 | 1.00 | 1.63 | 2.30 | 3.00 | 3.73 | 4.47 | 5.23 |

#### 3.2.4. Land Suitability Modeling under Different Scenarios

In this study, the OWA method was used to model land suitability for the development of pistachio processing facilities in different scenarios. In past studies, this method was used in various applications such as potential of renewable energies [4,28], vulnerability and resilience [15,59], and modeling the physical growth of cities [60]. OWA is a multi-criteria integration operator developed by Yager [61]. This method has two types of weights [62]; one is a criterion weight and the other is an order weight. The criterion weight shows the relative importance of one criterion over another criterion and is the same for all its cells in an information layer. But the order weights are in the form of a set, the number of which is equal to the number of criteria layers. For each cell in the final raster layer, there is an ordered set of weights. These weights are obtained from the cell values of the existing criterion layers, which are arranged in descending order.

Since the set of sequential weights is $V = [v_1, v_2, \ldots, v_n]$, with $v_j \in [0,1]$ for $j = 1,2,3,\ldots,n$ and $\sum_{i=1}^{n} v = 1$, the OWA operator is defined as Equation (7):

$$OWA_i = \sum_{j=1}^{n} \left( \frac{u_j v_j}{\sum_{j=1}^{n} u_j v_j} \right) z_{ij} \quad (7)$$

where $z_{ij}$ is the value of the $i^{th}$ cell according to the $j^{th}$ criterion, $u_j$ is the weight of the $j^{th}$ criterion, which is according to the relationship between the $j^{th}$ criterion and decision-maker priorities, and $v_j$ is the order weight [63]. An important feature of the OWA operator is that, unlike many other aggregate operators, multiple answers can be created based on the mental characteristics of the decision maker by using a decision matrix [64]. The most important difference between this operator and other aggregation operators is that OWA provides the possibility of flexible aggregation in a range between minimum and maximum, and takes into account the level of risk-taking/risk-aversion of the decision maker [65]. Considering the sensitivity of decision making in high-risk and stressful jobs such as investing, OWA yields more accurate and logical results [66]. This is because it is possible to calculate the weight of each option and its ranking by considering different degrees of optimism for decision makers to determine the best option.

Ordinal weights are associated with tradeoffs and risk taking. The numbers obtained from these two concepts are between zero and one [67]. The riskiness value indicates how similar this operator (obtained from the conceptual quantifier) is to the logical OR operator. The closer this number is to 1, the more risk-taking it is. The risk taking of this operator is because it is like the set community operator and places the maximum number of input map pixels in the output map. This action causes the maximum values to appear in the output map, while not all criteria play a role in creating the final map. The unsuitable locations, which cause risk in decision making, are then identified. The tradeoff value can be expressed as the size of the dispersion of order weights. The tradeoff maximum is for when the order weights are equally distributed among the criteria. A tradeoff value of 0 indicates non-tradeoff and a tradeoff value of 1 indicates a complete tradeoff [68]. Equations (8) and (9), respectively, show how to calculate risk and tradeoff.

$$ORness = \sum_{j=1}^{n} \left(\frac{n-j}{n-1}\right) \lambda_j,\ 1 \leq ORness \ \leq 0 \quad (8)$$

$$TRADE - OFF = 1 - \sqrt{\frac{n\sum_j (\lambda_j - 1/n)^2}{n-1}} \quad (9)$$

where $n$ is the number of criteria, $j$ is the order of the criteria that are sorted in descending order, and $\lambda_j$ is the order weight of the $j^{th}$ criterion [64].

In this study, land suitability maps for the development of pistachio processing facilities for the study area were prepared in very pessimistic, pessimistic, intermediate, optimistic, and very optimistic scenarios based on the OWA method. The land suitability value obtained from the OWA method in each scenario ranges from 0 (low land suitability) to 1 (high land suitability). Also, land suitability maps for the construction of pistachio processing facilities in five classes (very low (0–0.2), low (0.2–0.4), moderate (0.4–0.6), high (0.6–0.8), and very high (1–0.8)) were classified. Then, the area of each class was calculated in different scenarios.

#### 3.2.5. Sensitivity Analysis

There are different methods to perform the sensitivity analysis of the weight of the criteria. One of the most used methods of a sensitivity analysis is the method of changing each input factor to observe its effect on the output. This method is known as the OAT method [69]. In this method, any change in output is clearly due to a change in a criterion. By changing one criterion, all other criteria can be kept constant and the results can be compared [70].

In the OAT method, there are three common methods for analyzing the sensitivity of criteria: (1) changing criteria weights, (2) changing criteria values, and (3) changing relative importance of criteria [71]. In this study, changing the weight of criteria was used in order to identify criteria that are sensitive to weight change and to quantify changes in the priority of criteria. In this method, the user must define three parameters: the main change criterion, range of percent change, and increment of percent change. The main change criterion is the criterion whose weight will change in current executions. The range of percent change can be a limited set of discrete percentage changes from the main weights of the criteria. In this method, one interval can be used for all criteria or a different interval can be used for each criterion. The increment of the percent change parameter also expresses the percentage by which the weight of each criterion changes, which should not exceed the percentage change range [72]. After changing the weight of the main variable criterion according to the increase in the percentage change, the weights of other criteria are also changed to establish the condition that the total weight of the criteria must be equal to one [4].

## 4. Results

The weight of the effective criteria in determining the optimal location for the construction of pistachio processing factories is shown in Table 3. The consistency rate of the determined weights for different criteria was 0.003, which shows the high consistency of experts in determining the weight of these criteria. The criteria of proximity to pistachio orchards and proximity to gas lines had the highest and lowest weights. Proximity to pistachio orchards, proximity to the city and village, proximity to the road network, and proximity to facilities had a higher weight than other criteria. The weights of elevation and slope criteria were 0.05 and 0.06, respectively. The influence of the impact type of all the criteria on determining optimal locations was minimal, except for the proximity to faults and proximity to landslide points.

**Table 3.** Weight of effective criteria calculated using the AHP method.

| Criteria | Weight | Type of Impact |
|---|---|---|
| Proximity to city | 0.12 | Minimum |
| Elevation | 0.05 | Minimum |
| Proximity to faults | 0.04 | Maximum |
| Proximity to gas lines | 0.02 | Minimum |
| Proximity to landslide points | 0.04 | Maximum |
| Proximity to pistachio orchards | 0.16 | Minimum |
| Proximity to power lines | 0.03 | Minimum |
| Proximity to railway stations | 0.05 | Minimum |
| Proximity to stream networks | 0.04 | Minimum |
| Proximity to road networks | 0.10 | Minimum |
| Proximity to industrial areas | 0.09 | Minimum |
| Slope | 0.06 | Minimum |
| Proximity to coldstores | 0.03 | Minimum |
| Proximity to terminals | 0.07 | Minimum |
| Proximity to village points | 0.10 | Minimum |

Maps of the criteria used in this study are shown in Figure 3. Spatial distribution of values of effective criteria in determining land suitability for the development of pistachio processing facilities is heterogeneous and different from each other. The least spatial changes in the studied area are related to the criteria of elevation and slope. According to these criteria, the areas located in the northeast of the studied area have a lower suitability than other areas for the construction of pistachio processing facilities. The areas located in the southwest of the study area have a higher suitability than other areas, especially the north and southeast of the study area, due to being located at a greater proximity to landslide points. In terms of criteria of proximity to a fault, proximity to pistachio orchards, proximity to stream networks, and proximity to gas lines, the north and northeast regions have a lower ratio than other regions. The spatial changes that are of proximity to the city, proximity to the village, proximity to coldstores, proximity to terminals, proximity to industrial areas, and proximity to landslide points are radial. But the spatial adjustments of the proximity to stream networks, proximity to road networks, proximity to gas and electricity lines, and proximity to pistachio orchards are linear. The city of Zarand has been studied in the center of the region; therefore, in terms of many criteria such as access to the urban area, coldstores, a terminal, industrial areas, etc., the central areas have a higher proportion than other areas. The values of the criteria are variable between zero and one. So, a value of one (red color) indicates a high potential and a value of zero (blue color) indicates a low potential for the establishment of pistachio processing facilities. The visual examination of the maps shows that the central parts of the study area are more suitable locations in terms of most of the effective criteria.

Some parts of the study area fall into the restricted category; that is, they are not suitable for the construction of a pistachio processing industry factory. Figure 4 shows the map of restrictions. Almost 33% of the studied area is considered as a restricted area with existing pistachio garden lands. In terms of slope and elevation, respectively, 3.6% and 5% of the studied area are classified in the restriction class. In terms of fault and landslide criteria, some of the areas located in the east, northeast, and north of the study area are classified as restriction classes. The development of pistachio processing facilities in these areas is limited in terms of safety against natural hazards. Also, 5.84% of the area under study is located within 300 m of the waterway network, which is one of the geographical limitations in the development of transformation facilities due to the risk of flooding. In total, 5.5, 2.48, and 5.75% of the area of the studied area are defined as restrictions due to being within the boundaries of the road and gas and power lines. According to different criteria, more than 60% of the study area is classified as restricted.

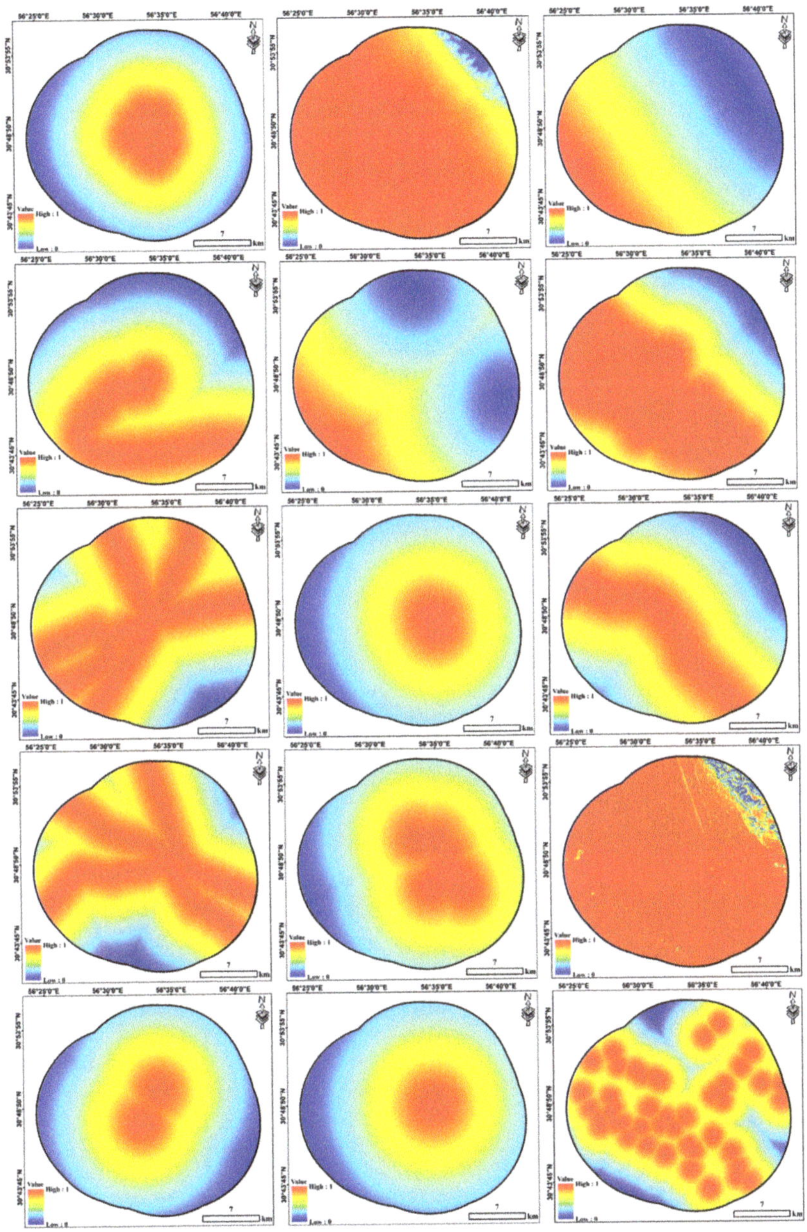

**Figure 3.** Standardized map of criteria (from left to right); proximity to city, elevation, proximity to faults, proximity to gas lines, proximity to landslide points, proximity to pistachio orchards, proximity to power lines, proximity to railway stations, proximity to stream networks, proximity to road networks, proximity to industrial areas, slope, proximity to coldstores, proximity to terminals, proximity to villages.

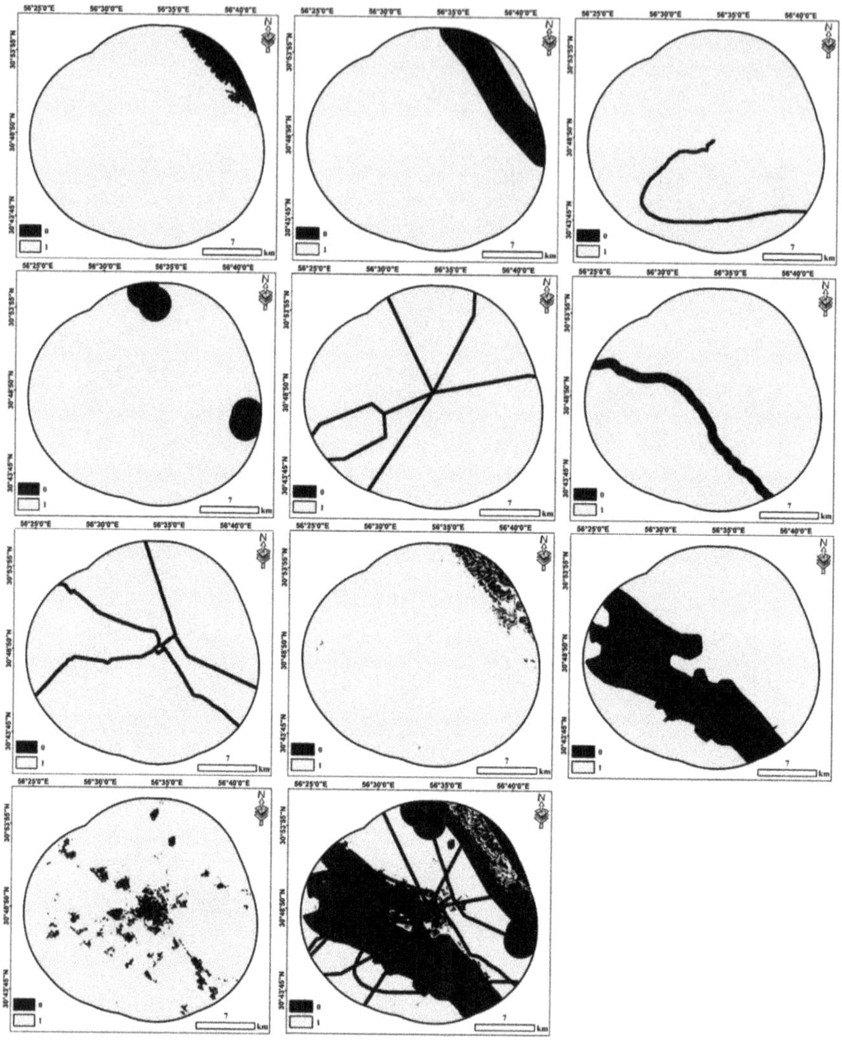

**Figure 4.** Maps of restrictions for the establishment of pistachio processing facilities (from left to right); elevation above 2200 m, proximity of 1000 m from faults, proximity of 200 m from gas lines, proximity of 1000 m from landslide points, proximity of 200 m from power lines, proximity of 300 m from stream networks, proximity of 200 m from road networks, slope above 35%, proximity of 300 m from pistachio orchards, proximity of 300 m from city, and 200 m from village; total restrictions for the establishment of pistachio processing facilities.

The potential maps for the construction of pistachio processing factories in different risk scenarios in decision making obtained from the OWA model are shown in Figure 5. With the increase in the degree of risk scenarios, the amount of areas with high potential and suitability for the construction of pistachio packaging factories increases. In the very pessimistic scenario map, many areas in the study area have potential values less than 0.5, while in the very optimistic scenario, almost the entire study area has potential values above 0.5. Due to the proximity to urban area, industrial areas, coldstores, a terminal, etc., the potential of the central areas in the study area is higher than the peripheral areas. But the difference between central and peripheral areas in very pessimistic scenarios is higher than in optimistic and very optimistic scenarios.

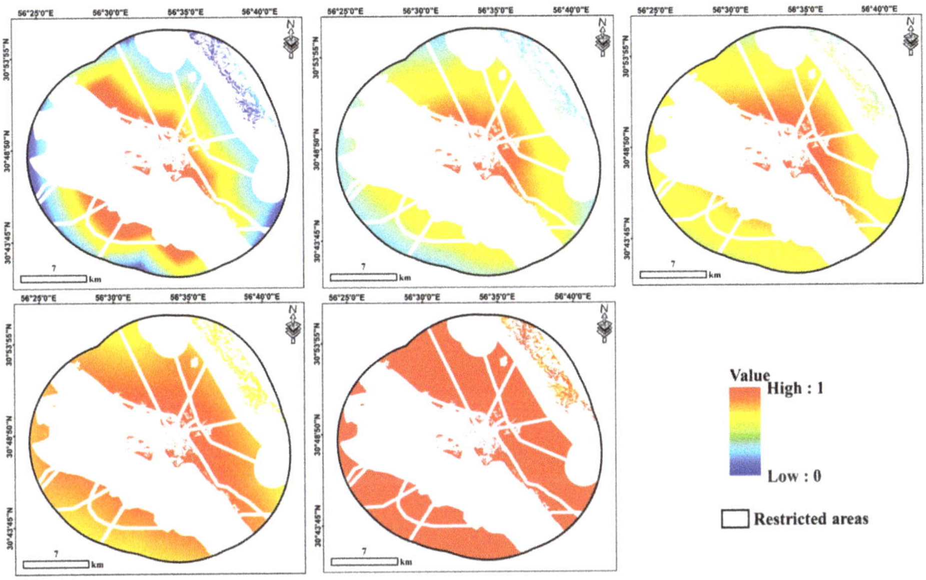


**Figure 5.** Potential maps for the construction of pistachio processing factories in different decision-making risk scenarios (from left to right; very pessimistic, pessimistic, intermediate, optimistic, and very optimistic).

The suitability values of the statistical parameters of the potential maps in different decision-making risk scenarios are shown in Table 4. The average values in very pessimistic, pessimistic, intermediate, optimistic, and very optimistic scenarios are 0.19, 0.47, 0.63, 0.78, and 0.97, respectively. These results show that with the increase in the degree of optimism, the amount of potential and suitability increases. The standard deviation values of these maps for very pessimistic, pessimistic, intermediate, optimistic, and very optimistic scenarios were 0.09, 0.13, 0.10, 0.07, and 0.02, respectively. The standard deviation values show that the maps based on pessimistic and intermediate scenarios have the highest heterogeneity of potential values. The range of changes in potential values calculated for the pixels of the study area in the very optimistic scenario was 0.26, which is less than in other scenarios. This value was 0.55, 0.66, 0.58, and 0.45 for very pessimistic, pessimistic, intermediate, and optimistic scenarios, respectively.

**Table 4.** Statistical parameters of risk scenarios.

| | Min | Max | Mean | STD |
|---|---|---|---|---|
| Very pessimistic | 0.01 | 0.56 | 0.19 | 0.09 |
| Pessimistic | 0.18 | 0.84 | 0.47 | 0.13 |
| Intermediate | 0.33 | 0.91 | 0.63 | 0.10 |
| Optimistic | 0.51 | 0.96 | 0.78 | 0.07 |
| Very optimistic | 0.74 | 1 | 0.97 | 0.02 |

Classified potential maps in different risk-based decision-making scenarios are shown in Figure 6. The spatial distribution of different potential classes varies in different decision scenarios. In the very pessimistic scenario, the potential map includes very low, low, and medium classes. Based on this scenario, only a small and limited area of the central areas of the study area has medium potential, which is located in the center of the study area. In the pessimistic scenario, the study area is classified into low-, medium-, high-, and very-high-potential classes. The obtained map based on this scenario for the study area has a very low class. Areas with low potential are marginal areas of the study area and areas with very high potential also cover a small part of the center of the study area. In other scenarios, the area of regions with high- and very-high-potential classes increases by increasing the degree of risk and optimism in decision making. In the optimistic scenario, the entire study area is classified into two classes of high and very high potential. Based on the very optimistic scenario, the entire study area was classified in a very high class. In other words, for the risk-taking investor, the entire study area, except for the restricted areas, is suitable for the development and construction of a pistachio processing factory.

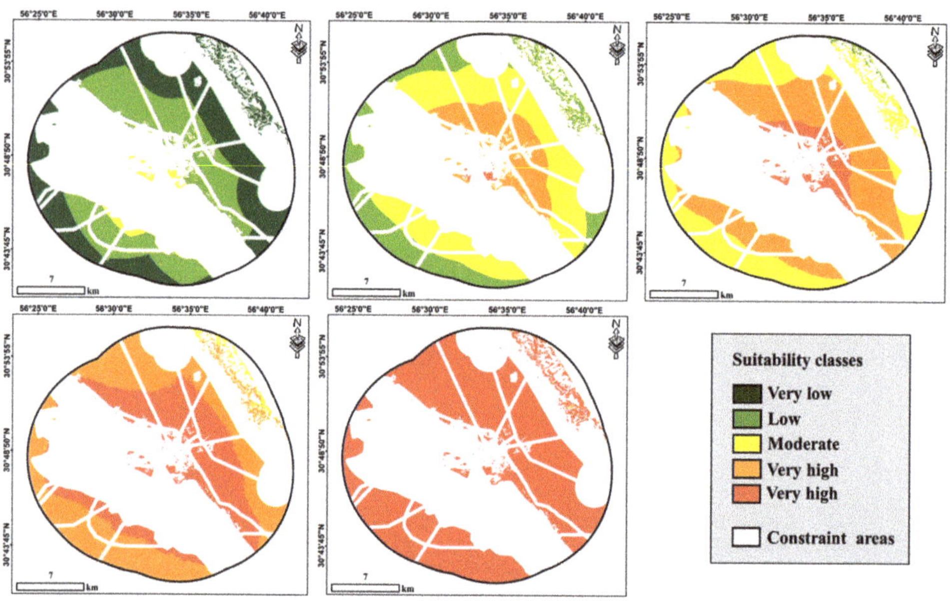


**Figure 6.** Suitability maps for the establishment of pistachio processing facilities in different risk scenarios (from left to right); very pessimistic, pessimistic, intermediate, optimistic, and very optimistic.

The area coverage percentage in the potential map classes under different risk scenarios is shown in Table 5. Here, 53.1% and 0.15% of the study area were classified in the very-low-potential class under the very pessimistic and pessimistic scenarios, respectively. The rest of the scenarios had no very-low-potential class. The coverage percentages of the high-potential class under very pessimistic, pessimistic, intermediate, optimistic, and very optimistic scenarios were 0, 19.47, 59.54, 54.76, and 0.22%, respectively. These values for the very high class under these scenarios were 0, 0.41, 8.25, 39.64, and 99.78 percent, respectively. With the increase in the degree of risk aversion, the number of suitable areas decreased.

The sensitivity analysis results of the percentage of areas with high and very high suitability are shown in Figure 7. These values are highly sensitive to the change in the criteria of proximity to the city, proximity to pistachio orchards, and proximity to road networks. By increasing the weight of these criteria, the percentage of areas with high and very high suitability increased significantly. Additionally, increasing the weights of the criteria for the distance from pistachio gardens, distance from cold storage, and distance from the road network up to 0.6 results in a significant increase in the percentage of areas with high and very high potential. For weights higher than 0.6, the changes in the percentage of these classes are limited. The percentage of areas with a high and very high suitability degree had a low sensitivity to the change in the weight of proximity to gas and power lines and proximity to terminals. Increasing the weight of proximity to a fault, slope, elevation, and proximity to stream networks caused a decrease in the percentage of areas with high and very high suitability. Increasing the weight of the fault distance criterion up to 0.4 leads to a noticeable reduction in the percentage of areas with high and very high potential.

**Table 5.** Coverage percentage of different classes under different risk scenarios.

| | Very Low | Low | Moderate | High | Very High |
|---|---|---|---|---|---|
| Very pessimistic | 53.10 | 45.13 | 1.77 | 0.00 | 0.00 |
| Pessimistic | 0.15 | 31.13 | 48.85 | 19.47 | 0.41 |
| Intermediate | 0.00 | 0.99 | 36.17 | 54.59 | 8.25 |
| Optimistic | 0.00 | 0.00 | 1.59 | 58.76 | 39.64 |
| Very optimistic | 0.00 | 0.00 | 0.00 | 0.22 | 99.78 |

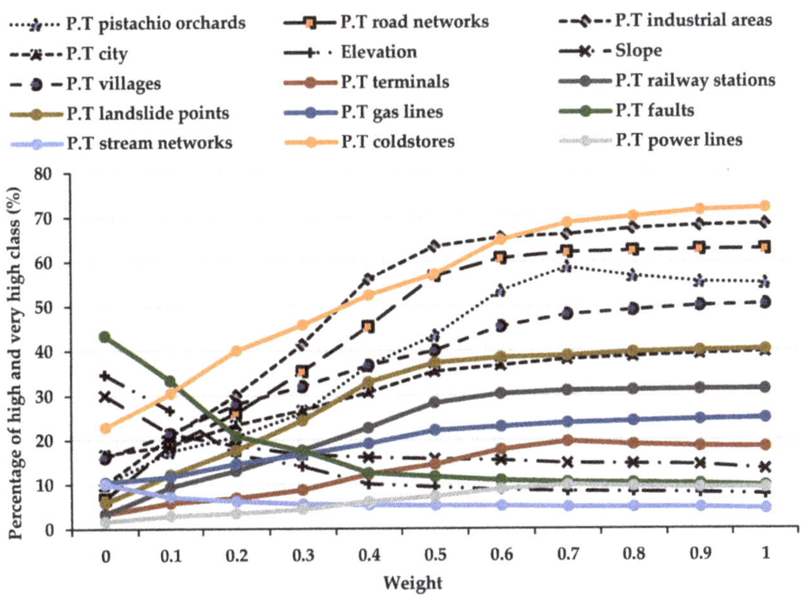


**Figure 7.** Sensitivity analysis of the percentage of areas with high and very high suitability to changes in criteria weights (Note: P.T (proximity to)).

## 5. Discussion

Determining the optimal location for the establishment of horticultural and agricultural product processing facilities is of great importance. A suitable location can prevent future economic, environmental, etc., costs and is a major effective factor in achieving local and regional sustainable development [10]. This study was the first attempt to determine potential and optimal locations for the development of pistachio processing facilities. In several previous studies, the optimal measurement potential was evaluated for some other industries such as healthcare [73,74], coffee [75], photovoltaics [76], steel [77], and compost food waste [78,79].

Determining the optimal location for the development of pistachio processing facilities is inherently a spatial decision and depends on various criteria, including economic, accessibility, and environmental criteria. Therefore, in this study, a GIS-MCDA model was used to assess the potential of different areas for the development of pistachio processing facilities. The accuracy and efficiency of GIS-MCDA models depend on the selected effective criteria, criterion weights, and the efficiency of aggregation models [80]. These criteria must enjoy certain conditions such as comprehensiveness and compatibility and can be selected from previous studies and expert opinions. Due to a shortage of studies on the topic addressed here, this study used the opinion of experts for this purpose.

The accuracy of criteria weights has a direct effect on the accuracy of the final model output. The methods of determining criteria weights can be divided into data-based, knowledge-based, and integrated methods [58]. Data-based methods are useful when information and past events are available [81]. This study employed the best-worst method, which falls into the knowledge-based category. In this method, experts determine the criteria weights. Considering that there is a difference of opinion among experts regarding the validity of different criteria, it is vital to calculate the compatibility rate. The condition for the acceptance of the weight calculated for different criteria based on the best-worst method is an acceptable compatibility rate [56]. Some of the advantages of the best-worst method are (i) generating more reliable criteria results, (ii) requiring less original pairwise comparison information, and (iii) smoothly fusing with other methodologies to manipulate uncertain information in the decision process [82].

In this study, the OWA model was used to aggregate the standardized values as well as the weights of the criteria. This method is more flexible than other methods such as WLC. The output and results of the OWA model can be used by a wide range of stakeholders, including managers, planners, and investors. Stakeholders with pessimistic attitudes are usually more strict in determining the optimal locations. Therefore, in this case, the selected locations must have suitable conditions in terms of a large number of effective criteria, whereas for optimistic stakeholders, a location can be optimal if it has suitable conditions in terms of a small number of effective criteria. In the OWA model, the ORness parameter regulates the degree of optimism and pessimism. Therefore, by changing the value of this parameter, it can produce different results that are suitable for stakeholders with different attitudes. The results of ORness = 0 are suitable for very pessimistic stakeholders and the results of ORness = 1 are suitable for very optimistic stakeholders. The OWA model has been used in several previous studies on measuring the potential of renewable energies [4,26,83], vulnerability and resilience [15,80], assessment of land suitability [84], and modeling the physical growth of cities [60].

For areas where there are pistachio processing facilities, the method proposed in this study can be used to evaluate the suitability of the existing situation and suggest new optimal areas. In some past studies, the outputs of GIS-MCDA models have been used to evaluate existing uses in an area such as restaurants [85]. However, the accuracy and uncertainty of the results of these models should be considered. In this study, the uncertainty caused by changing the weight of the criteria was investigated. The output of the model has a high sensitivity to the change in the weight of some criteria, such as the proximity to coldstores and the proximity to industrial areas, and a low sensitivity to the change in the weight of some other criteria, such as the proximity to power lines and proximity to stream networks. However, the efficiency of aggregation models is also effective on the accuracy and uncertainty of the final results of GIS-MCDA models, which is recommended to be paid attention to in future studies. Also, the use of algorithms based on artificial intelligence in modeling and analyses can also reduce uncertainty [86–90].

## 6. Conclusions

Measuring the spatial potential of processing facilities can be of great importance. The city of Zarand, Kerman, is one of the hubs of pistachio production in the world. This highlights the importance of determining the optimal location for the development of pistachio processing factories in this region. Different investors have different mental attitudes. Some of them are risk-taking and some are risk-averse. Therefore, it is of great importance to present the results of such spatial potential assessments based on different intellectual scenarios of investors. Including the concept of risk in the decision-making process is an important research gap in assessing land suitability for the development of pistachio processing facilities. This study presented for the first time a multi-criteria decision-making system based on risk in making decisions for the development of pistachio processing facilities. In this study, a scenario-based multi-criteria decision-making system was presented for measuring the spatial potential of developing pistachio processing factories. This system can prepare potential maps based on the attitude of different groups of investors, including very pessimistic, pessimistic, intermediate, optimistic, and very optimistic. Then, the OWA model was used to develop pistachio processing facilities in different decision-making scenarios. The research results showed that the effective criteria considering location selection have different weights. Proximity to pistachio orchards and proximity to gas transmission lines had the highest and lowest weights, respectively. With the increase in the degree of risk taking by investors, suitable locations increase. The results of the proposed system can be useful and practical for a wide range of stakeholders, including investors, managers, planners, and policy makers with different intellectual attitudes and risk tolerance. One of the remaining gaps in this study is the evaluation of uncertainty caused by different sources in the final result. It is suggested to use large-group decision-making methods in future studies to reduce the uncertainty in weights due to a low number of experts. Also, the development and use of a scenario to evaluate the accuracy of suitability maps for the establishment of pistachio processing facilities can be considered in future studies. Future studies can also use the system proposed in this study to flexibly locate suitable locations for different facilities based on different stakeholder attitudes.

**Author Contributions:** Conceptualization, M.E.S., E.T.Z. and N.M.; methodology, M.E.S., N.M., A.E., J.K. and R.S.; software, J.K., R.A., N.M. and R.S.; data curation, M.E.S., A.E. and J.K.; writing—original draft preparation, M.E.S., E.T.Z., J.K., A.E., R.A., R.S. and N.M.; writing—review and editing, J.J.A. All authors have read and agreed to the published version of the manuscript.

**Funding:** This research received no external funding.

**Informed Consent Statement:** Not applicable.

**Data Availability Statement:** The data used to support the findings of this study are available from the corresponding author upon reasonable request.

**Conflicts of Interest:** The authors declare no conflict of interest.

## Abbreviations

| | |
|---|---|
| GIS | Geographic Information System |
| MCDA | Multi-criteria Decision Analysis |
| OWA | Ordered Weighted Averaging |
| AHP | Analytic Hierarchy Process |
| TOPSIS | Technique for Order of Preference by Similarity to Ideal Solution |
| VIKOR | VlseKriterijumska Optimizcija I Kaompromisno Resenje |
| ANP | Analytical Network Process |
| ELECTRE | Elimination and Choice Translating Reality |
| PROMETHEE | Preference Ranking Organization Method for Enrichment Evaluation |
| DEM | Digital Elevation Model |
| BWM | Best-Worst Method |
| OAT | One-At-a-Time |

## References

1. Aburas, M.M.; Abdullah, S.H.; Ramli, M.F.; Asha'ari, Z.H. Land suitability analysis of urban growth in Seremban Malaysia, using GIS based analytical hierarchy process. *Procedia Eng.* **2017**, *198*, 1128–1136. [CrossRef]
2. Kordi, F.; Yousefi, H. Crop classification based on phenology information by using time series of optical and synthetic-aperture radar images. *Remote Sens. Appl. Soc. Environ.* **2022**, *27*, 100812. [CrossRef]

3. Lin, K.; Zhao, H. The Impact of Green Finance on the Ecologicalization of Urban Industrial Structure—Based on GMM of Dynamic Panel System. *J. Artif. Intell. Technol.* **2022**, *2*, 123–129. [CrossRef]
4. Shorabeh, S.N.; Firozjaei, M.K.; Nematollahi, O.; Firozjaei, H.K.; Jelokhani-Niaraki, M. A risk-based multi-criteria spatial decision analysis for solar power plant site selection in different climates: A case study in Iran. *Renew. Energy* **2019**, *143*, 958–973. [CrossRef]
5. Shorabeh, S.N.; Varnaseri, A.; Firozjaei, M.K.; Nickravesh, F.; Samany, N.N. Spatial modeling of areas suitable for public libraries construction by integration of GIS and multi-attribute decision making: Case study Tehran, Iran. *Libr. Inf. Sci. Res.* **2020**, *42*, 101017. [CrossRef]
6. Mohimi, A.; Esmaeily, A. Spatiotemporal analysis of urban sprawl using a multi-technique approach and remote sensing satellite imagery from 1990 to 2020: Kerman/Iran. *Environ. Dev. Sustain.* **2023**, 1–36. [CrossRef]
7. Shacheri, F.; Nikou, N.; Ziaei, A.; Saeedi, M. Surface dense discharge from rectangular and trapezoidal channels. *Flow Meas. Instrum.* **2022**, *87*, 102213. [CrossRef]
8. Pilevar, A.R.; Matinfar, H.R.; Sohrabi, A.; Sarmadian, F. Integrated fuzzy, AHP and GIS techniques for land suitability assessment in semi-arid regions for wheat and maize farming. *Ecol. Indic.* **2020**, *110*, 105887. [CrossRef]
9. Zolekar, R.B.; Bhagat, V.S. Multi-criteria land suitability analysis for agriculture in hilly zone: Remote sensing and GIS approach. *Comput. Electron. Agric.* **2015**, *118*, 300–321. [CrossRef]
10. Ramya, S.; Devadas, V. Integration of GIS, AHP and TOPSIS in evaluating suitable locations for industrial development: A case of Tehri Garhwal district, Uttarakhand, India. *J. Clean. Prod.* **2019**, *238*, 117872. [CrossRef]
11. Ostovari, Y.; Honarbakhsh, A.; Sangoony, H.; Zolfaghari, F.; Maleki, K.; Ingram, B. GIS and multi-criteria decision-making analysis assessment of land suitability for rapeseed farming in calcareous soils of semi-arid regions. *Ecol. Indic.* **2019**, *103*, 479–487. [CrossRef]
12. Hassan, I.; Javed, M.A.; Asif, M.; Luqman, M.; Ahmad, S.R.; Ahmad, A.; Akhtar, S.; Hussain, B. Weighted overlay based land suitability analysis of agriculture land in Azad Jammu and Kashmir using GIS and AHP. *Pak. J. Agric. Sci.* **2020**, *57*, 1509–1519.
13. Akıncı, H.; Özalp, A.Y.; Turgut, B. Agricultural land use suitability analysis using GIS and AHP technique. *Comput. Electron. Agric.* **2013**, *97*, 71–82. [CrossRef]
14. Nezhadkian, M.; Azimi, S.M.; Ferro, A.; Nafei, A.H. A model for new product development in business companies based on grounded theory approach and fuzzy method. *J. Comput. Cogn. Eng.* **2023**, *2*, 124–132. [CrossRef]
15. Afsari, R.; Nadizadeh Shorabeh, S.; Kouhnavard, M.; Homaee, M.; Arsanjani, J.J. A spatial decision support approach for flood vulnerability analysis in urban areas: A case study of Tehran. *ISPRS Int. J. Geo-Inf.* **2022**, *11*, 380. [CrossRef]
16. Shahpari Sani, D.; Heidari, M.T.; Tahmasebi Mogaddam, H.; Nadizadeh Shorabeh, S.; Yousefvand, S.; Karmpour, A.; Jokar Arsanjani, J. An Assessment of Social Resilience against Natural Hazards through Multi-Criteria Decision Making in Geographical Setting: A Case Study of Sarpol-e Zahab, Iran. *Sustainability* **2022**, *14*, 8304. [CrossRef]
17. Mijani, N.; Alavipanah, S.K.; Hamzeh, S.; Firozjaei, M.K.; Arsanjani, J.J. Modeling thermal comfort in different condition of mind using satellite images: An Ordered Weighted Averaging approach and a case study. *Ecol. Indic.* **2019**, *104*, 1–12. [CrossRef]
18. Kiavarz, M.; Jelokhani-Niaraki, M. Geothermal prospectivity mapping using GIS-based Ordered Weighted Averaging approach: A case study in Japan's Akita and Iwate provinces. *Geothermics* **2017**, *70*, 295–304. [CrossRef]
19. Erbaş, M.; Kabak, M.; Özceylan, E.; Çetinkaya, C. Optimal siting of electric vehicle charging stations: A GIS-based fuzzy Multi-Criteria Decision Analysis. *Energy* **2018**, *163*, 1017–1031. [CrossRef]
20. Sarkar, A.; Biswas, A.; Kundu, M. Development of q-rung orthopair trapezoidal fuzzy Einstein aggregation operators and their application in MCGDM problems. *J. Comput. Cogn. Eng.* **2022**, *1*, 109–121. [CrossRef]
21. Maihulla, A.S.; Yusuf, I.; Bala, S.I. Reliability and performance analysis of a series-parallel system using Gumbel–Hougaard family copula. *J. Comput. Cogn. Eng.* **2022**, *1*, 74–82. [CrossRef]
22. Tashayo, B.; Honarbakhsh, A.; Akbari, M.; Eftekhari, M. Land suitability assessment for maize farming using a GIS-AHP method for a semi-arid region, Iran. *J. Saudi Soc. Agric. Sci.* **2020**, *19*, 332–338. [CrossRef]
23. Tercan, E.; Dereli, M.A. Development of a land suitability model for citrus cultivation using GIS and multi-criteria assessment techniques in Antalya province of Turkey. *Ecol. Indic.* **2020**, *117*, 106549. [CrossRef]
24. Qureshi, S.; Shorabeh, S.N.; Samany, N.N.; Minaei, F.; Homaee, M.; Nickravesh, F.; Firozjaei, M.K.; Arsanjani, J.J. A new integrated approach for municipal landfill siting based on urban physical growth prediction: A case study mashhad metropolis in Iran. *Remote Sens.* **2021**, *13*, 949. [CrossRef]
25. Moghaddam, M.H.R.; Sedighi, A.; Fasihi, S.; Firozjaei, M.K. Effect of environmental policies in combating aeolian desertification over Sejzy Plain of Iran. *Aeolian Res.* **2018**, *35*, 19–28. [CrossRef]
26. Firozjaei, M.K.; Nematollahi, O.; Mijani, N.; Shorabeh, S.N.; Firozjaei, H.K.; Toomanian, A. An integrated GIS-based Ordered Weighted Averaging analysis for solar energy evaluation in Iran: Current conditions and future planning. *Renew. Energy* **2019**, *136*, 1130–1146. [CrossRef]
27. Moghaddam, H.A.; Shorabeh, S.N. Designing and implementing a location-based model to identify areas suitable for multi-renewable energy development for supplying electricity to agricultural wells. *Renew. Energy* **2022**, *200*, 1251–1264. [CrossRef]
28. Shorabeh, S.N.; Firozjaei, H.K.; Firozjaei, M.K.; Jelokhani-Niaraki, M.; Homaee, M.; Nematollahi, O. The site selection of wind energy power plant using GIS-multi-criteria evaluation from economic perspectives. *Renew. Sustain. Energy Rev.* **2022**, *168*, 112778. [CrossRef]

29. Shorabeh, S.N.; Samany, N.N.; Minaei, F.; Firozjaei, H.K.; Homaee, M.; Boloorani, A.D. A decision model based on decision tree and particle swarm optimization algorithms to identify optimal locations for solar power plants construction in Iran. *Renew. Energy* **2022**, *187*, 56–67. [CrossRef]
30. Hassan, S.A.; Amlan, H.A.; Alias, N.E.; Abd Kadir, M.A.; Sukor, N.S.A. Vulnerability of road transportation networks under natural hazards: A bibliometric analysis and review. *Int. J. Disaster Risk Reduct.* **2022**, *83*, 103393. [CrossRef]
31. Afsari, R.; Nadizadeh Shorabeh, S.; Bakhshi Lomer, A.R.; Homaee, M.; Arsanjani, J.J. Using Artificial Neural Networks to Assess Earthquake Vulnerability in Urban Blocks of Tehran. *Remote Sens.* **2023**, *15*, 1248. [CrossRef]
32. Carey, M.; Huggel, C.; Bury, J.; Portocarrero, C.; Haeberli, W. An integrated socio-environmental framework for glacier hazard management and climate change adaptation: Lessons from Lake 513, Cordillera Blanca, Peru. *Clim. Chang.* **2012**, *112*, 733–767. [CrossRef]
33. Ma, Y.; Lyu, D.; Sun, K.; Li, S.; Zhu, B.; Zhao, R.; Zheng, M.; Song, K. Spatiotemporal Analysis and War Impact Assessment of Agricultural Land in Ukraine Using RS and GIS Technology. *Land* **2022**, *11*, 1810. [CrossRef]
34. Withanage, W.; Gunathilak, M.; Mishra, P.K.; Wijesinghe, W.; Tripathi, S. Indexing habitat suitability and human-elephant conflicts using GIS-MCDA in a human-dominated landscape. *Geogr. Sustain.* **2023**, *4*, 343–355. [CrossRef]
35. Mijani, N.; Shahpari Sani, D.; Dastaran, M.; Karimi Firozjaei, H.; Argany, M.; Mahmoudian, H. Spatial modeling of migration using GIS-based multi-criteria decision analysis: A case study of Iran. *Trans. GIS* **2022**, *26*, 645–668. [CrossRef]
36. Luan, C.; Liu, R.; Peng, S. Land-use suitability assessment for urban development using a GIS-based soft computing approach: A case study of Ili Valley, China. *Ecol. Indic.* **2021**, *123*, 107333. [CrossRef]
37. Parry, J.A.; Ganaie, S.A.; Bhat, M.S. GIS based land suitability analysis using AHP model for urban services planning in Srinagar and Jammu urban centers of J&K, India. *J. Urban Manag.* **2018**, *7*, 46–56.
38. Kumar, A.; Pramanik, M.; Chaudhary, S.; Negi, M.S. Land evaluation for sustainable development of Himalayan agriculture using RS-GIS in conjunction with analytic hierarchy process and frequency ratio. *J. Saudi Soc. Agric. Sci.* **2021**, *20*, 1–17. [CrossRef]
39. Roy, S.; Singha, N.; Bose, A.; Basak, D.; Chowdhury, I.R. Multi-influencing factor (MIF) and RS–GIS-based determination of agriculture site suitability for achieving sustainable development of Sub-Himalayan region, India. *Environ. Dev. Sustain.* **2023**, *25*, 7101–7133. [CrossRef]
40. Zhang, F.; Cao, N. Application and research progress of Geographic Information System (GIS) in agriculture. In Proceedings of the 2019 8th International Conference on Agro-Geoinformatics (Agro-Geoinformatics), Istanbul, Turkey, 16–19 July 2019; pp. 1–5.
41. Benaissa, F.T.; Khalfallah, B. Industrial Activity Land Suitability Assessment Using Delphi and AHP to Control Land Consumption: The Case Study of Bordj Bouarreridj, Algeria. *Eng. Technol. Appl. Sci. Res.* **2021**, *11*, 7738–7744. [CrossRef]
42. Baghel, A. GIS-based Industrial Land Suitability Analysis for locating Industrial Parks in Raipur and Nava Raipur. In Proceedings of the IOP Conference Series: Earth and Environmental Science, Hanoi, Vietnam, 25–27 February 2022; p. 012024.
43. Radiom, S. Estimation of actual evapotranspiration in pistachio orchards using SEBAL algorithm in three irrigation system. *J. RS GIS Nat. Resour.* **2021**, *12*, 72–90.
44. Alijani, B.; Saligheh, M. Spatial Analysis of Pistachio Cultivates (Growth Period) in Iran. *J. Appl. Res. Geogr. Sci.* **2022**, *22*, 1–24.
45. Pour-taheri, M.; Roknoddin-eftkhari, A.-R.; Rahbari, M. The Analysis of Social–Economic Impacts of Pistachio Cultivation upon Development of Damghan. *Economy* **2013**, *2*, 69–86.
46. Farajnia, A.; Moravej, K.; Alamdari, P.; Eslahi, M. Application of FAO Agro-ecological Model for Locating Areas Prone to Pistachio Cultivation in East Azerbaijan Province. *Water Soil* **2021**, *35*, 703–717.
47. Bakhshi Lomer, A.R.; Rezaeian, M.; Rezaei, H.; Lorestani, A.; Mijani, N.; Mahdad, M.; Raeisi, A.; Arsanjani, J.J. Optimizing emergency shelter selection in earthquakes using a risk-driven large group decision-making support system. *Sustainability* **2023**, *15*, 4019. [CrossRef]
48. Al Garni, H.Z.; Awasthi, A. Solar PV power plant site selection using a GIS-AHP based approach with application in Saudi Arabia. *Appl. Energy* **2017**, *206*, 1225–1240. [CrossRef]
49. Tegou, L.-I.; Polatidis, H.; Haralambopoulos, D.A. Environmental management framework for wind farm siting: Methodology and case study. *J. Environ. Manag.* **2010**, *91*, 2134–2147. [CrossRef] [PubMed]
50. Baskurt, Z.M.; Aydin, C.C. Nuclear power plant site selection by Weighted Linear Combination in GIS environment, Edirne, Turkey. *Prog. Nucl. Energy* **2018**, *104*, 85–101. [CrossRef]
51. Lysaniuk, B.; Cely-García, M.F.; Giraldo, M.; Larrahondo, J.M.; Serrano-Calderón, L.M.; Guerrero-Bernal, J.C.; Briceno-Ayala, L.; Cruz Rodriguez, E.; Ramos-Bonilla, J.P. Using gis to estimate population at risk because of residence proximity to asbestos processing facilities in Colombia. *Int. J. Environ. Res. Public Health* **2021**, *18*, 13297. [CrossRef]
52. Boloorani, A.D.; Kazemi, Y.; Sadeghi, A.; Shorabeh, S.N.; Argany, M. Identification of dust sources using long term satellite and climatic data: A case study of Tigris and Euphrates basin. *Atmos. Environ.* **2020**, *224*, 117299. [CrossRef]
53. Boloorani, A.D.; Shorabeh, S.N.; Samany, N.N.; Mousivand, A.; Kazemi, Y.; Jaafarzadeh, N.; Zahedi, A.; Rabiei, J. Vulnerability mapping and risk analysis of sand and dust storms in Ahvaz, IRAN. *Environ. Pollut.* **2021**, *279*, 116859. [CrossRef] [PubMed]
54. Ramík, J. Pairwise comparisons method. *Lect. Notes Econ. Math. Syst.* **2020**, *690*. [CrossRef]
55. Saaty, T. The analytic hierarchy process (AHP) for decision making. In *Quantitative Assessment in Arms Control*; Springer: Boston, MA, USA, 1980; p. 69.
56. Rezaei, J. Best-worst multi-criteria decision-making method. *Omega* **2015**, *53*, 49–57. [CrossRef]

57. Rezaei, J. Best-worst multi-criteria decision-making method: Some properties and a linear model. *Omega* **2016**, *64*, 126–130. [CrossRef]
58. Vanolya, N.M.; Jelokhani-Niaraki, M. The use of subjective–Objective weights in GIS-based multi-criteria decision analysis for flood hazard assessment: A case study in Mazandaran, Iran. *GeoJournal* **2019**, *86*, 379–398. [CrossRef]
59. Huang, B.; Zha, R.; Chen, S.; Zha, X.; Jiang, X. Fuzzy evaluation of ecological vulnerability based on the SRP-SES method and analysis of multiple decision-making attitudes based on OWA operators: A case of Fujian Province, China. *Ecol. Indic.* **2023**, *153*, 110432. [CrossRef]
60. Karimi Firozjaei, M.; Sedighi, A.; Jelokhani-Niaraki, M. An urban growth simulation model based on integration of local weights and decision risk values. *Trans. GIS* **2020**, *24*, 1695–1721. [CrossRef]
61. Yager, R.R. On ordered weighted averaging aggregation operators in multicriteria decisionmaking. *IEEE Trans. Syst. Man Cybern.* **1988**, *18*, 183–190. [CrossRef]
62. Malczewski, J. Integrating multicriteria analysis and geographic information systems: The ordered weighted averaging (OWA) approach. *Int. J. Environ. Technol. Manag.* **2006**, *6*, 7–19. [CrossRef]
63. Malczewski, J.; Rinner, C. *Multicriteria Decision Analysis in Geographic Information Science*; Springer: Berlin/Heidelberg, Germany, 2015.
64. Malczewski, J.; Chapman, T.; Flegel, C.; Walters, D.; Shrubsole, D.; Healy, M.A. GIS–multicriteria evaluation with ordered weighted averaging (OWA): Case study of developing watershed management strategies. *Environ. Plan. A* **2003**, *35*, 1769–1784. [CrossRef]
65. Mokarram, M.; Hojati, M. Using ordered weight averaging (OWA) aggregation for multi-criteria soil fertility evaluation by GIS (case study: Southeast Iran). *Comput. Electron. Agric.* **2017**, *132*, 1–13. [CrossRef]
66. Rezvani, M.; Nickravesh, F.; Astaneh, A.D.; Kazemi, N. A risk-based decision-making approach for identifying natural-based tourism potential areas. *J. Outdoor Recreat. Tour.* **2022**, *37*, 100485. [CrossRef]
67. Moradi, M.; Delavar, M.R.; Moshiri, B. A GIS-based multi-criteria decision-making approach for seismic vulnerability assessment using quantifier-guided OWA operator: A case study of Tehran, Iran. *Ann. GIS* **2015**, *21*, 209–222. [CrossRef]
68. Eldrandaly, K.A.; AL-Amari, M.A. An expert GIS-based ANP-OWA decision making framework for tourism development site selection. *Int. J. Intell. Syst. Appl.* **2014**, *6*, 1–11. [CrossRef]
69. Daniel, C. One-at-a-time plans. *J. Am. Stat. Assoc.* **1973**, *68*, 353–360. [CrossRef]
70. Chen, Y.; Yu, J.; Khan, S. Spatial sensitivity analysis of multi-criteria weights in GIS-based land suitability evaluation. *Environ. Model. Softw.* **2010**, *25*, 1582–1591. [CrossRef]
71. Khalid, K.; Ali, M.; Abd Rahman, N.; Mispan, M. Application on one-at-a-time sensitivity analysis of semi-distributed hydrological model in tropical watershed. *Int. J. Eng. Technol.* **2016**, *8*, 132–136. [CrossRef]
72. Yu, S.; Yun, S.-T.; Hwang, S.-I.; Chae, G. One-at-a-time sensitivity analysis of pollutant loadings to subsurface properties for the assessment of soil and groundwater pollution potential. *Environ. Sci. Pollut. Res.* **2019**, *26*, 21216–21238. [CrossRef]
73. Dutta, B.; Das, M.; Roy, U.; Das, S.; Rath, S. Spatial analysis and modelling for primary healthcare site selection in Midnapore town, West Bengal. *GeoJournal* **2021**, *87*, 4807–4836. [CrossRef]
74. Torkayesh, A.E.; Zolfani, S.H.; Kahvand, M.; Khazaelpour, P. Landfill location selection for healthcare waste of urban areas using hybrid BWM-grey MARCOS model based on GIS. *Sustain. Cities Soc.* **2021**, *67*, 102712. [CrossRef]
75. Mighty, M.A. Site suitability and the analytic hierarchy process: How GIS analysis can improve the competitive advantage of the Jamaican coffee industry. *Appl. Geogr.* **2015**, *58*, 84–93. [CrossRef]
76. Zambrano-Asanza, S.; Quiros-Tortos, J.; Franco, J.F. Optimal site selection for photovoltaic power plants using a GIS-based multi-criteria decision making and spatial overlay with electric load. *Renew. Sustain. Energy Rev.* **2021**, *143*, 110853. [CrossRef]
77. Feng, P.; Yi-peng, J.; Li-nan, D.; Zhan-xiu, S. Study on layout optimization of the iron and steel industry in the Bohai Sea rim area based on GIS. In Proceedings of the 2013 International Conference on Management Science and Engineering 20th Annual Conference Proceedings, Prague, Czech Republic, 21–24 April 2013; pp. 1183–1189.
78. Awasthi, M.K.; Pandey, A.K.; Bundela, P.S.; Wong, J.W.; Li, R.; Zhang, Z. Co-composting of gelatin industry sludge combined with organic fraction of municipal solid waste and poultry waste employing zeolite mixed with enriched nitrifying bacterial consortium. *Bioresour. Technol.* **2016**, *213*, 181–189. [CrossRef] [PubMed]
79. Yalcinkaya, S.; Uzer, S.; Kaleli, H.İ.; Doğan, F.; Kayalık, M. Compost plant site selection for food waste using GIS based multicriteria analysis. *Turk. J. Agric.-Food Sci. Technol.* **2021**, *9*, 1908–1914. [CrossRef]
80. Rezaei, H.; Macioszek, E.; Derakhshesh, P.; Houshyar, H.; Ghabouli, E.; Bakhshi Lomer, A.R.; Ghanbari, R.; Esmailzadeh, A. A Spatial Decision Support System for Modeling Urban Resilience to Natural Hazards. *Sustainability* **2023**, *15*, 8777. [CrossRef]
81. Li, X.; Hijazi, I.; Koenig, R.; Lv, Z.; Zhong, C.; Schmitt, G. Assessing essential qualities of urban space with emotional and visual data based on GIS technique. *ISPRS Int. J. Geo-Inf.* **2016**, *5*, 218. [CrossRef]
82. Ghoushchi, S.J.; Dorosti, S.; Khazaeili, M.; Mardani, A. Extended approach by using best–worst method on the basis of importance–necessity concept and its application. *Appl. Intell.* **2021**, *51*, 8030–8044. [CrossRef]
83. Nadizadeh Shorabeh, S.; Neisany Samany, N.; Abdali, Y. Mapping the potential of solar power plants based on the concept of risk Case study: Razavi Khorasan Province. *Sci.-Res. Q. Geogr. Data SEPEHR* **2019**, *28*, 129–147.

84. Nadizadeh Shorabeh, S.; Neysani Samani, N.; Jelokhani-Niaraki, M.R.J.-N. Determination of optimum areas for the landfill with emphasis on the urban expansion trend based on the combination of the Analytical Hierarchy Process and the Ordered Weighted Averaging model. *J. Nat. Environ.* **2017**, *70*, 949–969.
85. Vanolya, N.M.; Jelokhani-Niaraki, M.; Toomanian, A. Validation of spatial multicriteria decision analysis results using public participation GIS. *Appl. Geogr.* **2019**, *112*, 102061. [CrossRef]
86. Gasparin, A.; Lukovic, S.; Alippi, C. Deep learning for time series forecasting: The electric load case. *CAAI Trans. Intell. Technol.* **2022**, *7*, 1–25. [CrossRef]
87. Meng, J.; Li, Y.; Liang, H.; Ma, Y. Single-image dehazing based on two-stream convolutional neural network. *J. Artif. Intell. Technol.* **2022**, *2*, 100–110. [CrossRef]
88. Zheng, M.; Zhi, K.; Zeng, J.; Tian, C.; You, L. A hybrid CNN for image denoising. *J. Artif. Intell. Technol.* **2022**, *2*, 93–99. [CrossRef]
89. Wu, Z.; Luo, G.; Yang, Z.; Guo, Y.; Li, K.; Xue, Y. A comprehensive review on deep learning approaches in wind forecasting applications. *CAAI Trans. Intell. Technol.* **2022**, *7*, 129–143. [CrossRef]
90. Ma, L.; Wei, P.; Qu, X.; Bi, S.; Zhou, Y.; Shen, T. Apple grading method based on neural network with ordered partitions and evidential ensemble learning. *CAAI Trans. Intell. Technol.* **2022**, *7*, 561–569. [CrossRef]

sustainability

MDPI

Article

# Improving Runoff Prediction Accuracy in a Mountainous Watershed Using a Remote Sensing-Based Approach

**Solmaz Fathololoumi [1,2,*], Ali Reza Vaezi [1], Seyed Kazem Alavipanah [3], Ardavan Ghorbani [4], Mohammad Karimi Firozjaei [3] and Asim Biswas [2]**

1 Department of Soil Science, Faculty of Agriculture, University of Zanjan, Zanjan 45371-38791, Iran; vaezi.alireza@znu.ac.ir
2 School of Environmental Sciences, University of Guelph, Guelph, ON N1G 2W1, Canada; biswas@uoguelph.ca
3 Department of Remote Sensing & GIS, Faculty of Geography, University of Tehran, Tehran 14155-6465, Iran; salavipa@ut.ac.ir (S.K.A.)
4 Faculty of Agriculture and Natural Resources, University of Mohaghegh Ardabili, Ardabil 56199-11367, Iran; a_ghorbani@uma.ac.ir
* Correspondence: sfatholo@uoguelph.ca

**Abstract:** Due to the limited number and sparse distribution of meteorological and hydrometric stations in most watersheds, the runoff estimation based on these stations may not be accurate. However, the accurate determination of the Antecedent Soil Moisture (ASM) in watersheds can improve the accuracy of runoff forecasting. The objective of this study is to utilize the ASM derived from satellite imagery to enhance the accuracy of runoff estimation in a mountainous watershed. In this study, a range of Remote Sensing (RS) data, including surface biophysical and topographic features, climate data, hydrometric station flow data, and a ground-based measured SM database for the Balikhli-Chay watershed in Iran, were utilized. The Soil Conservation Service Curve Number (SCS-CN) method was employed to estimate runoff. Two approaches were used for estimating the ASM: (1) using the precipitation data recorded in ground stations, and (2) using the SM data obtained from satellite images. The accuracy of runoff estimation was then calculated for these two scenarios and compared. The mean Nash–Sutcliffe statistic was found to be 0.63 in the first scenario and 0.74 in the second scenario. The inclusion of ASM derived from the satellite imagery in the precipitation–runoff model resulted in a 51% increase in the accuracy of runoff estimation compared to using precipitation data recorded in ground stations. These findings have significant implications for improving the accuracy of ASM and runoff modeling in various applications.

**Keywords:** runoff; ASM; MODIS; HEC-HMS; SCS-CN method

## 1. Introduction

Runoff is a crucial factor in the hydrological cycle, and changes in runoff have significant effects on the ecology, water, and soil storage resources in a watershed, particularly downstream [1–3]. The management of runoff in many watersheds is of utmost importance globally. The process of runoff production at the watershed scale is complex due to various factors such as precipitation patterns and surface characteristics [4,5]. Therefore, it is useful to determine and recognize runoff distribution patterns in order to develop appropriate control methods [6].

One important feature of watersheds that has been increasingly considered in runoff estimation systems is the Antecedent Soil Moisture (ASM) conditions prior to flooding [7–10]. As the ASM increases and the stability of aggregates decreases, there is a sharp increase in runoff and sediment production [11]. Consequently, the underestimation and overestimation of runoff discharge peaks in hydrological modeling, especially in runoff operational forecasts, is a persistent problem due to errors in estimating the ASM [12]. Incorporating observational ASM data into hydrological models is crucial for reducing errors in runoff estimation [13,14]. Therefore, the accurate measurement of this variable across watersheds will improve the performance of runoff forecasting, as rainfall–runoff models are sensitive to ASM changes [6,15].

The Soil Conservation Service Curve Number (SCS-CN) method is a common method used for runoff simulation [16–18], and it is supported by the Hydrologic Modeling System (HEC-HMS) runoff model. In the SCS-CN method, the ASM factor, along with land use and the hydrological group, is used to determine the Curve Number (CN) values. These variables are primarily calibrated using local data. Also, incorporating satellite data into hydrological models may help to overcome the limitations and uncertainties of hydrological modeling [19,20].

In several previous studies, the impact of ASM on the accuracy of runoff estimation was evaluated. Zhang et al. [8] assessed the effects of ASM on runoff generation in a semi-arid environment. The results of their study demonstrate the provision of concepts for runoff modeling based on the process. The study results show that a 1% change in the ASM resulted in a 0.05 mm change in the runoff. Shi et al. [21] presented a physical formula for estimating the Antecedent Soil Moisture. The study results show that their proposed method increased the model's efficiency by 88% in both calibration and validation, surpassing the conventional method based on the recorded precipitation data at ground stations. Shi and Wang [17] evaluated the influence of three variables, the slope, ASM, and storm duration, on the SCS-CN method to improve the accuracy of runoff estimation. Their study results show that the impact of the ASM variable on improving the runoff estimation accuracy was higher than the two variables of storm duration and slope. Mishra et al. [22] improved the accuracy of runoff estimation in 234 watersheds in the United States by incorporating the ASM data into the SCS-CN method. Tramblay et al. [9] evaluated the performance of several soil moisture indicators in runoff estimation, including the local Time Domain Reflectometry (TDR) measurements of the SM, and modeled the SM through the Interaction between Soil, Biosphere, and Atmosphere (ISBA) component of the SIM model (Mto-France), the antecedent precipitation, and the base flow. Their study results show that the TDR measurements had a higher accuracy in runoff estimation, while the antecedent-precipitation-based method had less efficiency. Brocca et al. [23] evaluated the performance of two scenarios for estimating antecedent conditions, including the Antecedent Precipitation Index (API) and the Base Flow Index (BFI), in runoff estimation. The results indicate that considering the ASM variable in the rainfall–runoff model helped to predict both the volume and peak discharge well, with a Nash–Sutcliffe efficiency of over 90%. The BFI-based scenario showed a higher accuracy in the runoff estimation compared to the API-based scenario in the rainfall–runoff model.

In previous studies, meteorological station information has commonly been used to determine the ASM. However, the accuracy of ASM determination is dependent on the number and distribution of stations in the region, as well as the accuracy of their recorded information. In many watersheds, particularly in mountainous and steep areas, there is a scarcity of meteorological stations, resulting in a limited access to sufficient data for precipitation and ASM determination [24–26]. Additionally, the spatial resolution of precipitation data for the entire area is low due to the limited number of rainfall recording stations. Typically, only one meteorological station provides rainfall and SM data for the entire area. However, the SM is a dynamic characteristic influenced by various factors such as surface biophysical and topographic factors, soil properties, and climate. At the watershed scale, SM exhibits heterogeneity [27]. Therefore, relying on a single value of SM for the entire area makes accurate runoff prediction challenging.

Previous studies have attempted to enhance the accuracy of runoff estimation by utilizing satellite data to calculate the SM [10,20,28,29]. However, the effectiveness of satellite data in improving the runoff estimation accuracy varies depending on the conditions. Consequently, it is necessary to evaluate the effectiveness of satellite data separately in each region. Estimating the runoff in mountainous and semi-arid areas presents greater challenges due to the heterogeneity of topographic and biophysical conditions.

The objective of this study is to utilize the ASM derived from satellite imagery to enhance the accuracy of runoff estimation in a mountainous watershed. To address these challenges and improve the runoff estimation accuracy, two approaches were employed for runoff modeling: (1) utilizing meteorological station precipitation data to determine ASM, and (2) utilizing satellite-imagery-derived SM. Finally, the results of the two approaches were compared to determine the most effective method.

## 2. Study Area

The Balikhli-Chay watershed is a sub-basin of the Qarahsu watershed, located in the northwest of Iran. This watershed is situated upstream of the Yam-chi Hydrometric Station. The highest elevation in the watershed is 4365 m, which is associated with the Sabalan Mountains in the northeast. The lowest elevation is 1550 m, located at the exit of the watershed on the northeast side. The average height of the watershed is 2109 m, and it covers an area of 567 $km^2$, with rainfed agriculture accounting for 34.3% of the total area.

Bozgoosh Mountain, located in the southern part, has an elevation of 2672 m above sea level. The Bagherloodagh Mountains can be found in the eastern and southeastern parts of the watershed. The output of the watershed is a flat plain known as the Ardabil Plain. This river is an important tributary of the Aras River in northwestern Iran, flowing in a south–north direction and collecting water from sub-tributaries in the east and south of the watershed. The Balikhli-Chay River is particularly significant in the region as it flows through the center of Ardabil, providing water for both drinking and agriculture. It is also surrounded by industrial and recreational areas.

The study area, characterized by surface condition heterogeneity such as slope, elevation, land use, and vegetation, is vulnerable to floods and erosion. Over a statistical period of 20 years (from 1999 to 2019), the average annual rainfall recorded by the Ardabil Meteorological Department is 360 mm. The high elevation of the study area greatly influences the type of rainfall, evapotranspiration, vegetation, and ultimately, the runoff. The average daily flow is 1.46 $m^3 \cdot s^{-1}$, with a base flow of 0.62 $m^3 \cdot s^{-1}$. The main drainage slope of the watershed is 3.4%. The slopes of both the main river and the watershed itself play crucial roles in the occurrence and severity of floods in the study area, as rivers and tributaries follow the slopes until they reach the main river. The average slope in the watershed is 17.2%, further contributing to the occurrence of floods. The Balikhli-Chay watershed (Figure 1) was chosen as the most suitable test watershed for this study based on the available recorded data from hydrometric and rain gauge stations.

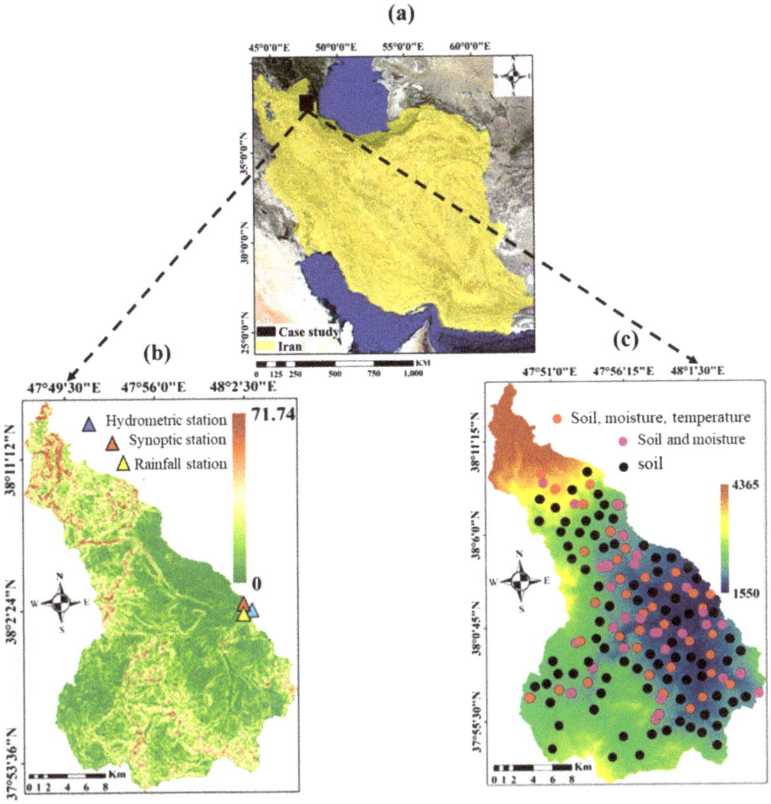


**Figure 1.** Geographical location of the study area in Iran (**a**), locations of hydrometric, synoptic, and rainfall stations on the slope map (**b**), and geographical locations of sampling points on the DEM map of the study area (**c**).

## 3. Materials and Methods

### *3.1. Data*

To evaluate the efficiency of the ASM in runoff modeling, the data used include in situ-measured soil properties, climatic and meteorological data, and a set of satellite images.

#### 3.1.1. Terrestrial Data

In this study, the effect of soil texture on the accuracy of SM modeling was investigated [30]. A total of 225 sampling points were selected in the area (Figure 1c) at a depth of 0–30 cm, and samples were collected in June 2018. The minimum and maximum distances between sampling points were 450 and 660 m, respectively. All samples were taken to a laboratory, air dried, and passed through a 2 mm sieve for texture analysis.

To model the SM and evaluate the accuracy of the model, ground measurements of the SM were taken at 148 points out of the 225 points. These measurements were taken simultaneously with the satellite overpass time in June, July, August, and September 2018. The reduction in the number of points for SM ground measurement was due to time limitations in relation to the size of the study area during the satellite overpass time. The SM was measured using an SM meter model, T SM150, at a depth of 0–30 cm. This portable device provides quick and accurate measurements of volumetric SM with an accuracy of 0.03 $m^3 \cdot m^{-3}$.

For runoff modeling, climatic and environmental information including daily and hourly rainfall (including intensity and duration) was obtained from the Meteorological

Organization of Ardabil Province. Flow and flood hydrograph data were obtained from the Ardabil Regional Water Company. Data on air temperature, rainfall, and air pressure were collected from meteorological stations. Eight rainfall–runoff events were used to simulate runoff, as shown in Table 1. The events are referred to as numbers 1 to 8 in the main text.

**Table 1.** The dates of the selected events numbered 1 to 8.

| Number of Events | 1 | 2 | 3 | 4 | 5 | 6 | 7 | 8 |
|---|---|---|---|---|---|---|---|---|
| Date of events | 15 March 2004 | 8 November 2007 | 27 September 2011 | 1 January 2014 | 10 March 2016 | 30 March 2015 | 26 April 2016 | 14 April 2017 |

3.1.2. Remote Sensing (RS) Data

In this study, the Advanced Spaceborne Thermal Emission and Reflection Radiometer (ASTER) Digital Elevation Model (DEM) was used to extract various surface topographic features. Spectral indices were extracted from Landsat 8 satellite images, which can be downloaded from https://earthexplorer.usgs.gov (accessed on 18 September 2019). The dates of the images used are presented in Table 2. These data were used on four different dates to determine the optimal model for calculating the SM in the study area. Additionally, in this stage, the average water vapor in the study area was obtained from MOD07 with a spatial resolution of 5000 m. These data were used to estimate the Land Surface Temperature (LST) based on the Landsat 8 image thermal bands. To determine the ASM in runoff estimation using the best model obtained in the first part, daily Moderate Resolution Imaging Spectroradiometer (MODIS) images from five days before and eight rainfall–runoff events were used (Table 2). Furthermore, in the process of generating the SM map based on the MODIS data, the LST was used as one of the spectral features to calculate the ASM. The daily MODIS LST product (MOD11A1) with a spatial resolution of 1000 m was used to determine the LST in the ASM estimation.

**Table 2.** RS data used in the study.

| Name | Dates |
|---|---|
| ASTER-DEM | - |
| Landsat 8 | 15 June, 1 July, 18 August, and 3 September, 2018 |
| MODIS | (14, 13, 12, 11, and 10 March 2004), (7, 6, 5, 4, and 3 November 2007), (26, 25, 24, 23, and 22 September 2011), (31, 30, 29, 28, and 27 December 2014), (29, 28, 27, 26, and 25, March 2015), (9, 8, 7, 6, and 5 March 2016), (25, 24, 23, 22, and 21 April 2016), (13, 12, 11, 10, and 9 April 2017) |
| MOD07 | The same as Landsat image dates |
| MOD11A1 | The same as MODIS image dates |

*3.2. Methods*

In this study, several steps were performed to evaluate the efficiency of the SM used in the runoff modeling (Figure 2). First, the SM of the study area was calculated based on Landsat 8 satellite images and DEM using different models. The accuracy of the SM maps, modeled based on real ground data, was evaluated to determine the best model for SM modeling. Then, based on the best model, the SM of the study area for five days before the events was calculated using MODIS satellite images and DEM. Secondly, the ASM of the study area was modeled based on two strategies using precipitation data and the SM obtained from satellite images. In the third step, the study area runoff values at different events were calculated based on two strategies using the ASM based on the rainfall data and the SM obtained from satellite images in the SCS-CN model. The efficiency of the two strategies in runoff modeling was evaluated based on the actual peak flow data and flood volume. Finally, the effect of the SM on the amount of runoff was investigated.

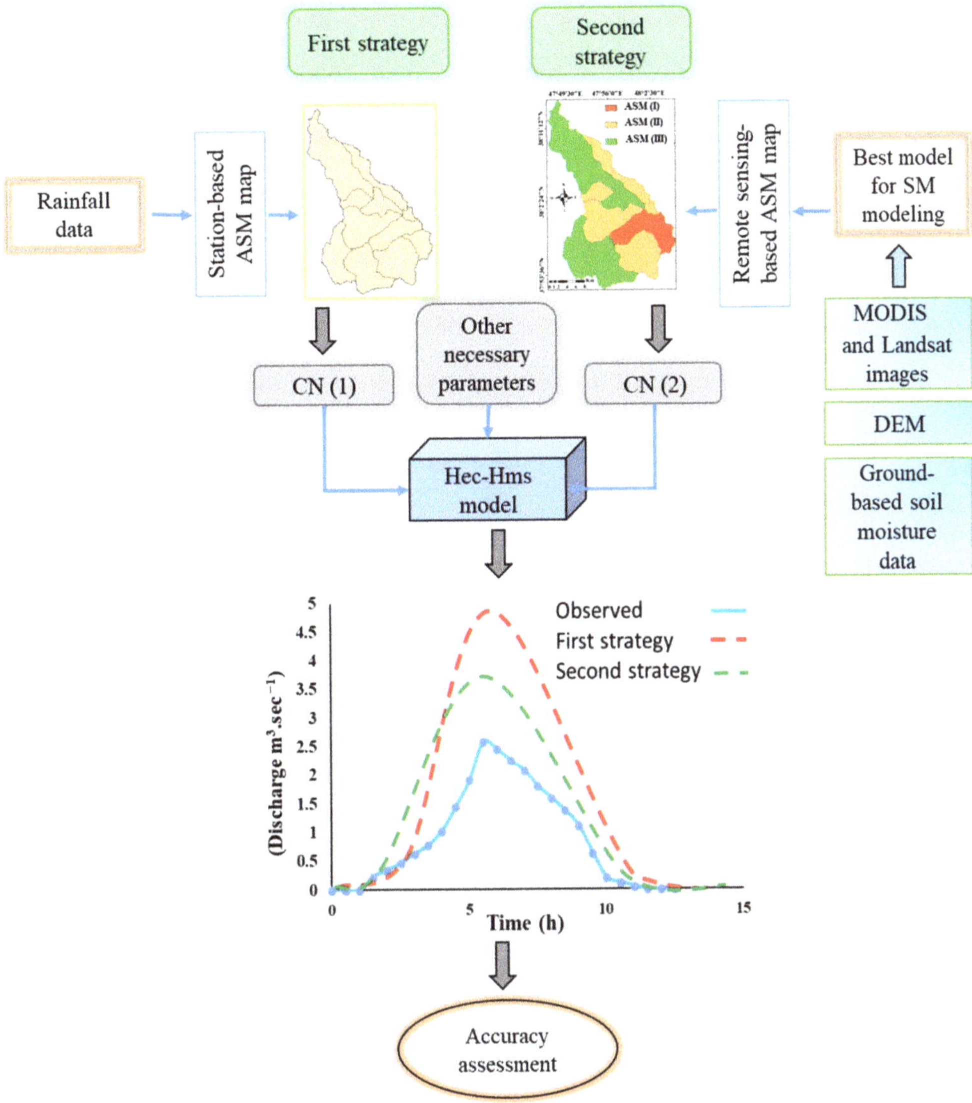


**Figure 2.** The flowchart of the study.

3.2.1. SM Estimation

Before calculating and extracting different spectral indices (Table 2) from satellite images, radiometric and atmospheric corrections of their different bands were necessary. The Fast Line-of-sight Atmospheric Analysis of Spectral Hypercubes (FLAASH) model was used for the atmospheric correction of reflective bands [31]. According to this model, spectral values resulting from water vapor and aerosol retrieval, as well as the adjacency effect, are corrected based on the spectral bands of satellite images. This model utilizes the moderate resolution atmospheric transmission (MODTRAN) for atmospheric correction, taking into account ancillary information such as satellite overpass time, sensor altitude, geographical location of the area, solar radiation angle, and regional atmospheric model.

The Indices required for SM modeling were extracted from Landsat, MODIS, and DEM images. In the first stage, Landsat and ASTER data (with a spatial resolution of 30 m) were used to determine the optimal model for calculating moisture in the study area. In the second stage, the optimized model was applied to MODIS data (1000 m) and ASTER data (1000 m resampled using the Nearest Neighbor method), and moisture maps for the days prior to the occurrence of runoff were prepared. In the first stage, after preparing the Landsat and DEM images and extracting all the indices, the Random Forest (RF) model was used to model the SM. The following indices, including albedo, brightness, greenness, wetness, Normalized Difference Vegetation Index (NDVI), Normalized Difference Water Index (NDWI), Normalized Difference Built-up Index (NDBI), Land Surface Temperature (LST), elevation, slope, and local incident angle, were used as predictive variables in the SM calculation process [32–34]. A supervised learning algorithm, RF, employing a regression tree structure [35], was employed for the prediction of the SM. RF comprises an ensemble of predictor trees that were grown without pruning [36], with the model's stability relying on the diversity of these trees [37]. In this research, tree diversity was achieved through the utilization of a range of RS datasets, encompassing surface biophysical characteristics and DEM-derived surface terrain indices as predictive variables. Concurrently, in situ SM data measured during satellite overpass time were integrated into the model. These predictors played a pivotal role in the creation of decision trees within the nodes. The construction of trees was carried out in parallel, with these variables being randomly selected to facilitate the process, and no interplay existing between them [33]. Additionally, the significance of variables in SM modeling was evaluated through the extraction of Variable Importance (VI) factors from the RF model. Noteworthy advantages of RF encompass (a) its high accuracy, rendering it suitable for various types of datasets; (b) its capacity to handle large datasets without necessitating data removal; (c) its ability to generate an unbiased prediction of generalization error, akin to a progressing forest; and (d) its efficacy in accurately predicting missing data, even when a substantial amount of data is absent [34,35,38].

After confirming the efficiency of the model, the best model was used to estimate ASM based on MODIS images and DEM in order to estimate the runoff in the SCS-CN model.

### 3.2.2. Antecedent Soil Moisture (ASM)

In this study, in the first runoff estimation strategy, based on previous studies, the average rainfall information of five days before the event, obtained from the meteorological station, was used to consider the ASM conditions (Table 3). As a result, the accuracy of determining the ASM depends on the number and distribution of stations in the region, as well as the accuracy of their recorded information. In many areas and watersheds, particularly in mountainous and steep regions, there is a limited number of meteorological stations, resulting in limited access to sufficient data for determining precipitation and ASM.

**Table 3.** ASM condition classification in watershed based on the first strategy [16].

| The Amount of Precipitation of the Previous Five Days | | |
|---|---|---|
| **ASM Class** | **Dormancy Season (mm)** | **Growing Season (mm)** |
| I | <12.5 | <35 |
| II | 12.5–27.5 | 35–52.5 |
| III | >27.5 | >52.5 |

In the second strategy, the ASM conditions were estimated using MODIS satellite images from the previous five days and the average of the five-day SM from Table 4. The best SM model was utilized to calculate the ASM based on MODIS images, as discussed in the previous sections. The average SM for each sub-watershed was calculated, and the corresponding SM classes for each sub-watershed were determined based on Table 4 (there were a total of 11 sub-watersheds in the study area (SW1 to SW11)).

**Table 4.** Classification of SM conditions in the watershed based on the second strategy.

| ASM Class | The SM of the Previous Five Days (Volumetric Percentage) |
|---|---|
| I | <10 |
| II | 10–30 |
| III | >30 |

#### 3.2.3. Runoff Modeling

This study employed the HEC-HMS rainfall–runoff model and the SCS-CN method to convert the rainfall–runoff relationship in the Balkhli-Chay watershed. The calculations included determining the amount of precipitation losses, converting excess precipitation into runoff, and calculating the baseline flow. Further details on the calculation of each parameter can be found in [16].

Model Calibration and Validation

Calibration is a process that involves adjusting the values of the variables in the model to achieve results that align with the observational data using an objective function. The objective function measures the variation between the observed and simulated hydrographs. The HEC-HMS model automatically minimizes the objective function and optimizes the values of the variables using a search method. In this study, the Nash–Sutcliffe criterion (CNS) based on Equation (1) was used as the objective function to optimize the CN variable and lag time:

$$C_{NS} = 1 - \frac{\sum_{i=1}^{n}(Q_{si} - Q_{oi})^2}{\sum_{i=1}^{n}(Q_{oi} - Q_{o})^2} \quad (1)$$

where $Q_{si}$ is the simulated discharge, $Q_{oi}$ is the observed discharge, and $Q_o$ is the mean observed discharge.

For calibration and validation, five and three events were randomly selected, respectively. After running the model with the corresponding rainfall–runoff events, the events were randomly divided into two categories. The variables for the five calibration events were optimized using the CNS objective function. The mean values of the optimized calibrated variables were then used for validation, and the model was run for the remaining three events. The results of the validation were compared with the observed values to test the accuracy of the data obtained from the calibration. To validate the model, the feature space was formed between the observed and estimated values. It was then analyzed based on the CNS index (Equation (1)), line 1:1, the absolute mean error (MAE) (Equation (2)), and the root mean square error (RMSE) (Equation (3)) between the observed and estimated values.

$$MAE = \frac{1}{n}\sum_{i=1}^{n}|Q_{si} - Q_{oi}| \quad (2)$$

$$RMSE = \sqrt{\frac{1}{n}\sum_{i=1}^{n}(Q_{si} - Q_{oi})^2} \quad (3)$$

where $Q_{si}$ and $Q_{oi}$ are the simulated and observed discharge, respectively.

### 3.3. *The Impact of SM on Runoff*

The amount of rainfall is the most important factor in determining the amount of runoff in a watershed. Increased rainfall can lead to higher peak discharge and flood volume in a basin. However, factors such as CN also play a role in determining the amount of runoff in a field. The temporal variation of runoff in a watershed depends on both rainfall and the ASM. To investigate the effect of the ASM on the runoff, the regression relationship between the discharge and measured flood volume in each event was obtained, along with the recorded rainfall at the meteorological station and the measured flow and flood volume estimated using RS. The relationship between the flow and measured flood volume was obtained simultaneously for both the station-measured rainfall and estimated SM.

## 4. Results and Discussion

### 4.1. The Effect of Soil Texture on the Accuracy of SM Modeling

The mean error of the modeled SM obtained from the RF model for the sampling points with fine, medium, and coarse textures was 1.6%, 2.1%, and 2.6%, respectively. Furthermore, the $R^2$ (RMSE) values between the measured and modeled SM for the sampling points with fine, medium, and coarse soil textures were 0.82 (1.5%), 0.73 (2.2%), and 0.60 (2.7%), respectively. These results indicate that the accuracy of SM prediction based on surface biophysical and topographic characteristics increases in fine-textured soils. In fine-textured soils, increased soil suction allows for more moisture retention. The higher water-holding capacity of fine-textured soils leads to a lower soil temperature and an improved SM content from an RS perspective. On the other hand, coarse-textured soils have a lower water-holding capacity and lose moisture faster than fine-textured soils. As the SM increases, the reflection in the red band increases, while the reflection in the infrared band decreases significantly, resulting in better quality information from RS images in fine-textured soil conditions [39]. Prakash et al. [40] also demonstrated the significant role of soil texture in SM retrieval.

### 4.2. Determination of ASM

The ASM status of events 1 to 8 were III, I, I, I, II, I, I, and II, respectively. In this study, in addition to the usual method (referred to as the first strategy), another strategy was used for the first time. The second strategy utilized satellite imagery and the best model for SM prediction to estimate the ASM. This allowed for the calculation of the average SM per sub-watershed instead of an overall estimate for the entire watershed. The ASM information for each sub-watershed was then used to determine the CN value of the watershed and its sub-watersheds. There was a total of 11 sub-watersheds in the area (SW1 to SW11).

In the first strategy, the ASM for the entire watershed was determined based on the precipitation information from a single meteorological station. Since there was only one data point, the precipitation value was the same for the entire area, resulting in the same ASM value for the whole area. This affected the spatial variation of the calculated CN for the sub-watersheds, which was influenced by the hydrological groups of soil and land cover. On the other hand, the second strategy took into account the amount of ASM in addition to the aforementioned factors, resulting in different CN values for different parts of the watershed. This is important for accurate runoff forecasting, as the CN plays a crucial role in runoff production. The proposed method in this study can be particularly useful in watersheds that lack sufficient and appropriate data, allowing for a more accurate determination of the CN. The ASM, being a hydrological variable, has a non-linear relationship with the runoff, and its values significantly impact runoff production. The spatial and temporal dynamics of the ASM are commonly used to describe the internal state of a watershed, indicating the sensitivity of the watershed to the production of surface runoff [7]. The spatial distribution of the ASM conditions for different events at the sub-watershed scale is shown in Figure 3.

### 4.3. Runoff Modeling with HEC-HMS Model

After preparing the model input variables and physically modeling the Balikhli-Chay watershed in the HEC-HMS model, the model was implemented for eight corresponding rainfall–runoff events based on two different strategies. The observational and modeled flow hydrographs based on the model implementation using two different strategies are presented in Figure 4.

A visual examination of the hydrographs shows that the hydrographs modeled with the second strategy, in which RS is used to estimate the ASM, are closer to the (real) observational values. This result demonstrates the efficiency of the RS method as a more accurate estimation of the SM at the watershed, and it is a better result than the one obtained when using the meteorological station data.

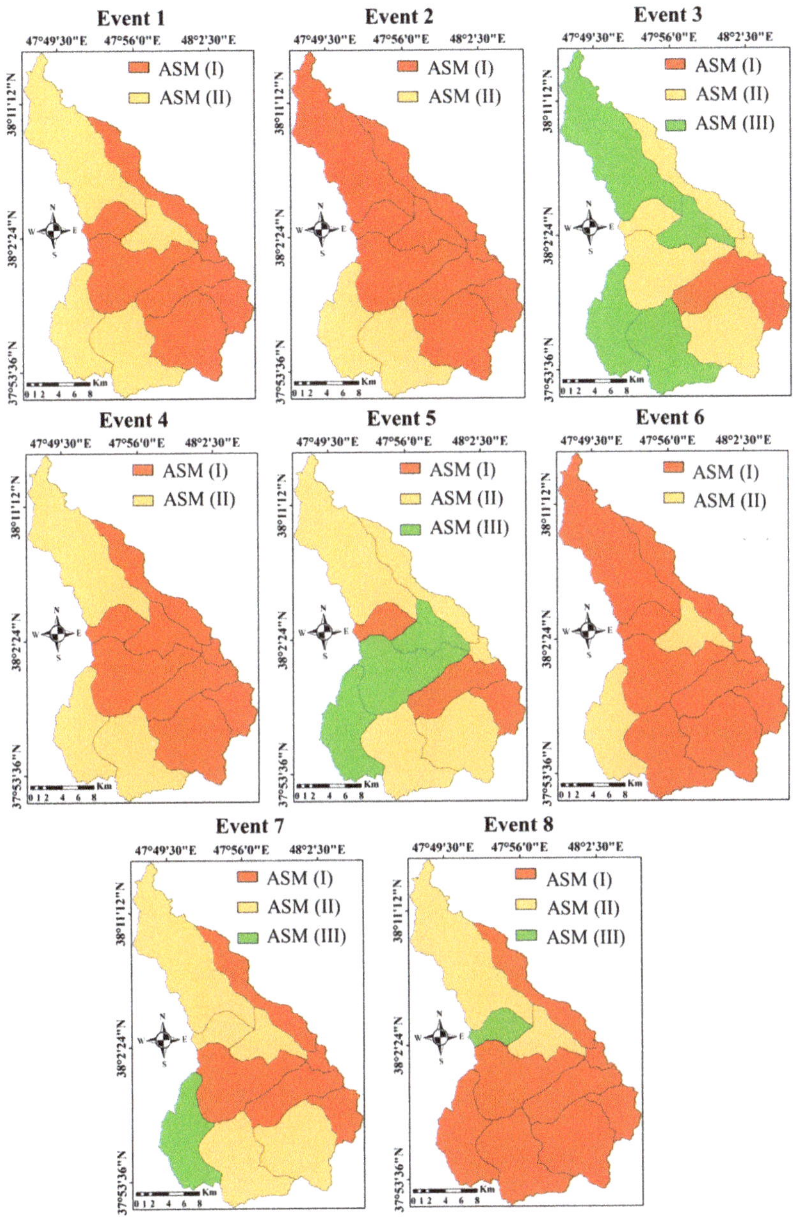


Figure 3. Spatial distribution of ASM conditions for different events at the sub-watershed scale.

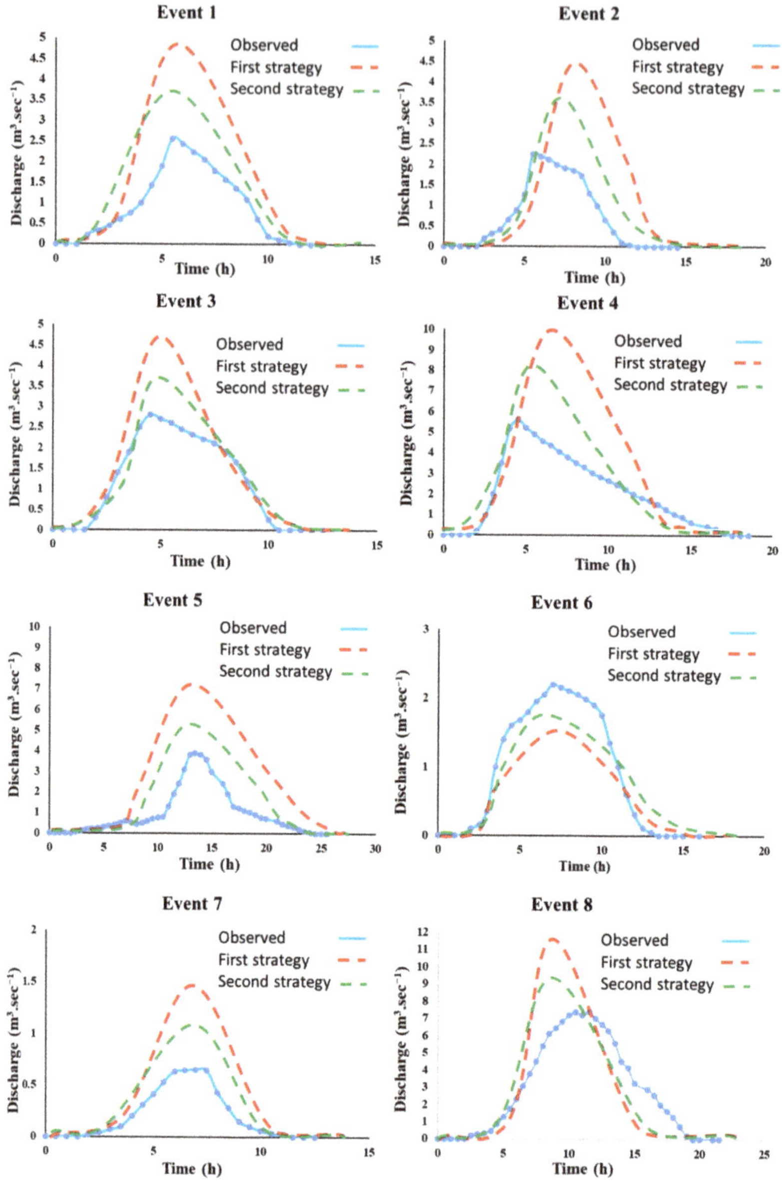


**Figure 4.** Observational and modeled flow hydrographs based on the two strategies used in the study for the corresponding rainfall–runoff events on 8 different dates before model calibration.

To compare the efficiency of the runoff prediction strategies, the peak discharge and volume of the hydrograph flood in both strategies were calculated and compared to the observational mode (Table 5). An examination of these values showed that the results of the runoff modeling with the second strategy (using RS to estimate the ASM) were closer to the observational values than the first strategy (using meteorological station data to estimate the ASM). The mean values of the observed and modeled peak discharge for the eight events, using the first strategy and the second strategy, were 3.40, 5.72, and 4.50 $m^3 \cdot s^{-1}$, respectively, and the mean flood values for these conditions were 74.63,

124.70, and 91.47 thousand $m^3$, respectively. Based on these values, the average differences between the modeled and observed peak flows using the first and second strategies were 2.32 and 1.10 $m^3 \cdot s^{-1}$, respectively, and the differences between the flood volume were 50.07 and 16.84 thousand $m^3$, respectively. In both cases, the difference between the modeled and observed values using the RS method (second strategy) was greater than the meteorological station data (first strategy).

**Table 5.** Peak flow and flood volume of observed and modeled hydrographs in 8 corresponding rainfall–runoff events.

| Event | Peak Discharge ($m^3 \cdot sec^{-1}$) | | | Flood Value (1000 $m^3$) | | |
|---|---|---|---|---|---|---|
| | Observed | Model 1 | Model 2 | Observed | Model 1 | Model 2 |
| 1 | 2.5 | 4.9 | 3.8 | 34.8 | 70.4 | 48.5 |
| 2 | 2.2 | 4.4 | 3.5 | 28.6 | 58.3 | 40.7 |
| 3 | 2.8 | 4.7 | 3.7 | 38.3 | 63.4 | 47.5 |
| 4 | 5.6 | 9.8 | 8.1 | 152.1 | 276.8 | 196.4 |
| 5 | 3.9 | 7.5 | 5.0 | 60.4 | 280.2 | 160.6 |
| 6 | 2.2 | 1.5 | 1.7 | 39.6 | 27.5 | 33.1 |
| 7 | 0.65 | 1.5 | 1.1 | 8.7 | 17.4 | 13.6 |
| 8 | 7.4 | 11.5 | 9.1 | 199.8 | 203.6 | 191.4 |
| Mean | 3.40 | 5.72 | 4.50 | 74.63 | 124.70 | 91.47 |

Calibration and Validation of HEC-HMS Model

The average optimal values of the CN and lag time were calculated based on the calibration of the HEC-HMS model in rainfall–runoff events. These values were determined under three ASM conditions and are presented in Table 6.

**Table 6.** Average optimized values of CN variable and lag time in three ASM conditions: I, II, and III.

| Sub-Watershed | CN | | | Lag Time (h) | | |
|---|---|---|---|---|---|---|
| | I | II | III | I | II | III |
| SW1 | 62.10 | 79.60 | 89.97 | 213.45 | 131.70 | 91.40 |
| SW2 | 58.57 | 77.10 | 88.56 | 304.79 | 185.47 | 126.23 |
| SW3 | 56.94 | 75.90 | 87.86 | 112.35 | 67.97 | 45.87 |
| SW4 | 61.09 | 78.90 | 89.58 | 156.78 | 96.34 | 66.48 |
| SW5 | 55.09 | 74.50 | 87.04 | 211.69 | 127.27 | 85.08 |
| SW6 | 60.67 | 78.60 | 89.41 | 117.81 | 72.27 | 49.75 |
| SW7 | 63.12 | 80.30 | 90.36 | 187.13 | 115.96 | 80.95 |
| SW8 | 63.27 | 80.40 | 90.41 | 110.15 | 68.30 | 47.72 |
| SW9 | 59.68 | 77.90 | 89.01 | 106.93 | 65.34 | 44.74 |
| SW10 | 68.48 | 83.80 | 92.24 | 122.40 | 77.81 | 56.15 |
| SW11 | 56.54 | 75.60 | 87.69 | 64.31 | 38.85 | 26.16 |
| Whole watershed | 60.67 | 78.60 | 89.41 | 224.77 | 137.88 | 94.91 |

During the validation stage, the calibrated variables were entered into the model, and it was run for three other events. The results of the model implementation in the validation stage are shown as observed, and the modeled flow hydrographs are shown in Figure 5. A visual evaluation of the observed and simulated hydrographs in both the implementation and validation stages demonstrated the accuracy of the model in simulating the flow hydrograph. Based on these results, it can be concluded that the model performance is good. Additionally, a visual comparison of the results of the two strategies in the hydrograph simulation revealed that the second strategy produced more accurate results. This means that the hydrographs related to the second strategy in all three events were closer to the observed hydrographs. Therefore, using the RS method to determine the ASM in the watershed leads to more accurate results in predicting the runoff.

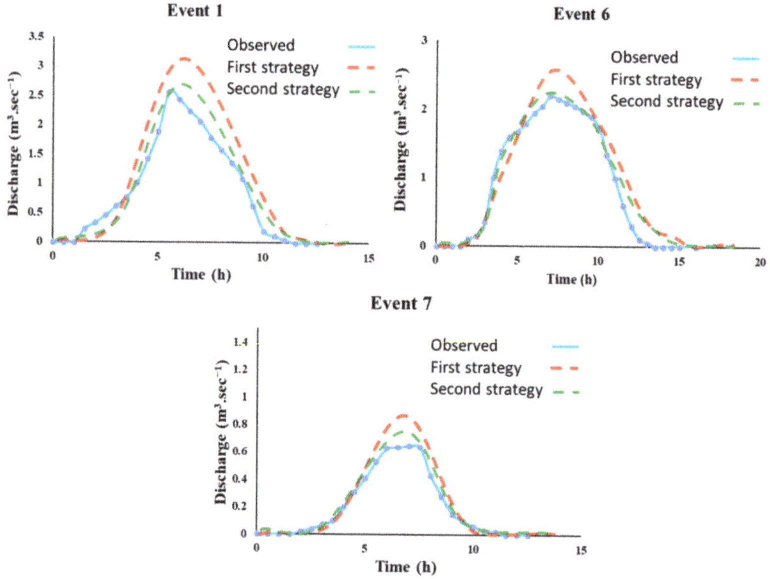


**Figure 5.** Observed and modeled flow hydrographs based on the two strategies used in the research for the corresponding rainfall–runoff events at 3 different dates after model calibration.

The results of the flood volume and peak discharge of the modeled and observed hydrographs in the model validation stage for the three corresponding rainfall–runoff events are presented in Table 7. The peak flow rate and the flood volume of the hydrograph simulated with the second strategy were closer to the actual values. For example, in event 1, the differences between the modeled peak discharge values using the first and second strategies and the observed values were 0.6 and 0.1 $m^3{\cdot}s^{-1}$, respectively. Similarly, the differences between the modeled flood volume using the first and second strategies and the observational values were 12.8 and 3.8 thousand $m^3$, respectively. These results indicate a better performance using the second strategy, which was based on RS.

**Table 7.** Peak flow rates and flood volume of observational and modeled hydrographs using two different strategies in three corresponding rainfall–runoff events.

| Event | Peak Discharge ($m^3{\cdot}s^{-1}$) | | | Flood Volume (1000 $m^3$) | | |
|---|---|---|---|---|---|---|
| | **Observed** | **First Strategy** | **Second Strategy** | **Observed** | **First Strategy** | **Second Strategy** |
| 1 | 2.5 | 3.1 | 2.6 | 34.8 | 47.6 | 38.6 |
| 6 | 2.2 | 2.6 | 2.3 | 39.6 | 49.8 | 45.5 |
| 7 | 0.65 | 0.85 | 0.75 | 8.7 | 9.6 | 9.1 |

In the next step, the accuracy of the HEC-HMS model in simulating the flow hydrograph was evaluated using the CNS, MAE, and RMSE evaluation indices (Table 8). Additionally, scatterplots and 1:1 lines are presented in Figure 6. The mean CNS, MAE, and

RMSE statistical criteria for the three events in the validation stage indicate the ability of the HEC-HMS model to simulate the flow hydrograph in the Balikhli-Chay watershed. The mean values of the evaluation indices (CNS, MAE, and RMSE) for the three events (first strategy and second strategy) were (0.63, 0.74), (0.27, 0.09), and (0.47, 0.23), respectively. Based on these results, it can be concluded that the HEC-HMS model has the ability to simulate the flow hydrograph in the Balikhli-Chay watershed. Previous studies have also shown that this model can provide satisfactory performance in estimating the runoff. However, this model typically includes a significant number of conceptual variables that are difficult to measure directly. Uncertainty in runoff estimation is often related to the choice of hydro-meteorological input for the hydrological model. The ASM is one of the known sources of uncertainty in runoff estimation in rainfall–runoff models [5]. For this reason, many studies have performed rainfall–runoff simulations without including ASM data in the model. However, various studies have also demonstrated that utilizing ASM information obtained from physical methods, RS, etc., can significantly enhance the accuracy of runoff estimation [8–10,17,21,22].

**Table 8.** Values of runoff accuracy assessment indices modeled based on different strategies.

| Event | Assessment Indices | | | | | |
|---|---|---|---|---|---|---|
| | $C_{NS}$ | | MAE ($m^3 \cdot s^{-1}$) | | RMSE ($m^3 \cdot s^{-1}$) | |
| | Strategy 1 | Strategy 2 | Strategy 1 | Strategy 2 | Strategy 1 | Strategy 2 |
| 1 | 0.41 | 0.56 | 0.38 | 0.13 | 0.66 | 0.35 |
| 6 | 0.69 | 0.78 | 0.31 | 0.11 | 0.55 | 0.23 |
| 7 | 0.80 | 0.89 | 0.13 | 0.05 | 0.21 | 0.11 |
| Mean | 0.63 | 0.74 | 0.27 | 0.09 | 0.47 | 0.23 |

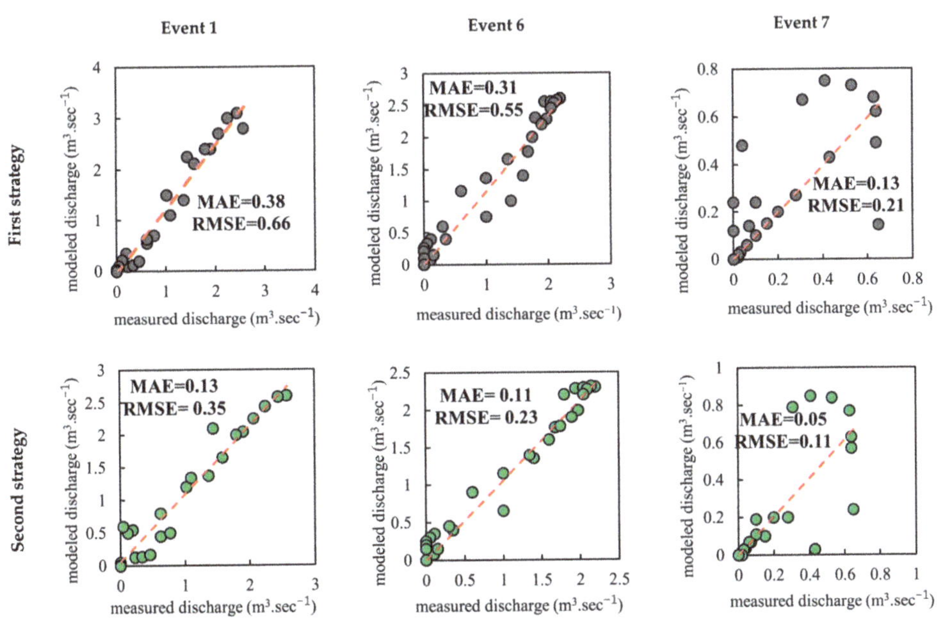


**Figure 6.** Validation of the model using the absolute mean error (MAE) and the root mean square error (RMSE) between the measured and modeled discharge values based on the validation events.

### 4.4. The Impact of ASM on Runoff

The amount of runoff in a watershed is primarily influenced by the amount of rainfall. Increased rainfall in a basin can lead to a higher peak discharge and flood volume. However, factors such as the CN also affect the amount of runoff in a field, in addition to the amount of rainfall. The temporal variation in the runoff for a watershed depends on the amount of rainfall and the ASM. To investigate the effect of SM on the runoff, the regression relationship between the discharge and measured flood volume was first determined for each event, using the recorded rainfall at the meteorological station and the measured flow and flood volume, while estimating the SM using RS. Additionally, the relationship between the flow and measured flood volume was simultaneously obtained for both the station-measured rainfall and the estimated SM. The $R^2$ value between the flow rate and the modeled flood was calculated for each of the three mentioned cases for all eight events (Figure 7).

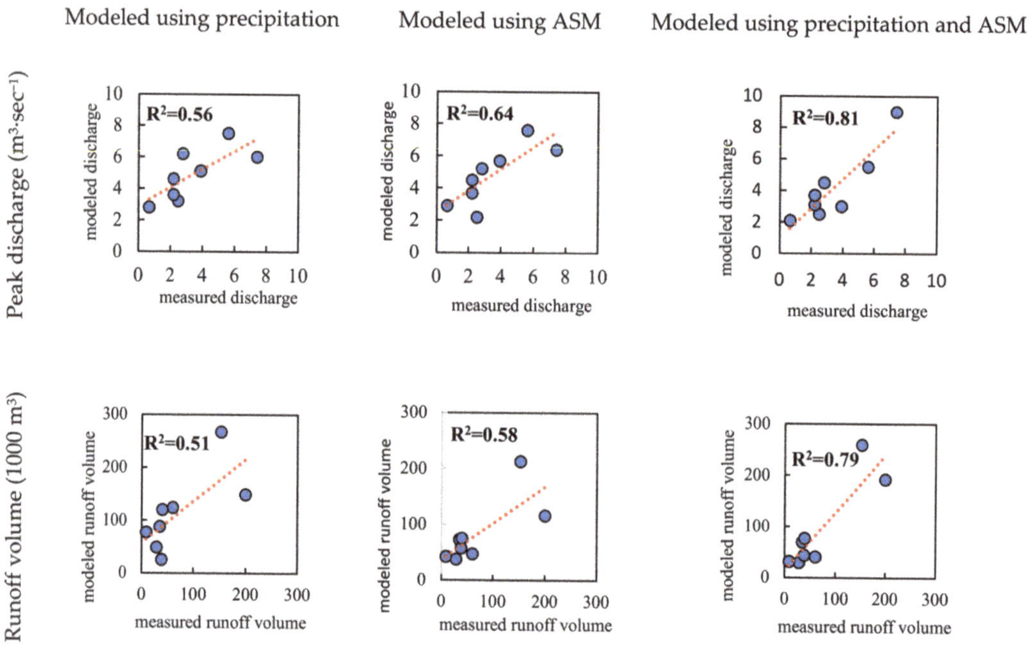


**Figure 7.** Determination coefficient ($R^2$) between measured and modeled peak flow and flood volume in three different modes ($n = 8$).

The results indicate that the ASM has a significant impact on the amount of runoff, with the runoff increasing as the ASM increases. The $R^2$ value between the observed and modeled peak flow rate (flood volume) was 0.56 (0.51) based on precipitation and 0.64 (0.58) based on the ASM. Considering both factors (precipitation and the ASM) resulted in an $R^2$ value of 0.81 (0.79), which represents an increase of approximately 0.25 (0.28) in the $R^2$ value compared to using precipitation information alone, and an increase of about 0.17 (0.21) compared to using ASM information alone. Zhang et al. [8] demonstrated that a 1% change in the ASM led to a 0.05 mm change in the runoff. Additionally, Brocca et al. [23] showed that incorporating the ASM variable in the rainfall–runoff model improved the accuracy of estimating both the volume and peak discharge of the runoff.

## 5. Conclusions

Effective water runoff management is crucial for the hydrological cycle, ecological conditions, water resource conservation, soil preservation, flood management, and other aspects of a watershed. Estimating the runoff in mountainous and semi-arid areas presents greater challenges due to the heterogeneity of topographic and biophysical conditions on the surface. In these areas, the ASM is one of the influential variables affecting the runoff. Most conventional methods for estimating the ASM rely on precipitation data collected at ground stations. However, watersheds located in mountainous regions often lack a sufficient and dense network of rain gauge stations. Therefore, the objective of this study is to utilize the ASM derived from satellite imagery to enhance the accuracy of runoff estimation in a mountainous watershed. In conclusion, this study demonstrates the advantages of using RS data and satellite imagery for estimating the runoff at the watershed scale. The following conclusions can be made based on the findings of this study:

- There is a correlation between the SM and soil properties, such as soil texture.
- The correlation between the ASM and runoff is higher than the correlation between precipitation and runoff. When considering both factors simultaneously, the correlation is even higher than their individual effects on the runoff.
- The results demonstrate that the use of RS techniques, which provide better temporal and spatial coverage of an area, can offer more accurate and comprehensive information about the ASM.
- The model's efficiency in predicting the runoff using RS-derived SM is superior to using precipitation information from meteorological stations.
- The results indicate a strong correlation between the SM in the watershed and the runoff.

**Author Contributions:** Conceptualization, S.F., A.R.V., A.G. and S.K.A.; methodology, S.F., A.R.V. and M.K.F.; software, S.F. and M.K.F.; data curation, S.F.; writing—original draft preparation, S.F.; writing—review and editing, S.K.A., A.G. and A.B. All authors have read and agreed to the published version of the manuscript.

**Funding:** This research received no external funding.

**Institutional Review Board Statement:** Not applicable.

**Informed Consent Statement:** Not applicable.

**Data Availability Statement:** The data used to support the findings of this study are available from the corresponding authors upon reasonable request.

**Acknowledgments:** The authors would like to thank the University of Zanjan for data support and for helping during the field experimental set-up and data collection.

**Conflicts of Interest:** The authors declare no conflict of interest.

## References

1. Feng, Z.-K.; Niu, W.-J.; Tang, Z.-Y.; Jiang, Z.-Q.; Xu, Y.; Liu, Y.; Zhang, H.-R. Monthly runoff time series prediction by variational mode decomposition and support vector machine based on quantum-behaved particle swarm optimization. *J. Hydrol.* **2020**, *583*, 124627. [CrossRef]
2. Pappas, E.; Smith, D.; Huang, C.; Shuster, W.; Bonta, J. Impervious surface impacts to runoff and sediment discharge under laboratory rainfall simulation. *Catena* **2008**, *72*, 146–152. [CrossRef]
3. Xu, C.; Rahman, M.; Haase, D.; Wu, Y.; Su, M.; Pauleit, S. Surface runoff in urban areas: The role of residential cover and urban growth form. *J. Clean. Prod.* **2020**, *262*, 121421. [CrossRef]
4. Ali, G.; Tetzlaff, D.; McDonnell, J.J.; Soulsby, C.; Carey, S.; Laudon, H.; McGuire, K.; Buttle, J.; Seibert, J.; Shanley, J. Comparison of threshold hydrologic response across northern catchments. *Hydrol. Process.* **2015**, *29*, 3575–3591. [CrossRef]
5. Chifflard, P.; Kranl, J.; zur Strassen, G.; Zepp, H. The significance of soil moisture in forecasting characteristics of flood events. A statistical analysis in two nested catchments. *J. Hydrol. Hydromech.* **2018**, *66*, 1–11. [CrossRef]
6. Abdulkareem, J.H.; Pradhan, B.; Sulaiman, W.N.A.; Jamil, N.R. Quantification of runoff as influenced by morphometric characteristics in a rural complex catchment. *Earth Syst. Environ.* **2018**, *2*, 145–162. [CrossRef]

7. Chifflard, P.; Moulding, D.; Petri, J.-T.; Zemke, J.J.; Reiss, M. Surface runoff of horse grazed pasture–a disregarded hydrological response unit in low mountain ranges. *DIE ERDE–J. Geogr. Soc. Berl.* **2018**, *149*, 76–85.
8. Zhang, Y.; Wei, H.; Nearing, M. Effects of antecedent soil moisture on runoff modeling in small semiarid watersheds of southeastern Arizona. *Hydrol. Earth Syst. Sci.* **2011**, *15*, 3171–3179. [CrossRef]
9. Tramblay, Y.; Bouvier, C.; Martin, C.; Didon-Lescot, J.-F.; Todorovik, D.; Domergue, J.-M. Assessment of initial soil moisture conditions for event-based rainfall–runoff modelling. *J. Hydrol.* **2010**, *387*, 176–187. [CrossRef]
10. Jadidoleslam, N.; Mantilla, R.; Krajewski, W.F.; Goska, R. Investigating the role of antecedent SMAP satellite soil moisture, radar rainfall and MODIS vegetation on runoff production in an agricultural region. *J. Hydrol.* **2019**, *579*, 124210. [CrossRef]
11. Liu, J.; Engel, B.A.; Wang, Y.; Wu, Y.; Zhang, Z.; Zhang, M. Runoff response to soil moisture and micro-topographic structure on the plot scale. *Sci. Rep.* **2019**, *9*, 2532. [CrossRef]
12. Zhuo, L.; Han, D. Could operational hydrological models be made compatible with satellite soil moisture observations? *Hydrol. Process.* **2016**, *30*, 1637–1648. [CrossRef]
13. Brocca, L.; Tullo, T.; Melone, F.; Moramarco, T.; Morbidelli, R. Catchment scale soil moisture spatial–temporal variability. *J. Hydrol.* **2012**, *422*, 63–75. [CrossRef]
14. Moradkhani, H. Hydrologic remote sensing and land surface data assimilation. *Sensors* **2008**, *8*, 2986–3004. [CrossRef]
15. Lei, W.; Dong, H.; Chen, P.; Lv, H.; Fan, L.; Mei, G. Study on Runoff and Infiltration for Expansive Soil Slopes in Simulated Rainfall. *Water* **2020**, *12*, 222. [CrossRef]
16. USDA, S. Soil Conservation Service (Chapters 4–10). In *National Engineering Handbook, Hydrology Section 4*; US Department of Agriculture: Washington DC, USA, 1972; Volume 4, pp. 4–10.
17. Shi, W.; Wang, N. An improved SCS-CN method incorporating slope, soil moisture, and storm duration factors for runoff prediction. *Water* **2020**, *12*, 1335. [CrossRef]
18. Soulis, K.; Valiantzas, J. Variation of runoff curve number with rainfall in heterogeneous watersheds. The Two-CN system approach. *Hydrol. Earth Syst. Sci.* **2012**, *16*, 1001–1015. [CrossRef]
19. Alexakis, D.D.; Mexis, F.-D.K.; Vozinaki, A.-E.K.; Daliakopoulos, I.N.; Tsanis, I.K. Soil moisture content estimation based on Sentinel-1 and auxiliary earth observation products. A hydrological approach. *Sensors* **2017**, *17*, 1455. [CrossRef]
20. Crow, W.T.; Bindlish, R.; Jackson, T.J. The added value of spaceborne passive microwave soil moisture retrievals for forecasting rainfall-runoff partitioning. *Geophys. Res. Lett.* **2005**, *32*, L18401. [CrossRef]
21. Shi, W.; Huang, M.; Gongadze, K.; Wu, L. A modified SCS-CN method incorporating storm duration and antecedent soil moisture estimation for runoff prediction. *Water Resour. Manag.* **2017**, *31*, 1713–1727. [CrossRef]
22. Mishra, S.K.; Jain, M.K.; Singh, V. Evaluation of the SCS-CN-based model incorporating antecedent moisture. *Water Resour. Manag.* **2004**, *18*, 567–589. [CrossRef]
23. Brocca, L.; Melone, F.; Moramarco, T.; Singh, V. Assimilation of observed soil moisture data in storm rainfall-runoff modeling. *J. Hydrol. Eng.* **2009**, *14*, 153–165. [CrossRef]
24. Behtari, M.; Vaezi, A. Effect of Initial Moisture on Runoff Generation and Soil Loss in Different Soil Textures under Simulated Rainfall Condition. *Iran. J. Watershed Manag. Sci. Eng.* **2018**, *11*, 11–21.
25. Ekwue, E.; Harrilal, A. Effect of soil type, peat, slope, compaction effort and their interactions on infiltration, runoff and raindrop erosion of some Trinidadian soils. *Biosyst. Eng.* **2010**, *105*, 112–118. [CrossRef]
26. Fu, X.; Zhang, L.; Wang, Y. Effect of Slope Length and Rainfall Intensity on Runoff and Erosion Conversion from Laboratory to Field. *Water Resour.* **2019**, *46*, 530–541. [CrossRef]
27. Fathololoumi, S.; Vaezi, A.R.; Alavipanah, S.K.; Ghorbani, A.; Biswas, A. Comparison of spectral and spatial-based approaches for mapping the local variation of soil moisture in a semi-arid mountainous area. *Sci. Total Environ.* **2020**, *724*, 138319. [CrossRef]
28. Fan, F.; Deng, Y.; Hu, X.; Weng, Q. Estimating composite curve number using an improved SCS-CN method with remotely sensed variables in Guangzhou, China. *Remote Sens.* **2013**, *5*, 1425–1438. [CrossRef]
29. Wadsworth, E.; Champagne, C.; Berg, A.A. Evaluating the utility of remotely sensed soil moisture for the characterization of runoff response over Canadian watersheds. *Can. Water Resour. J./Rev. Can. Des Ressour. Hydr.* **2020**, *45*, 77–89. [CrossRef]
30. Kemper, W.; Rosenau, R. Aggregate stability and size distribution. In *Methods of Soil Analysis: Part 1 Physical and Mineralogical Methods*; Wiley: Hoboken, NJ, USA, 1986; Volume 5, pp. 425–442.
31. Berk, A.; Conforti, P.; Kennett, R.; Perkins, T.; Hawes, F.; van den Bosch, J. MODTRAN® 6: A major upgrade of the MODTRAN® radiative transfer code. In Proceedings of the 2014 6th Workshop on Hyperspectral Image and Signal Processing: Evolution in Remote Sensing (WHISPERS), Lausanne, Switzerland, 24–27 June 2014; pp. 1–4.
32. Fathololoumi, S.; Vaezi, A.R.; Alavipanah, S.K.; Ghorbani, A.; Saurette, D.; Biswas, A. Effect of multi-temporal satellite images on soil moisture prediction using a digital soil mapping approach. *Geoderma* **2021**, *385*, 114901. [CrossRef]
33. Firozjaei, M.K.; Kiavarz, M.; Nematollahi, O.; Karimpour Reihan, M.; Alavipanah, S.K. An evaluation of energy balance parameters, and the relations between topographical and biophysical characteristics using the mountainous surface energy balance algorithm for land (SEBAL). *Int. J. Remote Sens.* **2019**, *40*, 5230–5260. [CrossRef]
34. Fathololoumi, S.; Firozjaei, M.K.; Biswas, A. Improving spatial resolution of satellite soil water index (SWI) maps under clear-sky conditions using a machine learning approach. *J. Hydrol.* **2022**, *615*, 128709. [CrossRef]

35. Were, K.; Bui, D.T.; Dick, Ø.B.; Singh, B.R. A comparative assessment of support vector regression, artificial neural networks, and random forests for predicting and mapping soil organic carbon stocks across an Afromontane landscape. *Ecol. Indic.* **2015**, *52*, 394–403. [CrossRef]
36. Ließ, M.; Glaser, B.; Huwe, B. Uncertainty in the spatial prediction of soil texture: Comparison of regression tree and Random Forest models. *Geoderma* **2012**, *170*, 70–79. [CrossRef]
37. Breiman, L. Random forests. *Mach. Learn.* **2001**, *45*, 5–32. [CrossRef]
38. Da Silva Chagas, C.; de Carvalho Junior, W.; Bhering, S.B.; Calderano Filho, B. Spatial prediction of soil surface texture in a semiarid region using random forest and multiple linear regressions. *Catena* **2016**, *139*, 232–240. [CrossRef]
39. Babaeian, E.; Sadeghi, M.; Franz, T.E.; Jones, S.; Tuller, M. Mapping soil moisture with the OPtical TRApezoid Model (OPTRAM) based on long-term MODIS observations. *Remote Sens. Env.* **2018**, *211*, 425–440. [CrossRef]
40. Prakash, R.; Singh, D.; Pathak, N.P. The effect of soil texture in soil moisture retrieval for specular scattering at C-band. *Prog. Electromagn. Res.* **2010**, *108*, 177–204. [CrossRef]

*sustainability*

MDPI

*Article*

# A Generalized Regression Neural Network Model for Accuracy Improvement of Global Precipitation Products: A Climate Zone-Based Local Optimization

**Saeid Mohammadpouri [1], Mostafa Sadeghnejad [2], Hamid Rezaei [3], Ronak Ghanbari [4], Safiyeh Tayebi [5], Neda Mohammadzadeh [2], Naeim Mijani [6], Ahmad Raeisi [7], Solmaz Fathololoumi [8,*] and Asim Biswas [8]**

1 Department of Remote Sensing and GIS, University of Tabriz, Tabriz 5166616471, Iran; saeid.m88a77a@gmail.com
2 Department of Geography and Geospatial Sciences, Kansas State University, 920 N17th Street, Manhattan, KS 66506-2904, USA; sadeghnejad@k-state.edu (M.S.); nedamohamadzadeh@k-state.edu (N.M.)
3 Department of Civil and Environmental Engineering, Florida International University, Miami, FL 33174, USA; hreza004@fiu.edu
4 Graduate Research in Remote Sensing, University of Iowa, Iowa City, IA 52242, USA; ruonak.ghanbari@gmail.com
5 Faculty of Geography, University of Tehran, Tehran 1417853933, Iran; tayebi.s@alumni.it.ac.ir
6 Department of Remote Sensing and GIS, University of Tehran, Tehran 1417853933, Iran; naeim.mijani@ut.ac.ir
7 Department of Electrical and Computer Engineering, University of Tehran, Tehran 1439957131, Iran; ahmadraeisi19@ut.ac.ir
8 School of Environmental Sciences, University of Guelph, Guelph, ON N1G 2W1, Canada; biswas@uoguelph.ca
* Correspondence: sfatholo@uoguelph.ca

**Abstract:** The ability to obtain accurate precipitation data from various geographic locations is crucial for many applications. Various global products have been released in recent decades for estimating precipitation spatially and temporally. Nevertheless, it is extremely important to provide reliable and accurate products for estimating precipitation in a variety of environments. This is due to the complexity of topographic, climatic, and other factors. This study proposes a multi-product information combination for improving precipitation data accuracy based on a generalized regression neural network model using global and local optimization strategies. Firstly, the accuracy of ten global precipitation products from four different categories (satellite-based, gauge-corrected satellites, gauge-based, and reanalysis) was assessed using monthly precipitation data collected from 1896 gauge stations in Iran during 2003–2021. Secondly, to enhance the accuracy of the modeled precipitation products, the importance score of effective and auxiliary variables—such as elevation, the Enhanced Vegetation Index (EVI), the Land Surface Temperature (LST), the Soil Water Index (SWI), and interpolated precipitation maps—was assessed. Finally, a generalized regression neural network (GRNN) model with global and local optimization strategies was used to combine precipitation information from several products and auxiliary characteristics to produce precipitation data with high accuracy. Global precipitation products scored higher than interpolated precipitation products and surface characteristics. Furthermore, the importance score of the interpolated precipitation products was considerably higher than that of the surface characteristics. SWI, elevation, EVI, and LST scored 53%, 20%, 15%, and 12%, respectively, in terms of importance. The lowest RMSE values were associated with IMERGFinal, TRMM3B43, PERSIANN-CDR, ERA5, and GSMaP-Gauge. For precipitation estimation, these products had Kling–Gupta efficiency (KGE) values of 0.89, 0.86, 0.77, 0.78, and 0.60, respectively. The proposed GRNN-based precipitation product with a global (local) strategy showed RMSE and KGE values of 9.6 (8.5 mm/mo) and 0.92 (0.94), respectively, indicating higher accuracy. Generally, the accuracy of global precipitation products varies depending on climatic conditions. It was found that the proposed GRNN-derived precipitation product is more efficient under different climatic conditions than global precipitation products. Moreover, the local optimization strategy based on climatic classes outperformed the global optimization strategy.

**Keywords:** global precipitation products; surface properties; climate-based local optimization; generalized regression neural network

## 1. Introduction

Various applications require high-precision precipitation data, including weather predictions, extreme climate event monitoring, water resources management, agricultural management, and urban planning [1–6]. Ground stations collect the most reliable and accurate precipitation data. However, ground station-derived precipitation data face challenges such as a limited topographical distribution and a relatively short time series. In addition, these stations are converting point-based precipitation data into area-based precipitation information.

A series of satellites equipped with various sensors monitor the earth from space at specific time intervals, each capable of obtaining unique data from the components of the earth system (atmosphere, lithosphere, hydrosphere, and biosphere) [7–11]. To date, remote sensing satellite sensors have been widely used in a variety of fields [12–18]. There is a wide range of meteorological phenomena recorded by meteorological satellites over a wide range of geographical regions that can provide both quantitative and numerical data. These data provide a more in-depth understanding of dynamic weather conditions and precipitation systems [19–23].

Satellite-based precipitation products have become popular due to their high spatial and temporal resolution as well as their global coverage. Previous studies have revealed the significant potential of these products in a diverse range of climate-related fields [24–27]. While providing high temporal and spatial resolution, satellite data can also be used to estimate precipitation in highlands and other difficult-access locations [19,28–34]. Nevertheless, a variety of studies have shown that satellite products often provide information about precipitation with uncertainty and significant errors [35–37]. Underestimation and overestimation can significantly affect the quality of satellite products [25,38]. A significant part of the effectiveness of natural hazard and climate change prediction depends on the accuracy of precipitation products both spatially and temporally [4,39]. In this regard, it is necessary to investigate the accuracy of the currently available precipitation products before utilizing them for various purposes. Additionally, a strategy should be developed to enhance the accuracy of precipitation data derived from satellites.

Previous studies on precipitation can be reviewed from several different perspectives. Several studies have focused on evaluating the accuracy of precipitation products [26,29,33,40–43]. In a study conducted by Liu, Aryastana, Liu, and Huang [34], the performance of three global precipitation datasets for Bali Island was evaluated at various elevations, rainfall intensity levels, and temporal scales during 2015–2017. Camici et al. [44] showed that the performance of global precipitation products varies in different parts of the world, making it necessary to evaluate their performance before operational applications.

Several studies have evaluated the accuracy of precipitation satellite products for regional, sub-regional, and climatic zones [25,31]. There can be no conclusive claim that any particular precipitation satellite product is always more or significantly less accurate than another in different conditions. In this respect, combining satellite-based products with ground and supplementary data can yield a more accurate precipitation product under a variety of conditions [43,45,46]. Global precipitation products can be divided into four categories: satellite-based, gauge-corrected satellites, gauge-based, and reanalysis products [25,43]. Due to the combination of several data sources in recent decades, different global precipitation products have been made available, such as Precipitation Estimation from Remotely Sensed Information using Artificial Neural Networks (PERSIANN) [35], Tropical Rainfall Measuring Mission (TRMM), US National Climate Service Forecasting Center (CMORPH) [47], Global Precipitation Measurement (GPM), Multi-Satellite Precipitation Analysis (TMPA) [48], Multi-Satellite Integrated Recovery for GPM (IMERG) [49], and the Naval Research Laboratory's (NRL) blended satellite (NRL-Blend) for precipitation estimates from laboratory NRL Naval Research [50]. Some studies have focused on developing methods and products based on gauge data, satellite data, or their combination [35,46,51]. According to Ceccherini et al. [46], a downscaling method has been developed that utilizes the Digital Elevation Model (DEM) and Enhanced Vegetation Index (EVI) to quantify Mean Annual Precipitation (MAP) using various precipitation products based on gauge data and satellite-based predictions. Additionally, studies such as Chen et al. (2020) and Bui et al. (2019) showed that satellite precipitation products or base models can be adjusted with ground precipitation data. This leads to the generation of a more precise precipitation product using the advantages of each dataset. Oliazadeh et al. [52] developed an algorithm for the optimal integration of different precipitation products to enable more precise estimations. They evaluated and combined the PERSIANN-CDR, TMPA-3B42, GPM-IMERG, and GSMaP MKV (SBPs) products. In previous studies, various models based on machine learning algorithms have been used to produce more accurate products by combining different precipitation products and other effective characteristics. These models had varying efficiency levels in preparing the combined precipitation product under different conditions.

This study proposes a multi-product information combination for improving precipitation data accuracy based on a neural network model using global and local optimization strategies. The accuracy of several available products was assessed for this purpose. A generalized regression neural network (GRNN) model was then employed to combine multiple global products, gauge station data, and surface characteristics to create a more accurate and reliable precipitation product. The rationale behind the model presented in this study is that a group of precipitation products can complement each other to provide additional information and improve results compared with a single product, which is supported by the collaborative decision-making principle [53]. Previous research has shown that the efficiency of different precipitation products varies under different climatic conditions. In line with the collaborative decision-making principle, this study focuses on using a group of precipitation products based on the fusion of multi-source information to improve their accuracy and performance compared with individual products.

## 2. Study Area

This study was performed in Iran. The characteristics of the study area and the location of the gauge stations are shown in Figure 1. Various criteria were considered for selecting the study area, including (1) the diversity of climatic conditions (five climates), (2) the diversity of topographic conditions (plain and mountainous areas), (3) the diversity of land cover types (built-up, bare land, agriculture, forest, grassland, shrubland, water body, etc.), and (4) the variation of weather conditions in different months. Located in the northwest of Asia (at latitude 25–40° N and longitude 44–64° E), Iran is the 18th-largest country in the world. It is connected to the Gulf of Oman and the Persian Gulf from the south and to the Caspian Sea from the north. Its area spans 1,648,195 km$^2$. The altitude of different geographical locations in Iran varies from −60 to 5590 m above sea level. The Alborz mountain range stretches from the northwest to the northeast, and the Zagros mountain range lies in the northwest-to-southwest direction. The central regions of Iran have arid climatic conditions and include the two large deserts of Kavir and Lut. Iran has a variety of climatic conditions, including humid, semi-arid, arid, and very arid.

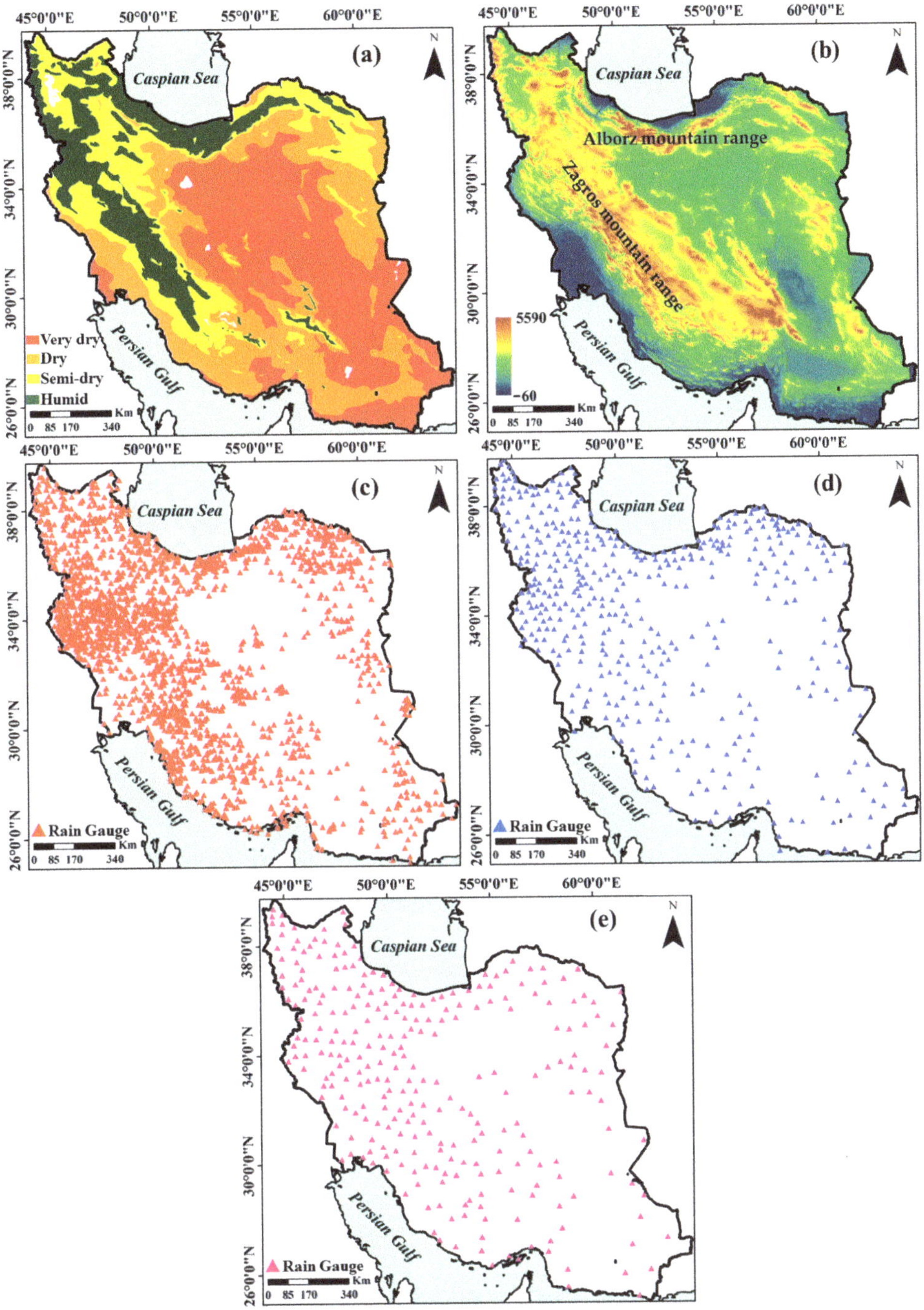


**Figure 1.** Maps of (**a**) climatic conditions, (**b**) DEM, and location of gauge stations in Iran that are used for (**c**) mapping precipitation using the interpolation method, (**d**) GRNN model calibration, and (**e**) GRNN model validation and validation of the global precipitation products.

## 3. Materials and Methods

### *3.1. Data*

This study used data from both ground and satellite sources, including precipitation data (monthly) collected by 1896 ground stations (https://www.irimo.ir/) (accessed on 18 August 2022), global precipitation products (https://giovanni.gsfc.nasa.gov/giovanni/) (accessed on 1 August 2022), the monthly Soil Water Index (SWI) (https://land.copernicus.eu/) (accessed on 27 August 2022), the monthly Enhanced Vegetation Index (EVI) (https://modis.gsfc.nasa.gov/data/dataprod/mod13.php) (accessed on 10 August 2022), the monthly Land Surface Temperature (LST) (https://modis.gsfc.nasa.gov/data/dataprod/mod11.php) (accessed on 22 August 2022), DEM (https://www.eorc.jaxa.jp/ALOS/en/dataset/aw3d30/aw3d30_e.htm) (accessed on 25 August 2022), and latitude data for the 2003–2021 period. Ground station information was collected by the Ministry of Energy and the Iranian Meteorological Organization (IRIMO). The initial number of stations was more than 4650. In the pre-processing step, station data were monitored for gaps, coordinates, and elevation. Afterwards, 1896 distinct stations with monthly records of precipitation were selected during the 2003–2021 period. Then, the precipitation stations were grouped into three categories as follows: (1) 1122 stations used for mapping precipitation using interpolation (Figure 1c); (2) 447 stations used for GRNN model calibration (Figure 1d); and (3) 327 stations used for both GRNN model validation and the validation of the precipitation products (Figure 1e). Random sampling was used to assign stations to different categories. To avoid cherry-picking, the user was not involved in selecting stations for each group.

Table 1 provides an overview of the global precipitation products used in this study. Although there are a wide variety of global precipitation products, this study focuses on those currently available. The MODIS MODIS11C3 products with a spatial resolution of 5000 m were used to assess the effect of surface temperature on precipitation modeling (accessible at the LAADS DAAC website). Additionally, Metop ASCAT's SWI product with a spatial resolution of 10,000 m was utilized to calculate soil moisture for precipitation modeling (accessible at the Copernicus global land service website). In the precipitation modeling process, an elevation variable was represented by the DEM of ALOS World 3D, which has a 30 m spatial resolution (accessible at the JAXA website). A cubic convolution interpolation method was used to convert the spatial resolutions of different datasets to 25 km.

**Table 1.** An overview of global precipitation products used in this study.

| Product | Type | Spatial Resolution | Temporal Resolution | Spatial Extent | Temporal Period | Reference |
|---|---|---|---|---|---|---|
| PRECL | Gauge-based | 0.5° × 0.5° | 1 mo | Global land | 1948–present | [54] |
| PERSIANN | Satellite-based | 0.25° × 0.25° | 1, 3, 6 h/1 d | 60° S–60° N | 2000–present | [35] |
| PERSIANN-CCS | | 0.04° × 0.04° | 1, 3, 6 h/1 d | 60° S–60° N | 2003–present | [55] |
| GSMaP-NRT | | 0.1° × 0.1° | 1 h/1 d | 60° S–60° N | 2000–present | [56] |
| PERSIANN-CDR | Gauge corrected satellites | 0.25° × 0.25° | 1 d/1 mo | 60° S–60° N | 1983-present | [57] |
| TRMM3B43 | | 0.25° × 0.25° | 3 h/1 d | 50° S–50° N | 1998–present | [48] |
| GSMaP-Gauge | | 0.1° × 0.1° | 1 h/1 d | 60° S–60° N | 2002–present | [58] |
| IMERGFinal | | 0.1° × 0.1° | 30 min | Global | Jun. 2000–present | [25] |
| ERA5 | Reanalysis | 31 km | 1 h/1 mo | Global | 1979–present | [59] |
| CHIRPS | | 0.05° × 0.05° | 1d | 50° S–50° N | 1981–present | [60] |

### *3.2. Methods*

Figure 2 illustrates the methodological process of the proposed strategy for generating improved precipitation products. Initially, the accuracy of the precipitation products was assessed at the study area and climatic region scales were assessed using point-to-pixel and pixel-to-pixel strategies. The next step involved assessing the importance of various factors that affect precipitation modeling accuracy, such as precipitation interpolation maps, EVI, LST, SWI, elevation, and latitude. Thirdly, a generalized regression neural network (GRNN) machine learning algorithm was utilized to combine information from several global precipitation products and supplementary data to produce an improved precipitation product.

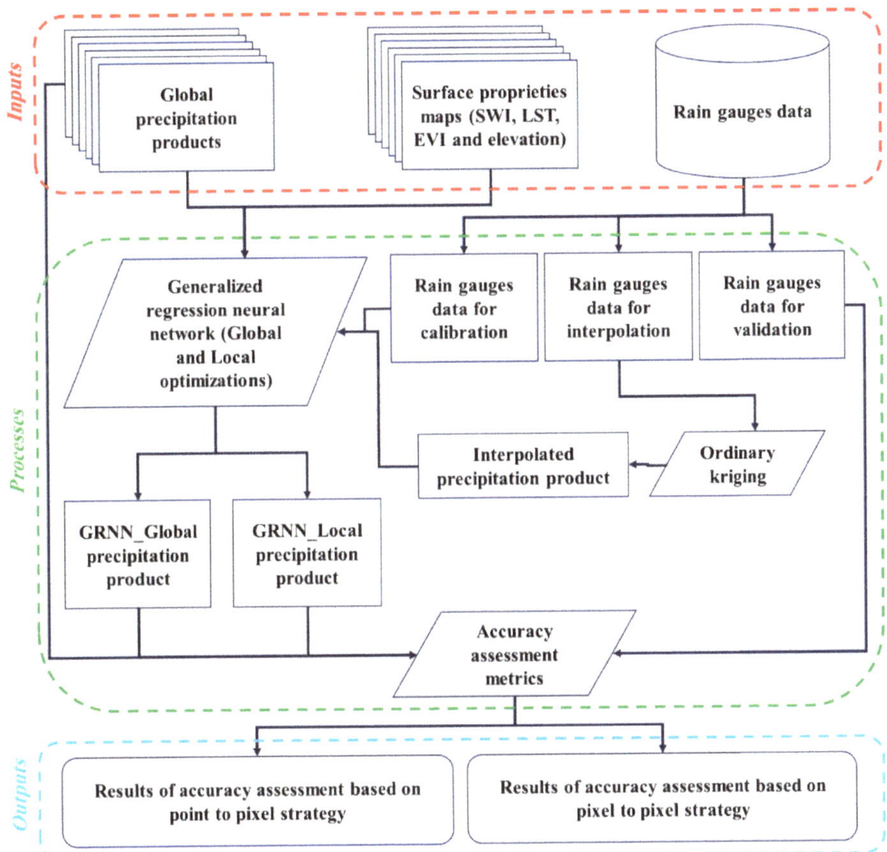


**Figure 2.** The methodology flowchart of the study.

### 3.2.1. Accuracy Assessment of Precipitation Products

To evaluate the accuracy of precipitation products, a variety of metrics and strategies based on ground data were employed in this study. The metrics included Relative Bias (RBias), Root Mean Squared Error (RMSE), Random Error (RE), Systematic Error (SE), Variability Ratio (VR), Correlation Coefficient (R), and Kling–Gupta Efficiency (KGE). Rbias measures the level of underestimation (negative values) or overestimation (positive values) of predictions. Errors are smaller when the index is close to zero. RMSE is used to calculate the magnitude of the average error. Lower RMSE values indicate higher performance. Pearson's correlation coefficient (R) is used to quantify the linear relationship between estimated and actual precipitation. The accuracy of the precipitation estimation increases as the R value increases. There are two types of errors in variable estimation: systematic and random errors. The term "random errors" refers to the unpredictable fluctuations in estimated precipitation as a result of measurements, while the term "systematic errors" refers to the predictable and repeatable errors in estimated precipitation. A KGE combines linear correlation, bias, and variability. In this case, the optimal value is 1, which can be calculated as follows:

$$\mathrm{KGE} = 1 - \sqrt{(\mathrm{r}-1)^2 + (\beta-1)^2 + (\gamma-1)^2} \tag{1}$$

where r is the Pearson correlation coefficient (optimal value = 1), β is the bias (optimal value = 1), and γ represents the variability ratio (optimal value = 1). The details of the calculation of these variables and evaluation metrics are presented in Saemian, Hosseini-Moghari, Fatehi, Shoarinezhad, Modiri, Tourian, Tang, Nowak, Bárdossy, and Sneeuw [25].

Two strategies were used to calculate each of these metrics: "point-to-pixel" and "pixel-to-pixel". A point-to-pixel approach was utilized to evaluate the accuracy of global precipitation products through ground data (from the validation map, Figure 1e). By using this strategy, accuracy assessment results were reported and displayed according to the ground station scale. A pixel-to-pixel approach was used to evaluate the accuracy of global precipitation products by comparing them with a precipitation map derived from ground data interpolation (Figure 1e). The Ordinary Kriging (OK) method was employed for interpolation [61]. The accuracy assessment results were reported and displayed at the pixel level using this strategy.

### 3.2.2. Climate Zone-Based Local Optimization Strategy

According to the proposed strategy, Equation (2) was used to estimate the amount of precipitation.

$$\begin{aligned}\mathrm{Precipitation}_{\mathrm{Proposed\ model}} \\ = \mathrm{f}(\mathrm{LST,\ Elevation,\ EVI,\ SWI,\ Global\ precipitation\ products, Interpolated\ precipitation\ map})\end{aligned} \tag{2}$$

The function f represents the relationship between the dependent variable (precipitation) and the independent variables. The GRNN model was utilized to calculate the function f.

The GRNN is based on radial basis functions and nonparametric regression analysis. With the help of the probability density function of the training data, the GRNN establishes a functional relationship between the dependent and independent variables [62,63]. GRNN, in addition to its nonlinear mapping ability and learning ability, can achieve convergence in regression analysis with a much larger sample size. The prediction output is generally very accurate when the sample size is small [64]. This study employed a K-fold cross-validation method to determine which GRNN parameters should be adjusted between the training and test data to achieve the lowest Mean Square Error (MSE). Five advantages of GRNN include (i) one-pass learning, which eliminates the need for backpropagation; (ii) a high accuracy in estimation since Gaussian functions are employed; (iii) the input noise can be handled; (iv) even with sparse data in a multidimensional measurement space, the model provides a smooth transition between observed values; and (v) it is a memory-based model. The GRNN model can be shown through Equation (3).

$$\mathrm{E}(\mathrm{Y}|\mathrm{X}) = \frac{\int_{-\infty}^{\infty} \mathrm{Y}\mathrm{f}(\mathrm{X.Y})\mathrm{dy}}{\int_{-\infty}^{\infty} \mathrm{f}(\mathrm{X.Y})\mathrm{dy}} \tag{3}$$

where X is the n-dimensional input vector, Y is the predicted value of the GRNN model, $\mathrm{E}(\mathrm{Y}|\mathrm{X})$ is the expected value of output Y according to the input vector X, and $\mathrm{f}(\mathrm{Y.X})$ is the joint probability density function of X and Y.

As part of its architecture, GRNN has four layers: input, pattern, aggregation, and output. After receiving information, the input layer stores the input vector X, which is equal to the number of neurons in the input vector. Afterward, neurons in the input layer feed data to the pattern layer. Input space to pattern space is transformed non-linearly by the pattern layer. Neurons in the pattern layer can memorize the relationship between input neurons and the appropriate response of the pattern layer. Moreover, the number of neurons is equal to the number of input variables. The Gaussian function ($p_i$) pattern is calculated using Equation (4).

$$p_i = \exp\left[-\frac{(X - X_i)^T(X - X_i)}{2\sigma^2}\right] \; (i = 1.2. \cdots .n) \tag{4}$$

where $\sigma$ is the smoothing parameter, X is the input variable of the network, and $X_i$ is the specific training vector of neuron i in the pattern layer. The aggregation layer has two addition operations: $S_s$ and $S_w$. The simple aggregation ($S_s$) performs the mathematical addition resulting from the pattern layer outputs, and its connection weight is equal to 1. In a weighted aggregation ($S_w$), the pattern layer outputs are summed together with their connection weight (w). The $S_s$ and $S_w$ can be determined from Equations (4) and (5).

$$S_s = \sum_{i=1} p_i \tag{5}$$

$$S_w = \sum_{i=1} w_i p_i \tag{6}$$

where $w_i$ is the weight of pattern neuron i which is connected to the aggregation layer. The number of neurons in the output layer is equal to the dimension k of the output vector Y. After aggregating the neurons in the aggregation layer, the Y output of the GRNN model can be calculated using Equation (7).

$$Y = \frac{S_s}{S_w} \tag{7}$$

In this study, a second group of ground data was used to calibrate the parameters of the GRNN model, which estimates precipitation (dependent variable) from independent variables (Figure 1d). Prior to implementing the GRNN model, an importance score was calculated for the variables, including global precipitation products, ground-based interpolated precipitation map, EVI, LST, SWI, and elevation. In order to reduce the processing volume and time, the global precipitation products with an importance degree of less than 5% were not included in the precipitation estimation process as required by the proposed model. In calibrating the proposed model, two scenarios were used: (1) a global scenario, where all ground data were used for determining the optimal structure of the GRNN; and (2) a local scenario, where ground data were used for determining the optimal structure of the GRNN based on the climate zone. In this scenario, the GRNN model was trained separately according to the climate zone (Figure 1a).

## 4. Results

### *4.1. Determining the Importance Score of Effective Variables*

The evaluation results of the importance scores for the effective variables in improving precipitation estimation accuracy are presented in Figure 3. The importance score for different variables was calculated based on the contribution of each one to reducing the MAE between the estimated and measured precipitation values (calibration data). The importance scores of the global precipitation products were higher than those of the interpolated precipitation product and the surface variables in generating a final precipitation product. Different climatic conditions resulted in varying importance scores for each variable. In semi-arid climates, the surface characteristics play a greater role in estimating precipitation than in other climates.

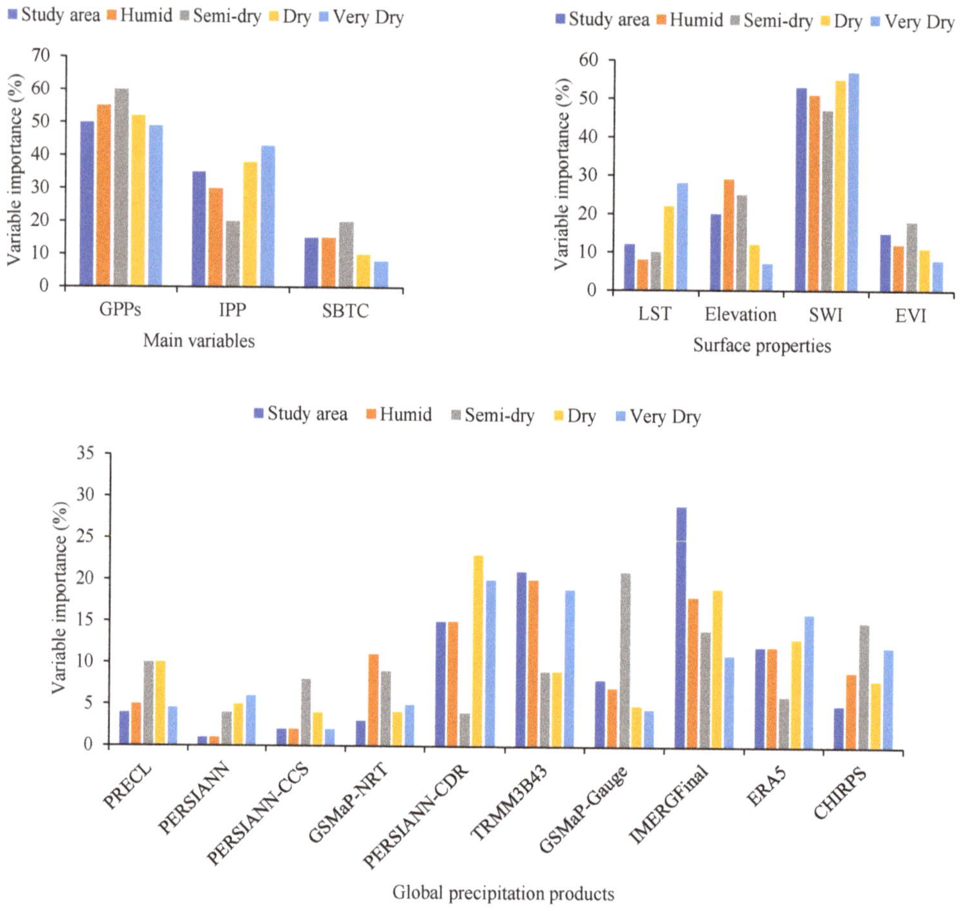


**Figure 3.** The effective variables and their relative importance in enhancing precipitation estimation accuracy (GPPs: global precipitation products; IPP: interpolated precipitation product; SPTC: surface biophysical and topographical characteristics; EVI: enhanced vegetation index; LST: land surface temperature; SWI: soil water index).

IMERGFinal, TRMM3B43, PERSIANN-CDR, and ERA5 had the highest importance scores concerning the final precipitation product across the study area. The contribution of global precipitation products to the GRNN model output varies by climate zone. In very arid, arid, semi-arid, and humid climates, TRMM3B43, GSMaP-Gauge, PERSIANN-CDR, and PERSIANN-CDR were the most significant, respectively. Global precipitation products with an importance degree of less than 5% were not used in the precipitation estimation process. Hence, PERSIANN, PERSIANN-CCS, and GSMaP-NRT products were not included in the estimation of precipitation based on the global strategy. Moreover, in the local strategy, PERSIANN and PERSIANN-CCS products (for humid climates), PERSIANN-CDR and PERSIANN products (for semi-arid climates), PERSIANN-CCS, GSMaP-NRT, and GSMaP-Gauge products (for dry climates), and PRECL, PERSIANN-CCS, and GSMaP-Gauge (for very dry climates) were ignored in the precipitation estimation process.

Compared with semi-arid and humid climates, LST had a higher importance score in very dry and arid climates. As for EVI, the highest importance score was achieved under semi-arid climatic conditions, while the lowest importance score was achieved under very dry conditions.

### 4.2. Evaluation of Global Precipitation Products by a Point-to-Pixel Approach

Table 2 presents the accuracy evaluation of different precipitation products using the data collected from the validation stations. Here, the dark green and dark red colors indicate the lower and higher values of various accuracy evaluation metrics, respectively. In terms of RMSE, PRECL had the lowest value and ERA5 the highest value. Among the evaluated precipitation products, the average RMSE values for the satellite-based, gauge-corrected satellites, gauge-based, and reanalysis products were 40.2, 34.4, 42.5, and 28.9 mm/mo, respectively. Reanalysis products and satellite-based products performed the best and the worst. Based on the proposed GRNN-Global (local) model, the precipitation estimate had an RMSE of 15.6 (13.2) mm/mo. The research results show that the developed product is more effective than existing products in providing precipitation estimates. In the proposed model, the local approach to optimization had a higher efficiency than the global approach. The highest and lowest RMSE values among the PERSIANN set of products were associated with PERSIANN-CDR and PERSIANN-CCR. A point-to-pixel evaluation revealed a lower precipitation estimation accuracy than a pixel-to-pixel evaluation for global precipitation products. There is a smaller variation in precipitation amounts derived from global products than actual precipitation amounts based on variability ratios (VRs) for all products. According to the Rbias statistic, precipitation values at PRECL, GSMaP-Gauge, and PERSIANN were less than actual precipitation amounts. While the precipitation values of CHIRPS, GSMaP-NRT, and PERSIANN-CCS were higher than the actual precipitation amounts recorded at ground stations. The KGE values calculated using a point-to-pixel strategy were lower than those calculated using a pixel-to-pixel strategy. A systematic error (SE) is typically greater than 0.5 in precipitation products; despite this, there were systematic errors (SE) of 0.28, 0.35, 0.33, and 0.40 for TRMM3B43, PERSIANN-CDR, ERA5, and IMERGFinal, respectively. The results of the proposed GRNN-Global and local products showed SE values of 0.18 and 0.15, respectively, indicating the developed product's remarkable reliability in estimating precipitation.

**Table 2.** Accuracy assessment results for the point-to-pixel approach (Different colors (from dark green to dark red) indicate different values (from lower to higher) of various accuracy evaluation metrics respectively).

| Product | KGE | R | Bias | VR | Rbias | RMSE | SE | RE |
|---|---|---|---|---|---|---|---|---|
| PRECL | 0.28 | 0.23 | 0.8 | 0.71 | −0.2 | 42.5 | 0.6 | 0.4 |
| PERSIANN | 0.08 | 0.2 | 0.67 | 0.73 | −0.33 | 41.5 | 0.77 | 0.23 |
| PERSIANN-CCS | −0.1 | 0.15 | 1.55 | 0.57 | 0.54 | 49.5 | 0.6 | 0.4 |
| GSMaP-NRT | 0.08 | 0.27 | 1.53 | 0.83 | 0.53 | 36.7 | 0.53 | 0.47 |
| PERSIANN-CDR | 0.1 | 0.29 | 1.01 | 0.81 | 0.15 | 35.9 | 0.35 | 0.65 |
| TRMM3B43 | 0.3 | 0.31 | 1.4 | 0.69 | 0.45 | 33.9 | 0.28 | 0.72 |
| GSMaP-Gauge | 0.25 | 0.3 | 0.94 | 0.83 | −0.06 | 33.6 | 0.51 | 0.49 |
| IMERGFinal | 0.32 | 0.38 | 1.03 | 0.71 | 0.03 | 28.5 | 0.4 | 0.47 |
| ERA5 | 0.21 | 0.46 | 1.33 | 0.53 | 0.33 | 25.5 | 0.33 | 0.47 |
| CHIRPS | 0.18 | 0.25 | 1.11 | 0.7 | 0.11 | 36 | 0.69 | 0.31 |
| GRNN_Global | 0.56 | 0.64 | 0.75 | 0.9 | 0.02 | 15.6 | 0.18 | 0.82 |
| GRNN_Local | 0.65 | 0.71 | 0.69 | 0.92 | 0.02 | 13.2 | 0.15 | 0.85 |

Figure 4 illustrates the geographic distribution of KGEs at validation sites over the study period. TRMM3B43, IMERG-Final, and PRECL offer better performance than others, with KGE averages of 0.30, 0.32, and 0.28. The PERSIANN set products exhibited low efficiency in estimating precipitation, with PERSIANN-CCS having negative KGE values at numerous stations, thus having the poorest performance among the set products. The precipitation estimation by the ERA5 product showed reasonable performance, especially in western Iran. However, this product performed poorly in the Alborz Mountains. Precipitation products generally performed better in western Iran, including the Zagros Mountains, than in other regions, especially in the Alborz Mountains and in the central and northwestern regions of the country. According to the proposed GRNN-Global (local) model, all the stations have KGE values greater than 0, with an average KGE of 0.56 (0.65), which indicates the superior performance of the proposed model.

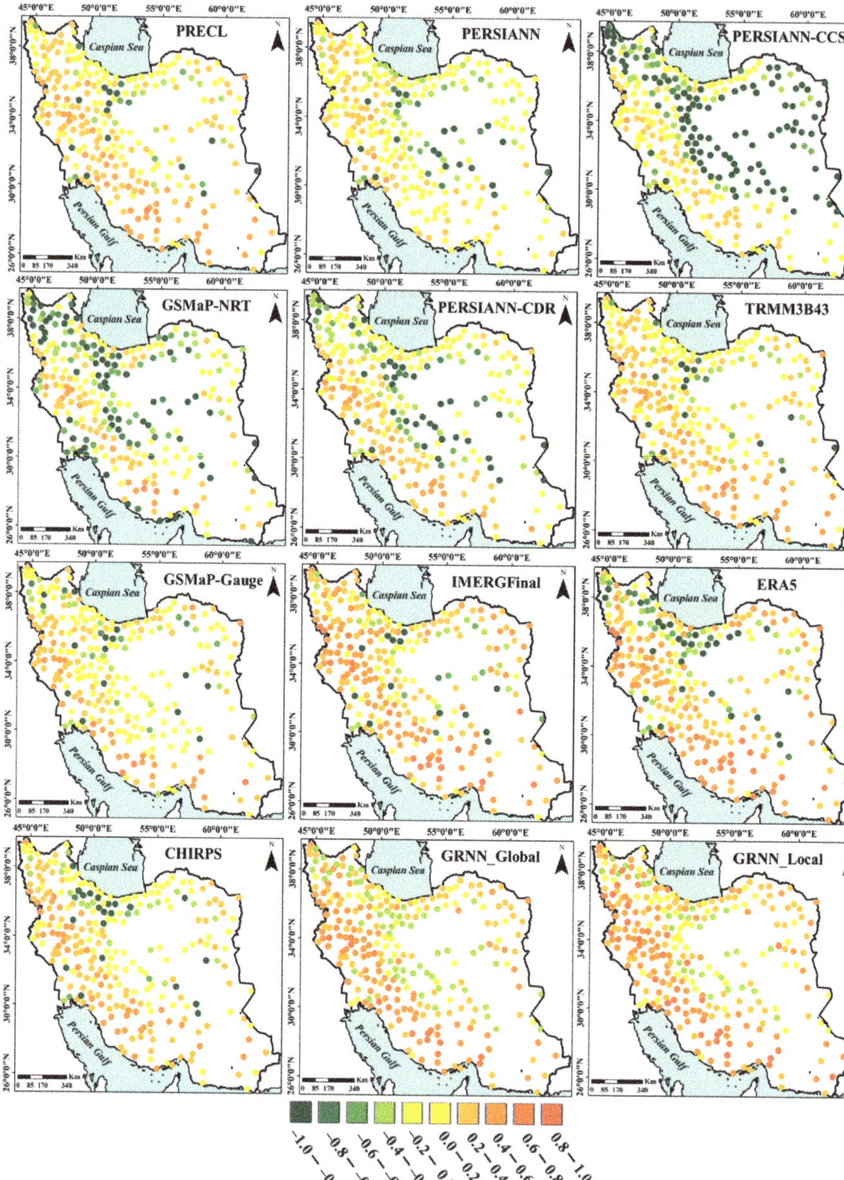


**Figure 4.** Geographic distribution of KGEs at validation sites based on point-to-pixel approach.

Figure 5 illustrates the KGE box plot of the evaluated precipitation products at the ground stations. IMERGFinal, TRMM3B43, PRECL, GSMaP-Gauge, and CHIRPS had higher KGE values than other products. GSMaP-NRT and PERSIANN-CCS had the lowest KGE values. In comparison with other global precipitation products, PRECL, TRMM3B43, and IMERGFinal had median values of 0.52, 0.30, and 0.41, respectively, which indicates their superior performance. The proposed products had a much narrower range of KGE values than the evaluated precipitation products. In addition, the median KGE values for the GRNN-Global and local products were 0.68 and 0.77, which are more accurate in estimating precipitation under different conditions than the evaluated precipitation products.

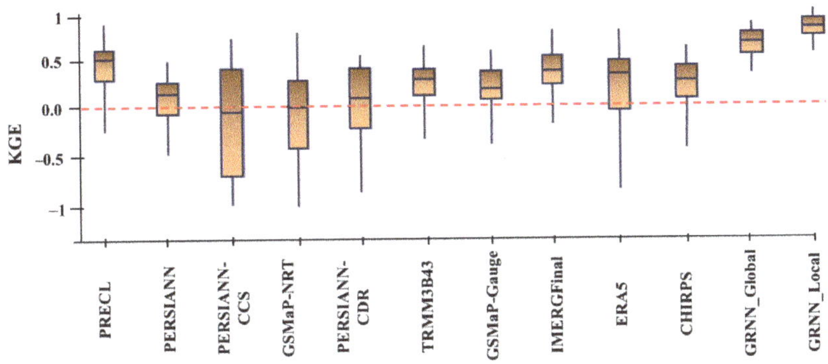


**Figure 5.** Graph of KGE values over Iran during the 2003–2021 period derived from a point-to-pixel evaluation approach.

### *4.3. Evaluation of Global Precipitation Products by a Pixel-to-Pixel Approach*

In this approach, the accuracy assessment of the precipitation products was conducted using the interpolated precipitation map produced by the OK method. Here, the dark green and dark red colors indicate the lower and higher values of various accuracy evaluation metrics, respectively. In terms of the RMSE values according to Table 3, the IMERGFinal (13.7), TRMM3B43 (15.6), PERSIANN-CDR (18.5), ERA5 (19.9), and GSMaP-Gauge (24.8 mm/mo) products had the lowest values. Across the entire study area, these products provided the most accurate precipitation estimates, with KGE values of 0.89, 0.86, 0.77, 0.78, and 0.60, respectively. As far as precipitation estimation accuracy is concerned, GSMaP-NRT and PERSIANN-CCS had the least accuracy. PRECL, PERSIANN, and GSMaP-Gauge overestimated, whereas GSMaP-NRT and PERSIANN-CCS underestimated, the precipitation amounts. According to the recorded amounts, the correlation coefficient between satellite-based and gauge-based products was low. Gauge-corrected satellite products were more accurate than other product groups.

**Table 3.** Accuracy assessment results for the pixel-to-pixel approach (Different colors (from dark green to dark red) indicate different values (from lower to higher) of various accuracy evaluation metrics respectively).

| Product | KGE | R | Bias | VR | Rbias | RMSE | SE | RE |
|---|---|---|---|---|---|---|---|---|
| PRECL | 0.47 | 0.55 | 0.8 | 0.81 | −0.2 | 27 | 0.6 | 0.4 |
| PERSIANN | 0.38 | 0.52 | 0.61 | 1.07 | −0.39 | 29.9 | 0.62 | 0.38 |
| PERSIANN-CCS | 0.14 | 0.5 | 1.68 | 0.83 | 0.71 | 43.6 | 0.17 | 0.83 |
| GSMaP-NRT | 0.46 | 0.58 | 1.3 | 1.16 | 0.36 | 41.1 | 0.03 | 0.97 |
| PERSIANN-CDR | 0.77 | 0.83 | 1.04 | 0.85 | 0.04 | 18.5 | 0.23 | 0.77 |
| TRMM3B43 | 0.86 | 0.9 | 1.1 | 0.99 | 0.1 | 15.6 | 0.02 | 0.98 |
| GSMaP-Gauge | 0.6 | 0.7 | 0.78 | 1.13 | −0.22 | 24.8 | 0.3 | 0.7 |
| IMERGFinal | 0.89 | 0.92 | 1.06 | 0.96 | 0.06 | 13.7 | 0.04 | 0.96 |
| ERA5 | 0.78 | 0.88 | 1.17 | 1.04 | 0.19 | 19.9 | 0.05 | 0.95 |
| CHIRPS | 0.71 | 0.74 | 1.11 | 0.97 | 0.11 | 25 | 0.08 | 0.92 |
| GRNN_Global | 0.92 | 0.92 | 1.02 | 1.02 | 0.02 | 9.6 | 0.05 | 0.95 |
| GRNN_Local | 0.94 | 0.93 | 1.01 | 1.01 | 0.01 | 8.5 | 0.04 | 0.96 |

Five of the evaluated precipitation products showed a similar variability to the recorded data (VRs between 0.90 and 1.10). Nevertheless, the other three products have different trends from the recorded data, such as PERSIANN-CCS, which had a VR of 1.68. Based on the research results, the RMSE for the proposed GRNN-Global (local) precipitation product was 9.6 (8.5 mm/mo), and the KGE value was 0.92 (0.94), indicating a higher level of accuracy for the developed product when compared with the global precipitation products. A total of 95 (96%) errors in the proposed GRNN-Global (local) products can be classified as random errors.

At the pixel level, Figure 6 illustrates the geographic distribution of KGEs over the study period. The evaluated precipitation products were more accurate in areas with very arid and arid climates than in semi-arid and humid climates. Nonetheless, there are some products that are less accurate in very arid climates than in semi-arid or arid climates. The accuracy of the products under evaluation in this study in the northern sub-basins of Iran was less than in other sub-basins. It is imperative to note that global precipitation products vary in accuracy from region to region. Persian-CCS had the highest KGE class area with values below zero. In more than 50% of regions, IMERGFinal, TRMM3B43, PERSIANN-CDR, and ERA5 all had a KGE of 0.8, indicating the high degree of accuracy with which these products can be used to estimate precipitation. The proportion of regions with KGE values greater than 0.8 in the proposed precipitation products was significantly higher compared with the evaluated precipitation products.

Figure 7 represents the KGE box plot for the evaluated precipitation products at the whole study area scale and under different climatic conditions. Considering the KGE median, product performance varies between climatic classes. Most products perform better in regions with arid climates, with only a few precipitation products, such as PERSIANN-CCS, performing poorly in these regions. The length of the box plot in regions with arid climates is shorter than in other climates, indicating low changes in precipitation products in these regions. In general, humid climates have lower KGE values than arid climates. Since the arid climate covers a large proportion of the study area, the accuracy of any product in these regions (precipitation is below 300 mm per year) is of significant importance. Since the study area contains a large area of arid climates, PERSIANN-CCS has the lowest accuracy in estimating precipitation. For regions with a humid climate, TRMM3B43, PERSIANN-CDR, and CHIRPS were more efficient than other products. At all scales, including the study area and climatic classes, the proposed model had a higher accuracy than the evaluated precipitation products. Furthermore, the developed product performed well in pixels of different climatic classes, indicating that it is stable and capable of accurately estimating precipitation under a variety of conditions. In the proposed model, the local approach had a higher efficiency than the global approach. PERSIANN-CCS and ERA5 products showed high variability in KGE values under different conditions, indicating poor performance.

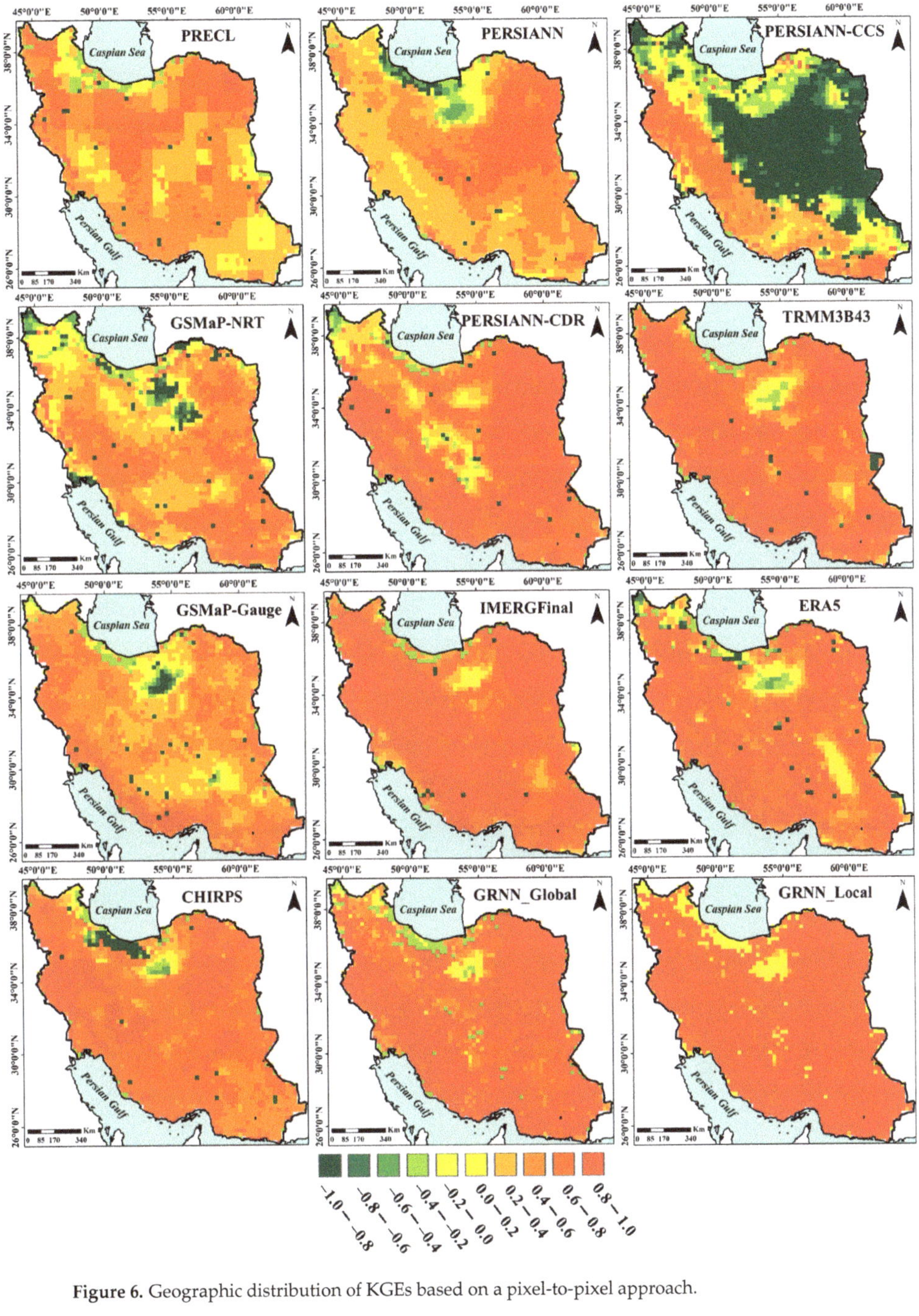


**Figure 6.** Geographic distribution of KGEs based on a pixel-to-pixel approach.

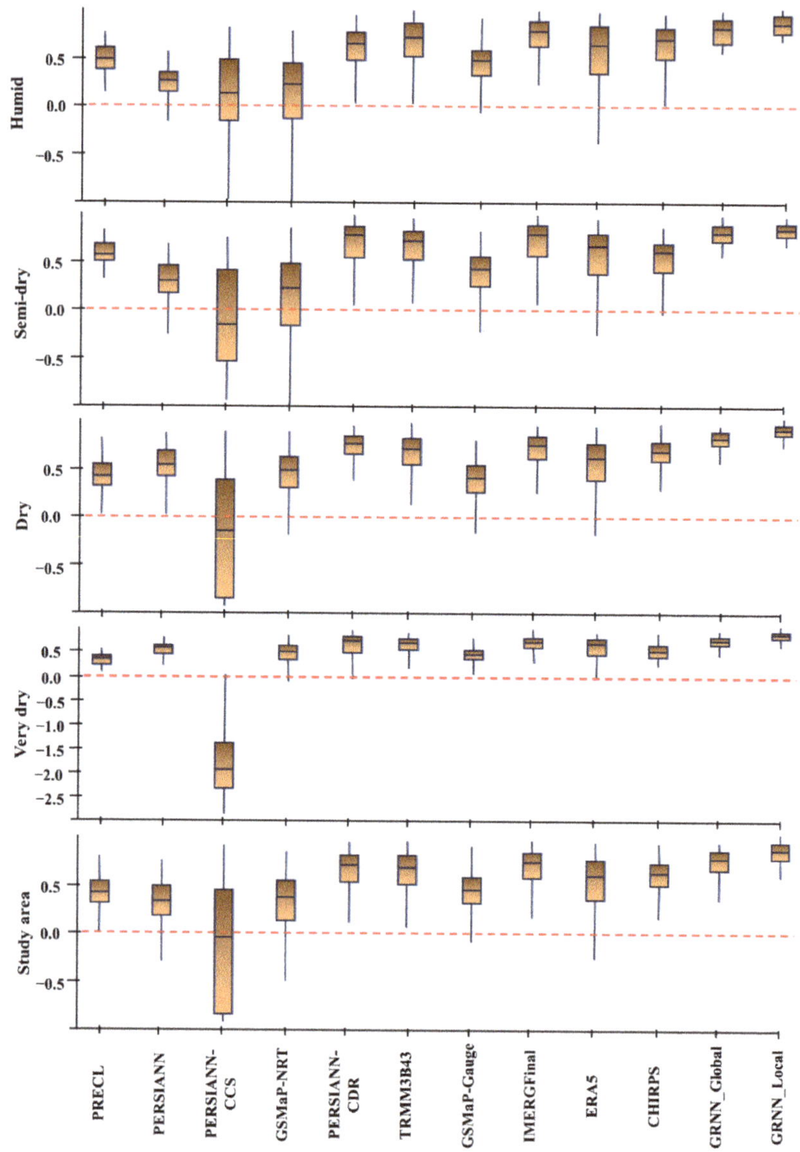


**Figure 7.** Graph of KGE values throughout Iran and climate zones during the 2003–2021 period derived from a pixel-to-pixel evaluation strategy.

## 5. Discussion

This study examined whether global precipitation products are efficient under a variety of conditions. Both point-to-pixel and pixel-to-pixel approaches were used to evaluate the accuracy of some global precipitation products [36,65]. The calculated accuracy values for the pixel-to-pixel approach were significantly higher for different products compared with the point-to-pixel approach. Nevertheless, both strategies ranked the products similarly. Thus, the two strategies can be employed interchangeably to rank products. The pixel-to-pixel approach can be more useful than the point-to-pixel approach for evaluating precipitation product accuracy because, based on this approach, precipitation products can also be evaluated in areas without ground stations.

Based on the accuracy assessment results, global precipitation products perform differently for each region. Here, IMERGFinal, TRMM3B43, and PERSIANN-CDR performed better than the others. This study showed that systematic error contributes to the total error in PRECL and PERSIANN. This conclusion is confirmed by Prakash [19]. On average, 95% of the error in the developed precipitation product was random error, which is one of the strengths of the proposed product compared with the existing precipitation products. The evaluation results indicate that the PERSIANN set products estimate precipitation at different accuracy levels. As shown in PERSIANN-CCS and PERSIANN, the precipitation amounts were overestimated and underestimated, respectively, compared with the ground measurements. A study conducted by Nguyen, Ombadi, Sorooshian, Hsu, AghaKouchak, Braithwaite, Ashouri, and Thorstensen [30] concluded that PERSIANN-CCS overestimates precipitation amounts. Given the high accuracy of PERSIANN-CDR, it is evident that a gauge-based adjustment is crucial for improving satellite estimation [24,26].

This study also indicates that gauge-corrected satellite products are more efficient than gauge-based, satellite-based, and reanalysis products. Gauge-corrected satellite products utilize data collected from gauge stations in the generation of their estimations, which can enhance their accuracy [27,66]. As a result of some previous studies [1,26,40], reanalysis products have been proven to be more accurate than other products in providing estimates of precipitation in warm and dry regions. Additionally, Xu, Chen, Moradkhani, Zhang, and Hu [21] indicated that reanalysis products are more accurate than satellite-based precipitation products in Australia, Europe, and North America. However, Chen, Chen, and Azorin-Molina [41] demonstrated that satellite-based products are more accurate than reanalysis products. Overall, previous studies have shown that the efficiency of different groups of precipitation products varies in different regions, so it is not possible to choose a suitable product for all conditions [67,68]. Despite their strengths and weaknesses, these products can be used together to achieve higher accuracy.

It is generally accepted that precipitation products vary in performance based on precipitation type, climate, and topography [2]. This study found that results are also affected by climatic conditions. Global precipitation products will differ in efficiency ranking depending on the conditions under which they are evaluated. Based on the research conducted by Hosseini-Moghari and Tang [29], global precipitation products vary in accuracy depending on precipitation type and climatic conditions. As a result, a machine learning method was employed in this study to combine information from several precipitation products, recorded precipitation data, and surface characteristics to develop a more accurate precipitation product. It was found that combining multi-product information improved precipitation estimates. Each dependent variable had a different importance score in terms of accuracy improvement. The combination of precipitation products and different surface characteristics based on the GRNN model resulted in a precipitation product with higher accuracy. Unlike the existing precipitation products, the precipitation product obtained by the proposed model had appropriate accuracy in all four climatic conditions: very arid, arid, semi-arid, and humid. The precipitation product of the proposed model showed similar levels of accuracy in all these climates, which shows its stability.

There were still some limitations to this study. First, the global precipitation products and correction variables, such as LST, EVI, SWI, and elevation, had different spatial resolutions. Second, there was some temporal inconsistency between the ground station data and the global precipitation products. Another, or perhaps the most significant, limitation of the proposed model is the time and volume of calculations required to produce a more accurate precipitation product. This is because the 10 existing precipitation products, along with biophysical and topographical characteristics, are combined based on the GRNN model. Of course, the number of these inputs can be reduced, which will affect the output accuracy. Depending on the application and the expected sensitivity to precipitation estimation accuracy, the volume and time of calculations can be reduced by reducing model inputs. On the other hand, other effective parameters can be considered as inputs. This increases the accuracy of the proposed model while also increasing computational costs.

## 6. Conclusions

It is important to assess the efficiency of global precipitation products under varying conditions. Various products have different levels of efficiency given the geographical, climatic, and topographic conditions, as demonstrated in this study. In order to create a product that is accurate in a variety of conditions, it has been proposed to combine existing global precipitation products with complementary data. TRMM3B43, IMERGFinal, PERSIANN-CDR, and PERSIANN products have the highest impact on very dry, arid, semi-arid, and humid conditions, respectively. In terms of importance, the interpolated precipitation map has higher significance than topography and surface biophysical variables. Among the surface biophysical variables, SWI has a higher importance score than LST and EVI. These results show that a precipitation product alone cannot be suitable for various geographical locations in a country such as Iran with a diverse climate. Even at a particular geographical location, the efficiency of different precipitation products differs at different times. As a result, combining the capabilities of different precipitation products can improve precipitation estimation accuracy in areas with diverse conditions. As a result of the accuracy evaluation, the developed product provides a more stable and accurate estimate of precipitation amounts than global precipitation products. In the proposed model, the local approach to optimization had a higher efficiency than the global approach. As opposed to existing precipitation products, the accuracy of precipitation estimation using the proposed model was close to each other under varied climate conditions, which indicates the stability of this model under different conditions. A basic concept of this study is to develop a model that uses the collaborative decision-making principle and integrates findings at the decision-making level to produce a more accurate precipitation product using existing precipitation products and their varying capabilities. As the proposed model uses multiple variables in the production process, it is limited in terms of time and the volume of calculations. Future studies should concentrate on determining effective and optimal variables (inputs) that can reduce the time and volume of the calculations required.

**Author Contributions:** Conceptualization, S.M., M.S., H.R., N.M. (Naeim Mijani) and S.F.; methodology, S.M., M.S., H.R., N.M. (Naeim Mijani), R.G., S.T. and N.M. (Neda Mohammadzadeh); software, S.M., M.S., H.R., N.M. (Naeim Mijani), R.G., S.T., N.M. (Neda Mohammadzadeh) and A.R.; data curation, N.M. (Naeim Mijani) and S.F.; writing—original draft preparation, S.M., M.S., H.R., N.M. (Naeim Mijani), R.G., S.T. and N.M. (Neda Mohammadzadeh); software, S.M., M.S., H.R., N.M. (Naeim Mijani), R.G., S.T., N.M. (Neda Mohammadzadeh) and A.R.; writing—review and editing, S.F. and A.B. All authors have read and agreed to the published version of the manuscript.

**Funding:** This work was supported by the Canada First Research Excellence Fund (CFREF)–Food from Thought project at the University of Guelph and the Natural Science and Engineering Research Council (NSERC) of Canada (RGPIN-2020-05017).

**Institutional Review Board Statement:** Not applicable.

**Informed Consent Statement:** Not applicable.

**Data Availability Statement:** The data used to support the findings of this study are available from the corresponding author upon reasonable request.

**Conflicts of Interest:** The authors declare no conflict of interest.

## References

1. Salman, S.A.; Shahid, S.; Ismail, T.; Al-Abadi, A.M.; Wang, X.-j.; Chung, E.-S. Selection of gridded precipitation data for Iraq using compromise programming. *Measurement* **2019**, *132*, 87–98. [CrossRef]
2. Seyyedi, H.; Anagnostou, E.N.; Beighley, E.; McCollum, J. Hydrologic evaluation of satellite and reanalysis precipitation datasets over a mid-latitude basin. *Atmos. Res.* **2015**, *164*, 37–48. [CrossRef]

3. Sheffield, J.; Goteti, G.; Wood, E.F. Development of a 50-year high-resolution global dataset of meteorological forcings for land surface modeling. *J. Clim.* **2006**, *19*, 3088–3111. [CrossRef]
4. Sun, W.; Ma, J.; Yang, G.; Li, W. Statistical and hydrological evaluations of multi-satellite precipitation products over Fujiang river basin in humid southeast China. *Remote Sens.* **2018**, *10*, 1898. [CrossRef]
5. Kordi, F.; Hamzeh, S.; Atarchi, S.; Alavipanah, S.K. Agricultural Product Classification for Optimal Water Resource Management Using the Data Time Series of Landsat8. *Iran. J. Ecohydrol.* **2018**, *5*, 1267–1283.
6. Kordi, F.; Yousefi, H.; Tajrishi, M. Estimation of water consumption in the downstream agricultural area of Hasanlu Dam using METRIC algorithm. *Water Irrig. Manag.* **2022**, *12*, 171–185.
7. Firozjaei, M.K.; Fathololoumi, S.; Alavipanah, S.K.; Kiavarz, M.; Vaezi, A.R.; Biswas, A. A new approach for modeling near surface temperature lapse rate based on normalized land surface temperature data. *Remote Sens Envrion.* **2020**, *242*, 111746. [CrossRef]
8. Firozjaei, M.K.; Sedighi, A.; Firozjaei, H.K.; Kiavarz, M.; Homaee, M.; Arsanjani, J.J.; Makki, M.; Naimi, B.; Alavipanah, S.K. A historical and future impact assessment of mining activities on surface biophysical characteristics change: A remote sensing-based approach. *Ecol. Indic.* **2021**, *122*, 107264. [CrossRef]
9. Weng, Q.; Firozjaei, M.K.; Sedighi, A.; Kiavarz, M.; Alavipanah, S.K. Statistical analysis of surface urban heat island intensity variations: A case study of Babol city, Iran. *GIScience Remote Sens.* **2019**, *56*, 576–604. [CrossRef]
10. Kordi, F.; Yousefi, H. Crop classification based on phenology information by using time series of optical and synthetic-aperture radar images. *Remote Sens. Appl. Soc. Environ.* **2022**, *27*, 100812. [CrossRef]
11. Kordi, F.; Yousefi, H.; Ghasemi, L.; Tajrishy, M. Investigation and comparison of land use map database in the Urmia lake basin. *Iran. J. Ecohydrol.* **2022**, *8*, 891–905.
12. Mijani, N.; Alavipanah, S.K.; Firozjaei, M.K.; Arsanjani, J.J.; Hamzeh, S.; Weng, Q. Modeling outdoor thermal comfort using satellite imagery: A principle component analysis-based approach. *Ecol. Indic.* **2020**, *117*, 106555. [CrossRef]
13. Firozjaei, M.K.; Alavipanah, S.K.; Liu, H.; Sedighi, A.; Mijani, N.; Kiavarz, M.; Weng, Q. A PCA–OLS model for assessing the impact of surface biophysical parameters on land surface temperature variations. *Remote Sens.* **2019**, *11*, 2094. [CrossRef]
14. Kiavarz, M.; Hosseinbeigi, S.B.; Mijani, N.; Shahsavary, M.S.; Firozjaei, M.K. Predicting spatial and temporal changes in surface urban heat islands using multi-temporal satellite imagery: A case study of Tehran metropolis. *Urban Clim.* **2022**, *45*, 101258. [CrossRef]
15. Boloorani, A.D.; Kazemi, Y.; Sadeghi, A.; Shorabeh, S.N.; Argany, M. Identification of dust sources using long term satellite and climatic data: A case study of Tigris and Euphrates basin. *Atmos. Environ.* **2020**, *224*, 117299. [CrossRef]
16. Zheng, Y.; Tang, L.; Wang, H. An improved approach for monitoring urban built-up areas by combining NPP-VIIRS nighttime light, NDVI, NDWI, and NDBI. *J. Clean. Prod.* **2021**, *328*, 129488. [CrossRef]
17. Tobías, A.; Carnerero, C.; Reche, C.; Massagué, J.; Via, M.; Minguillón, M.C.; Alastuey, A.; Querol, X. Changes in air quality during the lockdown in Barcelona (Spain) one month into the SARS-CoV-2 epidemic. *Sci. Total Environ.* **2020**, *726*, 138540. [CrossRef]
18. Yao, J.; Raffuse, S.M.; Brauer, M.; Williamson, G.J.; Bowman, D.M.; Johnston, F.H.; Henderson, S.B. Predicting the minimum height of forest fire smoke within the atmosphere using machine learning and data from the CALIPSO satellite. *Remote Sens. Environ.* **2018**, *206*, 98–106. [CrossRef]
19. Prakash, S. Performance assessment of CHIRPS, MSWEP, SM2RAIN-CCI, and TMPA precipitation products across India. *J. Hydrol.* **2019**, *571*, 50–59. [CrossRef]
20. Trinh-Tuan, L.; Matsumoto, J.; Ngo-Duc, T.; Nodzu, M.I.; Inoue, T. Evaluation of satellite precipitation products over Central Vietnam. *Prog. Earth Planet. Sci.* **2019**, *6*, 54. [CrossRef]
21. Xu, L.; Chen, N.; Moradkhani, H.; Zhang, X.; Hu, C. Improving global monthly and daily precipitation estimation by fusing gauge observations, remote sensing, and reanalysis data sets. *Water Resour. Res.* **2020**, *56*, e2019WR026444. [CrossRef]
22. Zhan, Y.; Ren, G.; Yang, S. Change in precipitation over the Asian continent from 1901-2016 based on a new multi-source dataset. *Clim. Res.* **2018**, *76*, 41–57. [CrossRef]
23. Zhou, Y.; Nelson, K.; Mohr, K.I.; Huffman, G.J.; Levy, R.; Grecu, M. A spatial-temporal extreme precipitation database from GPM IMERG. *J. Geophys. Res. Atmos.* **2019**, *124*, 10344–10363. [CrossRef]
24. Beck, H.E.; Pan, M.; Roy, T.; Weedon, G.P.; Pappenberger, F.; Van Dijk, A.I.; Huffman, G.J.; Adler, R.F.; Wood, E.F. Daily evaluation of 26 precipitation datasets using Stage-IV gauge-radar data for the CONUS. *Hydrol. Earth Syst. Sci.* **2019**, *23*, 207–224. [CrossRef]
25. Saemian, P.; Hosseini-Moghari, S.-M.; Fatehi, I.; Shoarinezhad, V.; Modiri, E.; Tourian, M.J.; Tang, Q.; Nowak, W.; Bárdossy, A.; Sneeuw, N. Comprehensive evaluation of precipitation datasets over Iran. *J. Hydrol.* **2021**, *603*, 127054. [CrossRef]
26. Satgé, F.; Defrance, D.; Sultan, B.; Bonnet, M.-P.; Seyler, F.; Rouche, N.; Pierron, F.; Paturel, J.-E. Evaluation of 23 gridded precipitation datasets across West Africa. *J. Hydrol.* **2020**, *581*, 124412. [CrossRef]
27. Sun, Q.; Miao, C.; Duan, Q.; Ashouri, H.; Sorooshian, S.; Hsu, K.L. A review of global precipitation data sets: Data sources, estimation, and intercomparisons. *Rev. Geophys.* **2018**, *56*, 79–107. [CrossRef]
28. Ghozat, A.; Sharafati, A.; Hosseini, S.A. Long-term spatiotemporal evaluation of CHIRPS satellite precipitation product over different climatic regions of Iran. *Theor. Appl. Climatol.* **2021**, *143*, 211–225. [CrossRef]
29. Hosseini-Moghari, S.-M.; Tang, Q. Validation of GPM IMERG V05 and V06 precipitation products over Iran. *J. Hydrometeorol.* **2020**, *21*, 1011–1037. [CrossRef]

30. Nguyen, P.; Ombadi, M.; Sorooshian, S.; Hsu, K.; AghaKouchak, A.; Braithwaite, D.; Ashouri, H.; Thorstensen, A.R. The PERSIANN family of global satellite precipitation data: A review and evaluation of products. *Hydrol. Earth Syst. Sci.* **2018**, *22*, 5801–5816. [CrossRef]
31. Sharifi, E.; Steinacker, R.; Saghafian, B. Assessment of GPM-IMERG and other precipitation products against gauge data under different topographic and climatic conditions in Iran: Preliminary results. *Remote Sens.* **2016**, *8*, 135. [CrossRef]
32. Yuan, F.; Zhang, L.; Soe, K.M.W.; Ren, L.; Zhao, C.; Zhu, Y.; Jiang, S.; Liu, Y. Applications of TRMM-and GPM-era multiple-satellite precipitation products for flood simulations at sub-daily scales in a sparsely gauged watershed in Myanmar. *Remote Sens.* **2019**, *11*, 140. [CrossRef]
33. Eini, M.R.; Olyaei, M.A.; Kamyab, T.; Teymoori, J.; Brocca, L.; Piniewski, M. Evaluating three non-gauge-corrected satellite precipitation estimates by a regional gauge interpolated dataset over Iran. *J. Hydrol. Reg. Stud.* **2021**, *38*, 100942. [CrossRef]
34. Liu, C.-Y.; Aryastana, P.; Liu, G.-R.; Huang, W.-R. Assessment of satellite precipitation product estimates over Bali Island. *Atmos. Res.* **2020**, *244*, 105032. [CrossRef]
35. Hsu, K.-l.; Gao, X.; Sorooshian, S.; Gupta, H.V. Precipitation estimation from remotely sensed information using artificial neural networks. *J. Appl. Meteorol. Climatol.* **1997**, *36*, 1176–1190. [CrossRef]
36. Peng, F.; Zhao, S.; Chen, C.; Cong, D.; Wang, Y.; Ouyang, H. Evaluation and comparison of the precipitation detection ability of multiple satellite products in a typical agriculture area of China. *Atmos. Res.* **2020**, *236*, 104814. [CrossRef]
37. Satgé, F.; Ruelland, D.; Bonnet, M.-P.; Molina, J.; Pillco, R. Consistency of satellite-based precipitation products in space and over time compared with gauge observations and snow-hydrological modelling in the Lake Titicaca region. *Hydrol. Earth Syst. Sci.* **2019**, *23*, 595–619. [CrossRef]
38. Taghizadeh, E.; Ahmadi-Givi, F.; Brocca, L.; Sharifi, E. Evaluation of satellite/reanalysis precipitation products over Iran. *Int. J. Remote Sens.* **2021**, *42*, 3474–3497. [CrossRef]
39. Tekeli, A.E.; Fouli, H. Evaluation of TRMM satellite-based precipitation indexes for flood forecasting over Riyadh City, Saudi Arabia. *J. Hydrol.* **2016**, *541*, 471–479. [CrossRef]
40. Basheer, M.; Elagib, N.A. Performance of satellite-based and GPCC 7.0 rainfall products in an extremely data-scarce country in the Nile Basin. *Atmos. Res.* **2019**, *215*, 128–140. [CrossRef]
41. Chen, A.; Chen, D.; Azorin-Molina, C. Assessing reliability of precipitation data over the Mekong River Basin: A comparison of ground-based, satellite, and reanalysis datasets. *Int. J. Climatol.* **2018**, *38*, 4314–4334. [CrossRef]
42. Chen, J.; Wang, Z.; Wu, X.; Chen, X.; Lai, C.; Zeng, Z.; Li, J. Accuracy evaluation of GPM multi-satellite precipitation products in the hydrological application over alpine and gorge regions with sparse rain gauge network. *Hydrol. Res.* **2019**, *50*, 1710–1729. [CrossRef]
43. Li, C.; Tang, G.; Hong, Y. Cross-evaluation of ground-based, multi-satellite and reanalysis precipitation products: Applicability of the Triple Collocation method across Mainland China. *J. Hydrol.* **2018**, *562*, 71–83. [CrossRef]
44. Camici, S.; Ciabatta, L.; Massari, C.; Brocca, L. How reliable are satellite precipitation estimates for driving hydrological models: A verification study over the Mediterranean area. *J. Hydrol.* **2018**, *563*, 950–961. [CrossRef]
45. Xiao, S.; Xia, J.; Zou, L. Evaluation of multi-satellite precipitation products and their ability in capturing the characteristics of extreme climate events over the Yangtze River Basin, China. *Water* **2020**, *12*, 1179. [CrossRef]
46. Ceccherini, G.; Ameztoy, I.; Hernández, C.P.R.; Moreno, C.C. High-resolution precipitation datasets in South America and West Africa based on satellite-derived rainfall, enhanced vegetation index and digital elevation model. *Remote Sens.* **2015**, *7*, 6454–6488. [CrossRef]
47. Joyce, R.J.; Janowiak, J.E.; Arkin, P.A.; Xie, P. CMORPH: A method that produces global precipitation estimates from passive microwave and infrared data at high spatial and temporal resolution. *J. Hydrometeorol.* **2004**, *5*, 487–503. [CrossRef]
48. Huffman, G.J.; Bolvin, D.T.; Nelkin, E.J.; Wolff, D.B.; Adler, R.F.; Gu, G.; Hong, Y.; Bowman, K.P.; Stocker, E.F. The TRMM multisatellite precipitation analysis (TMPA): Quasi-global, multiyear, combined-sensor precipitation estimates at fine scales. *J. Hydrometeorol.* **2007**, *8*, 38–55. [CrossRef]
49. Huffman, G.J.; Bolvin, D.T.; Braithwaite, D.; Hsu, K.; Joyce, R.; Xie, P.; Yoo, S.-H. NASA global precipitation measurement (GPM) integrated multi-satellite retrievals for GPM (IMERG). *Algorithm Theor. Basis Doc. ATBD Version* **2015**, *4*, 26.
50. Turk, J.T.; Mostovoy, G.V.; Anantharaj, V. The NRL-blend high resolution precipitation product and its application to land surface hydrology. In *Satellite Rainfall Applications for Surface Hydrology*; Springer: Berlin/Heidelberg, Germany, 2010; pp. 85–104.
51. Chen, S.; Xiong, L.; Ma, Q.; Kim, J.-S.; Chen, J.; Xu, C.-Y. Improving daily spatial precipitation estimates by merging gauge observation with multiple satellite-based precipitation products based on the geographically weighted ridge regression method. *J. Hydrol.* **2020**, *589*, 125156. [CrossRef]
52. Oliazadeh, A.; Bozorg-Haddad, O.; Pakdaman, M.; Baghbani, R.; Loáiciga, H.A. Optimal merging of multi-satellite precipitation data in urban areas. *Theor. Appl. Climatol.* **2022**, *147*, 1697–1712. [CrossRef]
53. Kacprzyk, J.; Fedrizzi, M. *Multiperson Decision Making Models Using Fuzzy Sets and Possibility Theory*; Springer Science & Business Media: Berlin, Germany, 2012; Volume 18.
54. Chen, M.; Xie, P.; Janowiak, J.E.; Arkin, P.A. Global land precipitation: A 50-yr monthly analysis based on gauge observations. *J. Hydrometeorol.* **2002**, *3*, 249–266. [CrossRef]
55. Sorooshian, S.; Hsu, K.-L.; Gao, X.; Gupta, H.V.; Imam, B.; Braithwaite, D. Evaluation of PERSIANN system satellite-based estimates of tropical rainfall. *Bull. Am. Meteorol. Soc.* **2000**, *81*, 2035–2046. [CrossRef]

56. Tashima, T.; Kubota, T.; Mega, T.; Ushio, T.; Oki, R. Precipitation extremes monitoring using the near-real-time GSMaP product. *IEEE J. Sel. Top. Appl. Earth Obs. Remote Sens.* **2020**, *13*, 5640–5651. [CrossRef]
57. Ashouri, H.; Hsu, K.-L.; Sorooshian, S.; Braithwaite, D.K.; Knapp, K.R.; Cecil, L.D.; Nelson, B.R.; Prat, O.P. PERSIANN-CDR: Daily precipitation climate data record from multisatellite observations for hydrological and climate studies. *Bull. Am. Meteorol. Soc.* **2015**, *96*, 69–83. [CrossRef]
58. Ushio, T.; Sasashige, K.; Kubota, T.; Shige, S.; Okamoto, K.i.; Aonashi, K.; Inoue, T.; Takahashi, N.; Iguchi, T.; Kachi, M. A Kalman filter approach to the Global Satellite Mapping of Precipitation (GSMaP) from combined passive microwave and infrared radiometric data. *J. Meteorol. Soc. Japan. Ser. II* **2009**, *87*, 137–151. [CrossRef]
59. Hersbach, H.; Bell, B.; Berrisford, P.; Hirahara, S.; Horányi, A.; Muñoz-Sabater, J.; Nicolas, J.; Peubey, C.; Radu, R.; Schepers, D. The ERA5 global reanalysis. *Q. J. R. Meteorol. Soc.* **2020**, *146*, 1999–2049. [CrossRef]
60. Funk, C.; Peterson, P.; Landsfeld, M.; Pedreros, D.; Verdin, J.; Shukla, S.; Husak, G.; Rowland, J.; Harrison, L.; Hoell, A. The climate hazards infrared precipitation with stations—A new environmental record for monitoring extremes. *Sci. Data* **2015**, *2*, 150066. [CrossRef]
61. Wackernagel, H. *Multivariate Geostatistics: An Introduction with Applications*; Springer Science & Business Media: Berlin, Germany, 2003.
62. Specht, D.F. A general regression neural network. *IEEE Trans. Neural Netw.* **1991**, *2*, 568–576. [CrossRef]
63. Chen, S.; Cowan, C.; Grant, P. Orthogonal Least Squares Learning Algorithm for Radial. *IEEE Trans. Neural Netw.* **1991**, *2*, 303. [CrossRef]
64. Li, H.-z.; Guo, S.; Li, C.-j.; Sun, J.-q. A hybrid annual power load forecasting model based on generalized regression neural network with fruit fly optimization algorithm. *Knowl. Based Syst.* **2013**, *37*, 378–387. [CrossRef]
65. Xu, R.; Tian, F.; Yang, L.; Hu, H.; Lu, H.; Hou, A. Ground validation of GPM IMERG and TRMM 3B42V7 rainfall products over southern Tibetan Plateau based on a high-density rain gauge network. *J. Geophys. Res. Atmos.* **2017**, *122*, 910–924. [CrossRef]
66. Krakauer, N.Y.; Lakhankar, T.; Dars, G.H. Precipitation trends over the Indus basin. *Climate* **2019**, *7*, 116. [CrossRef]
67. Cai, Y.; Jin, C.; Wang, A.; Guan, D.; Wu, J.; Yuan, F.; Xu, L. Spatio-temporal analysis of the accuracy of tropical multisatellite precipitation analysis 3B42 precipitation data in mid-high latitudes of China. *PLoS ONE* **2015**, *10*, e0120026. [CrossRef] [PubMed]
68. Ebert, E.E.; Janowiak, J.E.; Kidd, C. Comparison of near-real-time precipitation estimates from satellite observations and numerical models. *Bull. Am. Meteorol. Soc.* **2007**, *88*, 47–64. [CrossRef]

*sustainability*

MDPI

*Review*

# Impacts of Tillage on Soil's Physical and Hydraulic Properties in Temperate Agroecosystems

**Md Nayem Hasan Munna * and Rattan Lal**

CFAES Rattan Lal Center for Carbon Management and Sequestration (Lal Carbon Center), School of Environment and Natural Resources, College of Food, Agricultural, and Environmental Sciences, The Ohio State University, Columbus, OH 43210, USA; lal.1@osu.edu

* Correspondence: munna.1@osu.edu

**Abstract**

Tillage practices critically influence soil's physical properties, which are fundamental to sustainable agriculture in temperate climates. This review evaluates how conventional tillage (CvT; e.g., moldboard and chisel plowing), reduced tillage (RT), and conservation tillage (CT), particularly no-tillage (NT), affect six key indicators: bulk density (BD), saturated hydraulic conductivity (Ks), wet aggregate stability (WAS), penetration resistance (PR), available water capacity (AWC), and soil organic carbon (SOC). Special emphasis is placed on differentiating topsoil and subsoil responses to inform climate-resilient land management. A total of 70 peer-reviewed studies published between 1991 and 2025 were analyzed. Data were extracted for BD, Ks, WAS, PR, AWC, and SOC across tillage systems. Depths were standardized into topsoil (0–10 cm) and composite (>10 cm) categories. Descriptive statistics were used to synthesize cross-study trends. NT showed lower mean BD in the topsoil ($1.32 \pm 0.08$ Mg/m$^3$) compared with moldboard plow ($1.33 \pm 0.09$) and chisel tillage ($1.39 \pm 0.12$); however, the effects of tillage on BD were not statistically significant, while BD was higher at composite depths under NT ($1.56 \pm 0.09$ Mg/m$^3$), indicating subsoil compaction. Ks improved under NT, reaching 4.2 mm/h with residue retention. WAS rose by 33.4%, and SOC increased by 25% under CT systems. PR tended to be elevated in deeper layers under NT. Overall, CT, particularly NT, improves surface soil's physical health and SOC accumulation in temperate agroecosystems; however, persistent subsoil compaction highlights the need for depth-targeted management strategies, such as controlled traffic, periodic subsoil alleviation, or deep-rooted cover crops, to sustain long-term soil functionality and climate-resilient production systems.

**Keywords:** conservation tillage; soil organic carbon; saturated hydraulic conductivity; aggregate stability; tillage management

## 1. Introduction

The escalating global challenges of climate change, population growth, and worsening soil degradation underscore the urgent need for sustainable agricultural practices, particularly in temperate regions [1–4]. Temperate regions, which are characterized by distinct seasonal variations, play a vital role in global food production [5–7]. These regions face unique challenges in balancing the demands of agricultural productivity with the imperative to maintain soil health and ecosystem services [3,4]. Among the diverse soil management practices employed in agriculture, tillage remains one of the most significant factors influencing soil's physical properties [8,9]. Its effects ripple through key soil



https://doi.org/10.3390/su18021083

attributes such as bulk density (BD), saturated hydraulic conductivity (Ks), wet aggregate stability (WAS), available water capacity (AWC), penetration resistance (PR), and soil organic carbon (SOC), directly impacting both crop productivity and environmental sustainability [10].

Historically, tillage has been widely employed for seedbed preparation, weed control, and soil aeration. However, these benefits come with trade-offs that include soil compaction, reduced organic matter content, and increased susceptibility to erosion, particularly under conventional tillage (CvT) systems. These trade-offs are exacerbated in temperate climates, where freeze–thaw cycles, precipitation variability, and temperature fluctuations influence soil dynamics and agricultural decision-making [8,11]. Consequently, understanding the nuanced effects of different tillage systems, including no-tillage (NT) and reduced tillage (RT), on soil's physical properties is essential for developing region-specific, sustainable agricultural practices.

Emerging research highlights the potential of conservation tillage (CT) practices in mitigating soil degradation and enhancing agricultural resilience. RT has been shown to improve SOC sequestration, aggregate stability, and water retention, contributing to better resistance against droughts and floods [9,12]. For example, NT systems have demonstrated the ability to increase water infiltration rates and reduce soil compaction, particularly in semi-arid and arid regions. However, the benefits of NT and RT systems in temperate climates are highly dependent on additional agronomic factors such as crop rotation, residue retention, and the use of cover crops [13].

Despite these promising findings, significant knowledge gaps persist, particularly regarding the long-term and region-specific impacts of tillage practices on soil's physical properties. For instance, while NT practices are often lauded for their ability to enhance SOC content, their effects on subsoil characteristics, such as deep compaction and water movement, remain poorly understood [10]. Similarly, while CT can improve surface soil health, its interaction with microbial dynamics and nutrient cycling at greater depths requires further investigation [14,15]. Furthermore, tillage practices have complex implications for greenhouse gas emissions. While RT may enhance carbon sequestration in some contexts, it can also lead to increased nitrous oxide emissions under specific soil conditions [16].

Another critical aspect of tillage research lies in its integration with broader climate adaptation strategies. With climate change projected to exacerbate soil degradation and water scarcity, understanding how tillage practices influence water-use efficiency, erosion control, and soil thermal dynamics is essential for ensuring the resilience of agricultural systems [17]. However, such insights must be contextualized within the socio-economic realities of farming communities, particularly in temperate regions where farm sizes, resource availability, and policy frameworks vary widely [18].

The overall goal of this review is to evaluate the effects of different tillage practices on soil's physical properties, with particular emphasis on BD, Ks, WAS, AWC, PR, and SOC. Accordingly, a comprehensive literature review was conducted using peer-reviewed studies that examined tillage impacts on soil's physical characteristics in temperate climates. Despite the extensive body of research on tillage impacts, existing review articles and meta-analyses have largely focused on individual soil properties or specific tillage systems, offering limited comparative synthesis across multiple physical indicators of soil in temperate agroecosystems. This review advances beyond previous syntheses by integrating structural, hydraulic, and carbon-related indicators within a unified comparative framework, while explicitly accounting for variability, uncertainty, and evidence gaps across regions and management contexts. This approach is particularly relevant for


https://doi.org/10.3390/su18021083

temperate agroecosystem management, as it supports context-specific decision-making rather than uniform tillage prescriptions.

The primary objectives of this study are to (i) comprehensively review and quantitatively assess recent literature on the effects of different tillage practices on specific physical properties of soil in temperate climates, and (ii) identify knowledge gaps and future research directions in the field of tillage practices for improving soil's physical properties. Based on the objectives of this review, we hypothesized that reported responses of key physical indicators of soil to tillage practices in temperate agroecosystems are indicator-specific and context-dependent, rather than uniform across systems. We further hypothesized that structural indicators exhibit more consistent directional responses to conservation tillage (CT) than hydraulic indicators, for which outcomes remain more variable and uncertain due to differences in soil type, management duration, and limited evidence density. These hypotheses emphasize comparative synthesis and unresolved variability, rather than assuming universally beneficial effects of specific tillage systems.

## 2. Approach

### 2.1. Literature Identification and Appraisal Methods

A review was conducted to assess the impact of tillage practices on soil's physical properties in temperate climates. A comprehensive literature search was performed using Web of Science, Scopus, Google Scholar, and PubMed, employing keywords such as "soil physical properties", "bulk density", "saturated hydraulic conductivity", "available water capacity", "wet aggregate stability", "penetration resistance", "soil organic carbon", "no-tillage", and "tillage". The use of multiple databases ensured broad coverage of soil science, agronomy, and environmental research, thereby minimizing disciplinary and publication bias. The search was limited to peer-reviewed articles published in English between 1991 and 2025, a period that captures both early foundational studies on conservation tillage and more recent research addressing long-term and depth-dependent physical responses of soil; inclusion was restricted to studies conducted in temperate climates, verified using reported site locations, Köppen climate classification, and climate descriptions provided in each article. After screening approximately 300 publications for relevance and methodological rigor, a final set of 70 studies was retained, representing a balanced range of soil types, tillage systems, management durations, and geographic regions within temperate agroecosystems. This review adopted selected structured literature-search and screening practices to enhance transparency, but it was not conducted as a formal PRISMA-guided systematic review. A non-PRISMA or semi-systematic approach was adopted because the objective of this review was comparative and interpretive rather than effect-size estimation, focusing on indicator-specific responses, depth-dependent patterns, and methodological variability across studies. This approach allows integration of heterogeneous datasets, long-term field experiments, and studies reporting non-uniform depth intervals or response metrics, which are often excluded from formal meta-analyses. However, this approach does not provide pooled effect sizes and may be influenced by uneven evidence density across indicators. These limitations are explicitly acknowledged and addressed through transparent reporting of study distribution, variability, and evidence gaps within the synthesis. Studies were included if they met the following criteria: (i) conducted in temperate climates, (ii) field experiments or long-term trials ($\geq$2 years), (iii) specified tillage methods, (iv) provided quantitative data on at least one physical property of soil, and (v) included a control or baseline for comparison. A minimum duration of two years was chosen because soil's physical properties such as BD, PR, and near-surface aggregation can respond within multiple cropping cycles, whereas longer study durations are required to evaluate stabilization and


https://doi.org/10.3390/su18021083

depth-dependent trends. Studies with durations $\geq$ 3 years were therefore given greater interpretive weight when discussing long-term physical responses of soil.

Although SOC is often classified as a chemical property, this review included it among the physical indicators because it strongly influences soil structural behavior. SOC enhances aggregation, reduces BD, increases macroporosity, and improves water retention; thus, it is frequently integrated into soil physical quality (SPQ) frameworks assessing tillage effects [9]. Excluded studies consisted of conference proceedings, book chapters, laboratory-based experiments without field validation, and those focusing solely on soil chemical or biological properties. The final selection comprised 70 studies after an initial screening of 300 publications for relevance and methodological rigor.

### 2.2. Data Management and Standardization

The identified studies were cataloged using Zotero (Version 7.0.24), an open-source reference management software. Key study parameters, including location, soil texture, mean annual precipitation (MAP), mean annual temperature (MAT), tillage methods, experiment duration, soil sampling depth, and main findings, were extracted and tabulated. If MAP or MAT data were not reported, values were retrieved from the NOAA Climate Data Online (https://www.ncei.noaa.gov/ accessed on 17 September 2025) or other reliable sources.

To ensure consistency across studies, soil depth variability was addressed by categorizing data into surface and subsoil layers. When studies reported soil depth intervals that differed from our target (e.g., 0–10 cm), data were harmonized by calculating thickness-weighted averages across the reported layers so that values represented the equivalent standard depth. Data presented in graphical formats were digitized using WebPlotDigitizer (https://automeris.io/WebPlotDigitizer, accessed on 11 August 2025) to obtain numerical values. A qualitative synthesis of the reviewed literature was conducted to identify common findings, trends, and discrepancies. This approach facilitated the evaluation of hypotheses based on cumulative evidence from multiple studies, providing a comprehensive understanding of tillage effects on soil's physical properties.

This harmonization approach, while necessary for cross study comparison, introduces conceptual limitations that warrant explicit consideration. Soil's physical properties often exhibit strong vertical gradients, and aggregating heterogeneous depth intervals may mask localized responses, particularly at transitional layers influenced by tillage induced stratification. Similarly, differences in measurement techniques, such as laboratory versus field-based methods for Ks or AWC, varying aggregate stability protocols, and contrasting PR instruments, contribute to methodological variability that cannot be fully normalized through standardization alone. In addition, experimental duration varied widely among studies, ranging from short term trials to multi decadal experiments, and soil's physical responses, especially those related to structure and hydraulic behavior, are known to evolve nonlinearly over time. As a result, harmonized comparisons emphasize general patterns and ranges rather than precise effect magnitudes. These limitations are addressed in this review by explicitly reporting variability, avoiding overgeneralization, and interpreting results within the context of evidence density, depth resolution, and study duration.

Given the variability in tillage terminology and reference systems across studies, tillage practices were categorized based on their relative soil disturbance intensity as described by the original authors. When moldboard plow (MP) or full inversion tillage was used as the control or baseline, systems involving less soil disturbance (e.g., chisel plow, disk harrow, etc.) were classified as RT. Conversely, in studies where a cultivator or shallow tillage served as the standard, even lower-intensity methods, such as strip-till or shallow disking, were categorized as RT. NT was frequently defined as the absence of primary tillage, with soil disturbance limited to seeding operations. In this study, minimum tillage is considered



https://doi.org/10.3390/su18021083

a form of RT because it represents a system with substantially lower soil disturbance than CvT. In cases where tillage intensity was not clearly labeled or systems overlapped, classification was guided by reported parameters such as tillage depth, frequency, and implement type. This context-sensitive yet systematic approach enabled consistent cross-study comparisons while respecting the agronomic context of each experiment. While we acknowledge that classification overlap (e.g., a cultivator being "reduced" in one study but "conventional" in another) was sometimes unavoidable, we prioritized internal consistency within each study and applied a transparent framework to support reliable synthesis.

To ensure consistency in comparing BD values across studies, soil depths were categorized into two groups: 'Topsoil' (0–10 cm) and 'Composite depth.' The latter included any BD value reported for depths greater than 10 cm or for extended sampling profiles such as 0–15 cm, 0–30 cm, or 0–50 cm, regardless of whether the original authors referred to these values as subsoil, deeper soil, or profile average. For visual clarity, only the terms 'Topsoil' and 'Composite depth' are used in the graph and corresponding analysis presented in this review.

## 3. Result and Synthesis

### 3.1. Distribution of Reviewed Studies and Insights into Soil Sampling Depths

The implementation of various tillage practices across diverse regions of the USA and Europe (Figure 1) significantly impacts soil's physical properties, influencing aspects like BD, WAS, and SOC. Findings from the USA highlight that NT and RT systems often increase SOC levels and decrease BD, thereby improving soil quality. For example, studies conducted in Ohio, USA, using silty clay loam soils demonstrate that NT enhances soil structure and increases water infiltration.

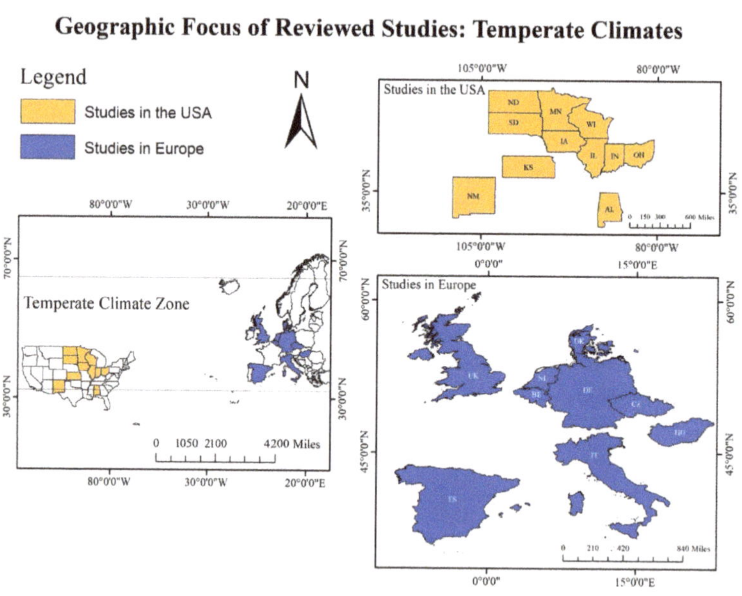


**Figure 1.** Locations of the reviewed studies in the USA (AL, Alabama; IN, Indiana; IA, Iowa; KS, Kansas; OH, Ohio; IL, Illinois; WI, Wisconsin; MN, Minnesota; SD, South Dakota; ND, North Dakota; NM, New Mexico) and Europe (IT, Italy; DK, Denmark; NL, The Netherlands; DE, Germany; ES, Spain; CZ, Czech Republic; UK, United Kingdom; BE, Belgium; and HU, Hungary) under temperate climate. The gray horizontal lines indicate the approximate boundaries of the temperate climate zone.


https://doi.org/10.3390/su18021083

Similarly, in Europe, such as in Denmark and the Netherlands, research indicates that RT tends to elevate PR in clay loam soils, pointing towards a denser soil structure under reduced disturbance. These observations suggest that the choice of tillage practice can have significant implications for soil quality and agricultural sustainability in different environmental and soil conditions.

Figure 2 illustrates the total number of studies on tillage focused on various physical properties of soil. WAS and SOC are the most studied properties, indicated by the highest counts of studies at 39 and 37, respectively. This is followed by BD with 35 studies, PR with 14 studies, Ks with 9 studies, and AWC with 5 studies. The figure highlights the prioritization of research areas within SPQ assessments under tillage conditions in temperate climates.

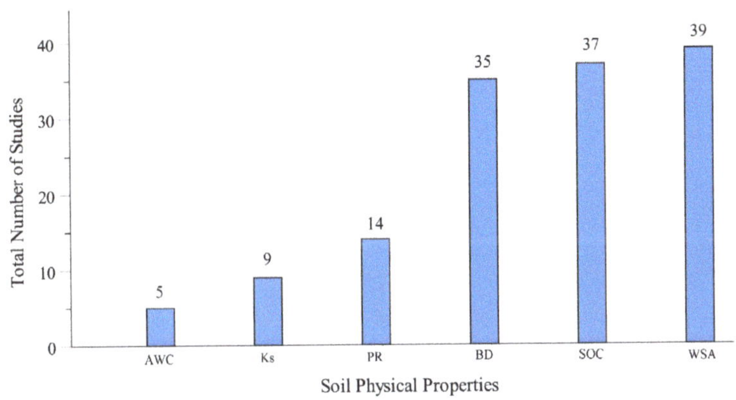


**Figure 2.** Distribution of reviewed studies (1991–2025) across physical properties of soil for tillage practices in the temperate climate zone (AWC: available water capacity; Ks: saturated hydraulic conductivity; PR: penetration resistance; BD: bulk density; SOC: soil organic carbon; WSA: water-stable aggregates).

Figure 3 illustrates the maximum soil sampling depths utilized in various tillage studies to investigate soil properties. The depth of soil sampling varied significantly across studies, ranging from as shallow as 5 cm to as deep as 60 cm. This variation highlights different research focuses and methodologies. For instance, deeper sampling depths, like those used in the studies by Fiorini et al. [19], and Zuber et al. [20], might be indicative of investigations into deeper soil layers' physical properties or nutrient profiles, which are crucial for understanding subsoil dynamics under different tillage practices. Conversely, studies with shallower sampling depths might focus on topsoil characteristics, which are directly influenced by tillage and are critical for seed germination and root development.

### 3.2. *Evaluation of Tillage Effects on Soil's Physical Properties*

Across indicators, responses to NT were not uniform but depended strongly on soil depth, texture, residue management, and duration of implementation. While NT generally improved surface structural indicators, responses of hydraulic and mechanical properties were more variable, particularly at depth.

#### 3.2.1. Trends in Soil Bulk Density Under Different Tillage Practices

Across temperate agroecosystems, BD responses to tillage practices show a clear depth-dependent pattern rather than uniform differences among systems (Table 1). CT, particularly NT and RT, generally results in lower BD in the surface soil, reflecting improved aggregation, residue retention, and reduced mechanical disturbance. Several studies


https://doi.org/10.3390/su18021083

consistently report surface BD reductions under NT relative to CvT, indicating enhanced structural stability and reduced compaction in the upper soil layer [21,22]. For example, Presley et al. [21] documented substantially lower BD under NT compared to CvT in the 0–10 cm depth, highlighting the role of long-term residue accumulation and biological activity in improving surface soil structure.

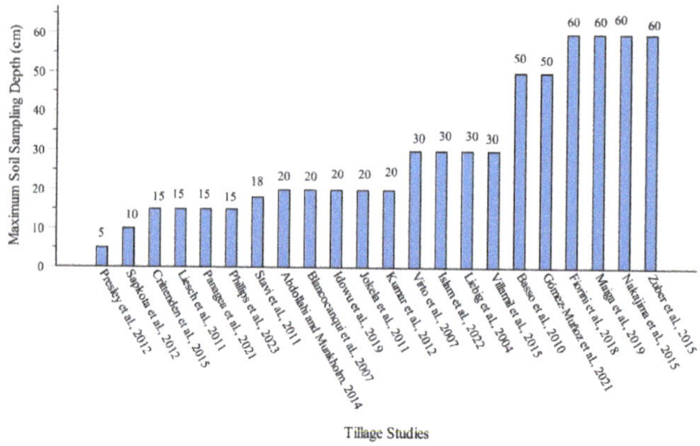


**Figure 3.** Maximum soil sampling depths reported in studies assessing the impact of tillage methods on soil's physical properties in temperate climate zones [11,19–39].

**Table 1.** Tillage effects on bulk density (BD) under different soil types, climates, soil sampling depth, and management durations. MAP = mean annual precipitation; MAT = mean annual temperature; NT = no-tillage; RT = reduced tillage; CvT = conventional tillage; MP = moldboard plow, NSE = No Significant Effect.

| Location | Texture | MAP (mm) | MAT (°C) | Tillage | Time (yr) | Depth (cm) | Major Findings | References |
|---|---|---|---|---|---|---|---|---|
| Kansas, USA | Silty clay loam | 780 | 13.1 | NT, CvT | Long-term | 0–5 | NT had lower BD | [21] |
| Ohio, USA | Silt loam | 940 | 10.6 | NT, Disk plow | 10 | 0–18 | CvT had higher BD | [26] |
| North Dacota, USA | Sandy loam | 366 | 5.4 | NT, Ripper | 4 | 0–40 | Ripper has higher BD | [40] |
| Kansas, USA | Silty clay loam | 580 | 12.2 | NT, RT | 33 | 0–7.5 | NT had lower BD | [41] |
| Illinois, USA | Silt loam | 1015 | 11.5 | NT, CvT | 15 | 0–60 | NT had higher BD | [20] |
| Ohio, USA | Silt loam | 1016 | 11.1 | NT, Chisel | 13 | 0–10 | NT had lower BD | [42] |
| Italy | Sandy loam | 850 | 14.7 | NT, CvT | 15 | 0–10 | NT had lower BD | [22] |
| North Dacota, USA | Loam | 477 | 5.5 | NT, CvT | 27 | 0–15 | NSE | [43] |
| South Dacota, USA | Silty clay loam | 627 | 8.6 | NT, CvT | 23 | 0–15 | NT had lower BD | [44] |
| Ohio, USA | Clay loam | 845 | 10.9 | NT, Chisel, MP | 47 | 0–10 | NT had lower BD | [31] |
| Germany | Loamy sand, Sandy loam | 498 | 9 | NT MP | 10+ | 0–35 | NT had higher BD | [45] |
| Denmark | Sandy loam | 765 | 8.3 | NT, CvT | 17 | 0–50 | NT had lower BD | [37] |
| Ohio, USA | Silt loam | 950 | 11.4 | NT, Chisel, MP | 43 | 0–15 | NT had lower BD | [46] |


https://doi.org/10.3390/su18021083

However, aggregated evidence indicates that these benefits are less consistent at greater depths. Figure 4 illustrates that differences in BD between NT and MP systems often diminish or overlap at composite or subsoil depths, suggesting that reduced disturbance alone may not alleviate compaction below the surface layer. This variability reflects differences in soil texture, legacy compaction, and the duration of NT implementation across studies. As a result, while NT frequently improves surface BD, its effectiveness in reducing deeper soil compaction remains context dependent.

Strategic tillage approaches, including OT and RT, further indicate that BD responses are influenced by both tillage intensity and the frequency of soil disturbance. Obour et al. [47] reported similarly low BD under strategic and reduced tillage, indicating that occasional soil disturbance can mitigate surface compaction without fully disrupting soil structure in certain systems. In contrast, Stavi et al. [26] observed higher BD under occasional tillage compared to continuous NT, emphasizing that the benefits of strategic tillage are not universal and depend on site-specific conditions. Collectively, these findings indicate that BD responses to tillage are controlled more by soil depth, management history, and disturbance timing than by tillage category alone, underscoring the need for context-specific interpretation of tillage effects on soil compaction.

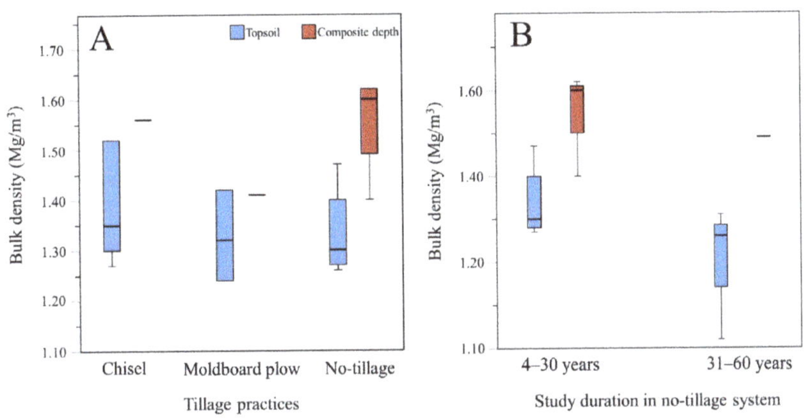


**Figure 4.** Bulk density under different tillage systems (**A**) and study durations in NT systems (**B**), categorized by soil depth. 'Topsoil' refers to 0–10 cm, while 'Composite depth' includes values from extended or unsegmented profiles (e.g., 0–15 cm, 0–50 cm). The short horizontal gray lines indicate cases where only a single observation was available (N = 1), and therefore no box or whiskers could be displayed [21,31,40–42,45,46,48–53].

The effects of tillage practices and study duration on BD at varying soil depths are presented in Figure 4. In the topsoil layer (0–10 cm), BD was slightly lower under NT (1.32 ± 0.08 Mg/m$^3$) compared with MP (1.33 ± 0.09 Mg/m$^3$) and chisel tillage (1.39 ± 0.12 Mg/m$^3$). However, two-way ANOVA indicated that tillage effects on BD were not statistically significant ($F = 1.39$, $p = 0.272$). Sample sizes (N) varied from 1 to 8 per group. The absence of whiskers reflects either limited sample size or lack of variability, while groups represented by a single value, such as chisel tillage at composite depth, do not have an associated standard deviation. The data were synthesized from published studies, where boxplots display the interquartile range and median, and whiskers represent the overall data range. The absence of statistically significant differences may be attributed to the high variability among soils and study sites, which can obscure management effects. BD is particularly sensitive to soil texture, organic matter content, and inherent structural heterogeneity, reducing statistical


https://doi.org/10.3390/su18021083

power even when apparent mean differences exist. Moreover, the limited number of studies and unequal replication across tillage categories likely constrained the ability to detect clear treatment effects.

This lack of significance may reflect high variability across soils and sites, which can mask management effects. BD is strongly influenced by soil texture, organic matter content, and natural structural heterogeneity, which can reduce statistical power even when mean differences appear evident. Additionally, the relatively small number of studies and unequal replication among tillage categories may have limited detection of clear treatment effects.

In contrast, soil depth exhibited a significant effect on BD (F = 12.49, $p = 0.002$), with greater compaction observed at deeper or composite layers. BD at composite depths (>10 cm) was higher under NT ($1.56 \pm 0.09$ Mg/m$^3$) and chisel systems (1.56 Mg/m$^3$, based on a single composite-depth observation), compared with MP (1.41 Mg/m$^3$, single reported value without standard deviation). Although single data points were used for some composite depths, they were retained to represent existing literature trends rather than for inferential comparison. These values provide descriptive context for depth-wise compaction under reduced soil disturbance. The interaction between tillage and depth was not significant ($p = 0.391$), indicating similar BD patterns across systems when comparing surface and subsurface layers.

Analysis of NT study duration revealed that BD in the topsoil layer tended to be higher in the 4–30-year group ($1.34 \pm 0.09$ Mg/m$^3$) than in the 31–60-year group ($1.21 \pm 0.13$ Mg/m$^3$), although the difference was not significant (F = 1.50, $p = 0.252$). For composite layers, BD was slightly higher in the 4–30-year group ($1.54 \pm 0.12$ Mg/m$^3$) than in the 31–60-year group (1.49 Mg/m$^3$), the latter derived from a single reported value. As before, these single data points were included descriptively to illustrate the long-term trend of gradual BD reduction with extended NT duration. The interaction between study duration and soil depth was also non-significant (F = 0.30, $p = 0.596$), suggesting consistent BD variation across depths regardless of NT duration. However, the main effect of soil depth remained significant (F = 10.18, $p = 0.011$), confirming that compaction generally increases with depth, independent of management duration.

These findings suggest that depth is the dominant factor influencing BD, rather than tillage system or duration alone. While NT shows potential for maintaining or slightly lowering surface BD over time, its effect is inconsistent and may not be sufficient to alleviate subsoil compaction. These results align with previous findings [54], emphasizing the need for integrated strategies, including deep-rooted cover crops, to mitigate subsoil compaction in long-term NT systems.

The broader implications of these findings are profound, indicating that while NT is generally beneficial for reducing BD and enhancing soil structure, it is not universally applicable. Specific conditions such as soil type, climatic conditions, and crop rotation must be considered. For instance, Fiorini et al. [19] observed a temporal decrease in BD under NT, which suggests that the benefits of NT accrue over time, enhancing soil's physical properties progressively.

However, the adoption of NT should be carefully managed. Wander et al. [55] and Weninger et al. [56] noted instances where NT induced higher BD compared to CvT, particularly in silt-dominated soils, potentially leading to increased soil compaction. This indicates that NT can have counterintuitive effects depending on the soil texture and environmental conditions. Moreover, Fornara and Higgins [57] demonstrated a positive correlation between BD and the frequency of tillage and reseeding events, which suggests that not all NT systems automatically reduce soil compaction, and strategic



https://doi.org/10.3390/su18021083

tillage might be necessary to mitigate some of the compaction issues associated with continuous NT.

3.2.2. Variations in Saturated Hydraulic Conductivity Across Tillage Systems

The adoption of CT practices such as NT and RT is generally associated with improved soil hydraulic properties that are critical for effective water management (Table 2). Across studies, NT consistently enhances Ks relative to CvT, reflecting improved soil structure, greater pore continuity, and reduced compaction in surface soils [21,58]. Reported increases in Ks under NT indicate more efficient water transmission and aeration within the soil profile, which are key functions of well-developed macropore networks. However, the magnitude of Ks improvement varies among studies and is influenced by soil texture, residue management, and the duration of CT implementation. While some studies report substantial increases in Ks under NT, others show more moderate responses, suggesting that hydraulic benefits are conditional rather than universal. Overall, these patterns indicate that Ks responses to CT are governed more by the preservation of soil pore structure and biological activity than by tillage category alone.

**Table 2.** Tillage effects on saturated hydraulic conductivity (Ks) under different soil types, climates, soil sampling depth, and management durations. MAP = mean annual precipitation; MAT = mean annual temperature; NT = no-tillage; RT = reduced tillage; CvT = conventional tillage; MP = moldboard plow; NIT = Non-inversion tillage, NSE = No Significant Effect.

| Location | Texture | MAP (mm) | MAT (°C) | Tillage | Time (yr) | Depth (cm) | Major Findings | References |
|---|---|---|---|---|---|---|---|---|
| North Dakota, USA | Sandy loam | 366 | 5.4 | NT | 10 | 0–40 | NSE | [40] |
| Ohio, USA | Silt loam | 1039 | 11.5 | NT, Chisel | Since 1994 | 0–60 | NT had higher Ks | [39] |
| Minnesota, USA | Silt loam | 652 | 5.3 | NT, CvT | Since 2007 | 0–5, 10–15 | NT had higher Ks than CvT | [24] |
| Iowa and Ohio, USA | Loam to clay loam | 886–1019 | 8–10 | NT, Chisel, MP | 6 to 28 | 0–7.5 | NT had higher Ks | [59] |
| Wisconsin, USA | Sandy | 832 | 6.7 | NT, MP | 5 | Surface | MP had higher Ks | [60] |
| Ohio, USA | Silt loam, clay loam | 1083 | 11.2 | NT | Since 2004 | 0–20 | Reduced Ks | [29] |
| Germany | Silt loam | 700 | 9.9 | RT, CvT | 10 | Surface | Reduced Ks for RT, CvT | [61] |
| The Netherlands | Clay loam | 825 | 9.7 | MP, NIT | 5 | 0–5, 10–15 | MP had higher Ks | [23] |
| Minnesota | Silt loam | 887 | 4.4 | NT, MP | 10 | surface | NT had higher Ks | [60] |

However, the benefits of NT on soil hydraulic properties are not uniform and can be influenced by management practices such as residue management. Blanco-Canqui et al. [28] highlighted a significant reduction in Ks from 4.2 to 0.6 mm/h with complete stover removal under NT, showcasing a potential downside to inappropriate residue management that can negate the benefits of conservation practices by impairing the soil's ability to transmit water.


https://doi.org/10.3390/su18021083

RT systems also contribute positively to soil hydraulic functions, although the extent can vary depending on specific practices and environmental conditions. Crittenden et al. [23] noted that non-inversion tillage (NIT), a form of RT, improved Ks in organic farming settings, particularly evident during autumn when soil compaction is less pronounced. This improvement is likely due to less soil disturbance compared to CvT, which preserves natural soil architecture and enhances pore connectivity.

Conversely, CvT often degrades these beneficial structures, leading to decreased Ks. Rasmussen [62] documented increased bulk density and reduced macroporosity under CvT, which adversely affects the soil's hydraulic capabilities. This degradation underscores the critical need for strategic management of tillage practices to avoid counterproductive outcomes on SPQ. Analogous challenges of pore blockage and hydraulic decline have been documented in engineered water systems, where maintaining pore continuity is critical for long-term functionality, reinforcing the importance of managing soil pore networks to sustain hydraulic conductivity under CT systems [63].

The implications of these findings are substantial for promoting the adoption of CT. They highlight the necessity for careful management of tillage depth and residue to optimize the hydraulic benefits of NT and RT systems. Such practices not only enhance the soil's water handling characteristics but also contribute to broader environmental conservation goals by improving water quality and reducing erosion. The integration of CT into modern agricultural practices offers a sustainable approach to managing soil and water resources effectively. However, the variable effects documented in different studies call for a nuanced application of these practices, tailored to specific soil types, climatic conditions, and cropping systems to maximize their benefits and mitigate potential drawbacks.

#### 3.2.3. Comparative Analysis of Available Water Capacity

Tillage practices exert complex and context-dependent effects on AWC (Table 3). Across studies, NT generally enhances AWC through improvements in soil structure and increased SOC, which promote favorable pore size distribution and greater water retention [31]. Similar improvements in AWC have also been reported under RT, where enhanced soil structure and higher SOC contribute to improved water holding capacity (WHC) relative to more intensive tillage systems [64]. Evidence from controlled incubation studies further indicates that optimized pore structure and surface charge can enhance WHC even under CvT conditions, as demonstrated by acid-modified pineapple biochar application in saline soils [65]. In contrast, intensive tillage practices often reduce AWC by degrading soil structure and lowering SOC. CvT has been associated with increased BD and reduced air capacity, which constrain both soil aeration and water retention, particularly in silt-dominated soils susceptible to compaction and erosion [56]. Collectively, these patterns indicate that AWC responses are governed by interactions among soil structure, SOC dynamics, and soil texture rather than tillage intensity alone.

Panagea et al. [11] noted that while SOC increases generally enhance water retention, the effects are more pronounced in coarser-textured soils with lower initial SOC content and less significant in finer-textured soils. CT practices, such as NT and RT, have frequently demonstrated measurable benefits by significantly increasing AWC. Similarly, long-term studies in Ohio showed that NT practices resulted in higher AWC in well-drained silt loam soils, providing evidence of its sustainable benefits in moisture retention and availability [31].


https://doi.org/10.3390/su18021083

**Table 3.** Tillage effects on available water capacity (AWC) under different soil types, climates, soil sampling depth, and management durations. MAP = mean annual precipitation; MAT = mean annual temperature; NT = no-tillage; RT = reduced tillage; CvT = conventional tillage; MP = moldboard plow; NSE = No Significant Effect.

| Location | Texture | MAP (mm) | MAT (°C) | Tillage | Time (yr) | Depth (cm) | Major Findings | References |
|---|---|---|---|---|---|---|---|---|
| Ohio, USA | Silt loam | 1016 | 11.1 | NT, Chisel | 13 | 0–10 | NSE | [42] |
| Spain | Sandy loam | 869 | 16 | NT, CvT | 8 | 0–5 | NT had higher AWC | [66] |
| Belgium, Czech Republic, Hungary, Italy, UK | Varies | Varies | Varies | NT, CvT, RT | 8 to 54 | 0–15 | Minimal effect on AWC | [11] |
| Illinois, USA | Silt loam | 949 | 11.7 | NT, Chisel, MP | 8 | 0–75 | NSE | [67] |
| Spain | Silt loam | 448 | 15.6 | NT, CvT | 8 | 0–30 | NT had higher AWC | [68] |

Variations in AWC among tillage systems arise from differences in soil structure, pore distribution, and organic carbon stabilization. NT enhances aggregation and mesoporosity, increasing water retention and reducing evaporative loss through improved structural stability [31,68]. In contrast, intensive tillage destroys aggregates and macropores, leading to reduced pore continuity and limited plant-available water [56]. RT exerts intermediate effects by loosening soil while maintaining SOC protection [63]. These findings collectively demonstrate that practices preserving soil structure and carbon are key to sustaining AWC in temperate agroecosystems.

#### 3.2.4. Evaluating Aggregate Stability for Assessing Soil Health

Different tillage practices exert pronounced effects on soil WAS, with responses reflecting both management intensity and soil structural preservation (Table 4). Across studies, conservation practices such as NT consistently enhance WAS by maintaining aggregate integrity and minimizing mechanical disruption. Improved WAS under NT is commonly attributed to greater aggregate cohesion driven by residue retention and increased SOC, resulting in higher proportions of stable aggregates compared to CvT systems [31,34]. RT systems further demonstrate that partial disturbance can still support aggregate stability when soil structure is largely preserved. Higher WAS and increased aggregate MWD under RT indicate improved soil structural condition relative to more intensive tillage, particularly in surface soils [69]. Similarly, combinations of organic management and RT have been shown to promote the formation of large macroaggregates and higher MWD in the upper soil profile, reinforcing the role of reduced disturbance and organic inputs in stabilizing soil aggregates [70]. Collectively, these patterns indicate that WAS responses are governed primarily by the degree of structural disturbance and organic matter inputs rather than tillage category alone.


https://doi.org/10.3390/su18021083

**Table 4.** Tillage effects on wet aggregate stability (WAS) under different soil types, climates, soil sampling depth, and management durations. MAP = mean annual precipitation; MAT = mean annual temperature; NT = no-tillage; RT = reduced tillage; CvT = conventional tillage; MP = moldboard plow; NSE = No Significant Effect.

| Location | Texture | MAP (mm) | MAT (°C) | Tillage | Time (yr) | Depth (cm) | Major Findings | References |
|---|---|---|---|---|---|---|---|---|
| Wisconsin, USA | Silt loam | 889 | 8.3 | CvT | 18 | 0–20 | Higher in pasture and alfalfa-based systems | [30] |
| Spain | Sandy loam | 389 | 14.7 | NT, Chisel, MP | 16 | 0–7.5 | Higher in NT | [71] |
| Illinois, USA | Silty clay loam | 965 | 11.5 | NT, CvT | 15 | 0–60 | Higher in NT | [20] |
| South Dakota, USA | Loam | 654 | 7.8 | NT, Chisel, MP | 2–3 | 0–15 | Higher in NT | [72] |
| Ohio, USA | Silt loam | 940 | 10.6 | NT, Disk plow | 10 | 0–12 | Higher in continuous NT | [26] |
| Kansas, USA | Silt loam | 580 | 26.2 | NT, CvT | 45 | 0–2.5 | Higher in NT | [73] |
| Northern Great Plains, USA | Silt loam | 425 | 7.2 | NT, CvT | 17 and 8 years | 0–30 | Higher in NT | [34] |
| Denmark | Sandy loam | 558 | 8.5 | NT, MP | 12 | 0–20 | NSE | [27] |
| Germany | Clay loam | 700 | 9 | CvT, RT | 10 | 10–240 | Higher in RT | [69] |
| Spain | Clayey | 525 | 13.5 | NT, CvT | 10 | 0–30 | NSE with stubble burning | [32] |
| Ohio, USA | Silt loam | 1016 | 11 | NT | 22 | 0–10 | Higher WSA in NT, 68.7% | [52] |

Long-term NT practices, coupled with diversified crop rotations, also showed significant WAS improvements [38]. However, RT may increase soil BD and induce soil consolidation, as Bilibio et al. [69] noted at a 0.10–0.24 m depth range, where CvT systems had lower total pore space and higher BD compared to conventionally plowed treatments. Büchi et al. [74] highlighted substantial intra-system heterogeneity in soil properties, sometimes surpassing differences between tillage regimes. Moreover, Stavi et al. [26] found that even sporadic soil disturbance in long-term NT systems negatively impacted soil quality indicators, including a decrease in WAS and MWD. This suggests that while NT practices generally benefit WAS, the degree of improvement is strongly influenced by soil texture, initial organic carbon content, and baseline aggregation, with finer-textured, carbon-rich soils responding more positively than coarse or degraded soils. Climatic conditions, such as precipitation regime and freeze–thaw intensity, also shape outcomes: humid temperate regions tend to favor macroaggregate stabilization, whereas semi-arid or drought-prone environments may limit WAS gains due to moisture deficits. Occasional disturbance can further offset these benefits by disrupting aggregate stability and reducing MWD.

#### 3.2.5. Long-Term Variations in Soil Penetration Resistance

The effects of tillage practices on SPQ, particularly PR, show strong depth- and management-dependent responses with important implications for soil function (Table 5). PR is a key indicator of soil compaction and directly influences root growth and crop productivity. Across studies, NT frequently reduces PR in the topsoil, especially in long-term systems with residue retention and soil organic matter (SOM) accumulation. These reductions are commonly linked to minimal mechanical disturbance, improved soil structure, and increased biological activity in surface layers [19,29,31].



https://doi.org/10.3390/su18021083

In contrast, PR responses at greater soil depths are more variable. Several studies report higher PR under NT in subsoil layers, reflecting the persistence or accumulation of compaction in the absence of periodic mechanical loosening [25,31]. Elevated subsoil PR under NT has been associated with soil texture, historical compaction, and management duration. During the transition from CvT to NT, soils may initially exhibit increased PR due to residual compaction from previous tillage operations, with gradual structural adjustment occurring over time [23]. Collectively, these findings indicate that PR responses to NT are governed by interactions among soil depth, management history, and SOM dynamics rather than being uniformly beneficial.

**Table 5.** Tillage effects on penetration resistance (PR) under different soil types, climates, soil sampling depth, and management durations. MAP = mean annual precipitation; MAT = mean annual temperature; NT = no-tillage; ST = striptillage; CvT = conventional tillage; MP = moldboard plow; NIT = Non-inversion tillage.

| Location | Texture | MAP (mm) | MAT (°C) | Tillage | Time (yr) | Depth (cm) | Major Findings | References |
|---|---|---|---|---|---|---|---|---|
| Illinois, USA | Silt loam | 890–914 | 10.6–11.1 | NT, Chisel plow | 7 | 0–30 | Higher PR in NT with residue removal | [35] |
| Ohio, USA | Silt loam, Clay loam | 845–905 | 9.1–9.9 | NT, MP, Chisel plow | 47–49 | 0–20 | Higher PR in NT | [31] |
| New Mexico, USA | Sandy | 250 | 17 | NT, CvT, ST | 2 | 0–20 | Higher PR in NT | [29] |
| Ohio, USA | Silt loam | 940 | 10.6 | NT, Disk plow | 10 | 0–18 | Higher PR in NT | [26] |
| Iowa, USA | Silty clay loam | 733 | 10 | NT, Chisel plow | 13 | 0–15 | Higher PR in CP | [25] |
| Wisconsin, USA | Silt loam | 889 | 8.3 | CvT | 18 | 0–20 | Higher PR in alfalfa-based systems | [30] |
| The Netherlands | Clay loam | 825 | 9.7 | MP, NIT | 3 to 4 | 0–5, 10–15 | Higher PR under NIT | [23] |
| Denmark | Sandy loam | 626 | 7.3 | NT, MP | 10 | 0–20 | Higher PR in NT | [27] |
| Spain | Clay loam | 525 | 13.5 | NT, MP | 10 | 0–10 | Higher PR in NT | [32] |
| Italy | Silty clay loam | 890 | 12.2 | NT, CvT | 3 | 0–60 | Higher PR in NT | [19] |

CvT methods such as moldboard plow effectively reduce PR, particularly in the subsoil, by mechanically disrupting compacted layers, which facilitates root growth and water infiltration. While this immediate reduction in PR is advantageous for crop establishment and root penetration, it is not devoid of long-term disadvantages. Frequent use of CvT can lead to increased soil erosion and decreased SOM, presenting a trade-off between immediate benefits and sustainable soil health [75,76]. Therefore, the choice of tillage system should be a strategic decision influenced by specific soil conditions, crop requirements, and long-term sustainability goals. The varying impacts of NT and CvT on soil PR exemplify the intricate relationship between tillage practices and soil health. To optimize agricultural productivity and environmental sustainability, management strategies must carefully consider both the immediate and extended consequences of tillage on soil structure and function [77].

#### 3.2.6. Dynamics of Soil Organic Carbon Storage and Distribution

CT methods, particularly NT and RT, consistently promote SOC accumulation, which plays a critical role in soil fertility and climate change mitigation (Table 6). Across studies,


https://doi.org/10.3390/su18021083

higher SOC concentrations are commonly reported in surface soils under NT relative to CvT, reflecting the combined effects of reduced soil disturbance, residue retention, and enhanced organic matter stabilization [39,78]. These patterns indicate that minimal mechanical disruption under NT favors the preservation and accumulation of SOC, particularly in the upper soil layers. In addition to increasing SOC, long-term NT management is closely linked with improvements in soil's physical structure. Enhanced SOC under NT is frequently accompanied by lower BD and higher WSA, indicating stronger aggregation and improved structural stability [21]. Evidence from temperate agroecosystems further demonstrates that sustained organic inputs under NT support coupled gains in SOC and nitrogen sequestration while simultaneously improving aggregation and reducing BD over time [79]. Collectively, these findings highlight that SOC responses to CT are governed by interactions between organic matter inputs, disturbance intensity, and soil structural development, rather than tillage category alone.

**Table 6.** Tillage effects on soil organic carbon (SOC) under different soil types, climates, soil sampling depth, and management durations. MAP = mean annual precipitation; MAT = mean annual temperature; NT = no-tillage; RT = reduced tillage; CvT = conventional tillage.

| Location | Texture | MAP (mm) | MAT (°C) | Tillage | Time (yr) | Depth (cm) | Major Findings | References |
|---|---|---|---|---|---|---|---|---|
| South Dakota, USA | Silty | 508 | 8.3 | NT | 14–24 | 0–60 | 4-year rotation increased SOC | [38] |
| Kansas, USA | Silt loam | 440 | 17.2 | NT, RT | 19 | 0–5 | NT increased SOC | [80] |
| Illinois, USA | Silty clay loam | 965 | 11.7 | NT, CvT | 15 | 0–60 | NT increased SOC | [20] |
| Alabama, Indiana, Ohio, USA | Varies | 991–1405 | 10.9–17.8 | NT | 4 | 0–30 | NT increased SOC | [33] |

The efficacy of NT in increasing SOC is, however, variable and may diminish as SOC levels begin to stabilize over time, as noted by VandenBygaart et al. [81]. This variability underscores the complex interaction between tillage practices and soil carbon dynamics. Additionally, the adoption of NT can occasionally lead to soil compaction and reduced water infiltration, potentially impacting crop yields adversely over time [82]. Such negative outcomes highlight the need for careful management and consideration of local soil conditions when implementing NT practices. Additionally, the integration of biochar as a soil amendment within CT systems has been suggested to address challenges like soil compaction, though environmental and health considerations related to biochar production and application must be carefully managed [83].

Despite these challenges, the long-term benefits of NT, particularly when combined with diverse crop rotations and organic amendments, have been demonstrated to significantly enhance SOC, microbial biomass, and overall soil health. Maiga et al. [38] reported improvements in these parameters, reinforcing the potential of NT to enhance agricultural sustainability. In similar findings, Krauss et al. [84] observed a 25% increase in topsoil SOC under reduced tillage coupled with organic amendments over a 15-year period, suggesting that a combination of reduced tillage and organic matter addition can synergistically boost SOC levels.

Moreover, Büchi et al. [74] have further refined our understanding of how tillage impacts SOC. They noted that NT, especially when integrated with cover cropping in soybean systems, resulted in significantly higher SOC levels compared to RT, illustrating the added benefits of cover crops in carbon sequestration strategies. They emphasized that pedoclimatic factors alongside management practices significantly dictate the extent

 https://doi.org/10.3390/su18021083

of SOC improvement, with NT and organic systems showing higher SOC at the surface compared to conventionally tilled fields.

Given these findings, it is clear that while CT practices like NT and RT can enhance SOC levels and improve soil health, the outcomes are dependent on a multitude of factors that require careful consideration and management. Continuous monitoring and adaptive management strategies are essential to optimize the benefits of CT for SOC enhancement, ensuring that agricultural practices align with sustainability goals while adapting to specific local conditions.

#### 3.2.7. Cross-Indicator Integration and Trade-Offs

Across the reviewed literature, tillage-induced changes in soil's physical and hydraulic indicators are closely interlinked rather than independent (Figure 5). Improvements in SOC under CT, particularly NT and RT, are consistently associated with enhanced WAS and reduced BD in surface soils, which collectively support improved pore continuity and higher Ks [21,31,34,39,78]. Increased aggregation and lower BD under NT further contribute to improved AWC by promoting a favorable balance of storage pores and stable soil structure [31,64]. These synergistic responses highlight the central role of SOC in coupling structural stability with hydraulic functioning under reduced disturbance systems.

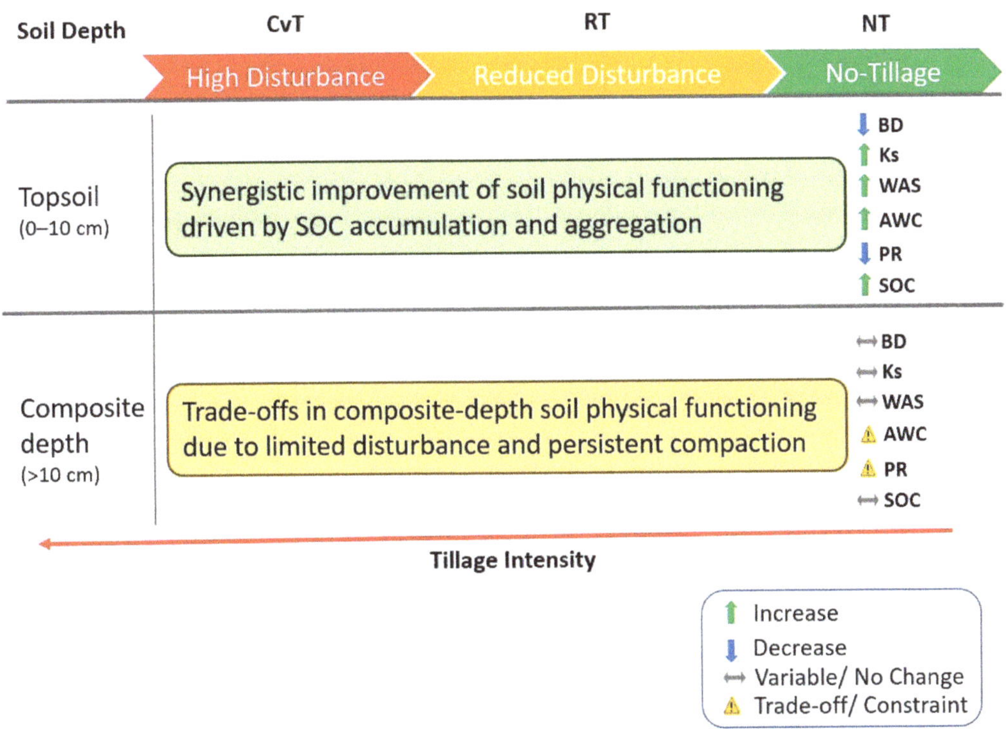


**Figure 5.** Conceptual framework showing soil's physical responses to tillage intensity across soil depths. Tillage systems range from conventional tillage (CvT) to reduced tillage (RT) and no-tillage (NT). Arrows indicate direction of change in soil properties across depths. (BD = bulk density; PR = penetration resistance; Ks = saturated hydraulic conductivity; WAS = wet aggregate stability; AWC = available water capacity; SOC = soil organic carbon). Arrows indicate the direction of change in soil properties: ↑ increase, ↓ decrease, ↔ variable or no change, and ⚠ trade-off or constraint.



https://doi.org/10.3390/su18021083

Clear trade-offs emerge across indicators between the surface soil and composite depth layers (Figure 5), highlighting the depth-dependent nature of tillage effects on soil's physical functioning. While NT frequently improves SOC, WAS, Ks, and AWC in the topsoil, several studies report elevated PR at greater depths, indicating persistent or accumulated subsoil compaction in the absence of periodic mechanical loosening [23,25,31]. This depth-dependent divergence suggests that gains in surface soil hydraulic and structural quality may coexist with mechanical constraints in the subsoil. Consequently, SPQ under CT reflects a balance between structural stabilization and compaction dynamics that varies with soil depth, management duration, and soil texture. These findings underscore the importance of evaluating tillage effects through integrated indicator responses rather than relying on single-property assessments.

## 4. Future Research Directions

Despite extensive research on the impacts of diverse tillage methods on soil's physical properties, several critical knowledge gaps remain. Long-term effects of CT practices, such as NT and RT, on soil structure and hydraulic properties across diverse soil types and climates need further study [9]. While CT and organic amendments have demonstrated clear benefits for SPQ, their long-term impacts under varying climatic and soil conditions require further investigation. Munna and Lal [85] highlight the need for region-specific management strategies that integrate organic inputs with reduced tillage to optimize soil water retention and structure. Future research should focus on multi-year trials to assess the interactive effects of organic amendments, tillage intensity, and residue management on soil hydraulic properties. Short-term studies show promising results in improved WAS and water infiltration rates, but the long-term sustainability and potential limitations, especially in the context of climate change, require more thorough evaluation [12]. Policies that promote sustainable tillage practices are essential for their widespread adoption. Kremen and Miles [86] argue that regulatory frameworks and incentive-based approaches can drive farmer behavior toward conservation practices. Further research is needed to evaluate the effectiveness of existing policies and design innovative approaches that combine financial, educational, and environmental incentives.

Long-term studies are essential for assessing the sustained effects of tillage practices on soil's physical properties. Short-term studies often fail to capture cumulative impacts on SOM, crop yields, and profitability [13]. Similar time-dependent trade-offs have been observed in land restoration systems, where vegetation introduction initially altered soil's physical and hydrological functioning but led to long-term recovery of soil structure and ecosystem services, underscoring the importance of temporal scale in evaluating soil–water management interventions [87]. Longitudinal research across diverse climatic and soil conditions is necessary to develop region-specific sustainable practices. Given climate change, tillage must adapt, integrating advanced models that incorporate climate variables to predict soil responses. Such models facilitate understanding interactions between tillage, precipitation, temperature fluctuations, and extreme weather [16], offering data-driven insights for resilient agricultural systems.

Emerging technologies such as remote sensing, AI, and PA provide high-resolution, real-time soil data, optimizing management strategies [88]. IoT-enabled soil sensors enhance predictive capabilities, improving on-farm decision-making. Comparative studies across ecological zones are crucial for identifying regional adaptations to tillage practices, considering variations in soil, climate, and cropping systems [89]. Such studies reveal global trends and localized solutions for enhancing SPQ.

Economic analyses of tillage should consider both direct and indirect costs. Evaluating long-term benefits of improved SPQ against short-term adoption costs can guide policy and



https://doi.org/10.3390/su18021083

decision-making [90]. Additionally, subsoil characteristics such as compaction and porosity are often overlooked despite their role in water movement and root development [10]. Depth-specific studies can enhance soil health management strategies. These research priorities highlight the need for an integrated, interdisciplinary approach to tillage, addressing agricultural sustainability and broader environmental challenges.

## 5. Conclusions

This review advances a synthesis-level understanding of how CT practices, particularly NT and RT, influence soil's physical and hydraulic functioning in temperate agroecosystems. Rather than indicating uniform benefits across all indicators, the evidence highlights depth-dependent and indicator-specific responses that reflect trade-offs between structural stabilization and compaction dynamics. Conceptually, this review integrates BD, Ks, WAS, AWC, PR, and SOC within a unified framework that emphasizes coupled physical processes in soil rather than isolated responses. Across studies, NT was associated with reduced BD in the topsoil and enhanced Ks, particularly under residue retention, although evidence for hydraulic responses remains more limited and variable than for structural indicators. Improvements in WAS under continuous NT reached 33.4%, while long-term RT practices incorporating organic amendments increased topsoil SOC by approximately 25%. These surface soil benefits are closely linked through SOC-driven aggregation and pore development. However, interpretations of SOC and AWC responses should be considered tentative, as these properties are supported by comparatively few studies. Despite these benefits, BD remained high at composite depths under NT, reflecting limited subsoil disturbance and ongoing compaction. Increased PR in deeper soil layers further indicates that NT systems may experience persistent physical constraints in the subsoil. From a practical perspective, these findings suggest that CT should be implemented with explicit consideration of soil depth, management duration, and complementary practices. The inclusion of deep-rooted cover crops may help alleviate subsoil PR and improve vertical soil structure under long-term NT. At the same time, this synthesis reveals clear knowledge gaps related to subsoil responses, regional variability, and the integration of structural and hydraulic indicators. Future studies should prioritize longer-term field experiments, standardized depth classifications, and improved reporting of variability and uncertainty. The integration of emerging tools such as artificial intelligence and precision agriculture may further support adaptive soil management strategies. When applied with contextual understanding, CT has strong potential to promote sustainable soil health and agricultural productivity under changing climatic conditions.

**Author Contributions:** M.N.H.M.: Literature search, synthesis of information, writing—original draft. R.L.: Conceptualization, supervision, validation, writing—review and editing, and guidance on structuring and refining the manuscript. All authors have read and agreed to the published version of the manuscript.

**Funding:** Financial support for this work was provided by the USDA-Agricultural Research Service and the Foundation for Food and Agriculture Research (FFAR) under Grant No. 22-000279, Enhanced Soil Carbon Farming as a Climate Solution, administered by The Ohio State University. The content of this publication is the sole responsibility of the author and does not necessarily reflect the official views of FFAR.

**Institutional Review Board Statement:** Not applicable.

**Informed Consent Statement:** Not applicable.

**Data Availability Statement:** The data supporting the findings of this study are available from the corresponding author upon reasonable request.



https://doi.org/10.3390/su18021083

**Conflicts of Interest:** The authors declare no conflicts of interest.

## Abbreviations

The following abbreviations are used in this manuscript:

| | |
|---|---|
| SPQ | Soil's Physical Quality |
| SOC | Soil Organic Carbon |
| SOM | Soil Organic Matter |
| BD | Bulk Density |
| Ks | Saturated Hydraulic Conductivity |
| WAS | Wet Aggregate Stability |
| WSAs | Water-stable Aggregates |
| AWC | Available Water Capacity |
| WHC | Water Holding Capacity |
| PR | Penetration Resistance |
| CvT | Conventional Tillage |
| NT | No-Tillage |
| RT | Reduced Tillage |
| MAP | Mean Annual Precipitation |
| MAT | Mean Annual Temperature |
| OT | Occasional Tillage |
| NIT | Non-Inversion Tillage |
| IoT | Internet of Things |
| CT | Conservation Tillage |
| MWD | Mean Weight Diameter |
| MP | Moldboard Plow |
| ST | Strip Tillage |
| DT | Deep Tillage |
| NSE | No Significant Effect |

## References

1. Godfray, H.C.J.; Beddington, J.R.; Crute, I.R.; Haddad, L.; Lawrence, D.; Muir, J.F.; Pretty, J.; Robinson, S.; Thomas, S.M.; Toulmin, C. Food security: The challenge of feeding 9 billion people. *Science* **2010**, *327*, 812–818. [CrossRef] [PubMed]
2. Smith, P.; Bustamante, M.; Ahammad, H.; Clark, H.; Dong, H.; Elsiddig, E.A.; Haberl, H.; Harper, R.; House, J.; Jafari, M.; et al. Agriculture, forestry and other land use (AFOLU). In *Climate Change 2014: Mitigation of Climate Change. Contribution of Working Group III to the Fifth Assessment Report of the Intergovernmental Panel on Climate Change*; Edenhofer, O., Pichs-Madruga, R., Sokona, Y., Minx, J.C., Farahani, E., Kadner, S., Seyboth, K., Adler, A., Baum, I., Brunner, S., et al., Eds.; Cambridge University Press: Cambridge, UK; New York, NY, USA, 2014; pp. 811–922.
3. Montanarella, L.; Pennock, D.J.; McKenzie, N.; Badraoui, M.; Chude, V.; Baptista, I.; Mamo, T.; Yemefack, M.; Aulakh Singh, M.S.; Yagi, K.; et al. World's soils are under threat. *Soil* **2016**, *2*, 79–82. [CrossRef]
4. Lal, R. Regenerative agriculture for food and climate. *J. Soil Water Conserv.* **2020**, *75*, 123A–124A. [CrossRef]
5. Tilman, D.; Cassman, K.G.; Matson, P.A.; Naylor, R.; Polasky, S. Agricultural sustainability and intensive production practices. *Nature* **2002**, *418*, 671–677. [CrossRef] [PubMed]
6. Foley, J.A.; Ramankutty, N.; Brauman, K.A.; Cassidy, E.S.; Gerber, J.S.; Johnston, M.; Zaks, D.P.M. Solutions for a cultivated planet. *Nature* **2011**, *478*, 337–342. [CrossRef]
7. Porter, J.R.; Xie, L.; Challinor, A.J.; Cochrane, K.; Howden, S.M.; Iqbal, M.M.; Lobell, D.B.; Travasso, M.I. Food security and food production systems. In *Climate Change 2014: Impacts, Adaptation, and Vulnerability. Part A: Global and Sectoral Aspects. Contribution of Working Group II to the Fifth Assessment Report of the Intergovernmental Panel on Climate Change*; Field, C.B., Barros, V.R., Dokken, D.J., Mach, K.J., Mastrandrea, M.D., Bilir, T.E., Chatterjee, M., Ebi, K.L., Estrada, Y.O., Genova, R.C., et al., Eds.; Cambridge University Press: Cambridge, UK; New York, NY, USA, 2014; pp. 485–533.
8. Derpsch, R.; Friedrich, T.; Kassam, A.; Hongwen, L. Current status of adoption of no-till farming in the world and some of its main benefits. *Int. J. Agric. Biol. Eng.* **2010**, *3*, 1–25. [CrossRef]
9. Blanco-Canqui, H.; Ruis, S.J. No-tillage and soil physical environment. *Geoderma* **2018**, *326*, 164–200. [CrossRef]


https://doi.org/10.3390/su18021083

10. Schneider, F.; Don, A.; Hennings, I.; Schmittmann, O.; Seidel, S.J. The effect of deep tillage on crop yield—What do we really know? *Soil Tillage Res.* **2017**, *174*, 193–204. [CrossRef]
11. Panagea, I.; Berti, A.; Cermak, P.; Diels, J.; Elsen, A.; Kusa, H.; Piccoli, I.; Poesen, J.; Stoate, C.; Tits, M.; et al. Soil water retention as affected by management-induced changes of soil organic carbon: Analysis of long-term experiments in Europe. *Land* **2021**, *10*, 362. [CrossRef]
12. Pittelkow, C.M.; Linquist, B.A.; Lundy, M.E.; Liang, X.; van Groenigen, K.J.; Lee, J.; van Gestel, N.; Six, J.; Venterea, R.T.; van Kessel, C. When does no-till yield more? A global meta-analysis. *Field Crops Res.* **2015**, *183*, 156–168. [CrossRef]
13. Jat, H.S.; Kumar, P.; Sutaliya, J.M.; Kumar, S.; Choudhary, M.; Singh, Y.; Jat, M.L. Conservation agriculture-based sustainable intensification of Basmati rice–wheat system in north-west India. *Arch. Agron. Soil Sci.* **2019**, *65*, 1370–1386. [CrossRef]
14. Six, J.; Ogle, S.M.; Breidt, F.J.; Conant, R.T.; Mosier, A.R.; Paustian, K. The potential to mitigate global warming with no-tillage management is only realized when practised in the long term. *Glob. Change Biol.* **2004**, *10*, 155–160. [CrossRef]
15. Van Der Heijden, M.G.A.; Bardgett, R.D.; Van Straalen, N.M. The unseen majority: Soil microbes as drivers of plant diversity and productivity in terrestrial ecosystems. *Ecol. Lett.* **2008**, *11*, 296–310. [CrossRef] [PubMed]
16. Powlson, D.S.; Stirling, C.M.; Thierfelder, C.; White, R.P.; Jat, M.L. Does conservation agriculture deliver climate change mitigation through soil carbon sequestration in tropical agro-ecosystems? *Agric. Ecosyst. Environ.* **2016**, *220*, 164–174. [CrossRef]
17. Verhulst, N.; Govaerts, B.; Verachtert, E.; Castellanos-Navarrete, A.; Mezzalama, M.; Wall, P.C.; Chocobar, A.; Deckers, J.; Sayre, K.D. Conservation agriculture: Improving soil quality for sustainable production systems? In *Food Security and Soil Quality*; Lal, R., Stewart, B.A., Eds.; CRC Press: Boca Raton, FL, USA, 2010; pp. 137–208. [CrossRef]
18. Giller, K.E.; Witter, E.; Corbeels, M.; Tittonell, P. Conservation agriculture and smallholder farming in Africa: The heretics' view. *Field Crops Res.* **2009**, *114*, 23–34. [CrossRef]
19. Fiorini, A.; Boselli, R.; Amaducci, S.; Tabaglio, V. Effects of no-till on root architecture and root–soil interactions in a three-year crop rotation. *Eur. J. Agron.* **2018**, *99*, 156–166. [CrossRef]
20. Zuber, S.M.; Behnke, G.D.; Nafziger, E.D.; Villamil, M.B. Crop rotation and tillage effects on soil physical and chemical properties in Illinois. *Agron. J.* **2015**, *107*, 971–978. [CrossRef]
21. Presley, D.; Sindelar, A.; Buckley, M.; Mengel, D. Long-term nitrogen and tillage effects on soil physical properties under continuous grain sorghum. *Agron. J.* **2012**, *104*, 749–755. [CrossRef]
22. Sapkota, T.; Mazzoncini, M.; Bàrberi, P.; Antichi, D.; Silvestri, N. Fifteen years of no-till increase soil organic matter, microbial biomass and arthropod diversity in cover crop-based arable cropping systems. *Agron. Sustain. Dev.* **2012**, *32*, 853–863. [CrossRef]
23. Crittenden, S.; Poot, N.; Heinen, M.; van Balen, D.; Pulleman, M. Soil physical quality in contrasting tillage systems in organic and conventional farming. *Soil Tillage Res.* **2015**, *154*, 136–144. [CrossRef]
24. Liesch, A.; Krueger, E.; Ochsner, T. Soil structure and physical properties under rye–corn silage double-cropping systems. *Soil Sci. Soc. Am. J.* **2011**, *75*, 1307–1314. [CrossRef]
25. Phillips, C.; Tekeste, M.; Ebrahimi, E.; Logsdon, S.; Malone, R.; O'Brien, P.; Emmett, B.; Karlen, D. Thirteen-year stover harvest and tillage effects on soil compaction in Iowa. *Agrosyst. Geosci. Environ.* **2023**, *6*, e20361. [CrossRef]
26. Stavi, I.; Lal, R.; Owens, L.B. On-farm effects of no-till versus occasional tillage on soil quality and crop yields in eastern Ohio. *Agron. Sustain. Dev.* **2011**, *31*, 475–482. [CrossRef]
27. Abdollahi, L.; Munkholm, L. Tillage system and cover crop effects on soil quality: I. Chemical, mechanical, and biological properties. *Soil Sci. Soc. Am. J.* **2014**, *78*, 262–270. [CrossRef]
28. Blanco-Canqui, H.; Lal, R.; Post, W.M.; Izaurralde, R.C.; Shipitalo, M.J. Soil hydraulic properties influenced by corn stover removal from no-till corn in Ohio. *Soil Tillage Res.* **2007**, *92*, 144–155. [CrossRef]
29. Idowu, O.; Sultana, S.; Darapuneni, M.; Beck, L.; Steiner, R. Short-term conservation tillage effects on corn silage yield and soil quality in an irrigated, arid agroecosystem. *Agronomy* **2019**, *9*, 455. [CrossRef]
30. Jokela, W.; Posner, J.; Hedtcke, J.; Balser, T.; Read, H. Midwest cropping system effects on soil properties and on a soil quality index. *Agron. J.* **2011**, *103*, 1552–1562. [CrossRef]
31. Kumar, S.; Kadono, A.; Lal, R.; Dick, W. Long-term no-till impacts on organic carbon and properties of two contrasting soils and corn yields in Ohio. *Soil Sci. Soc. Am. J.* **2012**, *76*, 1798–1809. [CrossRef]
32. Virto, I.; Imaz, M.; Enrique, A.; Hoogmoed, W.; Bescansa, P. Burning crop residues under no-till in semi-arid land, northern Spain—Effects on soil organic matter, aggregation, and earthworm populations. *Aust. J. Soil Res.* **2007**, *45*, 414–421. [CrossRef]
33. Islam, K.; Dick, W.; Watts, D.; Gonzalez, J.M.; Fausey, N.; Flanagan, D.; Reeder, R.; Vantoai, T.; Batte, M. Gypsum, crop rotation, and cover crop impacts on soil organic carbon and biological dynamics in rainfed transitional no-till corn–soybean systems. *PLoS ONE* **2022**, *17*, e0275198. [CrossRef]
34. Liebig, M.; Tanaka, D.; Wienhold, B. Tillage and cropping effects on soil quality indicators in the Northern Great Plains. *Soil Tillage Res.* **2004**, *78*, 131–141. [CrossRef]
35. Villamil, M.B.; Little, J.; Nafziger, E.D. Corn residue, tillage, and nitrogen rate effects on soil properties. *Soil Tillage Res.* **2015**, *151*, 61–66. [CrossRef]



https://doi.org/10.3390/su18021083

36. Basso, B.; Amato, M.; Bitella, G.; Rossi, R.; Kravchenko, A.; Sartori, L.; Carvalho, L.M.; Gomes, J. Two-dimensional spatial and temporal variation of soil physical properties in tillage systems using electrical resistivity tomography. *Agron. J.* **2010**, *102*, 440–449. [CrossRef]
37. Gómez-Muñoz, B.; Jensen, L.; Munkholm, L.; Olesen, J.; Hansen, E.; Bruun, S. Long-term effect of tillage and straw retention in conservation agriculture systems on soil carbon storage. *Soil Sci. Soc. Am. J.* **2021**, *85*, 1465–1478. [CrossRef]
38. Maiga, A.; Alhameid, A.; Singh, S.; Polat, A.; Singh, J.; Kumar, S.; Osborne, S. Responses of soil organic carbon, aggregate stability, carbon and nitrogen fractions to 15 and 24 years of no-till diversified crop rotations. *Soil Res.* **2019**, *57*, 149–163. [CrossRef]
39. Nakajima, T.; Lal, R.; Jiang, S. Soil quality index of a Crosby silt loam in Central Ohio. *Soil Tillage Res.* **2015**, *146*, 323–328. [CrossRef]
40. Jabro, J.D.; Iversen, W.M.; Stevens, W.B.; Evans, R.G.; Mikha, M.M.; Allen, B.L. Physical and hydraulic properties of a sandy loam soil under zero, shallow and deep tillage practices. *Soil Tillage Res.* **2016**, *159*, 67–72. [CrossRef]
41. Blanco-Canqui, H.; Klocke, N.L.; Schlegel, A.J.; Stone, L.R.; Rice, C.W. Impacts of deficit irrigation on carbon sequestration and soil physical properties under no-till. *Soil Sci. Soc. Am. J.* **2010**, *74*, 1301–1309. [CrossRef]
42. Abid, M.; Lal, R. Tillage and drainage impact on soil quality: I. Aggregate stability, carbon and nitrogen pools. *Soil Tillage Res.* **2008**, *100*, 89–98. [CrossRef]
43. Chatterjee, A.; Teboh, J.; Nelson, S.; Aberle, E.; Schatz, B.G.; Zilahi-Sebess, S. Long-term effect of nitrogen and tillage management on soil carbon pools in the semiarid Northern Great Plains. *Commun. Soil Sci. Plant Anal.* **2017**, *48*, 730–740. [CrossRef]
44. Alhameid, A.; Ibrahim, M.; Kumar, S.; Sexton, P.; Schumacher, T.E. Soil organic carbon changes impacted by crop rotational diversity under no-till farming in South Dakota, USA. *Soil Sci. Soc. Am. J.* **2017**, *81*, 868–877. [CrossRef]
45. Mueller, L.; Kay, B.D.; Deen, B.; Hu, C.; Zhang, Y.; Wolff, M.; Eulenstein, F.; Schindler, U. Visual assessment of soil structure: Part II. Implications of tillage, rotation and traffic on sites in Canada, China and Germany. *Soil Tillage Res.* **2009**, *103*, 188–196. [CrossRef]
46. Ussiri, D.A.N.; Lal, R.; Jarecki, M.K. Nitrous oxide and methane emissions from long-term tillage under a continuous corn cropping system in Ohio. *Soil Tillage Res.* **2009**, *104*, 247–255. [CrossRef]
47. Obour, A.; Holman, J.; Simon, L.; Schlegel, A. Strategic tillage effects on crop yields, soil properties, and weeds in dryland no-tillage systems. *Agronomy* **2021**, *11*, 662. [CrossRef]
48. Filho, O.G.; Blanco-Canqui, H.; Da Silva, A.P. Least limiting water range of the soil seedbed for long-term tillage and cropping systems in the Central Great Plains, USA. *Geoderma* **2013**, *207–208*, 99–110. [CrossRef]
49. Fernández, F.G.; Sorensen, B.A.; Villamil, M.B. A comparison of soil properties after five years of no-till and strip-till. *Agron. J.* **2015**, *107*, 1339–1346. [CrossRef]
50. Nawaz, A.; Lal, R.; Shrestha, R.K.; Farooq, M. Mulching affects soil properties and greenhouse gas emissions under long-term no-till and plough-till systems in Alfisol of Central Ohio. *Land Degrad. Dev.* **2017**, *28*, 673–681. [CrossRef]
51. Tormena, C.A.; Karlen, D.L.; Logsdon, S.; Cherubin, M.R. Corn stover harvest and tillage impacts on near-surface soil physical quality. *Soil Tillage Res.* **2017**, *166*, 122–130. [CrossRef]
52. Kahlon, M.S.; Lal, R.; Ann-Varughese, M. Twenty-two years of tillage and mulching impacts on soil physical characteristics and carbon sequestration in Central Ohio. *Soil Tillage Res.* **2013**, *126*, 151–158. [CrossRef]
53. Kumar, S.; Kadono, A.; Lal, R.; Dick, W. Long-term tillage and crop rotations for 47–49 years influence hydrological properties of two soils in Ohio. *Soil Sci. Soc. Am. J.* **2012**, *76*, 2195–2207. [CrossRef]
54. Wang, H.; Wang, L.; Ren, T. Long-term no tillage alleviates subsoil compaction and drought-induced mechanical impedance. *Int. Agrophys.* **2022**, *36*, 297–308. [CrossRef]
55. Wander, M.M.; Bidart, M.G.; Aref, S. Tillage impacts on depth distribution of total and particulate organic matter in three Illinois soils. *Soil Sci. Soc. Am. J.* **1998**, *62*, 1704–1711. [CrossRef]
56. Weninger, T.; Kreiselmeier, J.; Chandrasekhar, P.; Julich, S.; Feger, K.; Schwärzel, K.; Bodner, G.; Schwen, A. Effects of tillage intensity on pore system and physical quality of silt-textured soils detected by multiple methods. *Soil Res.* **2019**, *57*, 703–711. [CrossRef]
57. Fornara, D.; Higgins, A. Tillage and reseeding effects on soil carbon stocks: Evidence from 500 agricultural grasslands in the UK. *Agron. Sustain. Dev.* **2022**, *42*, 71. [CrossRef]
58. Bagnall, D.; Morgan, C. SLAKES and 3D scans characterize management effects on soil structure in farm fields. *Soil Tillage Res.* **2021**, *208*, 104893. [CrossRef]
59. Benjamin, J.G. Tillage effects on near-surface soil hydraulic properties. *Soil Tillage Res.* **1993**, *26*, 277–288. [CrossRef]
60. Reynolds, W.D.; Gregorich, E.G.; Curnoe, W.E. Characterisation of water transmission properties in tilled and untilled soils using tension infiltrometers. *Soil Tillage Res.* **1995**, *33*, 117–131. [CrossRef]
61. Buczko, U.; Bens, O.; Hüttl, R.F. Tillage effects on hydraulic properties and macroporosity in silty and sandy soils. *Soil Sci. Soc. Am. J.* **2006**, *70*, 1998–2007. [CrossRef]


 https://doi.org/10.3390/su18021083

62. Rasmussen, K. Impact of ploughless soil tillage on yield and soil quality: A Scandinavian review. *Soil Tillage Res.* **1999**, *53*, 3–14. [CrossRef]
63. Wang, L.; Zhang, J.-G.; Sha, H.-Y.; Wang, Y.-R.; Wang, H.-Y.; Zhu, G.-C.; Lu, Y.-Z. Lanthanum-quaternized chitosan-modified zeolite for long-lasting operation of constructed wetlands: A bifunctional strategy for simultaneous phosphorus removal and microbial clogging mitigation. *Water Res.* **2026**, *288*, 124688. [CrossRef]
64. Blanco-Canqui, H.; Lal, R. No-tillage and soil-profile carbon sequestration: An on-farm assessment. *Soil Sci. Soc. Am. J.* **2008**, *72*, 693–701. [CrossRef]
65. Munna, M.N.H.; Tanu, F.Z.; Mia, S.; Shapna, N.A.; Hakim, A.; Lal, R. Reclaiming saline soil by using acid-modified pineapple biochar. *Arch. Agron. Soil Sci.* **2025**, *71*, 1–16. [CrossRef]
66. Gómez-Paccard, C.; Hontoria, C.; Mariscal-Sancho, I.; Pérez, J.; León, P.; González, P.; Espejo, R. Soil–water relationships in the upper soil layer in a Mediterranean Palexerult as affected by no-tillage under excess water conditions—Influence on crop yield. *Soil Tillage Res.* **2015**, *146*, 303–312. [CrossRef]
67. Hussain, I.; Olson, K.R.; Ebelhar, S.A. Impacts of tillage and no-till on production of maize and soybean on an eroded Illinois silt loam soil. *Soil Tillage Res.* **1999**, *52*, 37–49. [CrossRef]
68. Fernández-Ugalde, O.; Virto, I.; Bescansa, P.; Imaz, M.J.; Enrique, A.; Karlen, D.L. No-tillage improvement of soil physical quality in calcareous, degradation-prone, semiarid soils. *Soil Tillage Res.* **2009**, *106*, 29–35. [CrossRef]
69. Bilibio, C.; Uteau, D.; Horvat, M.; Rosskopf, U.; Junge, S.; Finckh, M.; Peth, S. Impact of ten years conservation tillage in organic farming on soil physical properties in a loess soil—Northern Hesse, Germany. *Agriculture* **2023**, *13*, 133. [CrossRef]
70. Puerta, V.; Pereira, E.; Wittwer, R.; van der Heijden, M.; Six, J. Improvement of soil structure through organic crop management, conservation tillage and grass-clover ley. *Soil Tillage Res.* **2018**, *180*, 96–104. [CrossRef]
71. Santín-Montanyá, M.I.; Martín-Lammerding, D.; Zambrana, E.; Tenorio, J.L. Management of weed emergence and weed seed bank in response to different tillage, cropping systems and selected soil properties. *Soil Tillage Res.* **2016**, *161*, 38–46. [CrossRef]
72. Khakural, B.R.; Lemme, G.D.; Schumacher, T.E.; Lindstrom, M.J. Effects of tillage systems and landscape on soil. *Soil Tillage Res.* **1992**, *25*, 43–52. [CrossRef]
73. Blanco-Canqui, H.; Mikha, M.M.; Presley, D.R.; Claassen, M.M. Addition of cover crops enhances no-till potential for improving soil physical properties. *Soil Sci. Soc. Am. J.* **2011**, *75*, 1471–1482. [CrossRef]
74. Büchi, L.; Walder, F.; Banerjee, S.; Colombi, T.; van der Heijden, M.; Keller, T.; Charles, R.; Six, J. Pedoclimatic factors and management determine soil organic carbon and aggregation in farmer fields at a regional scale. *Geoderma* **2022**, *409*, 115632. [CrossRef]
75. Lal, R. Residue management, conservation tillage and soil restoration for mitigating greenhouse effect by $CO_2$-enrichment. *Soil Tillage Res.* **1997**, *43*, 81–107. [CrossRef]
76. Pimentel, D.; Harvey, C.; Resosudarmo, P.; Sinclair, K.; Kurz, D.; McNair, M.; Crist, S.; Shpritz, L.; Fitton, L.; Saffouri, R.; et al. Environmental and economic costs of soil erosion and conservation benefits. *Science* **1995**, *267*, 1117–1123. [CrossRef] [PubMed]
77. Montgomery, D.R. Soil erosion and agricultural sustainability. *Proc. Natl. Acad. Sci. USA* **2007**, *104*, 13268–13272. [CrossRef]
78. Dachraoui, M.; Sombrero, A. Soil organic carbon accumulation and carbon dioxide emissions during a 6-year study in irrigated continuous maize under two tillage systems in semiarid Mediterranean conditions. *Span. J. Agric. Res.* **2021**, *19*, e1102. [CrossRef]
79. Munna, M.N.H.; Lal, R. Long-term organic inputs effects on soil carbon and nitrogen sequestration in a temperate agroecosystem. *Soil Use Manag.* **2026**, *42*, e70161. [CrossRef]
80. Blanco-Canqui, H.; Stone, L.R.; Schlegel, A.J.; Lyon, D.J.; Vigil, M.F.; Mikha, M.M.; Stahlman, P.W.; Rice, C.W. No-till induced increase in organic carbon reduces maximum bulk density of soils. *Soil Sci. Soc. Am. J.* **2009**, *73*, 1871–1879. [CrossRef]
81. VandenBygaart, A.J.; Gregorich, E.G.; Angers, D.A. Influence of agricultural management on soil organic carbon: A compendium and assessment of Canadian studies. *Can. J. Soil Sci.* **2003**, *83*, 363–380. [CrossRef]
82. Melero, S.; Panettieri, M.; Madejón, E.; Macpherson, H.; Moreno, F.; Murillo, J. Implementation of chiselling and mouldboard ploughing in soil after 8 years of no-till management in SW Spain: Effect on soil quality. *Soil Tillage Res.* **2011**, *112*, 107–113. [CrossRef]
83. Mia, S.; Bristy, S.Y.; Jindo, K.; Munna, M.N.H.; Rahman, M.M.; Uddin, K.; Kasim, S.; Rahman, M.S. Potential health risks associated with biochar—From production to field application. In *Biochar Amendments for Environmental Remediation*; Nidheesh, P.V., Vithanage, M., Sreedharan, V., Bolan, N., Gao, B., Bhatnagar, A., Eds.; CRC Press: Boca Raton, FL, USA, 2024; pp. 293–304. [CrossRef]
84. Krauss, M.; Berner, A.; Perrochet, F.; Frei, C.; Niggli, M.; Mäder, P. Enhanced soil quality with reduced tillage and solid manures in organic farming—A synthesis of 15 years. *Sci. Rep.* **2020**, *10*, 4403. [CrossRef]
85. Munna, M.N.H.; Lal, R. Impacts of cover cropping and organic amendments on soil physical quality under temperate climate. *Cogent Food Agric.* **2025**, *11*, 2467452. [CrossRef]
86. Kremen, C.; Miles, A. Ecosystem services in biologically diversified versus conventional farming systems: Benefits, externalities, and trade-offs. *Ecol. Soc.* **2012**, *17*, 40. [CrossRef]



https://doi.org/10.3390/su18021083

87. Zhang, E.; Meng, C.; Qu, J.; Zhu, Z.; Niu, J.; Wang, L.; Song, N.; Yin, Z. Dual effects of Caragana korshinskii introduction on herbaceous vegetation in Chinese desert areas: Short-term degradation and long-term recovery. *Plant Soil* **2025**. [CrossRef]
88. Kumari, K.; Mirzakhani Nafchi, A.; Mirzaee, S.; Abdalla, A. AI-driven future farming: Achieving climate-smart and sustainable agriculture. *AgriEngineering* **2025**, *7*, 89. [CrossRef]
89. Steponavičienė, V.; Žiūraitis, G.; Rudinskienė, A.; Jackevičienė, K.; Bogužas, V. Long-term effects of different tillage systems and their impact on soil properties and crop yields. *Agronomy* **2024**, *14*, 870. [CrossRef]
90. Brevik, E.C. Soil health and productivity. In *Plant Growth and Crop Production, Encyclopedia of Life Support Systems (EOLSS)*; Verheye, W., Ed.; EOLSS Publishers: Oxford, UK, 2009; pp. 106–137.





https://doi.org/10.3390/su18021083

sustainability

MDPI

Review

# Environmental and Public Health Impacts of Mining Tailings in Chañaral, Chile: A Narrative Case-Based Review

Sandra Cortés [1,2,3,*], Pablo González [4], Cinthya Leiva [1], Yendry Vargas [5], Alejandra Vega [3] and Pablo Pastén [3,6]

1 School of Public Health, Faculty of Medicine, Pontificia Universidad Católica de Chile, Diagonal Paraguay 362, Santiago 8330077, Chile; ccleiva@uc.cl
2 Advanced Center for Chronic Diseases (ACCDIS), Sergio Livingstone 1007, Santiago 8380492, Chile
3 Center for Sustainable Urban Development (CEDEUS), El Comendador 1916, Santiago 7520245, Chile; asvega@uc.cl (A.V.); ppasten@uc.cl (P.P.)
4 National School of Public Health, Sérgio Arouca, Oswaldo Cruz Foundation, Av. Brasil, 4365, Rio de Janeiro 21040900, Brazil; pablo.castilho@aluno.fiocruz.br
5 Institute of Social Studies in Population, National University of Costa Rica, Heredia 86-3000, Costa Rica; yendryvatre@gmail.com
6 Department of Hydraulic and Environmental Engineering, Faculty of Engineering, Pontificia Universidad Católica de Chile, Av. Vicuña Mackenna 4860, Santiago 7820436, Chile
* Correspondence: scortesn@uc.cl

**Abstract**

This narrative case-based review describes the environmental and public health impacts in Chañaral, a town in northern Chile affected by the accumulation of copper mining tailings for the past 80 years. The review included 34 scientific articles published between 1978 and 2025. The keywords used were "mining tailings" and "Chañaral", without year limits, and covering disciplines such as ecology, public health, environmental history, and territorial studies. The scientific evidence demonstrates the negative impacts on the ecosystem and the human population exposed to toxic metals and arsenic. Geomorphological and biogeochemical alterations have been found on the Chañaral coast, affecting marine biodiversity and water quality. In addition, epidemiological studies indicate exposure to toxic metals measured in street dust and urine, raising concerns on respiratory health in children and metabolic conditions in adults. According to the social sciences, the lack of environmental monitoring and human exposure data contributes to the high health risk perception in the population, posing the need to strengthen environmental monitoring, raise awareness on the risks of exposure to toxic metals, and promote mitigation and restoration strategies. These measures will contribute to sustainable conditions for the Chañaral community through the improvement of comprehensive public policies.

**Keywords:** Chañaral; Chile; mining tailings; toxic metals; biochemical disruption; public health

## 1. Introduction

Mining represents one of the largest volumetric waste streams in the world [1], where less than 1% of the extracted ore is economically relevant, and the remaining fraction is considered waste. Such waste can cause serious damage to the ecosystem and place human health at risk due to toxic metal contamination [2–4], in addition to tailings dam incidents. Case reports in the Americas include countries such as the United States (US) (98 cases from 1910 to 1999 and 17 cases between 2000 and 2020), Canada (19 cases before 2000 and 11 cases after 2000), and Chile (32 cases before 2000 and 9 cases after 2000). Regarding global-scale impact, tailings dam incidents from copper mines have been the most frequent ($n = 86$), mostly originating from Chile ($n = 35$), mainly due to earthquakes. Chile, the US, Peru and the Philippines account for roughly 77% of all cases of copper mine failure. During the study period (1915–2020), the countries reporting the highest numbers of human deaths were China ($n = 516$), Bulgaria ($n = 488$), Brazil ($n = 331$), Mexico ($n = 306$), Italy ($n = 269$), and Chile ($n = 258$) [5].

Chile stands out as one of the world's leading copper producers. In 2022, copper production accounted for 10.9% of the country's Gross Domestic Product (GDP) [6]. Despite its economic relevance, the boom in mega-mining has led to interventions in mining territories with limited surveillance on environmental and human impacts. Insufficient surveillance and regulation over the past decades have generated various consequences, such as the presence of mining tailings and environmental liabilities, which constitute a permanent risk for nearby populations, such as in the municipality of Chañaral [7].

In this sense, socio-environmental inequality, also called environmental injustice, occurs when territorial, national, or regional development, driven by economic ventures or public policy actions or inactions, exposes parts of the population to disproportionate environmental risks. Additionally, it includes situations where access to natural resources in areas inhabited and worked in by local populations is restricted, infringing upon numerous human rights such as the right to a clean environment, safe living and working conditions, and the enjoyment of cultural practices [8]. This form of injustice arises from a societal disregard for the value, autonomy, and territory of affected communities, manifesting in various ways. These include economic valuations that prioritize profit over well-being, biased environmental regulations, limited participation opportunities for impacted communities, the exclusion of local economic practices, surveillance gaps in institutional monitoring, inadequacies in health services, and biases in research on environmental and health issues across time [8].

Environmental issues and the effects on inhabitants' health in Chañaral have been addressed by a wide range of professionals, from ecology and geochemistry to public health experts, generating evidence on socio-environmental issues. However, this evidence has not been considered in urban planning to promote sustainable development that protects the life and health of the population in the territory. Therefore, the purpose of this article is to analyze the scientific evidence on the environmental, social, and health impacts of exposure to toxic metals from mining tailings in the city of Chañaral. This research will contribute elements for a debate on public environmental, health, and urban planning policies that address the socio-environmental problem of Chañaral, taking into consideration the fact that Chañaral is identified as a "potential sacrifice zone" according to the Human Rights Council, United Nations [9].

## 2. Materials and Methods

### 2.1. Study Area

Chañaral is a small coastal city located in the Atacama region, 167 km northwest of Copiapó (regional capital) and 963 km north of Santiago, Chile (national capital) (Figure 1). According to the 2017 census, Chañaral housed 12,219 inhabitants [10]. Of these, 11,705 inhabitants were registered in the state health service (National Health Fund, FONASA) (95.8% of the total population). The main economic activities are metal manufacturing and the exploitation of mines and quarries, which represent 10.5% and 42% of the city's economic activity, respectively [10]. This is consistent with the regional situation, as the mining industry is the basis of the economy in the Atacama region, representing 48% of the region's Gross National Product (GNP) [11]. In Chañaral, 10 million tons of tailings and a similar amount of sterile waste were produced in 2017 by Codelco's Salvador Division [12], which is the main economic actor.

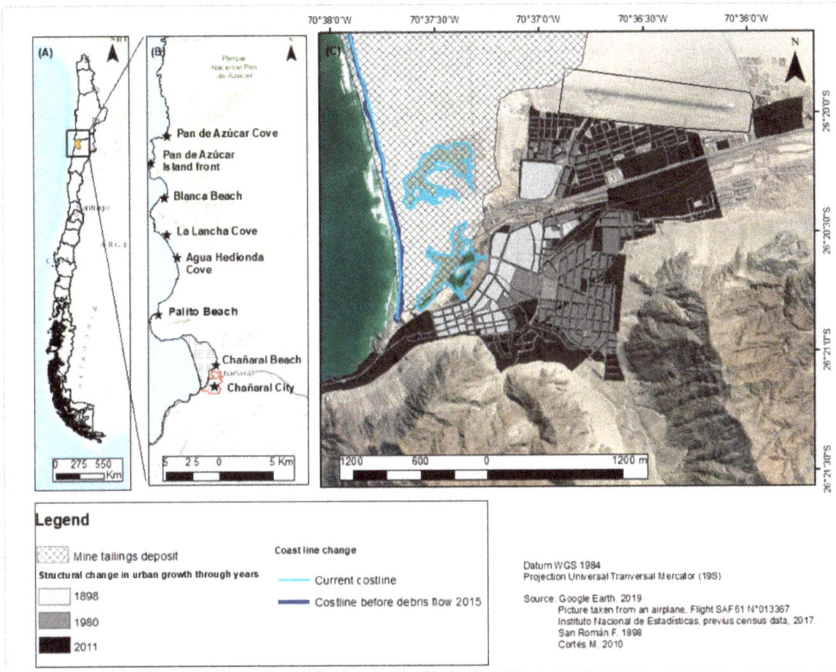


**Figure 1.** Location of (**A**) Chañaral and (**B**) relevant bay areas. (**C**) Temporary evolution of the urban area of Chañaral. It is important to consider that the years correspond to years of available information, and do not refer to a particular event in the territory. Red line show the urban area. Source: Own elaboration [13,14].

Mining activity and the problems associated with waste and tailings are not recent. Since the mid-20th century, the population of Chañaral has been exposed to mining tailings dumped into the Salado River by the Potrerillos and El Salvador mines, managed by the US company Andes Copper Mining Company and, since 1970, by the state-owned Codelco Chile. It is estimated that between 1938 and 1990, more than 300 million tons of toxic waste, containing arsenic, lead, and other toxic metals, as well as approximately 850 million tons of sewage, were deposited in the Chañaral Bay [15]. The tailings dam exceeded its capacity in 1938, so the mine tailings were initially discharged into the "Los Patos" estuary and subsequently dumped into the bay [15].

For more than three decades, tailings were uninterruptedly deposited on the coast of Chañaral until the Ministry of Public Works diverted their dumping towards the area of Caleta Palito, 12 km north of Chañaral (Figure 1, letter b) in 1974 through canalization to "avoid the progressive sand banking of the bay of Chañaral and the consequent disablement of its port facilities" [16]. All these deposits with more than 300 metric tons of copper tailings [17] caused the modification of the coastline, so that today a large part of the area used as a beach is mostly made up of mining tailings deposits (Figure 1). These tailings have generated serious environmental problems due to their repercussions on the terrestrial and marine environment, as well as on people's health [18].

Mining activity has also influenced the development of Chañaral. Since the first half of the 20th century, copper mining and port activity have played a key role in the development of the municipality's urban infrastructure. Although the unsafe disposal of mining tailings exposes the population to serious health risks, the urban fabric continued to grow to the north and south of Route 5 due to the gradual increase in population. By 1980, the population had increased to 7000 inhabitants and, by 2011, it expanded further south with an additional 3000 inhabitants, reaching as far as the Chañaral aerodrome (Figure 1) [19]. According to the 2017 census, 90.07% of the municipality's population resides in urban areas [7], mainly highlighting the urban area of Chañaral and, to a lesser extent, the locality of Barquito. This expansion poses challenges such as the lack of space for housing and the need to expand urbanization coverage. In addition, no remediation work has been carried out, maintaining the presence of toxic substances in an urbanized area [20].

### 2.2. Methods

A narrative review of the scientific literature on toxic metal cases in Chañaral was conducted using databases such as PubMed, Scopus, Web of Science, and Jstor. The time frame was between the years 1978 and 2025.

Inclusion criteria comprised peer-reviewed articles in Spanish and English in indexed journals. Scientific research articles, technical reports, and dissertations were included.

A total of 34 articles were selected from indexed journals published between 1978 and 2022 and analyzed. The keywords used were Chañaral, tailings, metal exposure, metal contamination of soils, particulate matter exposure effects, seaweed, and health effects. Each article was given an identification code to facilitate systematization and organization. The first step toward organization involved major themes of scientific production, that is, environmental, health, and social research, following a chronological order. Through a descriptive analysis, the content of the publications was categorized according to study population, methodological approach, the toxic metals involved, and scientific evidence. In addition, using the same methodological strategy, several postgraduate research studies in the fields of public health, law and urban studies were identified and are presented as Supplementary Materials (Table S1), classified into studies with online access and those without online access.

## 3. Results

Considering a timeline spanning 1978 through 2025, different types of research were conducted in Chañaral, Chile. The results were categorized by decades, marking significant events and corresponding studies. In the period 1922–1980, events such as the "1922 earthquake/tsunami" and "1972 debris flows" prompted primarily environmental studies, including geomorphological impacts (1978), and biological degradation in vegetation, invertebrates, and birds (1980). After 2010, a shift toward integrated studies emerged as the methodology diversified, featuring social studies (2010, 2011, 2022), including citizen mobilization against mining (2022); atmospheric environmental assessments (2012, 2013, 2014); and epidemiological studies (2016, 2018), covering toxic metals in humans and metabolic conditions in adults. Finally, socio-environmental research, integrating disciplines, covered the period 2021–2025. Interestingly, debris flows reappeared in 2015 and 2017, highlighting their critical role in cumulative environmental degradation. The results of this case-based review of the scientific literature on the environmental, health, and social impacts of the presence of mining tailings are classified by thematic issues and chronological order.

### 3.1. Environmental Impacts

The interest in determining the environmental consequences of tailings deposition in the Chañaral Bay has involved the participation of a variety of experts, including ecologists, geochemists, chemical engineers, and biologists (Table 1). The first studies were carried out in the 1970s to investigate the geomorphological impact and identify the mechanisms of contamination and their interactions on the seabed. There is now strong scientific evidence showing how mining activity and the release of toxic metals affect marine sediment composition, biological communities, and water quality in this bay.

Studies carried out by Castilla and Nealler [21] revealed that the low light penetrability values in Chañaral Bay are due to the presence of accumulated sediments. On the other hand, Dold [22] and Lee and Correa [23] (2006) point out that seawater is mainly affected by As, Mo, Cu, and Zn contamination through infiltration and colloidal transport from tailings and the Rio Salado.

Similarly, Andrade et al. [24] concluded that the highest concentrations of Cu, both as solid particles and dissolved in water, are found in sediments from mining activities that have been agitated or suspended in the water column. Lee et al. [25] found high concentrations of metals in interstitial water, indicating the possible presence of contaminants that may affect aquatic organisms. This underlines the importance of taking into account microbial activity and sedimentation processes in the dynamics of contaminants in the aquatic environment.

Since the late 1990s, several studies have provided valuable evidence on the biological effects of metals from mining tailings on the diversity and abundance of species present in Chañaral Bay, such as algae and birds [26,27]. These studies indicate that adult and juvenile *Lessonia nigrescens* algae can show normal growth and regeneration even in the presence of relatively high levels of copper in the water (150 $\mu g\ L^{-1}$) [26,28]. However, some minor effects on growth may be present. Such data could be used as a detoxification strategy, as algae may contribute to metal attenuation through their uptake and accumulation [29]. Such a procedure could be implemented as part of a tool for the ecological restoration of Chañaral Bay.

Likewise, parallel studies carried out by Medina et al. [30] determined that dissolved copper at levels of 8.72–25.74 μg/L, affects the diversity of sessile organisms (organisms that do not move from their location), suggesting that the presence of copper in water may have negative implications on the community of organisms fixed or attached to the substrate. De la Iglesia et al. [31] also highlighted that the presence of copper has a strong effect (statistically significant R-value 0.49) on the structure of epilithic bacterial communities, but that this effect is reversible and returns to its original state after 48 h of copper removal. These results show how metals can have an impact on biological communities but also indicate that some effects may be temporary and mitigable.

Studies carried out by Fariña and Castilla [32] indicate that the low diversity and abundance of algae at present are strongly influenced by herbivory, but the chronic effects of contaminant discharges cannot be ruled out as a possible factor that also affects the diversity and abundance of algae. All these studies highlight the importance of understanding the combined effects of herbivory and copper contamination in the marine ecosystem of Chañaral Bay. The interaction of these factors may have implications for the biological diversity and stability of marine habitats in the impacted area.

In the case of birds, Valladares et al. [27] found an accumulation of cadmium in their kidneys, at concentrations of 1.05 μg/g (range 0.044 to 9.86 μg/g), values considered within the normal range, suggesting a possible effect by this metal. Although lead concentration levels were considered normal and no significant effects were found, there is evidence that cadmium may be negatively affecting these birds compared to others with a similar ecological role. Therefore, these findings show that the presence of copper and other contaminating metals can have negative effects on biological communities, both in algae and birds. In addition, bacterial communities can also be affected by the presence of copper, although its impact appears to be temporary and reversible. Such biological effects are important considerations for the conservation and management of the marine environment of Chañaral Bay.

**Table 1.** Studies on adaptation and ecological alteration due to exposure to toxic metals, Chañaral Bay, Chile (1978–2025).

| ID | Year | Authors | Population/Matrix | Pollutant(s) | Method | Reported Evidence | Ref. |
|---|---|---|---|---|---|---|---|
| 1 | 1978 | Castilla, J. and Nealler, E. | Sub-littoral organisms | Not applicable | Observations and samplings | Low values of light penetrability reflected | [21] |
| 2 | 1983 | Castilla, J.C. | Marine invertebrates | Molybdenite and chalcopyrite | Sandy beach morphometric and sediment analysis | Geomorphological coastal modifications | [33] |
| 3 | 1997 | Riquelme, C, et al. | Green algae *Enteromorpha compressa* | Cu | Laboratory incubation and isolation of epiphyte bacteria | High copper tolerance of epiphytic bacteria | [34] |
| 4 | 1998 | Correa, J.A. et al. | Algae *Lessonia nigrescens* | Cu | Three alternative hypotheses were tested on copper tolerance | Sea water mixed with affluent was not lethal to algae | [26] |
| 5 | 1999 | Correa, J.A. et al. | Algae *Enteromorpha compressa* | Cu | Monitoring macroalgae, sessile, and mobile invertebrates | Low algae diversity influenced by herbivory | [28] |
| 6 | 2001 | Lee, M.R. et al. | Nematode, copepod | Trace metals | Laboratory analysis | Extent of metal impacts on meiofauna densities | [35] |
| 7 | 2001 | Fariña, J.M. and Castilla, J.C. | Seawater | Cu, Zn, Cd, inorganic particles | Standard methods designed | Variation in sessile species in rocky intertidal benthic communities | [32] |
| 8 | 2002 | Lee, M.R. et al. | Cores of porewater samples | Cu, Al, As, Cr, Mn, Ni, Zn | Diffusive gradients in thin-film (DGT) techniques | High concentrations of metals in porewater | [25] |
| 9 | 2005 | Medina, M. et al. | Algae and sessile invertebrates | Cu | Measurement of total dissolved copper concentration | Dissolved copper affects the diversity of sessile organisms | [30] |
| 10 | 2005 | Ramírez, M. et al. | Sessile species | Cd, Cu, Fe, Mn, Ni, Pb, Zn | Sequential chemical extraction method | Mining activity leads to the presence of copper | [36] |
| 11 | 2005 | Lee, M.R. and Correa, J.A. | Meiofaunal assemblages | Cu | Diffusion gel technique (DGT) | Copper found in seawater is connected to mining tailings | [37] |
| 12 | 2006 | Correa, J.A. et al. | Kelp *Lessonia nigrescens (Phaeophyceae)* | Not applicable | Experimental transplants of the large kelp *Lessonia nigrescens (Phaeophyceae)* | Transplanted plants regenerated | [38] |
| 13 | 2006 | Dold, B. | Tailing deposits | Cu, Ni, Zn, Pb, Mo, As | Measurement of parameters in the tailings deposit | The sea is affected by As, Mo, Cu, and Zn contamination. The population of Chañaral is mainly exposed to high concentrations of Cu and minor concentrations of Ni and Zn | [22] |

**Table 1.** *Cont.*

| ID | Year | Authors | Population/Matrix | Pollutant(s) | Method | Reported Evidence | Ref. |
|---|---|---|---|---|---|---|---|
| 14 | 2006 | Andrade, S. et al. | Intertidal ecosystem | Cu | Total dissolved copper concentration was determined by ASV | Resuspended mine-derived sediments have the highest value of particulate and dissolved copper | [24] |
| 15 | 2010 | Bea, S.A. et al. | Efflorescent salt in coastal mine tailing deposits | Halite (NaCl) and eriochalcite ($CuCl_2 \cdot 2H_2O$) | Reactive transport modeling | The reduction in evaporation is a key factor that prevents the formation of salt deposits and a specific mineral (eriochalcite) | [39] |
| 16 | 2012 | De la Iglesia et al. | Epilithic intertidal bacterial | Cu | Culture-independent molecular approach, field sampling, and laboratory microcosm experiments | Copper significantly affects bacterial communities, but its impact can be temporary and reversible | [31] |
| 17 | 2012 | Ramos-Jiliberto, R. et al. | Intertidal community | Cu | Retrospective qualitative analysis of ecological networks | Four initiators were identified as the primary drivers of observed community structure changes in the intertidal system | [40] |
| 18 | 2012 | Koski, R. | Mining activities in Spain (Gulf of Cádiz, Portman Bay), Chile (Chañaral Bay), the Philippines (Marinduque Island), and the United States (Callahan mine, Maine; Ellamar and Beatson mines, Prince William Sound, Alaska) | Cu, Pb, Zn | The dispersal of metals is described along a series of pathways leading from source rocks to the marine biosphere | The pathway approach considers a number of physical and chemical processes that influence the dispersion of metals in the continuum from the metal source to bioreceptor | [41] |
| 19 | 2013 | Valladares, P. et al. | Wild turkey vulture, Cathartes aura | Cd, Pb | Analysis of Cd and Pb using an atomic absorption spectrophotometer and an autosampler | High cadmium accumulation in the kidney compared to similar birds | [27] |
| 20 | 2013 | Korehi H et al. | Microorganisms | Cu | Scanning electron microscopy | Extreme mine tailings provide an environment that supports the growth of prokaryotes | [42] |
| 21 | 2014 | Besaury L et al. | Archaeal and bacterial communities | Cu | Molecular tools such as the 16S rRNA gene Q-PCR and RT-Q-PCR | Bacteria are the dominant group within the prokaryotic community | [43] |

**Table 1.** *Cont.*

| ID | Year | Authors | Population/Matrix | Pollutant(s) | Method | Reported Evidence | Ref. |
|---|---|---|---|---|---|---|---|
| 22 | 2018 | Bonnail E et al. | Water fog composition | Cu, As | Fog water collectors. Volumes oscillate between 0 and 9000 L per month | High acidity and high concentrations of Cu and As in water collected on top of the mountain were found | [44] |

### 3.2. Impacts on Human Health (Exposure and Damage)

Research has established an association between the aeolian action of particulate matter from mining tailings deposited in the Chañaral Bay and adverse health effects on the surrounding populations. Korehi et al. [42] provide evidence of how microbial activity in extreme mining waste contributes to the release of acid and metals into water and air, which can lead to atmospheric contamination in areas near mining deposits. Together with the studies of Dold [22], a general context of exposure to metal-enriched air-toxic metals in the marine environment of the bay was formed, revealing possible impacts through both aquatic and atmospheric routes.

In the field of population health, a quantitative approach was used to establish the presence of metals in urine and blood samples and changes in lung function in residents of Chañaral (Table 2) [45–49].

Cortés et al. [45] found that the levels of total arsenic, inorganic arsenic, nickel, copper, mercury, and lead in the urine of people in this area exceeded the values found in other regions of Chile and international measurements. The high percentages of people with detectable levels of each metal suggest the recent contact of a significant part of the population with these metals, and potential long-term damage, especially from simultaneous exposure to these metals.

**Table 2.** Studies on evidence of exposure and health effects in Chañaral Bay, Chile (1978–2025).

| ID | Year | Authors | Population (Sample) | Pollutant(s) | Method | Reported Evidence | Ref. |
|---|---|---|---|---|---|---|---|
| 23 | 2016 | Cortés, S et al. | Adults 18 to 65 years old. Sample: 205 individuals | Cu, Hg, Ni, Pb, As | A questionnaire and a urine sample | The study revealed elevated urinary levels of total and inorganic arsenic and lead in this area, surpassing international standards. These levels were higher than those in other regions of Chile, indicating potential environmental health concerns for the population. | [45] |
| 24 | 2018 | Mesías Monsalve, S. et al. | School spaces. Sample: 10 schools | Na, Cl, S, Ca, Fe, K, Mn, Ti, Si, MP2.5 | Trace element concentrations were measured using X-ray fluorescence on two consecutive days in both summer and winter of 2012 and 2013 | Indoor school spaces, particularly classrooms, had higher trace element levels compared to outdoor environments. The most abundant elements detected were Na, Cl, S, Ca, Fe, K, Mn, Ti, and Si, associated with the Earth's crust. | [46] |

**Table 2.** *Cont.*

| ID | Year | Authors | Population (Sample) | Pollutant(s) | Method | Reported Evidence | Ref. |
| --- | --- | --- | --- | --- | --- | --- | --- |
| 25 | 2019 | Moya, P.M. et al. | Sample of 158 participants, 18 to 65 years old | As, Cu, Ni, Pb, Mn | Cross-sectional epidemiological study | High concentrations of arsenic and copper were found in street dust; arsenic concentration in street dust was correlated with the distance to the mine tailings; urinary levels of metals were low and did not significantly correlate with the socioeconomic level (SES) of the population; there was no clear relationship between SES and urinary levels of metals. | [47] |
| 26 | 2021 | Cortés S et al. | Sample: 25 volunteers, 45 to 65 years old | As and its metabolites, Cu, Ni, Cr, and Pb | Cross-sectional study. Inductively coupled plasma mass spectrometry (ICP-MS) to measure metals. Questionnaire for demographic, lifestyle, metal exposure, and health status | Urinary Ni levels are positively linked to glycemia and IL-6. AsBet levels are positively linked to total cholesterol, while inorganic arsenic is linked to glycemia. A negative association was found between AsBet and 8-OHdG. | [48] |
| 27 | 2021 | Cortés S et al. | Adults no Sample: 205 | Cu, Hg, Ni, Pb, As | Questionnaire about health risk perception of environmental hazards | The final questionnaire is a straightforward, reliable, and valuable tool for assessing environmental health risk perceptions in Latin American countries. Environmental health risk perception is higher in women. | [49] |

Monsalve et al. [46] also identified high levels of trace elements in the air (PM2.5) and soil (Na, Cl, S, Ca, Fe, K, Mn, Ti, and Si) from indoor spaces of schools ($n = 10$), especially in classrooms, compared to outdoor environments. The most abundant elements found in school dust are associated with the Earth's crust. This indicates possible indoor air pollution in schools, which may have implications for the health of students and staff.

Similar results were found by Moya et al. [47], who highlight the presence of arsenic, nickel, and copper levels in the urine of people living near the tailings in the bay. The authors also argue that there is contamination of the urban environment due to high levels of copper and arsenic found in street dust near the tailings. Identifying differences in environmental exposure according to socioeconomic status within Chañaral suggests that different population groups may be exposed to different levels of contamination.

Cortés et al. [48] describe how urinary metal exposure levels are related to metabolic alterations. In particular, urinary nickel levels are directly associated with significant increases in total cholesterol, while increases in inorganic arsenic levels are associated with increases in glycemia.

There is an additional study on the perception of environmental risks and their consequences on health, which has shown the usefulness of quantitative instruments for this purpose. In general terms, the people of Chañaral show a high perception of health risks associated with the characteristics of the environment and exposure to chemical agents [49].

These results are in agreement with the notion that people living near the mine tailings in Chañaral Bay are exposed to toxic metals, which can have chronic detrimental effects on respiratory and cardiovascular health in adults and children. The presence of high levels of metals in urine and street dust, as well as the differences in exposure according to socioeconomic level, emphasizes the importance of taking measures to reduce contamination and protect the health of the population near the mine tailings in Chañaral Bay due to the exposure scenario described above.

### 3.3. Analysis from the Social Sciences

In the social sciences, studies carried out by Vergara [15], González [20,50,51], Schorr [52], Aedo [53], and Quintana [54] share a common interest in addressing the pollution and environmental impact caused by large-scale copper mining in Chañaral, from a historical and socio-environmental perspective (Table 3). These investigations explore diverse topics, such as mitigation measures, community resistance, and socio-cultural aspects related to environmental suffering. In addition, they employ a variety of methodologies, including historical analysis, qualitative approaches and case studies, combining written sources, interviews, and public health data. Through their findings, they highlight the importance of approaching the environmental problem from multiple perspectives to better understand its complexity and to find effective solutions.

**Table 3.** Studies on the social evidence and historical reconstruction of impact and citizen mobilization in Chañaral Bay, Chile (1978–2025).

| ID | Year | Reference | Subject of Study | Social Group | Method | Reported Evidence | Ref. |
|---|---|---|---|---|---|---|---|
| 28 | 2011 | Vergara, A. | Impact of large copper mining in the Province of Chañaral between 1938 and 1990 | Not applicable | Reconstruction of environmental history through documentary consultation | The disposal of tailings into the riverbed and sea has changed the coastline, devastated bay plants and marine life, and impacted the health of local residents | [15] |
| 29 | 2018 | Schorr, B. | Analysis of the environmental conflict in Chañaral and the population's opposition to mining | Environmental activists and mining companies | Documentary consultation | Activists in Chañaral struggle to gain support in holding the mining company accountable and seeking compensation for the damage caused | [52] |
| 30 | 2018 | González, P. | Study of the background of the environmental problem at Chañaral until the legalization and termination of tailings disposal in 1990 | Mining companies and population | Documentary consultation | Failed solutions and unsuccessful projects demonstrate socio-ecological inequalities, worsening the environmental impact, and leading to urban lethargy in Chañaral | [20] |

**Table 3.** *Cont.*

| ID | Year | Reference | Subject of Study | Social Group | Method | Reported Evidence | Ref. |
|---|---|---|---|---|---|---|---|
| 31 | 2019 | González, P. | Territorial production in the coastal town of Chañaral; changes were promoted since the 1920s | Mining companies and population | Analysis of written, oral, and cartographic sources | Contamination remains a significant concern, creating a physical and symbolic burden and adversely affecting urban quality of life. Hence, the socio-environmental conflict endures in a state of latency | [50] |
| 32 | 2019 | Aedo, M. | The affections and knowledge of women leaders in Chañaral regarding the search for justice and reparation of the territory | Women leaders of social organizations | Interviews | Due to the denial of justice and reparations by governments, women have established support systems encompassing health, food, child and elderly care, as well as education. These systems aim to repair the damage caused by coming together, uplifting their community's spirits, and creating spaces for enjoyment and connection | [53] |
| 33 | 2021 | González, P. | The socio-environmental situation of Chañaral as an environmental, biomedical and sociocultural phenomenon | Mining companies and population | Analysis of qualitative information and reports on public health | Daily suffering and uncertainties about toxic risks create environmental distress due to the lack of comprehensive measures to address the issue effectively | [51] |
| 34 | 2022 | Quintana-Muñoz, J. | Analysis of territory meanings | Mining companies and population | Interviews and speech analysis | The population demands the right to stay in a safe territory and shape it according to their needs | [54] |

For example, Vergara [15] reconstructs the history of mining tailings dumped into the Salado River and the Pacific Ocean by mining companies between 1938 and 1990. He also highlights the lack of action by the authorities and how local inhabitants denounced the contamination early on. Following the same line of analysis from environmental history, González [20] reviews the environmental problems of Chañaral up to 1990. He aims to explain the socio-ecological inequality and how mitigation measures have been ineffective, leading to urban lethargy and latent conflict in the collective memory of the inhabitants. For his part, Gonzalez's [50] argument consists of the analysis of the territorial production of the coastal town of Chañaral and the changes caused by large-scale copper exploitation. In his opinion, the actors in dispute within the community take different positions, which complicates the analysis regarding the opposition between the company and the affected community.

Studies by Schorr [52], Aedo [53], and González [51] address the impact of pollution on the daily lives of the inhabitants of Chañaral. Schorr [52] highlights that many inhabitants have naturalized pollution and consider it an intrinsic part of their lives, which could affect their participation in protests and mobilizations against the mining companies responsible for the pollution. On the other hand, Aedo [53] highlights the strength of resistance in Chañaral, focusing on the emotional experiences and responses of the community, particularly women leaders. Her study is based on environmental suffering and highlights the importance of female resistance against contamination, positioning women as key actors in the search for justice and reparation for the territory affected by mining. In this sense, Aedo [53] provides a deeper and more detailed look at the experiences of women in the community.

Quintana [54] explores the meanings that the population gives to the territory from an emotional dimension and sets out to understand how people have built a territorial demand based on the "right to stay in a safe territory" and the "right to make territory". Through a qualitative interpretative–hermeneutic methodology, Quintana focuses on the construction of meaning in the social world, analyzing emotions, and exploring the feelings and meanings associated with the territory of Chañaral.

González [51] also analyzes environmental suffering in Chañaral, relating it to uncertainty about toxic risk and the lack of concrete measures available to address the problem. His approach focuses on the complex relationship between the daily experiences of the inhabitants, the incidence of biomedical knowledge on the scientific legitimacy of the environmental problem, and the need to consider sociocultural aspects in the urban-environmental planning of territories affected by contamination.

These studies show that interdisciplinary research is essential to comprehensively address Chañaral's socio-environmental problem and to find solutions that protect the health of the population and restore the quality of the environment. Only through a joint effort in which the voices of the community are heard and the complex socio-environmental dynamic is considered, is it possible to achieve a significant change in the current situation and guarantee a sustainable future for Chañaral and its inhabitants.

## 4. Discussion

Evidence on the environmental, ecological, and public health impacts in Chañaral, as a result of the dispersion of mining tailings along the bay, has led to a scenario of socio-environmental inequality that embodies what Gouveia et al. [55] describe as environmental injustice. From the environmental sciences, there is valuable scientific evidence that shows that the accumulation of tailings has caused an alteration of the geomorphology and biogeochemical dynamics of the coast, as well as the retreat of the sea and solidification of the bay (Table 1). In terms of public health, it is reported that people are directly impacted by the particulate matter (PM) lifted by wind action, transporting metal-rich particles to the city of Chañaral in the absence of containment measures. This leads to clear exposure to toxic metals, as evidenced by biomonitoring studies, and respiratory and metabolic alterations, which require further investigation (Table 2). In the social sciences, studies have highlighted the importance of approaching the environmental problem from different perspectives, exploring mitigation measures, community resilience, and sociocultural aspects related to environmental suffering. Research has highlighted the need to understand the emotional and affective connections that people establish within the territory and how these perceptions influence their perception of the inhabited space (Table 3). Similar findings were reported by Cacciuttolo et al. (2023) [17], who concluded that mine tailings are potentially toxic to both communities and the environment, posing a considerable health risk, and cannot be considered as inert and innocuous materials.

They reported chemical compounds present in copper mine tailings, including metals such as copper, manganese, vanadium, metalloids, such as As, and other non-metals such as chlorides and sulfates, as also reported by Dold (2006) [22].

In the Chañaral Case, health risk factors for humans are linked to the water and contaminant-containing acid rock drainages; to the soil, due potential instability underlied by seismic events; and to the alteration of the soil's chemical quality due to mine tailings and land use [17]. Additionally, air exposure is related to the generation of dust enriched in toxic metals [22]. Our analysis supported the findings that changes in the quality of soil, water, the atmosphere, and locally-produced food caused by the presence of toxic metals can cause changes in the structure and physiology of living organisms, including human populations in Chañaral. Also, it is interesting that the timeline of the research made in Chañaral reflects a historically grounded, interdisciplinary approach, marked by a gradual integration of physical, biomedical, and social sciences, which highlights direct links between environmental crises and scientific responses, and an evolution from descriptive accounts to human-centered and policy-relevant analysis.

These studies show that pollution in Chañaral is a complex and multidimensional problem that requires a comprehensive and collaborative response. Despite all this scientific evidence, the Chañaral municipality does not have urban plans in place which adequately addresses this problem [51]. This generates an important challenge for its urban sustainability, in which visions must converge between environmental studies, environmental management and urban planning to reduce the health risk to the population. Not addressing this problem in a comprehensive manner deepens the environmental suffering of inhabitants [51–54].

Therefore, the broad health impacts on children, adults, and the elderly should be taken into consideration when inhabitants are exposed to a mix of toxic metals and metalloids [56]. It is advisable to promote public policies that seek decontamination under appropriate environmental governance criteria and that use scientific knowledge to develop environmental restoration experiences. It is important to include the historical and sociocultural dimensions of pollution in a sustainable development plan and in environmental education programs. This will make it possible to recognize and address the problem in a preventive manner, thus contributing to the health of both humans and the environment, understanding that both entities are intrinsically related. To achieve this, it is necessary to have the collaboration of urban-environmental planners, ecologists, experts in public health and environmental epidemiology, as well as educators and social scientists. These actors must work together to carry out an environmental remediation process that encompasses the different dimensions of pollution.

According to Gouveia et al.'s ([55] concept of environmental justice, defined as *"addressing both unequal exposure to environmental hazards and the resulting health disparities"* in Chañaral, residents are disproportionately exposed to pollution due to mining activities, which exacerbates public health inequities among the most vulnerable, including children. Thus, the situation reflects environmental injustice by underscoring how industrial waste and socio-ecological inequalities intersect to disproportionately impact the health and environment of these communities. Environmental actions that meet the characteristics mentioned by Porto et al. [8] align with the perspectives of both Gouveia et al. [55] and Villasana et al. [57], who call for structural changes to address the root causes of environmental and health inequalities. While Gouveia et al. stress the need to mitigate unequal exposure and health effects among vulnerable populations, Villasana et al. advocate for a constitutional and political framework that ensures dignity, health, and environmental protection. Together, these approaches support a comprehensive development plan for Chañaral, which can protect both human and environmental health by addressing systemic inequalities and fostering resilience through policies centered on social and environmental equity.

The lack of urban planning that addresses environmental issues in Chañaral reveals a marked disconnect between urban planning policies and public health decisions, both of which are unaligned with the scientific evidence. Although environmental legislation has advanced in Chile over the last 30 years, environmental liabilities are not systematically considered in urban planning. Law No. 19,300 on the General Bases of the Environment, enacted in 1994, specifically determined that urban development projects that generate socio-environmental alterations must declare their impacts through the Environmental Impact Assessment System. Subsequent modifications established the guidelines for Strategic Environmental Assessment (SEA), a mechanism that allows the incorporation of environmental vision into the Territorial Planning Instruments in the early stages of the decision-making process. Specifically, it states that "the environmental factors of sustainable development must be considered during the formulation of public policies and plans of a general normative nature that have an impact on the environment or sustainability".

Therefore, the SEA is indicated as mandatory for intermunicipal and municipal regulatory plans. In practice, however, these regulations do not have clear commitments at the local level, so the relationship between environmental problems, public health, and urban planning instruments is insufficiently considered [58–61]. Although Article 2.1.17 of the General Ordinance of Urbanism and Construction (OGUC), enacted in 1992, establishes areas with the presence of contaminants due to anthropogenic activities as "risk zones" [58], the areas where mining tailings are located are not classified as risk zones due to the outdatedness of communal regulatory plans and the lack of relevant soil studies [60,61]. In addition, the closure of mining sites and facilities was only regulated in Chile in 2011 by Law 20,551. The objective of the law is to mitigate the negative effects of the mining industry and to protect the life, health, and safety of people and the environment. However, this law does not have a retroactive effect on previously inactive mining operations. Therefore, Chañaral's environmental problems would not be regulated or mitigated, since they predate the law.

In addition, this study proposes that it is fundamental to address the socio-environmental inequality in the design and implementation of such policies. Environmental justice implies taking actions based on the acknowledgement of unequal exposure to toxic risks affecting the health, human rights and dignity of communities exposed to the chemical and socio-cultural effects of pollution [8]. Joint work between experts from different disciplines and community participation is essential to find effective solutions that protect the health of the population through planning processes leading to mitigating actions, achieving a significant change in the current situation and guaranteeing a sustainable future for Chañaral and its inhabitants. Decision-making based on scientific evidence and the consideration of social and emotional aspects is key to addressing the challenge of the contamination in this area and promoting the well-being of the population and the environment. Boyd's report (2024) establishes that *"Chile should apply a human rights-based approach to laws, regulations, policies, and actions governing the production, import, sale, use, release, and disposal of substances that may harm human health or the environment, to eliminate the negative impacts on human rights. A rights-based approach should also govern clean-up, remediation, restoration, and compensation. A rights-based approach clarifies the obligations of Governments and the responsibilities of businesses, prioritizes the most disadvantaged, and catalyses ambitious action. Immediate action must be taken to eliminate residents' exposure to environmental hazards. Putting economic considerations ahead of human rights is a fundamentally flawed form of decision-making, as the Inter-American Commission on Human Rights recently clarified."* [9].

In short, actionable recommendations emerging from different disciplines must consider the following as a minimum [56,62]:

(a) Remediation and management of mining tailing, integrating environmental, geochemical, and hydraulic aspects. Efforts are required in the control of particulate matter with metals emitted from the tailings.
(b) Research on toxicity, biodegradability, and bioaccumulation in biota and living organisms. It may be necessary to evaluate phytoremediation measures for soil detoxification.
(c) Community-based trans-generational participatory health research to improve the understanding of chronic exposure to toxic metals; efforts must include mental and physical health, neurotoxicity and respiratory damage in children, cardiovascular and cognitive disorders in the elderly, and metabolic and respiratory damage in adults, related to chronic exposure to toxic metals and arsenic.
(d) Promotion of community environmental monitoring together with local and regional governance models, integrating public participation.

Complementing the above, other authors suggest the need to reassess environmental definitions and criteria for mining tailing storage facilities and site selection, among other technical and regulatory changes involving governmental agencies, such as SERNAGEOMIN, the General Directorate of Waters (DGA), the Ministry of Health (MINSAL), and the Environmental Assessment Service (SEA), and to establish a regulatory framework for submarine deep-sea tailings disposal, considering the historical and recent mine tailing production in Chile, and the location of all large mining projects [17]. Along the same lines, it could be useful to consider ecological indicators and sustainability metrics for improving integrated surveillance systems [63].

## 5. Conclusions

The scientific evidence from 1978 to date demonstrates the negative impacts on the ecosystem and the human population exposed to toxic metals and arsenic. Geomorphological and biogeochemical alterations have been found on the Chañaral coast, affecting marine biodiversity and water quality. In humans, chronic exposure to toxic metals at low levels measured in street dust and urine raises health concerns in children and adults. In the social sciences, it is argued that the lack of environmental monitoring and data on human exposure contributes to an elevated health risk perception in the population.

In concordance with other authors, the improvement of the final disposal of mining tailings relies on the lessons learned from the past in many parts of the world. In fact, in Chañaral, as well as in other Chilean areas and globally, the improper management of mine tailings generates socio-environmental impacts on marine and terrestrial ecosystems that persist to this day. Considering the lack of information and practical experience on the effects of these practices, it is necessary to apply ethical approaches based on human rights and state-of-the-art available technologies, concerning society and ecosystems, to avoid repeating the mistakes of the past in present and future tailings management operations [17].

In conclusion, it is important to conceive the urban environment as a space where health must be prioritized and the lives of the inhabitants of Chañaral must be protected. Based on notions of social-environmental justice, it is considered fundamental to establish the minimum axes of analysis to seek solutions in public policies.

**Supplementary Materials:** The following supporting information can be downloaded at https://www.mdpi.com/article/10.3390/su17177732/s1. Table S1: Postgraduate research in public health, law, and urban studies.

**Author Contributions:** Conceptualization, S.C., P.G., C.L. and Y.V.; methodology, S.C., P.G., C.L. and Y.V.; formal analysis, P.G., C.L., Y.V. and A.V.; investigation, Y.V.; resources, S.C., P.G. and C.L.; writing—original draft preparation, P.G. and S.C.; writing—review and editing, C.L., Y.V. and A.V.; visualization, S.C. and P.G.; supervision, A.V. and P.P.; project administration, S.C.; funding acquisition, S.C. All authors have read and agreed to the published version of the manuscript.

**Funding:** This research was funded by Agencia Nacional de Investigación y Desarrollo (ANID) through the grants Fondo de Financiamiento de Centros de Investigación en Áreas Prioritarias (FONDAP) 1523A0008 and FONDAP 1523A0004.

**Institutional Review Board Statement:** Not applicable.

**Informed Consent Statement:** Not applicable.

**Data Availability Statement:** The data are contained within the article.

**Acknowledgments:** The authors would like to thank the community of Chañaral for their support and interest in sharing information and stories relevant to our study, as well as Héctor Adaros Marambio of the "Jeronimo Mendez Chañaral Hospital", for his constant collaboration and interest in raising awareness on the case of Chañaral, to improve the quality of life of its inhabitants.

**Conflicts of Interest:** The authors declare no conflicts of interest.

## References

1. Hudson-Edwards, K.A.; Jamieson, H.E.; Lottermoser, B.G. Mine Wastes: Past, Present, Future. *Elements* **2011**, *7*, 375–380. [CrossRef]
2. Castillo, E.; Cortes-Maramba, N.P.; Reyes, J.P.; Makalinao, I.; Dioquino, C.; Francisco-Rivera, A.T.; Timbang, R. Health and environmental assessment of the impact of mine tailings spillage in the Philippines. *J. Phys. IV* **2003**, *107*, 275–279. [CrossRef]
3. Csavina, J.; Field, J.; Taylor, M.P.; Gao, S.; Landázuri, A.; Betterton, E.A.; Sáez, A.E. A review on the importance of metals and metalloids in atmospheric dust and aerosol from mining operations. *Sci. Total Environ.* **2012**, *433*, 58–73. [CrossRef] [PubMed]
4. Doumas, P.; Munoz, M.; Banni, M.; Becerra, S.; Bruneel, O.; Casiot, C.; Sappin-Didier, V. Polymetallic pollution from abandoned mines in Mediterranean regions: A multidisciplinary approach to environmental risks. *Reg. Environ. Change* **2016**, *18*, 677–692. [CrossRef]
5. Islam, K.; Murakami, S. Global-scale impact analysis of mine tailings dam failures: 1915–2020. *Glob. Environ. Change* **2021**, *70*, 102361. [CrossRef]
6. SERNAGEOMIN. *Anuario de la Minería de Chile*; Servicio Nacional de Geología y Minería: Santiago, Chile, 2022.
7. Pérez, G. Pasivos ambientales mineros en Chile: Lineamientos para priorización y remediación. In *Avances y Recomendaciones en la Gestión de Pasivos Ambientales Mineros (PAM) y Cierre de Minas en los Países Andinos*; Ministerio de Energía y Minas: Lima, Peru, 2019. Available online: https://www.cepal.org/sites/default/files/events/files/11_grecia_perez_caso_pams_chile_pdf (accessed on 17 March 2025).
8. Porto, M.; Ferreira, D.; Finamore, R. Health as dignity: Political ecology, epistemology and challenges to environmental justice movements. *J. Polit. Ecol.* **2017**, *24*, 110–124.
9. Boyd, D.R. A/HRC/55/43/Add.1: Visit to Chile Report of the Special Rapporteur on the Issue of Human Rights Obligations Relating to the Enjoyment of a Safe, Clean, Healthy and Sustainable Environment 2024. Updated 3 January 2024. Available online: https://www.ohchr.org/en/documents/country-reports/ahrc5543add1-visit-chile-report-special-rapporteur-issue-human-rights (accessed on 18 July 2025).
10. Biblioteca del Congreso Nacional de Chile. Reporte Comunal Chañaral. 2024. Available online: https://www.bcn.cl/siit/reportescomunales/comunas_v.html?anno=2024&idcom=3201 (accessed on 17 March 2024).
11. Delegación Presidencial Provincial de Chañaral. Conozca la Provincia. 2021. Available online: https://dppchanaral.dpp.gob.cl/conozca-la-provincia/ (accessed on 17 March 2025).
12. Codelco. Reporte de Sustentabilidad. 2017. Available online: https://www.codelco.com/prontus_codelco/site/artic/20180614/asocfile/20180614122058/reporte_sustentabilidad_2017_codelco.pdf (accessed on 17 March 2025).
13. San Román, F.J. Plano del Puerto de Chañaral [material cartográfico]. Santiago: Lit. e Imp. "La Nueva Alemana", Santiago. 1898.
14. Cortés, M. La Muerte Gris de Chañaral. El Libro Negro de la División Salvado del CODELCO Chile [Internet]. 2010. Available online: https://www.semillasdeagua.cl/wp-content/uploads/LA-MUERTE-GRIS-DE-CHA%C3%91ARALPDF.pdf (accessed on 17 August 2025).

15. Vergara, A. Cuando el río suena, piedras trae: Relaves de cobre en la bahía de Chañaral, 1938–1990. *Cuad. Hist.* **2011**, *135*, 135–151. [CrossRef]
16. Ministerio de Obras Públicas (MOP). *Memoria Anual*; Dirección de Planeamiento, Departamento de Estadística y Control: Santiago, Chile, 1974.
17. Cacciuttolo, C.; Cano, D.; Custodio, M. Socio-Environmental Risks Linked with Mine Tailings Chemical Composition: Promoting Responsible and Safe Mine Tailings Management Considering Copper and Gold Mining Experiences from Chile and Peru. *Toxics* **2023**, *11*, 462. [CrossRef]
18. Cortés, S. Percepción y Medición del Riesgo a Metales en una Población Expuesta a Residuos Mineros. Universidad de Chile. 2009. Available online: http://repositorio.uchile.cl/handle/2250/116517 (accessed on 17 March 2025).
19. Ilustre Municipalidad de Chañaral. Plan de Desarrollo Comunal Chañaral. 2019. Available online: https://www.portaltransparencia.cl/PortalPdT/documents/10179/62801/Informe+Final+PLADECO+2019-2026.pdf/3699b935-370e-48ae-a42f-d5b70473f99d (accessed on 17 March 2025).
20. González, P. Chañaral, un problema ambiental insoslayable. Intentos de solución en una ciudad bajo letargo. *Rev. Estud. Urbano-Reg.* **2018**, *1*, 1–2.
21. Castilla, J.C.; Nealler, E. Marine environmental impact due to mining activities of El Salvador Copper Mine, Chile. *Mar. Pollut. Bull.* **1978**, *9*, 67–70. [CrossRef]
22. Dold, B. Element flows associated with marine shore mine tailings deposits. *Environ. Sci. Technol.* **2006**, *40*, 752–758. [CrossRef]
23. Lee, M.R.; Correa, J.A.; Seed, R. A sediment quality triad assessment of the impact of copper mine tailings disposal on the littoral sedimentary environment in the Atacama region of northern Chile. *Mar. Pollut. Bull.* **2006**, *52*, 1389–1395. [CrossRef] [PubMed]
24. Andrade, S.; Moffett, J.; Correa, J.A. Distribution of dissolved species and suspended particulate copper in an intertidal ecosystem affected by copper mine tailings in Northern Chile. *Mar. Chem.* **2006**, *101*, 203–212. [CrossRef]
25. Lee, M.R.; Correa, J.A.; Zhang, H. Effective metal concentrations in porewater and seawater labile metal concentrations associated with copper mine tailings disposal into the coastal waters of the Atacama region of northern Chile. *Mar. Pollut. Bull.* **2002**, *44*, 956–961. [CrossRef]
26. Correa, J.A.; Roman, D.A.; Rivera, L.; Ramírez, M.A.; De, J.P.; Harpe, L.A.; Román, D.; Lidia, R. Copper, copper mining effluents, and grazing as potential determinants of algal abundance and diversity in Northern Chile. *Environ. Monit. Assess.* **1998**, *61*, 267–283. [CrossRef]
27. Valladares, P.; Alvarado, S.; Urra, C.; Abarca, J.; Inostroza, J.; Codoceo, J.; Ruz, M. Cadmium and Lead content in Liver and Kidney tissues of Wild Turkey Vulture Cathartes aura (Linneo, 1758) from Chañaral, Atacama desert, Chile. *Gayana* **2013**, *77*, 97–104. [CrossRef]
28. Correa, J.A.; Castilla, J.C.; Ramirez, M.; Varas, M.; Lagos, N.; Vergara, S.; Moenne, A.; Román, D.; Brown, M.T. Copper, copper mine tailings and their effect on marine algae in Northern Chile. *J. Appl. Phycol.* **1999**, *11*, 57–67. [CrossRef]
29. Gobierno Regional de Atacama. Evaluación de una Estrategia de Detoxificación de Efluentes Contaminados con Metales Pesados Utilizando Algas Marinas Chilenas. Informe Final. Copiapó, Chile. 2015. Available online: https://goreatacama.gob.cl/wp-content/uploads/Informe-final-Evaluacion-de-una-estrategia-de-detoxificacion-contaminados-con-metales-pesados-utilizando-algas-marinas-chilenas.pdf (accessed on 25 August 2025).
30. Medina, M.; Andrade, S.; Faugeron, S.; Lagos, N.; Mella, D.; Correa, J.A. Biodiversity of rocky intertidal benthic communities associated with copper mine tailing discharges in northern Chile. *Mar. Pollut. Bull.* **2005**, *50*, 396–409. [CrossRef]
31. De la Iglesia, R.; Valenzuela-Heredia, D.; Andrade, S.; Correa, J.; González, B. Composition dynamics of epilithic intertidal bacterial communities exposed to high copper levels. *FEMS Microbiol. Ecol.* **2012**, *79*, 720–727. [CrossRef]
32. Fariña, J.M.; Castilla, J.C. Temporal variation in the diversity and cover of sessile species in rocky intertidal communities affected by copper mine tailings in Northern Chile. *Mar. Pollut. Bull.* **2001**, *42*, 554–568. [CrossRef]
33. Castilla, J.C. Environmental impact in sandy beaches of copper mine tailings at Chañaral, Chile. *Mar. Pollut. Bull.* **1983**, *14*, 459–464. [CrossRef]
34. Riquelme, C.; Rojas, A.; Florest, V.; Correa, J.A. Epiphytic bacteria in a copper-enriched environment in Northern Chile. *Mar. Pollut. Bull.* **1997**, *34*, 816–820. [CrossRef]
35. Lee, M.R.; Correa, J.A.; Castilla, J.C. An assessment of the potential use of the nematode to copepod ratio in the monitoring of metals pollution. The Chañaral case. *Mar. Pollut. Bull.* **2001**, *42*, 691–701. [CrossRef]
36. Ramírez, M.; Massolo, S.; Frache, R.; Correa, J.A. Metal speciation and environmental impact on sandy beaches due to El Salvador copper mine, Chile. *Mar. Pollut. Bull.* **2005**, *50*, 62–72. [CrossRef]
37. Lee, M.R.; Correa, J.A. Effects of copper mine tailings disposal on littoral meiofaunal assemblages in the Atacama region of northern Chile. *Mar. Environ. Res.* **2005**, *59*, 1–18. [CrossRef]
38. Correa, J.A.; Lagos, N.A.; Medina, M.H.; Castilla, J.C.; Cerda, M.; Ramírez, M.; Martínez, E.; Faugeron, S.; Andrade, S.; Pinto, R.; et al. Experimental transplants of the large kelp *Lessonia nigrescens* (Phaeophyceae) in high-energy wave-exposed rocky intertidal habitats of northern Chile: Experimental, restoration and management applications. *Hydrobiologia* **2006**, *335*, 13–18. [CrossRef]

39. Bea, S.A.; Ayora, C.; Carrera, J.; Saaltink, M.W.; Dold, B. Geochemical and environmental controls on the genesis of soluble efflorescent salts in coastal mine tailings deposits: A discussion based on reactive transport modeling. *J. Contam. Hydrol.* **2010**, *111*, 65–82. [CrossRef]
40. Ramos-Jiliberto, R.; Garay-Narváez, L.; Medina, M.H. Retrospective qualitative analysis of ecological networks under environmental perturbation: A copper-polluted intertidal community as a case study. *Ecotoxicology* **2012**, *21*, 234–243. [CrossRef]
41. Koski, R. Metal Dispersion Resulting from Mining Activities in Coastal Environments: A Pathways Approach. *Oceanography* **2012**, *25*, 170–183. [CrossRef]
42. Korehi, H.; Blöthe, M.; Sitnikova, M.A.; Dold, B.; Schippers, A. Metal mobilization by iron- and sulfur-oxidizing bacteria in multiple extreme mine tailings in the Atacama Desert, Chile. *Environ. Sci. Technol.* **2013**, *47*, 2189–2196. [CrossRef]
43. Besaury, L.; Ghiglione, J.F.; Quillet, L. Abundance, activity, and diversity of archaeal and bacterial communities in both uncontaminated and highly copper-contaminated marine sediments. *Mar. Biotechnol.* **2014**, *16*, 230–242. [CrossRef]
44. Bonnail, E.; Cunha Lima, R.; Martínez Turrieta, G. Trapping fresh sea breeze in desert? Health status of Camanchaca, Atacama's fog. *Environ. Sci. Pollut. Res.* **2018**, *25*, 18204–18212. [CrossRef] [PubMed]
45. Cortés, S.; Molina, L.; Burgos, S.; Adaros, H.; Ferreccio, C. Urinary metal levels in a Chilean community 31 years after the dumpings of mine tailings. *J. Health Pollut.* **2016**, *6*, 19–27. [CrossRef] [PubMed]
46. Mesías Monsalve, S.; Martínez, L.; Yohannessen Vásquez, K.; Alvarado Orellana, S.; Klarián Vergara, J.; Martín Mateo, M.; Costilla Salazar, R.; Fuentes Alburquenque, M.; Cáceres Lillo, D.D. Trace element contents in fine particulate matter (PM2.5) in urban school microenvironments near a contaminated beach with mine tailings, Chañaral, Chile. *Environ. Geochem. Health* **2018**, *40*, 1077–1091. [CrossRef] [PubMed]
47. Moya, P.M.; Arce, G.J.; Leiva, C.; Vega, A.S.; Gutiérrez, S.; Adaros, H.; Muñoz, L.; Pastén, P.A.; Cortés, S. An integrated study of health, environmental and socioeconomic indicators in a mining-impacted community exposed to metal enrichment. *Environ. Geochem. Health* **2019**, *41*, 2505–2519. [CrossRef]
48. Cortés, S.; Zúñiga-Venegas, L.; Pancetti, F.; Covarrubias, A.; Ramírez-Santana, M.; Adaros, H.; Muñoz, L. A positive relationship between exposure to heavy metals and development of chronic diseases: A case study from Chile. *Int. J. Environ. Res. Public Health* **2021**, *18*, 1419. [CrossRef]
49. Cortés, S.; Burgos, S.; Adaros, H.; Lucero, B.; Quirós-Alcalá, L. Environmental health risk perception: Adaptation of a population-based questionnaire from Latin America. *Int. J. Environ. Res. Public Health* **2021**, *18*, 8600. [CrossRef]
50. González, P. Historia Ambiental de Chañaral. Intrusión de Relaves Mineros, Transformación Territorial Y Conflicto de Contenido Ambiental. Available online: https://www.researchgate.net/publication/343615788_Historia_ambiental_de_Chanaral_Intrusion_de_relaves_mineros_transformacion_territorial_y_conflicto_de_contenido_ambiental (accessed on 19 August 2025).
51. González, P. Living among tailing sands. Sanitary uncertainty and environmental suffering in Chañaral (Chile). *Rev. INVI* **2021**, *36*, 83–108.
52. Schorr, B. Oportunidades desiguales: Empresas y Estado en conflictos sobre la minería en Chile. *Estud. Atacameños* **2018**, *57*, 239–255. [CrossRef]
53. Aedo, M. Afectos y resistencias de las mujeres de Chañaral frente a los impactos de la minería estatal en Chile. *Sustainability* **2019**, *10*, 87–103.
54. Quintana-Muñoz, J. Making (Un)Safe Territory in the Midst of Socio-Environmental Disasters: Territory Meanings and Emotions in Chañaral, Chile. *Rev. Austral Cienc. Sociales* **2022**, *2022*, 107–128. [CrossRef]
55. Gouveia, N.; Slovic, A.D.; Kanai, C.M.; Soriano, L. Air pollution and environmental justice in Latin America: Where are we and how can we move forward? *Curr. Environ. Health Rep.* **2022**, *9*, 152–164. [CrossRef]
56. Fuller, R.; Landrigan, P.J.; Balakrishnan, K.; Bathan, G.; Bose-O'Reilly, S.; Brauer, M.; Caravanos, J.; Chiles, T.; Cohen, A.; Corra, L.; et al. Pollution and health: A progress update. *Lancet Planet Health.* **2022**, *6*, e535–e547. [CrossRef]
57. Villasana López, P.E.; Dörner Paris, A.P.; Estay Sepúlveda, J.G.; Moreno Leiva, G.M.; Monteverde Sanchez, A. Zonas de sacrificio y justicia ambiental en Chile. Una mirada crítica desde los Objetivos de Desarrollo Sostenible 2030. *Hist. Ambient. Latinoam. Caribeña* **2020**, *10*, 342–365. [CrossRef]
58. Ministerio de Vivienda y Urbanismo. Decreto N°47. Fija Nuevo Texto de la Ordenanza General de la Ley General de Urbanismo y Construcciones. Ley Chile. 1992. Available online: https://www.bcn.cl/leychile/navegar?idNorma=8201 (accessed on 17 March 2025).
59. Romero, H.; Vásquez, A. La comodificación de los territorios urbanizables y la degradación ambiental en Santiago de Chile. *Scripta Nova* **2005**, *IX*, 1–68.
60. Barton, J.R. Sustentabilidad urbana como planificación estratégica. *EURE* **2006**, *32*, 27–45. [CrossRef]
61. Hervé, D. Noción y elementos de la justicia ambiental: Directrices para su aplicación en la planificación territorial y en la evaluación ambiental estratégica. *Rev. Derecho* **2010**, *23*, 9–36.

62. Cacciuttolo, C.; Atencio, E. Past, Present, and Future of Copper Mine Tailings Governance in Chile (1905–2022): A Review in One of the Leading Mining Countries in the World. *Int. J. Environ. Res. Public Health* **2022**, *19*, 13060. [CrossRef] [PubMed]
63. Copaja, S.V.; Muñoz, F. Ecological risk assessment of trace elements in the tailings from Andacollo city, Northern Chile. *J. Chil. Chem. Soc.* **2022**, *67*, 5674–5681. [CrossRef]

MDPI AG
Grosspeteranlage 5
4052 Basel
Switzerland
Tel.: +41 61 683 77 34

*Sustainability* Editorial Office
E-mail: sustainability@mdpi.com
www.mdpi.com/journal/sustainability

www.ingramcontent.com/pod-product-compliance
Lightning Source LLC
Chambersburg PA
CBHW061619220926
51850CB00005B/549
* 9 7 8 3 7 2 5 8 7 3 5 6 2 *